Methods in Enzymology

Volume 213
CAROTENOIDS
Part A
Chemistry, Separation, Quantitation, and Antioxidation

METHODS IN ENZYMOLOGY

EDITORS-IN-CHIEF

John N. Abelson   Melvin I. Simon

DIVISION OF BIOLOGY
CALIFORNIA INSTITUTE OF TECHNOLOGY
PASADENA, CALIFORNIA

FOUNDING EDITORS

Sidney P. Colowick and Nathan O. Kaplan

Methods in Enzymology

Volume 213

Carotenoids

Part A
Chemistry, Separation, Quantitation, and Antioxidation

EDITED BY

Lester Packer

DEPARTMENT OF MOLECULAR AND CELL BIOLOGY
UNIVERSITY OF CALIFORNIA, BERKELEY
BERKELEY, CALIFORNIA

ACADEMIC PRESS, INC.
Harcout Brace Jovanovich Publishers
San Diego   New York   Boston
London   Sydney   Tokyo   Toronto

This book is printed on acid-free paper. ∞

Copyright © 1992 by ACADEMIC PRESS, INC.
All Rights Reserved.
No part of this publication may be reproduced or transmitted in any form or by any means, electronic or mechanical, including photocopy, recording, or any information storage and retrieval system, without permission in writing from the publisher.

Academic Press, Inc.
1250 Sixth Avenue, San Diego, California 92101-4311

United Kingdom Edition published by
Academic Press Limited
24–28 Oval Road, London NW1 7DX

Library of Congress Catalog Number: 54-9110

International Standard Book Number: 0-12-182114-5

PRINTED IN THE UNITED STATES OF AMERICA
92 93 94 95 96 97   MM   9 8 7 6 5 4 3 2 1

Table of Contents

CONTRIBUTORS TO VOLUME 213   ix

PREFACE xiii

VOLUMES IN SERIES. xv

Section I. Chemistry: Synthesis, Properties, and Characterization

1. Carotenoids: An Overview — HANSPETER PFANDER — 3

2. Synthesis and Characterization of Carotenoids by Different Methods — MASAYOSHI ITO, YUMIKO YAMANO, AND YUKO SHIBATA — 13

3. Structure and Characterization of Carotenoids from Various Habitats and Natural Sources — TAKAO MATSUNO — 22

4. Surface-Enhanced Raman Scattering Spectroscopy of Photosynthetic Membranes and Complexes — MICHAEL SEIBERT, RAFAEL PICOREL, JAE-HO KIM, AND THERESE M. COTTON — 31

5. Synthesis of Carotenoids Specifically Labeled with Isotopic Carbon and Tritium — ARNOLD A. LIEBMAN, WALTER BURGER, SATISH C. CHOUDHRY, AND JOSEPH CUPANO — 42

6. Synthesis of Deuterated β-Carotene — H. ROBERT BERGEN, III — 49

7. Formation of Volatile Compounds by Thermal Degradation of Carotenoids — J. CROUZET AND P. KANASAWUD — 54

8. Analysis of the Formation of Carotene Chromophore and Ionone Rings in *Narcissus pseudonarcissus* L. Chromoplast Membranes — PETER BEYER AND HANS KLEINIG — 62

9. C_{50} Bicyclic Carotenoids: Sarcinaxanthin Synthesis — J. P. FÉRÉZOU — 75

10. Synthesis of Carotenoporphyrin Models for Photosynthetic Energy and Electron Transfer — DEVENS GUST, THOMAS A. MOORE, ANA L. MOORE, AND PAUL A. LIDDELL — 87

11. Use of Borohydride Reduction Methods to Probe Carotenoid–Protein Binding — R. GÓMEZ AND J. C. G. MILICUA — 100

12. Gas-Phase Reaction Mass Spectrometric Analysis of Carotenoids — WILLIAM R. LUSBY, FREDERICK KHACHIK, GARY R. BEECHER, AND JAMES LAU — 111

13. Formation of Nonvolatile Compounds by Thermal Degradation of β-Carotene: Protection by Antioxidants — CLAUDETTE BERSET AND CLAIRE MARTY — 129

14. Natural Sources of Carotenoids from Plants and Oils — AUGUSTINE S. H. ONG AND E. S. TEE — 142

15. Distribution of Carotenoids — T. W. GOODWIN — 167

Section II. Separation and Quantitation

16. Stability of β-Carotene under Different Laboratory Conditions — GIORGIO SCITA — 175

17. Carotenoid Reversed-Phase High-Performance Liquid Chromatography Methods: Reference Compendium — NEAL E. CRAFT — 185

18. Separation and Quantification of Carotenoids in Human Plasma — FREDERICK KHACHIK, GARY R. BEECHER, MUDLAGIRI B. GOLI, WILLIAM R. LUSBY, AND CHARLES E. DAITCH — 205

19. Measurement of Carotenoids in Human and Monkey Retinas — GARRY J. HANDELMAN, D. MAX SNODDERLY, ALICE J. ADLER, MARK D. RUSSETT, AND EDWARD A. DRATZ — 220

20. Isolation of Fucoxanthin and Peridinin — JARLE ANDRÉ HAUGAN, TORUNN AAKERMANN, AND SYNNØVE LIAAEN-JENSEN — 231

21. Preparative High-Performance Liquid Chromatography of Carotenoids — GEORGE W. FRANCIS — 246

22. Profiling and Quantitation of Carotenoids by High-Performance Liquid Chromatography and Photodiode Array Detection — A. P. DE LEENHEER AND H. J. NELIS — 251

23. Mammalian Metabolism of Carotenoids Other Than β-Carotene — MICHELINE M. MATHEWS-ROTH AND NORMAN I. KRINSKY — 265

24. Extraction and Analysis by High-Performance Liquid Chromatography of Carotenoids in Human Serum — ARUN B. BARUA AND HAROLD C. FURR — 273

25. Analysis of Apocarotenoids and Retinoids by Capillary Gas Chromatography–Mass Spectrometry — HAROLD C. FURR, ANDREW J. CLIFFORD, AND A. DANIEL JONES — 281

26. Separation of Isomers of cis-β-Carotenes — KIYOSHI TSUKIDA — 291

27. Proton Nuclear Magnetic Resonance and Raman Spectroscopies of *cis–trans*-Carotenoids from Pigment–Protein Complexes	YASUSHI KOYAMA	298
28. Electron Paramagnetic Resonance Studies of Carotenoids	HARRY A. FRANK	305
29. Identification of Carotenoid Pigments in Birds	JOCELYN HUDON AND ALAN H. BRUSH	312
30. Fast-Atom Bombardment and Continuous-Flow Fast-Atom Bombardment Mass Spectrometry in Carotenoid Analysis	HAROLD H. SCHMITZ, RICHARD B. VAN BREEMEN, AND STEVEN J. SCHWARTZ	322
31. High-Resolution Analysis of Carotenoids in Human Plasma by High-Performance Liquid Chromatography	GARRY J. HANDELMAN, BINGHUA SHEN, AND NORMAN I. KRINSKY	336
32. Separation and Quantitation of Carotenoids in Foods	FREDERICK KHACHIK, GARY R. BEECHER, MUDLAGIRI B. GOLI, AND WILLIAM R. LUSBY	347
33. Distribution of Macular Pigment Components, Zeaxanthin and Lutein, in Human Retina	RICHARD A. BONE AND JOHN T. LANDRUM	360
34. Carotenoid Glycoside Ester from *Rhodococcus rhodochrous*	SHINICHI TAKAICHI AND JUN-ICHI ISHIDSU	366
35. Characterization of Carotenoids in Photosynthetic Bacteria	SHINICHI TAKAICHI AND KEIZO SHIMADA	374
36. Enhancement and Determination of Astaxanthin Accumulation in Green Alga *Haematococcus pluvialis*	SAMMY BOUSSIBA, LU FAN, AND AVIGAD VONSHAK	386
37. Simultaneous Quantitation and Separation of Carotenoids and Retinol in Human Milk by High-Performance Liquid Chromatography	A. R. GIULIANO, E. M. NEILSON, B. E. KELLEY, AND L. M. CANFIELD	391

Section III. Antioxidation and Singlet Oxygen Quenching

38. Antioxidant Effects of Carotenoids *in Vivo* and *in Vitro*: An Overview	PAOLA PALOZZA AND NORMAN I. KRINSKY	403
39. Efficiency of Singlet Oxygen Quenching by Carotenoids Measured by Near-Infrared Steady-State Luminescence	ESTHER OLIVEROS, PATRICIA MURASECCO-SUARDI, ANDRÉ M. BRAUN, AND HANS-JÜRGEN HANSEN	420
40. Assay of Lycopene and Other Carotenoids as Singlet Oxygen Quenchers	PAOLO DI MASCIO, ALFRED R. SANDQUIST, THOMAS P. A. DEVASAGAYAM, AND HELMUT SIES	429

41. Biosynthesis of β-Carotene in *Dunaliella* — AVIV SHAISH, AMI BEN-AMOTZ, AND MORDHAY AVRON — 439

42. Carotenoid Free Radicals — MICHAEL G. SIMIC — 444

43. Lipid Hydroperoxide Assay for Antioxidant Activity of Carotenoids — JUNJI TERAO, AKIHIKO NAGAO, DONG-KI PARK, AND BOEY PENG LIM — 454

44. Antioxidant Radical-Scavenging Activity of Carotenoids and Retinoids Compared to α-Tocopherol — MASAHIKO TSUCHIYA, GIORGIO SCITA, HANS-JOACHIM FREISLEBEN, VALERIAN E. KAGAN, AND LESTER PACKER — 460

45. Epoxide Products of β-Carotene Antioxidant Reactions — DANIEL C. LIEBLER AND TODD A. KENNEDY — 472

46. Techniques for Studying Photoprotective Function of Carotenoid Pigments — MICHELINE M. MATHEWS-ROTH — 479

AUTHOR INDEX 485

SUBJECT INDEX 503

Contributors to Volume 213

Article numbers are in parentheses following the names of contributors.
Affiliations listed are current.

TORUNN AAKERMANN (20), *The Norwegian Institute of Technology, Organic Chemistry Laboratories, The University of Trondheim, N-7034 Trondheim-Nth, Norway*

ALICE J. ADLER (19), *Eye Research Institute, Boston, Massachusetts 02114*

MORDHAY AVRON* (41), *Department of Biochemistry, The Weizmann Institute of Sciences, Rehovot 76100, Israel*

ARUN B. BARUA (24), *Department of Biochemistry and Biophysics, Iowa State University, Ames, Iowa 50011*

GARY R. BEECHER (12, 18, 32), *Nutrient Composition Laboratory, Beltsville Human Nutrition Research Center, U.S. Department of Agriculture, Agricultural Research Service, Beltsville, Maryland 20705*

AMI BEN-AMOTZ (41), *Israel Oceanographic and Limnological Research, Tel-Shikmona, Haifa 31080, Israel*

H. ROBERT BERGEN, III (6), *Department of Chemistry, University of Wisconsin—La Crosse, La Crosse, Wisconsin 54601*

CLAUDETTE BERSET (13), *Laboratoire Pigments Naturels et Couleur des Aliments, Département Science de l'Aliment, Ecole Nationale Supérieure des Industries, Agricoles et Alimentaires, 91305 Massy, France*

PETER BEYER (8), *Institut für Biologie II, Universität Freiburg, D-7800 Freiburg, Germany*

RICHARD A. BONE (33), *Department of Physics, Florida International University, Miami, Florida 33199*

SAMMY BOUSSIBA (36), *Micro-algal Biotechnology Laboratory, The Jacob Blaustein Institute for Desert Research, Ben Gurion University of the Negev, Sede-Boker 84990, Israel*

ANDRÉ M. BRAUN (39), *Département de Chimie, Institut de Chimie Physique, Ecole Polytechnique Fédérale de Lausanne, CH-1015 Lausanne, Switzerland*

ALAN H. BRUSH (29), *Department of Physiology and Neurobiology, University of Connecticut, Storrs, Connecticut 06268*

WALTER BURGER (5), *Roche Research Center, Hoffman–LaRoche Inc., Nutley, New Jersey 07110*

L. M. CANFIELD (37), *Department of Biochemistry, University of Arizona, Tucson, Arizona 85721*

SATISH C. CHOUDHRY (5), *Roche Research Center, Hoffman–LaRoche Inc., Nutley, New Jersey 07110*

ANDREW J. CLIFFORD (25), *Department of Nutrition, Univeristy of California, Davis, Davis, California 95616*

THERESE M. COTTON (4), *Department of Chemistry, Ames Laboratory, Iowa State University, Ames, Iowa 50011*

NEAL E. CRAFT (17), *Chemical Science and Technology Laboratory, Organic Analytical Research Division, National Institute of Standards and Technology, Gaithersburg, Maryland 20899*

J. CROUZET (7), *Centre de Génie et de Technologie Alimentaires Institut des Sciences de l'Ingenieur, Université de Montpellier II, F-34095 Montpellier Cedex 05, France*

JOSEPH CUPANO (5), *Roche Research Center, Hoffman–LaRoche Inc., Nutley, New Jersey 07110*

CHARLES E. DAITCH (18), *Nutrient Composition Laboratory, Beltsville Human Nutrition, Research Center, U.S. Department of Agriculture, Agricultural Research Service, Beltsville, Maryland 20705*

* Deceased.

A. P. DE LEENHEER (22), *Laboratoria voor Medische Biochemie en voor Klinische Analyse, Universiteit Gent, B-9000 Gent, Belgium*

THOMAS P. A. DEVASAGAYAM (40), *Biochemistry Division, Bhabha Atomic Research Center, Bombay, India*

PAOLO DI MASCIO (40), *Instituto de Quimica, University of Sao Paulo, San Paulo, Brazil*

EDWARD A. DRATZ (19), *Department of Chemistry and Biochemistry, Montana State University, Bozeman, Montana 59717*

LU FAN (36), *Micro-algal Biotechnology Laboratory, The Jacob Blaustein Institute for Desert Research, Ben Gurion University of the Negev, Sede-Boker 84990, Israel*

J. P. FÉRÉZOU (9), *Ecole Normale Supérieure, Laboratoire de Chimie, 75231 Paris Cedex 05, France*

GEORGE W. FRANCIS (21), *Department of Chemistry, University of Bergen, N-5007 Bergen, Norway*

HARRY A. FRANK (28), *Department of Chemistry, University of Connecticut, Storrs, Connecticut 06269*

HANS-JOACHIM FREISLEBEN (44), *Klinikum der Johann Goethe-Universität, Gustav-Embden-Zentrum der Biologischen Chemie, Abt. Mikrobiologische Chemie, 6000 Frankfurt/Main, Germany*

HAROLD C. FURR (24, 25), *Department of Nutritional Sciences, University of Connecticut, Storrs, Connecticut 06269*

A. R. GIULIANO (37), *Department of Biochemistry, University of Arizona, Tucson, Arizona 85721*

MUDLAGIRI B. GOLI (18, 32), *Center for Drug Design and Development, College of Pharmacy, University of Toledo, Toledo, Ohio 43606*

R. GÓMEZ (11), *Department of Biochemistry and Molecular Biology, University of the Basque Country, 48080 Bilbao, Spain*

T. W. GOODWIN (15), *Department of Biochemistry, University of Liverpool, Liverpool L69 3BX, England*

DEVENS GUST (10), *Department of Chemistry and Biochemistry, Center for Early Events in Photosynthesis, Arizona State University, Tempe, Arizona 85287*

GARRY J. HANDELMAN (19, 31), *Department of Nutrition, University of California, Davis, Davis, California 95616*

HANS-JÜRGEN HANSEN (39), *Organisch-Chemisches Institut, Universität Zürich, CH-8057 Zürich, Switzerland*

JARLE ANDRÉ HAUGAN (20), *The Norwegian Institute of Technology, Organic Chemistry Laboratories, The University of Trondheim, N-7034 Trondheim-Nth, Norway*

JOCELYN HUDON (29), *Department of Veterinary Anatomy, The University of Saskatchewan, Saskatoon, Saskatchewan, Canada S7N OWO*

JUN-ICHI ISHIDSU (34), *Biological Laboratory, Nippon Medical School, Kawasaki 211, Japan*

MASAYOSHI ITO (2), *Kobe Women's College of Pharmacy, Kobe 658, Japan*

A. DANIEL JONES (25), *Facility for Advanced Instrumentation, University of California, Davis, Davis, California 95616*

VALERIAN E. KAGAN (44), *Department of Molecular and Cell Biology, University of California, Berkeley, Berkeley, California 94720*

P. KANASAWUD (7), *Department of Chemistry, Chiang Mai University, Chiang Mai, Thailand 50002*

B. E. KELLY (37), *Department of Biochemistry, University of Arizona, Tucson, Arizona 85721*

TODD A. KENNEDY (45), *Department of Pharmacology and Toxicology, College of Pharmacy, University of Arizona, Tucson, Arizona 85721*

FREDERICK KHACHIK (12, 18, 32), *Nutrient Composition Laboratory, Beltsville Human Nutrition Research Center, U.S. Department of Agriculture, Agricultural Research Service, Beltsville, Maryland 20705*

JAE-HO KIM (4), *Department of Chemistry,*

Ames Laboratory, Iowa State University, Ames, Iowa 50011

HANS KLEINIG (8), *Institut Für Biologie II, Universität Freiburg, D-7800 Freiburg, Germany*

YASUSHI KOYAMA (27), *Faculty of Science, Kwansei Gakuin University, Uegahara, Nishinomiya 662, Japan*

NORMAN I. KRINSKY (23, 31, 38), *Department of Biochemistry, Tufts University School of Medicine, Boston, Massachusetts 02111*

JOHN T. LANDRUM (33), *Department of Chemistry, Florida International University, Miami, Florida 33199*

JAMES LAU (12), *Analytical Instruments Division, Hewlett-Packard, Paramus, New Jersey 07652*

SYNNØVE LIAAEN-JENSEN (20), *The Norwegian Institute of Technology, Organic Chemistry Laboratories, The University of Trondheim, N-7034 Trondheim-Nth, Norway*

PAUL A. LIDDELL (10), *Department of Chemistry and Biochemistry, Center for Early Events in Photosynthesis, Arizona State University, Tempe, Arizona 85287*

DANIEL C. LIEBLER (45), *Department of Pharmacology and Toxicology, College of Pharmacy, University of Arizona, Tucson, Arizona 85721*

ARNOLD A. LIEBMAN (5), *Roche Research Center, Hoffman-LaRoche Inc., Nutley, New Jersey 07110*

BOEY PENG LIM (43), *School of Chemical Sciences, Universiti Sains Malaysia, USM Penang 11800, Malaysia*

WILLIAM R. LUSBY (12, 18, 32), *Insect Neurobiology and Hormone Laboratory, U.S. Department of Agriculture, Agriculture Research Service, Beltsville, Maryland 20705*

CLAIRE MARTY (13), *Laboratoire Pigments Naturels et Couleur des Aliments, Département Science de l'Aliment, Ecole Nationale Supérieure des Industries, Agricoles et Alimentaires, 91305 Massy, France*

MICHELINE M. MATHEWS-ROTH (23, 46), *Department of Medicine, Channing Laboratory, Harvard Medical School, Boston, Massachusetts 02115*

TAKAO MATSUNO (3), *Department of Natural Products Research, Kyoto Pharmaceutical University, Kyoto 607, Japan*

J. C. G. MILICUA (11), *Department of Biochemistry and Molecular Biology, University of the Basque Country, 48080 Bilbao, Spain*

ANA L. MOORE (10), *Department of Chemistry and Biochemistry, Center for Early Events in Photosynthesis, Arizona State University, Tempe, Arizona 85287*

THOMAS A. MOORE (10), *Department of Chemistry and Biochemistry, Center for Early Events in Photosynthesis, Arizona State University, Tempe, Arizona 85287*

PATRICIA MURASECCO-SUARDI (39), *Departement de Chimie, Institut de Chimie Physique, Ecole Polytechnique Fédérale de Lausanne, CH-1015 Lausanne, Switzerland*

AKIHIKO NAGAO (43), *National Food Research Institute, Ministry of Agriculture, Forestry and Fishery, Tsukuba, Ibaraki 305, Japan*

E. M. NEILSON (37), *Department of Biochemistry, University of Arizona, Tucson, Arizona 85721*

H. J. NELIS (22), *Laboratoria voor Farmaceutische Microbiologie, Universiteit Gent, B-9000 Gent, Belgium*

ESTHER OLIVEROS (39), *Département de Chimie, Institut de Chimie Physique, Ecole Polytechnique Fédérale de Lausanne, CH-1015 Lausanne, Switzerland*

AUGUSTINE S. H. ONG (14), *Malaysian Palm Oil Promotion Council, 50450 Kuala Lumpur, Malaysia*

LESTER PACKER (44), *Department of Molecular and Cell Biology, University of California, Berkeley, Berkeley, California 94720*

PAOLA PALOZZA (38), *Istituto di Patologia Generale, Universita Cattolica del Sacro Cuore, Roma 01168, Italy*

DONG-KI PARK (43), *Department of Biochemistry, Kon-Kok University, Chung-Ju, Korea*

HANSPETER PFANDER (1), *Institute of Organic Chemistry, University of Bern, CH-3012 Bern, Switzerland*

RAFAEL PICOREL (4), *Department of Plant Nutrition, Estación Experimental Aula Dei, 50080 Zaragoza, Spain*

MARK D. RUSSETT (19), *Department of Biochemistry, Tufts University School of Medicine, Boston, Massachusetts 02111*

HAROLD H. SCHMITZ (30), *Department of Food Science, College of Agricultural and Life Sciences, North Carolina State University, Raleigh, North Carolina 27695*

STEVEN J. SCHWARTZ (30), *Department of Food Science, College of Agricultural and Life Sciences, North Carolina State University, Raleigh, North Carolina 27695*

GIORGIO SCITA (16, 44), *Department of Nutrition Sciences, University of California, Berkeley, Berkeley, California 94720*

MICHAEL SEIBERT (4), *Photoconversion Branch, National Renewable Energy Laboratory, Golden, Colorado 80401*

AVIV SHAISH (41), *Department of Biochemistry, The Weizmann Institute of Science, Rehovot 76100, Israel*

BINGHUA SHEN (31), *Department of Biochemistry, Tufts University School of Medicine, Boston, Massachusetts 02111*

YUKO SHIBATA (2), *Kobe Women's College of Pharmacy, Kobe 658, Japan*

KEIZO SHIMADA (35), *Department of Biology, Tokyo Metropolitan University, Hachiohji, Tokyo 192-03, Japan*

HELMUT SIES (40), *Institut für Physiologische Chemie 1, Universität Düsseldorf, D-4000 Düsseldorf 1, Germany*

MICHAEL G. SIMIC (42), *Department of Pharmacology and Toxicology, School of Pharmacy, University of Baltimore, Baltimore, Maryland 21201*

D. MAX SNODDERLY (19), *Eye Research Institute, Boston, Massachusetts 02114*

ALFRED R. SUNDQUIST (40), *Institut für Physiologische Chemie 1, Universität Düsseldorf, D-4000 Düsseldorf 1, Germany*

SHINICHI TAKAICHI (34, 35), *Biological Laboratory, Nippon Medical School, Kawasaki 211, Japan*

E. S. TEE (14), *Division of Human Nutrition, Institute for Medical Research, 50588 Kuala Lumpur, Malaysia*

JUNJI TERAO (43), *National Food Research Institute, Ministry of Agriculture, Forestry and Fishery, Tsukuba, Ibaraki 305, Japan*

MASAHIKO TSUCHIYA (44), *Department of Molecular and Cell Biology, University of California, Berkeley, Berkeley, California 94720*

KIYOSHI TSUKIDA (26), *Kobe Women's College of Pharmacy, Kobe 658, Japan*

RICHARD B. VAN BREEMEN (30), *Department of Chemistry, North Carolina State University, Raleigh, North Carolina 27695*

AVIGAD VONSHAK (36), *Micro-algal Biotechnology Laboratory, The Jacob Blaustein Institute for Desert Research, Ben Gurion University of the Negev, Sede-Boker 84990, Israel*

YUMIKO YAMANO (2), *Kobe Women's College of Pharmacy, Kobe 658, Japan*

Preface

Carotenoids are unique pigments synthesized in photosynthetic higher plants and photosynthetic microorganisms which serve essential functions in the protection against singlet oxygen-generated damage by photosensitized reactions. In the plant they participate in light-harvesting reactions and may act as a protective covering for certain higher plant species. In animals, carotenoids have attracted considerable interest, beginning over fifty years ago when it was discovered that they are precursors for vitamin A and its derivatives. Mammalian species do not synthesize carotenoids or vitamin A. These essential substances are therefore derived solely from plant carotenoids, which are absorbed and stored in tissues and subsequently metabolized.

From a biomedical and biotechnological viewpoint, the carotenoids have also received considerable attention recently. In numerous epidemiological studies, the dietary intake of foods rich in carotenoids, but not of preformed vitamin A, has been associated with a reduced risk of some types of cancer. Chemical studies have shown that the carotenoids are very powerful quenchers of singlet oxygen and other reactive oxygen species.

Carotenoids, which are considered to be safe and stable, are currently used primarily as colorants in food products such as margarine. However, there is increasing interest in their possible role in maintaining health particularly in regard to lowering the risk of cancer. Many multivitamin supplements now contain beta carotene.

In addition to being of widespread biomedical interest, the carotenoids and their biosynthetic pathways are also being considered as targets for precise inhibitors that would act as herbicides.

Studies of the genetics and molecular biology of carotenoids in microbial photosynthetic systems, or nonphotosynthetic organisms such as *Escherichia coli,* are leading to the identification of the gene sequence homologies and differences between microbial and higher plant systems. This area of genetics and molecular and cell biology is undergoing rapid development. It is suspected that carotenoids, or perhaps their metabolic products, may act as regulators of gene expression.

In addition to research on their natural occurrence, structure, and biological activity, the distribution of carotenoids in animal and plant tissues has also been receiving considerable attention.

These developments have led to an explosion of activity in this research area. This volume and its companion Volume 214 of *Methods in Enzymology* present a comprehensive and state-of-the-art compilation of the molecular and cellular methodologies needed for pursuing research with carotenoids.

LESTER PACKER

METHODS IN ENZYMOLOGY

VOLUME I. Preparation and Assay of Enzymes
Edited by SIDNEY P. COLOWICK AND NATHAN O. KAPLAN

VOLUME II. Preparation and Assay of Enzymes
Edited by SIDNEY P. COLOWICK AND NATHAN O. KAPLAN

VOLUME III. Preparation and Assay of Substrates
Edited by SIDNEY P. COLOWICK AND NATHAN O. KAPLAN

VOLUME IV. Special Techniques for the Enzymologist
Edited by SIDNEY P. COLOWICK AND NATHAN O. KAPLAN

VOLUME V. Preparation and Assay of Enzymes
Edited by SIDNEY P. COLOWICK AND NATHAN O. KAPLAN

VOLUME VI. Preparation and Assay of Enzymes (*Continued*)
Preparation and Assay of Substrates
Special Techniques
Edited by SIDNEY P. COLOWICK AND NATHAN O. KAPLAN

VOLUME VII. Cumulative Subject Index
Edited by SIDNEY P. COLOWICK AND NATHAN O. KAPLAN

VOLUME VIII. Complex Carbohydrates
Edited by ELIZABETH F. NEUFELD AND VICTOR GINSBURG

VOLUME IX. Carbohydrate Metabolism
Edited by WILLIS A. WOOD

VOLUME X. Oxidation and Phosphorylation
Edited by RONALD W. ESTABROOK AND MAYNARD E. PULLMAN

VOLUME XI. Enzyme Structure
Edited by C. H. W. HIRS

VOLUME XII. Nucleic Acids (Parts A and B)
Edited by LAWRENCE GROSSMAN AND KIVIE MOLDAVE

VOLUME XIII. Citric Acid Cycle
Edited by J. M. LOWENSTEIN

VOLUME XIV. Lipids
Edited by J. M. LOWENSTEIN

VOLUME XV. Steroids and Terpenoids
Edited by RAYMOND B. CLAYTON

VOLUME XVI. Fast Reactions
Edited by KENNETH KUSTIN

VOLUME XVII. Metabolism of Amino Acids and Amines (Parts A and B)
Edited by HERBERT TABOR AND CELIA WHITE TABOR

VOLUME XVIII. Vitamins and Coenzymes (Parts A, B, and C)
Edited by DONALD B. MCCORMICK AND LEMUEL D. WRIGHT

VOLUME XIX. Proteolytic Enzymes
Edited by GERTRUDE E. PERLMANN AND LASZLO LORAND

VOLUME XX. Nucleic Acids and Protein Synthesis (Part C)
Edited by KIVIE MOLDAVE AND LAWRENCE GROSSMAN

VOLUME XXI. Nucleic Acids (Part D)
Edited by LAWRENCE GROSSMAN AND KIVIE MOLDAVE

VOLUME XXII. Enzyme Purification and Related Techniques
Edited by WILLIAM B. JAKOBY

VOLUME XXIII. Photosynthesis (Part A)
Edited by ANTHONY SAN PIETRO

VOLUME XXIV. Photosynthesis and Nitrogen Fixation (Part B)
Edited by ANTHONY SAN PIETRO

VOLUME XXV. Enzyme Structure (Part B)
Edited by C. H. W. HIRS AND SERGE N. TIMASHEFF

VOLUME XXVI. Enzyme Structure (Part C)
Edited by C. H. W. HIRS AND SERGE N. TIMASHEFF

VOLUME XXVII. Enzyme Structure (Part D)
Edited by C. H. W. HIRS AND SERGE N. TIMASHEFF

VOLUME XXVIII. Complex Carbohydrates (Part B)
Edited by VICTOR GINSBURG

VOLUME XXIX. Nucleic Acids and Protein Synthesis (Part E)
Edited by LAWRENCE GROSSMAN AND KIVIE MOLDAVE

VOLUME XXX. Nucleic Acids and Protein Synthesis (Part F)
Edited by KIVIE MOLDAVE AND LAWRENCE GROSSMAN

VOLUME XXXI. Biomembranes (Part A)
Edited by SIDNEY FLEISCHER AND LESTER PACKER

VOLUME XXXII. Biomembranes (Part B)
Edited by SIDNEY FLEISCHER AND LESTER PACKER

VOLUME XXXIII. Cumulative Subject Index Volumes I–XXX
Edited by MARTHA G. DENNIS AND EDWARD A. DENNIS

VOLUME XXXIV. Affinity Techniques (Enzyme Purification: Part B)
Edited by WILLIAM B. JAKOBY AND MEIR WILCHEK

VOLUME XXXV. Lipids (Part B)
Edited by JOHN M. LOWENSTEIN

VOLUME XXXVI. Hormone Action (Part A: Steroid Hormones)
Edited by BERT W. O'MALLEY AND JOEL G. HARDMAN

VOLUME XXXVII. Hormone Action (Part B: Peptide Hormones)
Edited by BERT W. O'MALLEY AND JOEL G. HARDMAN

VOLUME XXXVIII. Hormone Action (Part C: Cyclic Nucleotides)
Edited by JOEL G. HARDMAN AND BERT W. O'MALLEY

VOLUME XXXIX. Hormone Action (Part D: Isolated Cells, Tissues, and Organ Systems)
Edited by JOEL G. HARDMAN AND BERT W. O'MALLEY

VOLUME XL. Hormone Action (Part E: Nuclear Structure and Function)
Edited by BERT W. O'MALLEY AND JOEL G. HARDMAN

VOLUME XLI. Carbohydrate Metabolism (Part B)
Edited by W. A. WOOD

VOLUME XLII. Carbohydrate Metabolism (Part C)
Edited by W. A. WOOD

VOLUME XLIII. Antibiotics
Edited by JOHN H. HASH

VOLUME XLIV. Immobilized Enzymes
Edited by KLAUS MOSBACH

VOLUME XLV. Proteolytic Enzymes (Part B)
Edited by LASZLO LORAND

VOLUME XLVI. Affinity Labeling
Edited by WILLIAM B. JAKOBY AND MEIR WILCHEK

VOLUME XLVII. Enzyme Structure (Part E)
Edited by C. H. W. HIRS AND SERGE N. TIMASHEFF

VOLUME XLVIII. Enzyme Structure (Part F)
Edited by C. H. W. HIRS AND SERGE N. TIMASHEFF

VOLUME XLIX. Enzyme Structure (Part G)
Edited by C. H. W. HIRS AND SERGE N. TIMASHEFF

VOLUME L. Complex Carbohydrates (Part C)
Edited by VICTOR GINSBURG

VOLUME LI. Purine and Pyrimidine Nucleotide Metabolism
Edited by PATRICIA A. HOFFEE AND MARY ELLEN JONES

VOLUME LII. Biomembranes (Part C: Biological Oxidations)
Edited by SIDNEY FLEISCHER AND LESTER PACKER

VOLUME LIII. Biomembranes (Part D: Biological Oxidations)
Edited by SIDNEY FLEISCHER AND LESTER PACKER

VOLUME LIV. Biomembranes (Part E: Biological Oxidations)
Edited by SIDNEY FLEISCHER AND LESTER PACKER

VOLUME LV. Biomembranes (Part F: Bioenergetics)
Edited by SIDNEY FLEISCHER AND LESTER PACKER

VOLUME LVI. Biomembranes (Part G: Bioenergetics)
Edited by SIDNEY FLEISCHER AND LESTER PACKER

VOLUME LVII. Bioluminescence and Chemiluminescence
Edited by MARLENE A. DELUCA

VOLUME LVIII. Cell Culture
Edited by WILLIAM B. JAKOBY AND IRA PASTAN

VOLUME LIX. Nucleic Acids and Protein Synthesis (Part G)
Edited by KIVIE MOLDAVE AND LAWRENCE GROSSMAN

VOLUME LX. Nucleic Acids and Protein Synthesis (Part H)
Edited by KIVIE MOLDAVE AND LAWRENCE GROSSMAN

VOLUME 61. Enzyme Structure (Part H)
Edited by C. H. W. HIRS AND SERGE N. TIMASHEFF

VOLUME 62. Vitamins and Coenzymes (Part D)
Edited by DONALD B. MCCORMICK AND LEMUEL D. WRIGHT

VOLUME 63. Enzyme Kinetics and Mechanism (Part A: Initial Rate and Inhibitor Methods)
Edited by DANIEL L. PURICH

VOLUME 64. Enzyme Kinetics and Mechanism (Part B: Isotopic Probes and Complex Enzyme Systems)
Edited by DANIEL L. PURICH

VOLUME 65. Nucleic Acids (Part I)
Edited by LAWRENCE GROSSMAN AND KIVIE MOLDAVE

VOLUME 66. Vitamins and Coenzymes (Part E)
Edited by DONALD B. MCCORMICK AND LEMUEL D. WRIGHT

VOLUME 67. Vitamins and Coenzymes (Part F)
Edited by DONALD B. MCCORMICK AND LEMUEL D. WRIGHT

VOLUME 68. Recombinant DNA
Edited by RAY WU

VOLUME 69. Photosynthesis and Nitrogen Fixation (Part C)
Edited by ANTHONY SAN PIETRO

VOLUME 70. Immunochemical Techniques (Part A)
Edited by HELEN VAN VUNAKIS AND JOHN J. LANGONE

VOLUME 71. Lipids (Part C)
Edited by JOHN M. LOWENSTEIN

VOLUME 72. Lipids (Part D)
Edited by JOHN M. LOWENSTEIN

VOLUME 73. Immunochemical Techniques (Part B)
Edited by JOHN J. LANGONE AND HELEN VAN VUNAKIS

VOLUME 74. Immunochemical Techniques (Part C)
Edited by JOHN J. LANGONE AND HELEN VAN VUNAKIS

VOLUME 75. Cumulative Subject Index Volumes XXXI, XXXII, XXXIV–LX
Edited by EDWARD A. DENNIS AND MARTHA G. DENNIS

VOLUME 76. Hemoglobins
Edited by ERALDO ANTONINI, LUIGI ROSSI-BERNARDI, AND EMILIA CHIANCONE

VOLUME 77. Detoxication and Drug Metabolism
Edited by WILLIAM B. JAKOBY

VOLUME 78. Interferons (Part A)
Edited by SIDNEY PESTKA

VOLUME 79. Interferons (Part B)
Edited by SIDNEY PESTKA

VOLUME 80. Proteolytic Enzymes (Part C)
Edited by LASZLO LORAND

VOLUME 81. Biomembranes (Part H: Visual Pigments and Purple Membranes, I)
Edited by LESTER PACKER

VOLUME 82. Structural and Contractile Proteins (Part A: Extracellular Matrix)
Edited by LEON W. CUNNINGHAM AND DIXIE W. FREDERIKSEN

VOLUME 83. Complex Carbohydrates (Part D)
Edited by VICTOR GINSBURG

VOLUME 84. Immunochemical Techniques (Part D: Selected Immunoassays)
Edited by JOHN J. LANGONE AND HELEN VAN VUNAKIS

VOLUME 85. Structural and Contractile Proteins (Part B: The Contractile Apparatus and the Cytoskeleton)
Edited by DIXIE W. FREDERIKSEN AND LEON W. CUNNINGHAM

VOLUME 86. Prostaglandins and Arachidonate Metabolites
Edited by WILLIAM E. M. LANDS AND WILLIAM L. SMITH

VOLUME 87. Enzyme Kinetics and Mechanism (Part C: Intermediates, Stereochemistry, and Rate Studies)
Edited by DANIEL L. PURICH

VOLUME 88. Biomembranes (Part I: Visual Pigments and Purple Membranes, II)
Edited by LESTER PACKER

VOLUME 89. Carbohydrate Metabolism (Part D)
Edited by WILLIS A. WOOD

VOLUME 90. Carbohydrate Metabolism (Part E)
Edited by WILLIS A. WOOD

VOLUME 91. Enzyme Structure (Part I)
Edited by C. H. W. HIRS AND SERGE N. TIMASHEFF

VOLUME 92. Immunochemical Techniques (Part E: Monoclonal Antibodies and General Immunoassay Methods)
Edited by JOHN J. LANGONE AND HELEN VAN VUNAKIS

VOLUME 93. Immunochemical Techniques (Part F: Conventional Antibodies, Fc Receptors, and Cytotoxicity)
Edited by JOHN J. LANGONE AND HELEN VAN VUNAKIS

VOLUME 94. Polyamines
Edited by HERBERT TABOR AND CELIA WHITE TABOR

VOLUME 95. Cumulative Subject Index Volumes 61–74, 76–80
Edited by EDWARD A. DENNIS AND MARTHA G. DENNIS

VOLUME 96. Biomembranes [Part J: Membrane Biogenesis: Assembly and Targeting (General Methods; Eukaryotes)]
Edited by SIDNEY FLEISCHER AND BECCA FLEISCHER

VOLUME 97. Biomembranes [Part K: Membrane Biogenesis: Assembly and Targeting (Prokaryotes, Mitochondria, and Chloroplasts)]
Edited by SIDNEY FLEISCHER AND BECCA FLEISCHER

VOLUME 98. Biomembranes (Part L: Membrane Biogenesis: Processing and Recycling)
Edited by SIDNEY FLEISCHER AND BECCA FLEISCHER

VOLUME 99. Hormone Action (Part F: Protein Kinases)
Edited by JACKIE D. CORBIN AND JOEL G. HARDMAN

VOLUME 100. Recombinant DNA (Part B)
Edited by RAY WU, LAWRENCE GROSSMAN, AND KIVIE MOLDAVE

VOLUME 101. Recombinant DNA (Part C)
Edited by RAY WU, LAWRENCE GROSSMAN, AND KIVIE MOLDAVE

VOLUME 102. Hormone Action (Part G: Calmodulin and Calcium-Binding Proteins)
Edited by ANTHONY R. MEANS AND BERT W. O'MALLEY

VOLUME 103. Hormone Action (Part H: Neuroendocrine Peptides)
Edited by P. MICHAEL CONN

VOLUME 104. Enzyme Purification and Related Techniques (Part C)
Edited by WILLIAM B. JAKOBY

VOLUME 105. Oxygen Radicals in Biological Systems
Edited by LESTER PACKER

VOLUME 106. Posttranslational Modifications (Part A)
Edited by FINN WOLD AND KIVIE MOLDAVE

VOLUME 107. Posttranslational Modifications (Part B)
Edited by FINN WOLD AND KIVIE MOLDAVE

VOLUME 108. Immunochemical Techniques (Part G: Separation and Characterization of Lymphoid Cells)
Edited by GIOVANNI DI SABATO, JOHN J. LANGONE, AND HELEN VAN VUNAKIS

VOLUME 109. Hormone Action (Part I: Peptide Hormones)
Edited by LUTZ BIRNBAUMER AND BERT W. O'MALLEY

VOLUME 110. Steroids and Isoprenoids (Part A)
Edited by JOHN H. LAW AND HANS C. RILLING

VOLUME 111. Steroids and Isoprenoids (Part B)
Edited by JOHN H. LAW AND HANS C. RILLING

VOLUME 112. Drug and Enzyme Targeting (Part A)
Edited by KENNETH J. WIDDER AND RALPH GREEN

VOLUME 113. Glutamate, Glutamine, Glutathione, and Related Compounds
Edited by ALTON MEISTER

VOLUME 114. Diffraction Methods for Biological Macromolecules (Part A)
Edited by HAROLD W. WYCKOFF, C. H. W. HIRS, AND SERGE N. TIMASHEFF

VOLUME 115. Diffraction Methods for Biological Macromolecules (Part B)
Edited by HAROLD W. WYCKOFF, C. H. W. HIRS, AND SERGE N. TIMASHEFF

VOLUME 116. Immunochemical Techniques (Part H: Effectors and Mediators of Lymphoid Cell Functions)
Edited by GIOVANNI DI SABATO, JOHN J. LANGONE, AND HELEN VAN VUNAKIS

VOLUME 117. Enzyme Structure (Part J)
Edited by C. H. W. HIRS AND SERGE N. TIMASHEFF

VOLUME 118. Plant Molecular Biology
Edited by ARTHUR WEISSBACH AND HERBERT WEISSBACH

VOLUME 119. Interferons (Part C)
Edited by SIDNEY PESTKA

VOLUME 120. Cumulative Subject Index Volumes 81–94, 96–101

VOLUME 121. Immunochemical Techniques (Part I: Hybridoma Technology and Monoclonal Antibodies)
Edited by JOHN J. LANGONE AND HELEN VAN VUNAKIS

VOLUME 122. Vitamins and Coenzymes (Part G)
Edited by FRANK CHYTIL AND DONALD B. MCCORMICK

VOLUME 123. Vitamins and Coenzymes (Part H)
Edited by FRANK CHYTIL AND DONALD B. MCCORMICK

VOLUME 124. Hormone Action (Part J: Neuroendocrine Peptides)
Edited by P. MICHAEL CONN

VOLUME 125. Biomembranes (Part M: Transport in Bacteria, Mitochondria, and Chloroplasts: General Approaches and Transport Systems)
Edited by SIDNEY FLEISCHER AND BECCA FLEISCHER

VOLUME 126. Biomembranes (Part N: Transport in Bacteria, Mitochondria, and Chloroplasts: Protonmotive Force)
Edited by SIDNEY FLEISCHER AND BECCA FLEISCHER

VOLUME 127. Biomembranes (Part O: Protons and Water: Structure and Translocation)
Edited by LESTER PACKER

VOLUME 128. Plasma Lipoproteins (Part A: Preparation, Structure, and Molecular Biology)
Edited by JERE P. SEGREST AND JOHN J. ALBERS

VOLUME 129. Plasma Lipoproteins (Part B: Characterization, Cell Biology, and Metabolism)
Edited by JOHN J. ALBERS AND JERE P. SEGREST

VOLUME 130. Enzyme Structure (Part K)
Edited by C. H. W. HIRS AND SERGE N. TIMASHEFF

VOLUME 131. Enzyme Structure (Part L)
Edited by C. H. W. HIRS AND SERGE N. TIMASHEFF

VOLUME 132. Immunochemical Techniques (Part J: Phagocytosis and Cell-Mediated Cytotoxicity)
Edited by GIOVANNI DI SABATO AND JOHANNES EVERSE

VOLUME 133. Bioluminescence and Chemiluminescence (Part B)
Edited by MARLENE DELUCA AND WILLIAM D. MCELROY

VOLUME 134. Structural and Contractile Proteins (Part C: The Contractile Apparatus and the Cytoskeleton)
Edited by RICHARD B. VALLEE

VOLUME 135. Immobilized Enzymes and Cells (Part B)
Edited by KLAUS MOSBACH

VOLUME 136. Immobilized Enzymes and Cells (Part C)
Edited by KLAUS MOSBACH

VOLUME 137. Immobilized Enzymes and Cells (Part D)
Edited by KLAUS MOSBACH

VOLUME 138. Complex Carbohydrates (Part E)
Edited by VICTOR GINSBURG

VOLUME 139. Cellular Regulators (Part A: Calcium- and CalmodulinBinding Proteins
Edited by ANTHONY R. MEANS AND P. MICHAEL CONN

VOLUME 140. Cumulative Subject Index Volumes 102–119, 121–134

VOLUME 141. Cellular Regulators (Part B: Calcium and Lipids)
Edited by P. MICHAEL CONN AND ANTHONY R. MEANS

VOLUME 142. Metabolism of Aromatic Amino Acids and Amines
Edited by SEYMOUR KAUFMAN

VOLUME 143. Sulfur and Sulfur Amino Acids
Edited by WILLIAM B. JAKOBY AND OWEN GRIFFITH

VOLUME 144. Structural and Contractile Proteins (Part D: Extracellular Matrix)
Edited by LEON W. CUNNINGHAM

VOLUME 145. Structural and Contractile Proteins (Part E: Extracellular Matrix)
Edited by LEON W. CUNNINGHAM

VOLUME 146. Peptide Growth Factors (Part A)
Edited by DAVID BARNES AND DAVID A. SIRBASKU

VOLUME 147. Peptide Growth Factors (Part B)
Edited by DAVID BARNES AND DAVID A. SIRBASKU

VOLUME 148. Plant Cell Membranes
Edited by LESTER PACKER AND ROLAND DOUCE

VOLUME 149. Drug and Enzyme Targeting (Part B)
Edited by RALPH GREEN AND KENNETH J. WIDDER

VOLUME 150. Immunochemical Techniques (Part K: *In Vitro* Models of B and T Cell Functions and Lymphoid Cell Receptors)
Edited by GIOVANNI DI SABATO

VOLUME 151. Molecular Genetics of Mammalian Cells
Edited by MICHAEL M. GOTTESMAN

VOLUME 152. Guide to Molecular Cloning Techniques
Edited by SHELBY L. BERGER AND ALAN R. KIMMEL

VOLUME 153. Recombinant DNA (Part D)
Edited by RAY WU AND LAWRENCE GROSSMAN

VOLUME 154. Recombinant DNA (Part E)
Edited by RAY WU AND LAWRENCE GROSSMAN

VOLUME 155. Recombinant DNA (Part F)
Edited by RAY WU

VOLUME 156. Biomembranes (Part P: ATP-Driven Pumps and Related Transport: The Na,K-Pump)
Edited by SIDNEY FLEISCHER AND BECCA FLEISCHER

VOLUME 157. Biomembranes (Part Q: ATP-Driven Pumps and Related Transport: Calcium, Proton, and Potassium Pumps)
Edited by SIDNEY FLEISCHER AND BECCA FLEISCHER

VOLUME 158. Metalloproteins (Part A)
Edited by JAMES F. RIORDAN AND BERT L. VALLEE

VOLUME 159. Initiation and Termination of Cyclic Nucleotide Action
Edited by JACKIE D. CORBIN AND ROGER A. JOHNSON

VOLUME 160. Biomass (Part A: Cellulose and Hemicellulose)
Edited by WILLIS A. WOOD AND SCOTT T. KELLOGG

VOLUME 161. Biomass (Part B: Lignin, Pectin, and Chitin)
Edited by WILLIS A. WOOD AND SCOTT T. KELLOGG

VOLUME 162. Immunochemical Techniques (Part L: Chemotaxis and Inflammation)
Edited by GIOVANNI DI SABATO

VOLUME 163. Immunochemical Techniques (Part M: Chemotaxis and Inflammation)
Edited by GIOVANNI DI SABATO

VOLUME 164. Ribosomes
Edited by HARRY F. NOLLER, JR. AND KIVIE MOLDAVE

VOLUME 165. Microbial Toxins: Tools for Enzymology
Edited by SIDNEY HARSHMAN

VOLUME 166. Branched-Chain Amino Acids
Edited by ROBERT HARRIS AND JOHN R. SOKATCH

VOLUME 167. Cyanobacteria
Edited by LESTER PACKER AND ALEXANDER N. GLAZER

VOLUME 168. Hormone Action (Part K: Neuroendocrine Peptides)
Edited by P. MICHAEL CONN

VOLUME 169. Platelets: Receptors, Adhesion, Secretion (Part A)
Edited by JACEK HAWIGER

VOLUME 170. Nucleosomes
Edited by PAUL M. WASSARMAN AND ROGER D. KORNBERG

VOLUME 171. Biomembranes (Part R: Transport Theory: Cells and Model Membranes)
Edited by SIDNEY FLEISCHER AND BECCA FLEISCHER

VOLUME 172. Biomembranes (Part S: Membrane Isolation and Characterization)
Edited by SIDNEY FLEISCHER AND BECCA FLEISCHER

VOLUME 173. Biomembranes [Part T: Cellular and Subcellular Transport: Eukaryotic (Nonepithelial) Cells]
Edited by SIDNEY FLEISCHER AND BECCA FLEISCHER

VOLUME 174. Biomembranes [Part U: Cellular and Subcellular Transport: Eukaryotic (Nonepithelial) Cells]
Edited by SIDNEY FLEISCHER AND BECCA FLEISCHER

VOLUME 175. Cumulative Subject Index Volumes 135–139, 141–167

VOLUME 176. Nuclear Magnetic Resonance (Part A: Spectral Techniques and Dynamics)
Edited by NORMAN J. OPPENHEIMER AND THOMAS L. JAMES

VOLUME 177. Nuclear Magnetic Resonance (Part B: Structure and Mechanism)
Edited by NORMAN N. OPPENHEIMER AND THOMAS L. JAMES

VOLUME 178. Antibodies, Antigens, and Molecular Mimicry
Edited by JOHN J. LANGONE

VOLUME 179. Complex Carbohydrates (Part F)
Edited by VICTOR GINSBURG

VOLUME 180. RNA Processing (Part A: General Methods)
Edited by JAMES E. DAHLBERG AND JOHN N. ABELSON

VOLUME 181. RNA Processing (Part B: Specific Methods)
Edited by JAMES E. DAHLBERG AND JOHN N. ABELSON

VOLUME 182. Guide to Protein Purification
Edited by MURRAY P. DEUTSCHER

VOLUME 183. Molecular Evolution: Computer Analysis of Protein and Nucleic Acid Sequences
Edited by RUSSELL F. DOOLITTLE

VOLUME 184. Avidin-Biotin Technology
Edited by MEIR WILCHEK AND EDWARD A. BAYER

VOLUME 185. Gene Expression Technology
Edited by DAVID V. GOEDDEL

VOLUME 186. Oxygen Radicals in Biological Systems (Part B: Oxygen Radicals and Antioxidents)
Edited by LESTER PACKER AND ALEXANDER N. GLAZER

VOLUME 187. Arachidonate Related Lipid Mediators
Edited by ROBERT C. MURPHY AND FRANK A. FITZPATRICK

VOLUME 188. Hydrocarbons and Methylotrophy
Edited by MARY E. LIDSTROM

VOLUME 189. Retinoids (Part A: Molecular and Metabolic Aspects)
Edited by LESTER PACKER

VOLUME 190. Retinoids (Part B: Cell Differentiation and Clinical Applications)
Edited by LESTER PACKER

VOLUME 191. Biomembranes (Part V: Cellular and Subcellular Transport: Epithelial Cells)
Edited by SIDNEY FLEISCHER AND BECCA FLEISCHER

VOLUME 192. Biomembranes (Part W: Cellular and Subcellular Transport: Epithelial Cells)
Edited by SIDNEY FLEISCHER AND BECCA FLEISCHER

VOLUME 193. Mass Spectrometry
Edited by JAMES A. MCCLOSKEY

VOLUME 194. Guide to Yeast Genetics and Molecular Biology
Edited by CHRISTINE GUTHRIE AND GERALD R. FINK

VOLUME 195. Adenylyl Cyclase, G Proteins, and Guanylyl Cyclase
Edited by ROGER A. JOHNSON AND JACKIE D. CORBIN

VOLUME 196. Molecular Motors and the Cytoskeleton
Edited by RICHARD B. VALLEE

VOLUME 197. Phospholipases
Edited by EDWARD A. DENNIS

VOLUME 198. Peptide Growth Factors (Part C)
Edited by DAVID BARNES, J. P. MATHER, AND GORDON H. SATO

VOLUME 199. Cumulative Subject Index Volumes 168–174, 176–194 (in preparation)

VOLUME 200. Protein Phosphorylation (Part A: Protein Kinases: Assays Purification, Antibodies, Functional Analysis, Cloning, and Expression)
Edited by TONY HUNTER AND BARTHOLOMEW M. SEFTON

VOLUME 201. Protein Phosphorylation (Part B: Analysis of Protein Phosphorylation, Protein Kinase Inhibitors, and Protein Phosphotases)
Edited by TONY HUNTER AND BARTHOLOMEW M. SEFTON

VOLUME 202. Molecular Design and Modeling: Concepts and Applications (Part A: Proteins, Peptides, and Enzymes)
Edited by JOHN J. LANGONE

VOLUME 203. Molecular Design and Modeling: Concepts and Applications (Part B: Antibodies and Antigens, Nucleic Acids, Polysaccharides, and Drugs)
Edited by JOHN J. LANGONE

VOLUME 204. Bacterial Genetic Systems
Edited by JEFFREY H. MILLER

VOLUME 205. Metallobiochemistry (Part B: Metallothionein and Related Molecules)
Edited by JAMES F. RIORDAN AND BERT L. VALLEE

VOLUME 206. Cytochrome P450
Edited by MICHAEL R. WATERMAN AND ERIC F. JOHNSON

VOLUME 207. Ion Channels
Edited by BERNARDO RUDY AND LINDA E. IVERSON

VOLUME 208. Protein–DNA Interactions
Edited by ROBERT T. SAUER

VOLUME 209. Phospholipid Biosynthesis
Edited by EDWARD A. DENNIS AND DENNIS E. VANCE

VOLUME 210. Numerical Computer Methods
Edited by LUDWIG BRAND AND MICHAEL L. JOHNSON

VOLUME 211. DNA Structures (Part A: Synthesis and Physical Analysis of DNA)
Edited by DAVID M. J. LILLEY AND JAMES E. DAHLBERG

VOLUME 212. DNA Structures (Part B: Chemical and Electrophoretic Analysis of DNA)
Edited by DAVID M. J. LILLEY AND JAMES E. DAHLBERG

VOLUME 213. Carotenoids (Part A: Chemistry, Separation, Quantitation, and Antioxidation)
Edited by LESTER PACKER

VOLUME 214. Carotenoids (Part B: Metabolism, Genetics, and Biosynthesis) (in preparation)
Edited by LESTER PACKER

VOLUME 215. Platelets: Receptors, Adhesion, Secretion (Part B)
Edited by JACEK J. HAWIGER

VOLUME 216. Recombinant DNA (Part G) (in preparation)
Edited by RAY WU

VOLUME 217. Recombinant DNA (Part H) (in preparation)
Edited by RAY WU

VOLUME 218. Recombinant DNA (Part I) (in preparation)
Edited by RAY WU

VOLUME 219. Reconstitution of Intracellular Transport (in preparation)
Edited by JAMES E. ROTHMAN

Section I

Chemistry: Synthesis, Properties, and Characterization

[1] Carotenoids: An Overview

By HANSPETER PFANDER

Introduction

Of the various classes of pigments in nature the carotenoids are among the most widespread and important ones, especially due to their most varied functions. In 1831 Wackenroder isolated carotene from carrots and in 1837 Berzelius named the yellow pigments from autumn leaves *xanthophylls*. This marks the beginning of carotenoid research and since then continuous developments have taken place. Because of their ubiquitous occurrence, different functions, and interesting properties carotenoids are the subject of interdisciplinary research in chemistry, biochemistry, biology, medicine, physics, and many other branches of science. The industrial production of carotenoids has also contributed to knowledge in this field.

This topic was first summarized in the book *Carotenoids* by Karrer and Jucker,[1] followed by *Carotenoids,* edited by Isler.[2] The biochemical aspects have been covered in *The Biochemistry of the Carotenoids* by Goodwin,[3] in *The Biochemistry of Natural Pigments* by Britton,[4] and, more recently, in *Plant Pigments,* edited by Goodwin.[5] The technical and nutritional applications of carotenoids have been treated in a book edited by Bauernfeind.[6] Advances in the entire field of carotenoid research are treated in the publications of the *International Symposium on Carotenoids.*[7-12]

---

[1] P. Karrer and E. Jucker, "Carotinoide." Birkhäuser, Basel, 1948.
[2] O. Isler (ed.), "Carotenoids." Birkhäuser, Basel, 1971.
[3] T. W. Goodwin, "The Biochemistry of the Carotenoids," 2nd Ed., Vols. 1 and 2. Chapman and Hall, London, 1980 and 1984.
[4] G. Britton, "The Biochemistry of Natural Pigments." Cambridge University Press, Cambridge, England, 1983.
[5] T. W. Goodwin (ed.), "Plant Pigments." Academic Press, San Diego, 1988.
[6] J. C. Bauernfeind (ed.), "Carotenoids as Colorants and Vitamin A Precursors: Technical and Nutritional Applications." Academic Press, New York, 1981.
[7] B. C. L. Weedon, "Carotenoids—4," Proc. 4th Int. Symp. Carotenoids. Pergamon, Oxford, 1976.
[8] T. W. Goodwin, "Carotenoids—5," Proc. 5th Int. Symp. Carotenoids. Pergamon, Oxford, 1979.
[9] G. Britton and T. W. Goodwin (eds.), "Carotenoid Chemistry and Biochemistry," Proc. 6th Int. Symp. Carotenoids. Pergamon, Oxford, 1982.
[10] 7th Int. Symp. Carotenoids, *Pure Appl. Chem.* **57,** 639 (1985).
[11] N. I. Krinsky, M. M. Mathews-Roth, and R. F. Taylor (eds.), "Carotenoids: Chemistry and Biology," Proc. 8th Int. Symp. Carotenoids. Plenum, New York, 1989.
[12] 9th Int. Symp. Carotenoids, *Pure Appl. Chem.* **63,** 1 (1991).

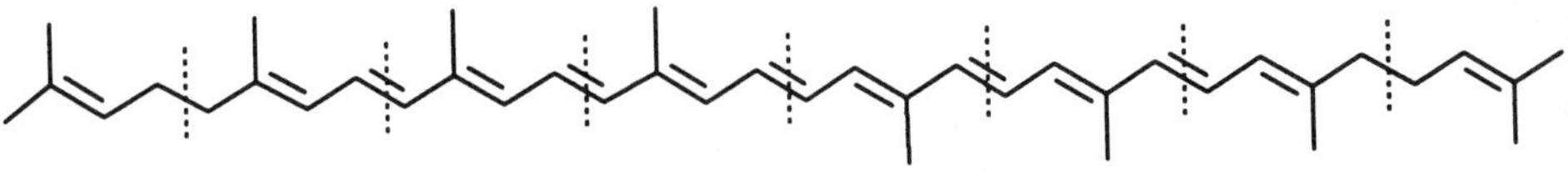

FIG. 1. Acyclic $C_{40}H_{56}$ structure (I).

Nomenclature and Structure

Carotenoids are a class of hydrocarbons (carotenes) and their oxygenated derivatives (xanthophylls). They consist of eight isoprenoid units joined in such a manner that the arrangement of isoprenoid units is reversed at the center of the molecule so that the two central methyl groups are in a 1,6-position relationship and the remaining nonterminal methyl groups are in a 1,5-position relationship. All carotenoids may be formally derived from the acyclic $C_{40}H_{56}$ structure (I) (see Fig. 1), having a long central chain of conjugated double bonds, by (1) hydrogenation, (2) dehydrogenation, (3) cyclization, or (4) oxidation, or any combination of these processes. The class also includes compounds that arise from certain rearrangements or degradations of the carbon skeleton (I), provided that the two central methyl groups are retained.

Rules for the nomenclature of carotenoids (semisystematic names) have been published by the International Union of Pure and Applied Chemistry (IUPAC) and IUPAC-International Union of Biochemists (IUB) Commissions on Nomenclature.[13] For the most common carotenoids trivial names are normally used. If these trivial names are used in a paper, the semisystematic name should always be given, in parentheses or in a footnote, at the first mention. All specific names are based on the stem name *carotene* (see Fig. 2), which corresponds to the structure and numbering in (II). The name of a specific compound is constructed by adding two Greek letters as prefixes (Table I) to the stem name carotene; the Greek letter prefixes are cited in alphabetical order.

The oxygenated carotenoids (xanthophylls) are names according to the usual rules of organic chemical nomenclature. The functions most frequently observed are hydroxy, methoxy, carboxy, oxo, aldehyde, and epoxy. In addition, carotenoids with triple bonds are also known. Important and characteristic carotenoids (Fig. 3) are lycopene (ψ,ψ-carotene) (I) and the compounds III–XIV shown in Fig. 3: β-carotene (β,β-carotene) (III), zeaxanthin [$(3R,3'R)$-β,β-carotene-3,3'-diol] (IV), lutein ["xanthophyll," $(3R,3'R,6'R)$-β,ε-carotene-3,3'-diol] (V), spirilloxanthin (1,1'-dimethoxy-3,4,3',4'-tetrahydro-1,2,1',2'-tetrahydro-ψ,ψ-carotene)

---

[13] *In* "Carotenoids" (O. Isler, ed.), p. 851. Birkhäuser, Basel, 1971.

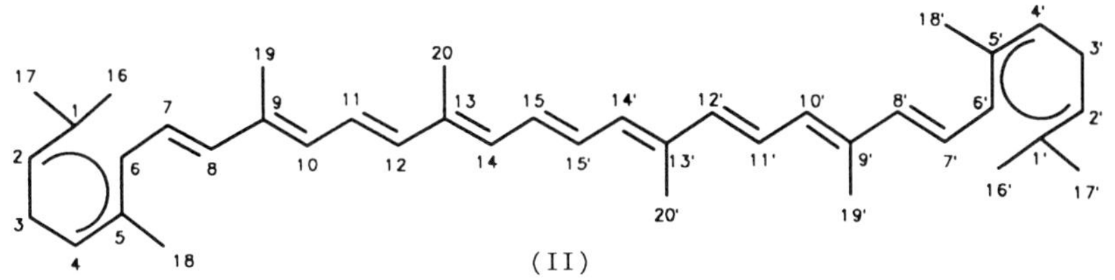

CAROTENOIDS: AN OVERVIEW

FIG. 2. Structure and numbering of stem name carotene (**II**).

TABLE I
PREFIXES TO STEM NAME CAROTENE

Greek letter	Structure	Greek letter	Structure
ψ	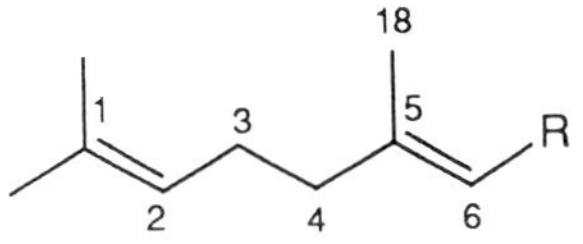 (acyclic)	κ	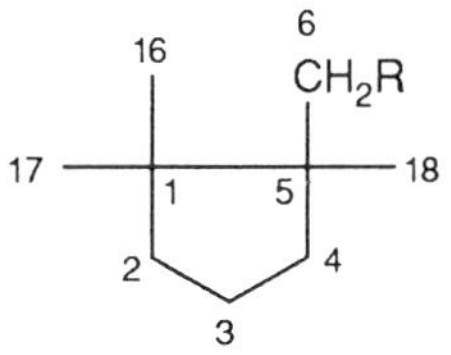 (cyclopentane with CH₂R)
β	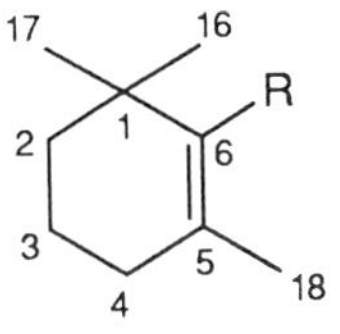 (cyclohexene, double bond at 5,6)	φ	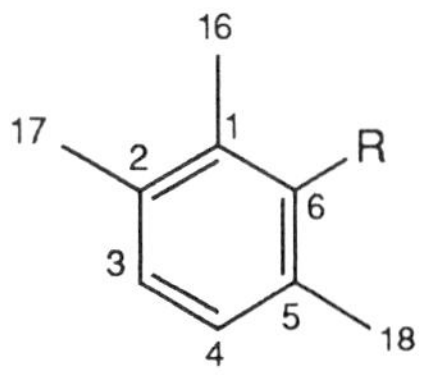 (aromatic ring)
ε	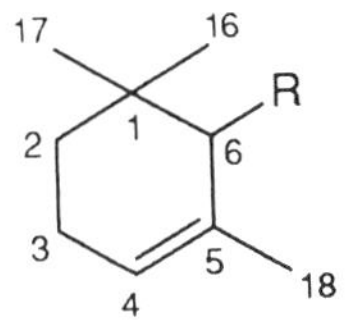 (cyclohexene, double bond at 4,5)	χ	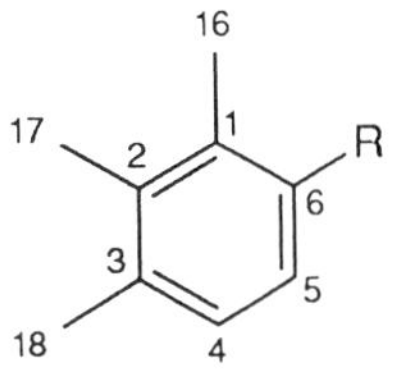 (aromatic ring)
γ	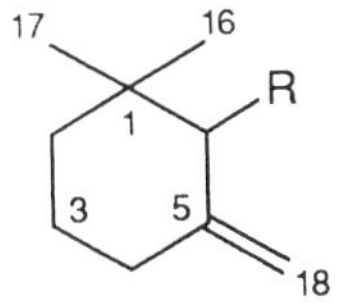 (cyclohexane with exocyclic methylene at 18)		

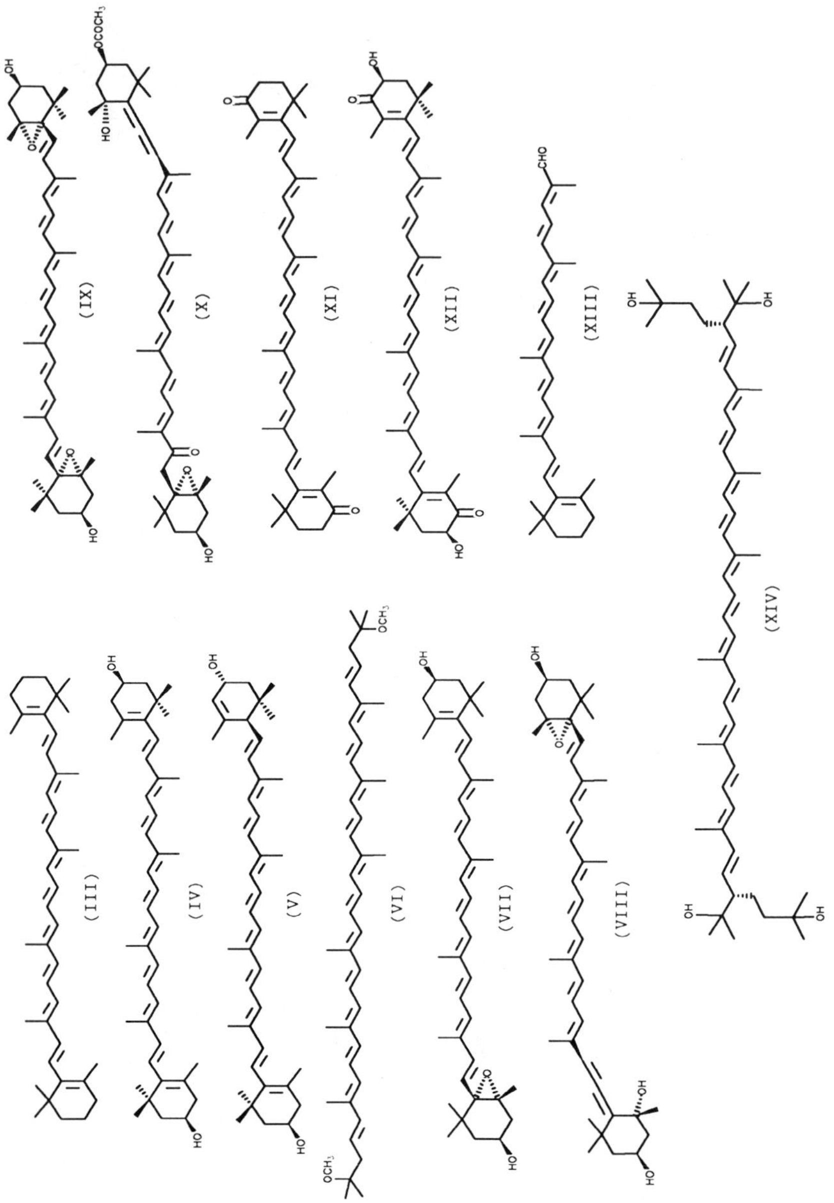

FIG. 3. Structures of important carotenoids (III–XIV).

(VI), antheraxanthin (5,6-epoxy-5,6-dihydro-β,β-carotene-3,3'-diol) (VII), neoxanthin [(3S,5R,6R,3'S,5'R,6'S)-5',6'-epoxy-6,7-didehydro-5,6,5',6'-tetrahydro-β,β-carotene-3,5,3'-triol] (VIII), violaxanthin [(3S,5R,6S,3'S,-5'R,6'S)-5,6,5',6'-diepoxy-5,6,5',6'-tetrahydro-β,β-carotene-3,3'-diol] (IX), fucoxanthin [(3S,5R,6S,3'S,5'R,6'R)-5,6-epoxy-3,3',5'-trihydroxy-6',7'-didehydro-5,6,7,8,5',6'-hexahydro-β,β-caroten-8-one 3'-acetate] (X), canthaxanthin (β,β-carotene-4,4'-dione) (XI), and astaxanthin [(3S,3'S)-3,3'-dihydroxy-β,β-carotene-4,4'-dione] (XII).

Derivatives in which the carbon skeleton has been shortened by the formal removal of fragments from one or both ends of a carotenoid are named apo- and diapocarotenoids, respectively, e.g., β-apo-8'-carotenal (8'-apo-β-caroten-8'-al) (XIII). An example of homocarotenoids (higher carotenoids), which contain more than eight isoprenoid units, is bacterioruberin[(2S,2'S)-2,2'-bis(3-hydroxy-3-methylbutyl)-3,4,3',4'-tetradehydro-1,2,1',2'-tetrahydro-ψ,ψ-carotene-1,1'-diol] (XIV).

About 600 carotenoids have been isolated from natural sources; they are listed with their trivial and semisystematic names in *Key to Carotenoids*,[14] which also includes literature references for their spectroscopic and other properties. It must be pointed out, however, that for many of the carotenoids listed the structure (this term includes the stereochemistry) is still uncertain and in all these cases a reisolation, followed by structural elucidation with all the modern spectroscopic methods [especially high-resolution nuclear magnetic resonance (NMR) spectroscopy] is absolutely necessary. About 370 of the naturally occurring carotenoids are chiral, bearing from 1 to 5 asymmetric carbon atoms. In most cases one carotenoid occurs only in one configuration. However, exceptions are known: whereas in higher plants lutein [(3R,3'R,6'R)-β,ε-carotene-3,3'-diol] was isolated, 3'-epilutein [the (3R,3'S,6'R)-isomer] was found in goldfish. Other examples are the different stereoisomers of astaxanthin [(3S,3'S)-, (3R,3'R)-, and (3R,3'S)-astaxanthin] from salmon and the different configurations of the 5,8-epoxides.

This cis–trans or (E/Z)-isomerism of the carbon–carbon double bonds is another interesting feature of the stereochemistry of the carotenoids. The literature in this field is extensive: the first comprehensive review of the cis–trans isomerism of carotenoids and vitamin A was published in 1962 by Zechmeister.[15] According to the number of double bonds a great number of (E/Z)-isomers exist for each carotenoid, e.g., 1056 for lycopene and 272 for β-carotene. In view of the (E/Z)-isomerism the double bonds of the polyene chain can be divided into two groups: (1) double bonds with no

---

[14] H. Pfander (ed.), "Key to Carotenoids," 2nd Ed. Birkhäuser, Basel, 1987.

[15] L. Zechmeister, "*cis–trans* Isomeric Carotenoids, Vitamin A, and Arylpolyenes." Springer-Verlag, Vienna, 1962.

steric hindrance of the (Z)-isomer (central 15,15'-double bond and the double bonds bearing a methyl group, such as the 9-, 9'-, 13-, and 13'-double bonds) and (2) double bonds with steric hindrance (7-, 7'-, 11-, and 11'-double bonds). Although isomers with sterically hindered (Z)-double bonds are known [(11Z)-retinal] the number of possible (Z)-isomers is in practice reduced considerably, e.g., for lycopene to 72. Normally carotenoids occur in nature as the (all-E)-isomer. However, exceptions are known, such as the (15Z)-phytoene isolated from carrots, tomatoes, and other organisms. Furthermore, modern analytical methods, especially high-performance liquid chromatography (HPLC), revealed many other (Z)-isomers, although mostly in minor amounts compared to the (all-E)-isomer. On the other hand, some carotenoids undergo isomerization very easily during workup; therefore many (Z)-isomers that are described in the literature as natural products are artifacts. For experimental work it must be kept in mind that (E/Z)-isomerization may occur when a carotenoid is kept in solution. Normally the percentage of the (Z)-isomers is rather low, but it is enhanced at higher temperature. Furthermore, the formation of (Z)-isomers is increased by exposure to light. Today HPLC is the method of choice for the separation of these isomers; however, it must be pointed out that careful selection of the conditions is necessary to avoid isomerization during separation. Most important for practical work is the influence of (E/Z)-isomerism on the electronic spectra (Fig. 4). In comparison with the (all-E)-isomer the absorption maxima of the (Z)-isomers are shifted toward shorter wavelengths and the fine structure of the spectrum is reduced. The cis peak that appears at ~ 142 nm below the maximum of the highest wavelength is characteristic.

Occurrence

As already mentioned, the carotenoids are a class of natural pigments that is very widespread. They are found throughout the plant kingdom, although their presence is often masked by chlorophyll, as, e.g., in green leaves. They are responsible for the beautiful colors of many fruits (pineapple, citrus fruits, tomatoes, paprika, rose hips) and flowers *(Eschscholtzia, Narcissus)*, as well as the colors of many birds (flamingo, cock of rock, ibis, canary), insects (lady bird), and marine animals (crustaceans, salmon). Normally carotenoids occur in low concentrations, but this varies enormously from one source to another. The total carotenoid production in nature has been estimated at about 10^8 tons a year. Most of this production is in the form of the four major carotenoids: fucoxanthin (**X**), the characteristic pigment of many marine algae, which is the most abundant natural carotenoid, and lutein (**V**), violaxanthin (**IX**), and neoxanthin (**VIII**), the carotenoids in green leaves.

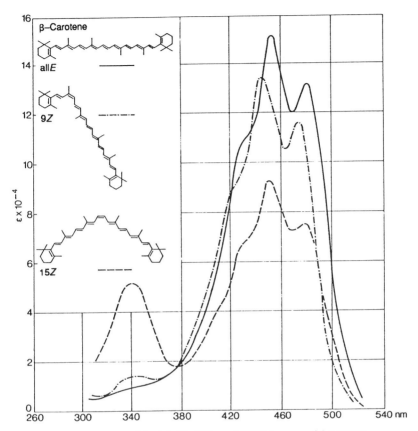

FIG. 4. Spectra of the (all-*E*), (9*Z*), and (15*Z*) isomers of β-carotene.

Industrial Production: Synthesis

Ever since the elucidation of the structure of β-carotene and other carotenoids much effort has been devoted to the synthesis of carotenoids and other polyenes. The first synthesis of β-carotene was reported independently by Karrer and Eugster,[16] Inhoffen and co-workers,[17] and Milas[18] in 1950. The Inhoffen synthesis was later developed into an industrial process and since 1954 β-carotene has been produced commercially. Today six synthetic carotenoids (Fig. 5) have become commercially important: β-apo-8′-carotenal (**XIII**) (C_{30}), β-apo-8′-carotenoic acid ethyl ester (ethyl 8′-apo-

---

[16] P. Karrer and C. H. Eugster, *Helv. Chim. Acta* **33**, 1172 (1950).
[17] H. H. Inhoffen, H. Pommer, and F. Bohlmann, *Ann. Chem.* **569**, 237 (1950).
[18] N. A. Milas, P. Davis, I. Belič, and D. Fleš, *J. Am. Chem. Soc.* **72**, 4844 (1950).

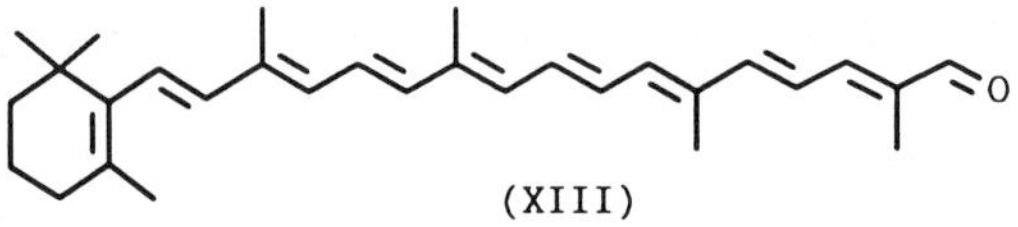

(XIII)

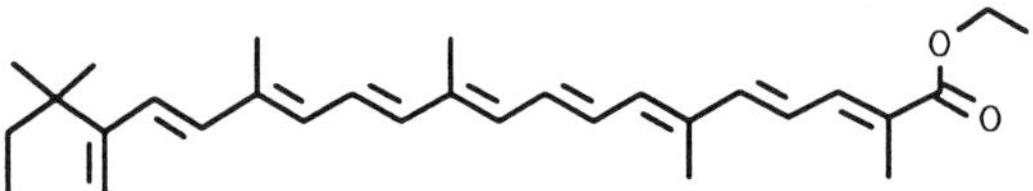

(XV)

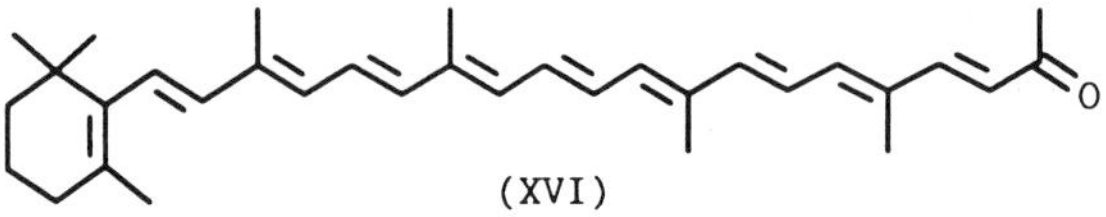

(XVI)

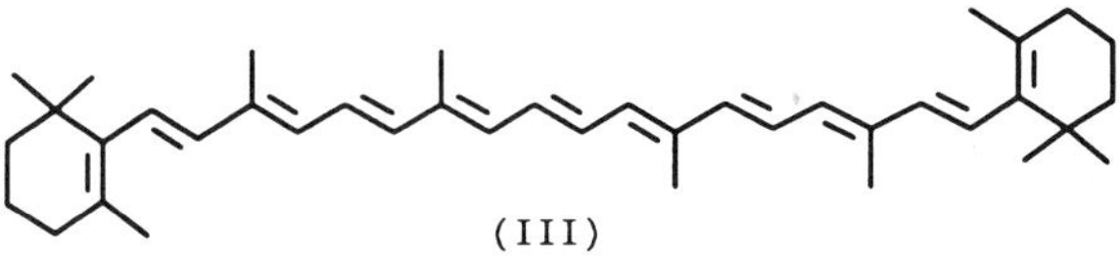

(III)

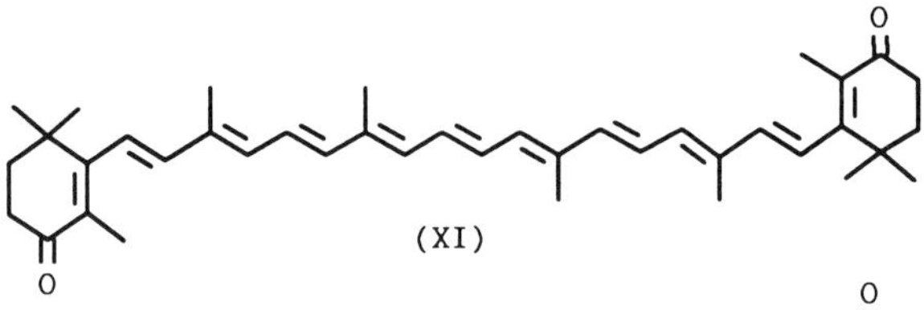

(XI)

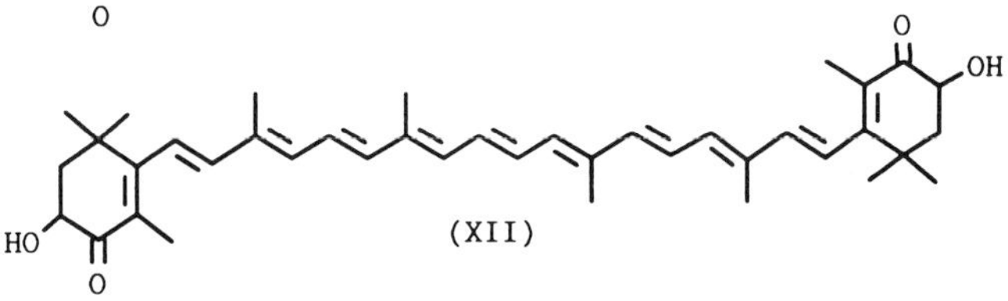

(XII)

FIG. 5. Structures of six synthetic carotenoids (**III, XI–XIII, XV,** and **XVI**).

β-caroten-8′-oate) **(XV)** (C_{32}), citranaxanthin (5′,6′-dihydro-5′-apo-18′-nor-β-caroten-6′-one) **(XVI)** (C_{30}), β-carotene **(III)** (C_{40}), canthaxanthin **(XI)** (C_{40}), and racemic astaxanthin **(XII)** (C_{40}).

Commercial synthetic carotenoids are mainly used as pigments for food (egg yolk, chickens, or farm-raised salmon) and for coloration of food products (margarine, cheese). Various methods have been developed for carotenoid application. A microcrystalline dispersion in an edible fat is used for coloring margarine, and in fruit juices a powder, in which β-carotene is in the form of a microdispersion in a hydrophilic protective colloid, is used. Current market prices for stabilized dispersible powders containing 5–10% active substance range from $600 for β-carotene, $900 for β-apo-8'-carotenoids, $1300 for canthaxanthin, to $2500 for astaxanthin per kilogram. It has been estimated that a production capacity for more than 500 tons of β-carotene per year is currently planned or under construction worldwide. Total sales of synthetic carotenoids are today about $300 million and may pass the $500 million mark in about 5 years.

Besides the above-mentioned carotenoids, which are of commercial interest, many other naturally occurring carotenoids have been synthesized. The main emphasis has been on the synthesis of carotenoids in optically active form, and today most of the important chiral end groups have been synthesized.

Chemical Properties

For practical work it is important to keep in mind that most carotenoids are unstable and are especially sensitive to light, oxygen, acid, and elevated temperature. All operations should be carried out in an inert atmosphere (nitrogen or vacuum), at low temperature (room temperature to 20°), in darkness or diffuse light, under acid-free conditions, and using freshly purified peroxide-free solvents. Carotenoids should be stored as solids under nitrogen and refrigerated. Methods for the isolation of carotenoids from natural sources are mainly determined by the nature of the biological material, the ease of solvent extraction, and the properties and quantities of the carotenoids. A general approach for the isolation of carotenoids is given by Liaaen-Jensen.[19]

Biosynthesis

Carotenoids are synthesized in nature by plants and many microorganisms. Animals can metabolize carotenoids in a characteristic manner, but they are not able to synthesize carotenoids.

Carotenoids, being terpenoids, are synthesized from the basic C_5 terpenoid precursor, isopentenyl pyrophosphate (IPP) (**XVII,** Fig. 6). This compound is converted to geranylgeranyl pyrophosphate (C_{20}) (**XVIII**). The dimerization of **XVIII** leads to phytoene (**XIX**) and the stepwise dehydrogenation via phytofluene (**XX**), ζ-carotene (**XXI**), and neurosporene (**XXII**) gives lycopene (**I**). Subsequent cyclizations, dehydrogenations, oxidations, etc., lead to the individual naturally occurring carotenoids.

[19] S. Liaaen-Jensen, *in* "Carotenoids" (O. Isler, ed.), p. 64. Birkhäuser, Basel, 1971.

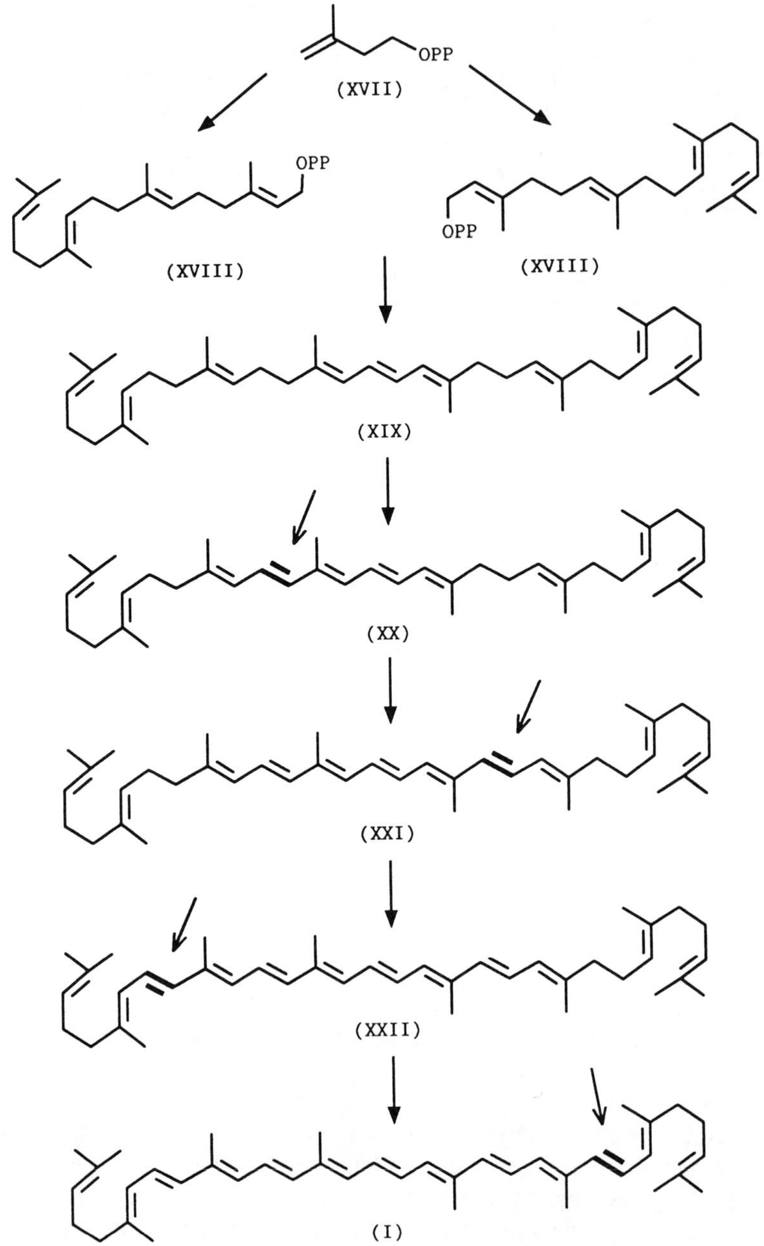

FIG. 6. Synthesis of carotenoids from isopentenyl pyrophosphate (XVII).

Functions

The best known function of carotenoids is the long-established role of β-carotene and other carotenoids with an unsubstituted β-ring as provitamin A. In Western countries the supply of vitamin A is not critical but in countries of the third world it is still a severe problem. According to an estimate of the World Health Organization (WHO), 250,000–500,000 children go blind every year due to a deficiency of vitamin A.

In photosynthetic organisms and tissues, e.g., algae and the chloroplasts of green plants, the carotenoids play essential roles as accessory light-harvesting pigments and, especially, in protection against damage by photosensitized oxidation. They are currently being investigated as antioxidants that may exert an important protective action against many diseases, including cancer. This protective effect is independent of the long-established role of β-carotene as provitamin A.

β-Carotene also enhances immunity and is used in livestock for enhancing fertility. Terms such as *chemoprevention* in precancerous states and *radioprotection* in conjunction with radiation treatment of cancer and *early prophylaxis in cardiovascular diseases* have been coined.

In summary, it is obvious that the carotenoids are a widely distributed and important class of natural pigments and a fascinating field for interdisciplinary research.

[2] Synthesis and Characterization of Carotenoids by Different Methods

By MASAYOSHI ITO, YUMIKO YAMANO, and YUKO SHIBATA

The various syntheses of carotenoids employ some key reactions for the formation of the carbon–carbon double bond, in particular, the aldol condensation, the Wittig condensation, the Emmons–Horner reaction, the Julia's method, and the addition of acetylides.[1] We have accomplished the total synthesis[2,3] of peridinin (**1**) (Fig. 1), which is representative of the butenolide carotenoids and is known as an auxiliary light-harvesting pig-

---

[1] H. Mayer and O. Isler, *in* "Carotenoids" (O. Isler, ed.), p. 325. Birkhäuser, Basel, 1971; F. Kienzle, *Pure Appl. Chem.* **47**, 183 (1976).
[2] M. Ito, Y. Hirata, Y. Shibata, and K. Tsukida, *J. Chem. Soc., Perkin Trans. 1* **1990**, 197 (1990).
[3] M. Ito, *Pure Appl. Chem.* **63**, 13 (1991).

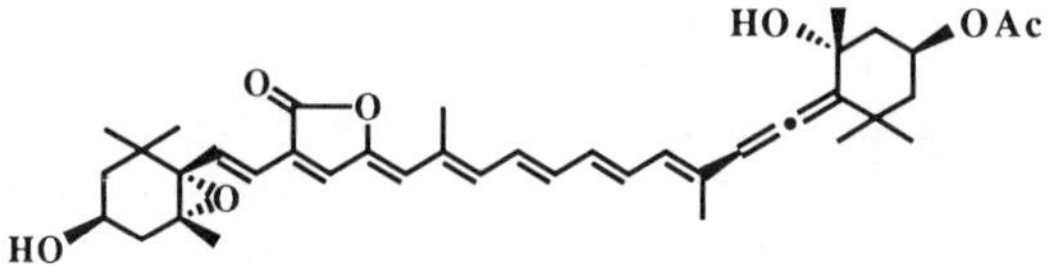

peridinin (1)

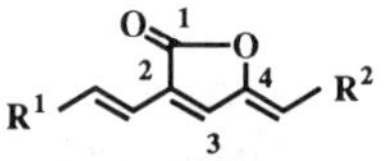

(2)

R^1=alkyl or polyene chain

R^2=polyene chain

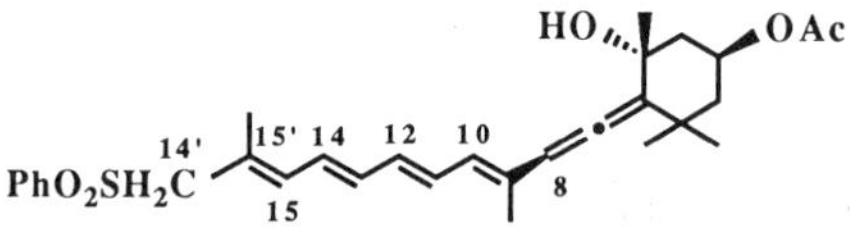

(3)

FIG. 1. Structures of peridinin (**1**), 4-alkylidenebutenolides (**2**), and the C_{22}-allenic sulfone (**3**).

ment for photosynthesis. Peridinin (**1**) is a unique C_{37}-tricyclic carotenoid containing a 4-alkylidenebutenolide structure carrying an allene function in the main polyene chain.[4]

In this total synthesis, three new reactions are included: (1) formation of a 4-alkylidenebutenolide structure (**2**) (Fig. 1) displaying extended conjugation at the C-2 position, (2) synthesis of a C_{22}-allenic sulfone (**3**) (Fig. 1) possessing the abnormal arrangement of an in-chain methyl group, and (3) palladium-catalyzed olefination of the vinyl triflate.

Synthesis of Alkylidenebutenolides

A new method[2,3] (a sulfone method, route C in Fig. 2) for the synthesis of carotenoidal alkylidenebutenolides such as (**4**) (Fig. 2) was developed

[4] H. H. Strain, W. A. Svec, K. Aitzetmüller, M. C. Grandolfo, J. J. Katz, H. Kjøsen, S. Norgård, S. Liaaen-Jensen, F. T. Haxo, P. Wegfahrt, and H. Rapoport, *J. Am. Chem. Soc.* **93,** 1823 (1971); H. H. Strain, W. A. Svec, P. Wegfahrt, H. Rapoport, F. T. Haxo, S. Norgård, H. Kjøsen, and S. Liaaen-Jensen, *Acta Chem. Scand., Ser. B* **B30,** 109 (1976); J. E. Johansen, W. A. Svec, S. Liaaen-Jensen, and F. T. Haxo, *Phytochemistry* **13,** 2261 (1974).

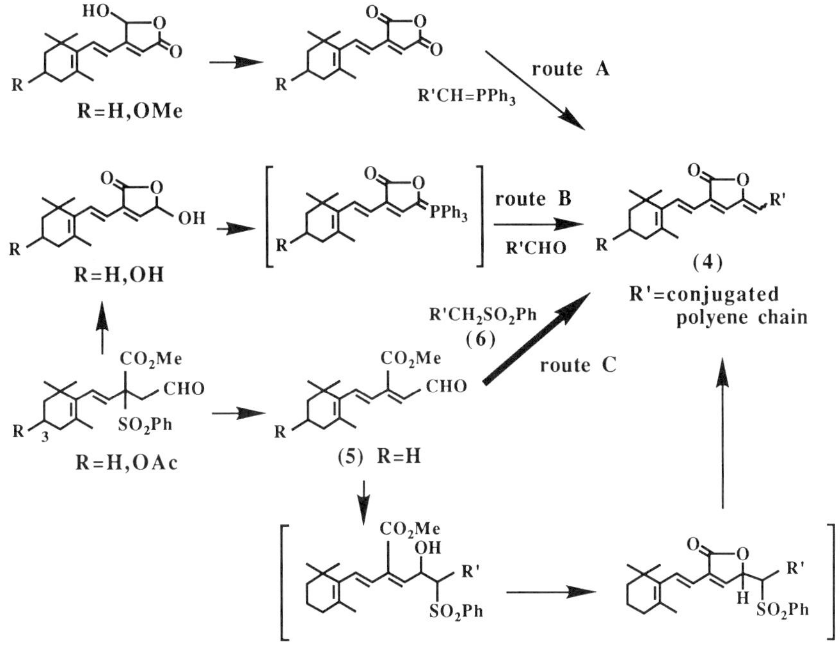

FIG. 2. Synthetic methodology for the conjugated alkylidenebutenolides (4).

that is more useful than other procedures[5] (the two Wittig methods, routes A and B, in Fig. 2).

Sulfone Method

The reaction of the conjugated formyl ester (5)[2] with various allylic sulfones (6)[6] in the presence of lithium diisopropylamide (LDA) at $-78°$ gives conjugated alkylidenebutenolides (4) in moderate yields as a mixture (about 1:1) of (Z) and (E) isomers about the ylidene double bond (Table I). In this reaction, addition of α-sulfonyl carbanion to the formyl group, cyclization of the resulting hydroxy ester, and elimination of the sulfonyl group take place successively in one vessel to give the expected products (4). Stereostructures of the products are determined from the high-field ^{1}H NMR data.

The configuration about the newly formed ylidene double bond is confirmed on the basis of an empirically determined rule[3]: in compounds

---

[5] M. Ito, T. Iwata, and K. Tsukida, *Chem. Pharm. Bull.* **32**, 1709 (1984); M. Ito, Y. Hirata, Y. Shibata, A. Sato, and K. Tsukida, *J. Nutr. Sci. Vitaminol.* **33**, 313 (1987).

[6] Allylic sulfones were prepared from corresponding allylic alcohols by the standard method.

TABLE I
Synthesis of Conjugated Alkylidenebutenolides

$$RCH_2SO_2Ph \xrightarrow[-78--50°]{\substack{LDA \\ THF-hexane\,(1:1)}} (4)$$

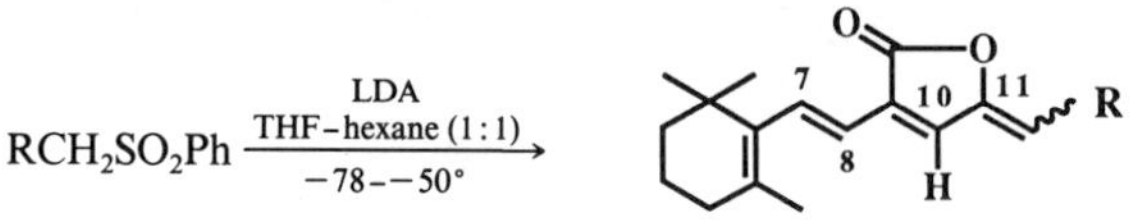

(6)   (5)   (4)

Sulfone	Product	Total yield (%) of E and Z isomers
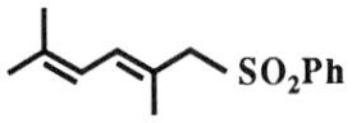 (6a)	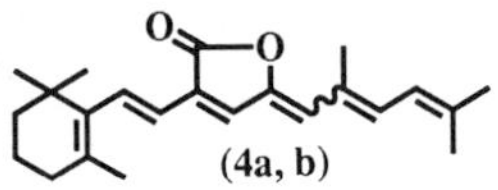 (4a, b)	56
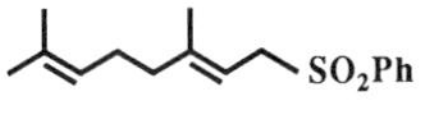	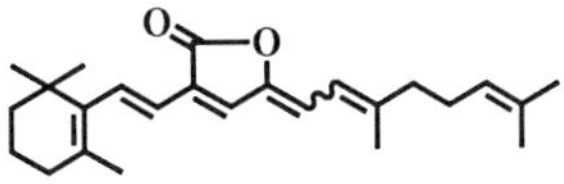	32
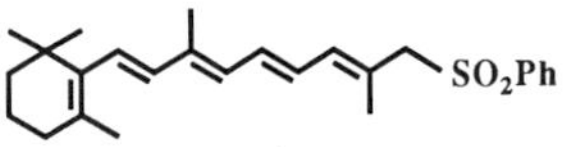	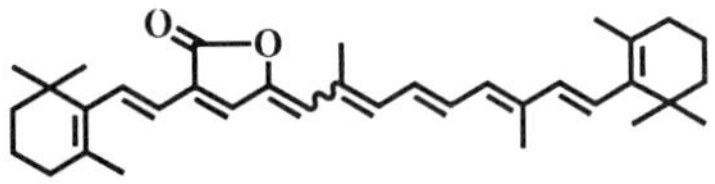	46
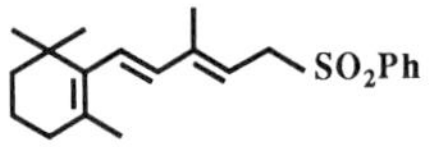	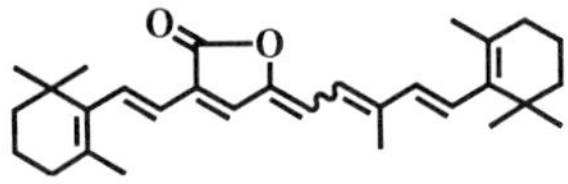	33
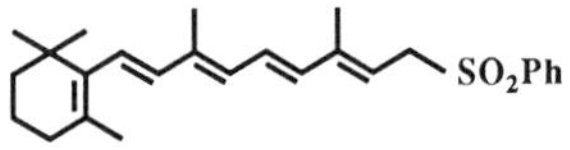	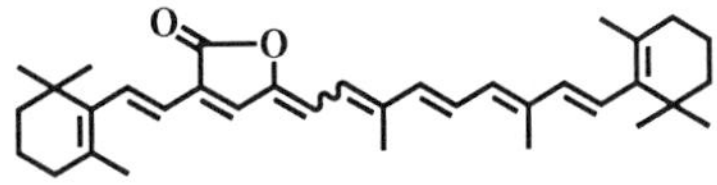	46

of this type, the NMR signal for 10-H[7] in the (11Z)-isomer is observed at δ 7.00 to δ 7.20, whereas the corresponding signal for the (11E)-isomer was found downfield below δ 7.40.

Preparation of Alkylidenebutenolide: General Procedure for Conjugated Alkylidenebutenolide Synthesis

A solution (0.36 ml, 0.57 mmol) of *n*-butyllithium (1.59 M in hexane) is added to a stirred solution of diisopropylamine (58 mg, 0.57 mmol) in dry tetrahydrofuran (THF) (1.5 ml) and hexane (1.5 ml) at $-78°$ under N_2

---

[7] We have employed the numbering system generally used for retinoids and carotenoids.

and the mixture is stirred for 30 min. To this LDA solution is added a solution of the sulfone (**6a**) (143 mg, 0.57 mmol) in a mixture (4 ml) of dry THF and hexane (1:1). After the addition is complete, the mixture is stirred for 30 min, then a solution of the formyl ester (**5**) (100 mg, 0.38 mmol) in dry THF (2 ml) and hexane (2 ml) is added dropwise at $-78°$. The reaction mixture is stirred at $-78°$ for 10 min before being allowed to warm to room temperature over about 20 min with stirring. The reaction is quenched with saturated aqueous NH_4Cl and extracted with ether. The extracts are washed with brine, dried with Na_2SO_4, and evaporated *in vacuo* to give an oil that is purified by silica gel short-column chromatography (ether–hexane, 1:9) under reduced pressure to afford **4a,b** (72 mg, 56%). Isomers (Z:E, ~1:1) are separated by preparative TLC (silica gel/benzene–hexane, 2:3) to give each pure specimen.

Compound **4a**: (11Z)^7-isomer, mp 127–130° (hexane); ultraviolet-visible (UV–VIS) λ_{max}^{EtOH} nm (ε): 237 (9000), 414 (57,000); infrared (IR) $\nu_{max}^{CHCl_3}$ cm^{-1}: 1741 (C=O), 1612 (C=C); ^{1}H NMR (200 MHz, CDCl$_3$) δ: 1.06 (6H, s, 1-*gem*-methyl), 1.76 (3H, s, 5-methyl), 1.86, 1.91 (each 3H, s, 16-*gem*-methyl), 2.20 (3H, s, 13-methyl), 5.72 (1H, s, 12-H), 6.20 (1H, d, $J = 16$ Hz, 8-H), 6.22 (1H, br d, $J = 12$ Hz, 15-H), 6.55 (1H, br d, $J = 12$ Hz, 14-H), 7.02 (1H, s, 10-H), 7.25 (1H, br d, $J = 16$ Hz, 7-H); high-resolution mass spectroscopy (MS) *m/z*: 338.224 (M$^+$, $C_{23}H_{30}O_2$ requires 338.224).

Compound **4b**: (11E)-isomer, mp 123–126° (hexane); UV–VIS λ_{max}^{EtOH} nm (ε): 238 (9000), 419 (57,000); IR $\nu_{max}^{CHCl_3}$ cm^{-1}: 1744 (C=O), 1608 (C=C); ^{1}H NMR (200 MHz, CDCl$_3$) δ: 1.07 (6H, s, 1-*gem*-methyl), 1.77 (3H, s, 5-methyl), 1.87, 1.92 (each 3H, s, 16-*gem*-methyl), 2.07 (3H, s, 13-methyl), 6.22 (1H, br d, $J = 12$ Hz, 15-H), 6.24 (1H, d, $J = 16$ Hz, 8-H), 6.39 (1H, s, 12-H), 6.55 (1H, br d, $J = 12$ Hz, 14-H), 7.37 (1H, br d, $J = 16$ Hz, 7-H), 7.43 (1H, s, 10-H); high-resolution MS *m/z*: 338.224 (M$^+$, $C_{23}H_{30}O_2$ requires 338.224).

Synthesis of C_{22}-Allenic Sulfone Possessing Abnormal Arrangement of In-Chain Methyl Group

The C_{22}-allenic sulfone (**3**) is synthesized (Scheme I) from C_{22}-allenic apocarotenals (**11, 12**), which are previously prepared[8] via a Wittig condensation of the C_{15}-allenic aldehyde (**10**) with the C_7-phosphonium bromide (**9a**). For this reaction the Wittig condensation is modified by use of the C_7-phosphonium chloride (**9b**) instead of **9a**.

---

[8] M. Ito, Y. Hirata, K. Tsukida, N. Tanaka, K. Hamada, R. Hino, and T. Fujiwara, *Chem. Pharm. Bull.* **36**, 3328 (1988).

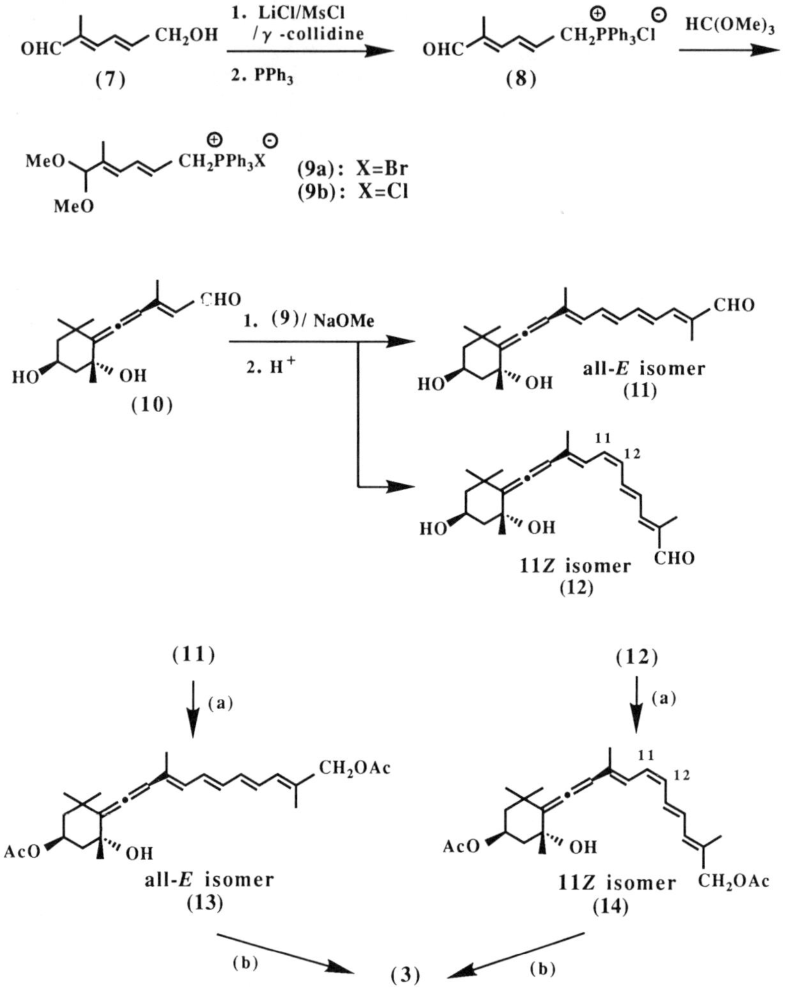

SCHEME I. Synthesis of the C_{22}-allenic sulfone (3) possessing the abnormal arrangement of the in-chain methyl group. Reagents for path (a): (1) $NaBH_4$, (2) Ac_2O/Py; for path (b): $PhSO_2Na$, 2-propanol–water, reflux.

Preparation of C_7-Phosphonium Chloride

To a stirred mixture of the formyl alcohol (7)[8] (1.2 g) and γ-collidine (1.4 ml, 1.1 Eq) is added a solution of LiCl (0.41 g, 1.0 Eq) in dry dimethylformamide (DMF) (3 ml) at an ice-cooled temperature under N_2. The mixture is stirred at the same temperature for a further 10 min. To this reaction mixture is added MsCl (0.81 ml, 1.1 Eq) and stirring is continued at the same temperature for a further 1 hr. The reaction mixture is poured into ice-water and extracted with ether. The organic layer is washed with 3% HCl saturated with NaCl, saturated aqueous $NaHCO_3$ solution, and then brine. Evaporation of the dried (Na_2SO_4) solvent *in vacuo* provides the residue, which is purified by silica gel short-column chromatography (ether-hexane, 2:3) under reduced pressure to afford the corresponding chloride (1.19 g). Subsequently, PPh_3 (2.05 g, 1.05 Eq) is added to a solution of the chloride (1.19 g) in CH_2Cl_2 (60 ml) and the mixture is refluxed for 22 hr under N_2. Evaporation of the solvent *in vacuo* gives the residue, which is washed with ether to provide the phosphonium chloride (**8**) (2.23 g, 55% from **7**). [IR 1680 cm^{-1} (conjugated CHO), UV 225, 275 nm.]

Preparation of C_{22}-Allenic Apocarotenals

To a solution of the C_7-phosphonium chloride (**8**) (1.1 g, 2.0 Eq) in methanol (5 ml) are added an acidic solution (1 ml) prepared from *p*-TsOH (150 mg) and H_3PO_4 (0.2 ml) in methanol (50 ml), and methyl orthoformate (1 ml). The reaction mixture is stirred at room temperature for 18 hr under N_2 and neutralized with NaOMe until just before the red color of an ylide appears to give a Wittig salt (**9b**) solution, to which a solution of the C_{15}-allenic aldehyde (**10**) (316 mg) in CH_2Cl_2 (15 ml) and NaOMe solution prepared from sodium (70 mg) and methanol (2 ml) are added. The reaction mixture is stirred at room temperature for 30 min, then poured into ice-water and extracted with ether. The organic layer is shaken with 3% HCl until the fine structure in the UV spectrum disappears, then washed with saturated aqueous $NaHCO_3$ solution and brine. Evaporation of the dried (Na_2SO_4) solvent *in vacuo* gives the residue, which is purified by silica gel short column chromatography (acetone-hexane, 3:7) under reduced pressure to afford an isomeric mixture (**11** and **12**). pHPLC [column, LiChrosorb Si-60 (7 μm), 2.5 × 25 cm; mobile phase, 2-propanol-THF-hexane, 1:35:64) separation provides the (all-E)-isomer (**11**) (168 mg, 41%) and the (11Z)-isomer (**12**) (164 mg, 38%), as orange-yellow solids, respectively.[8]

Preparation of C_{22}-Allenic Sulfone

$NaBH_4$ (16 mg) is added to an ice-cooled solution of (all-*E*)-allenic apocarotenal (**11**)[8] (290 mg) in methanol (12 ml). The mixture is stirred for 15 min, poured into ice-water, and extracted with ether. The extract is washed with brine. Evaporation of the dried (Na_2SO_4) extract *in vacuo* gives the triol, which without purification is dissolved in dry pyridine (11 ml) and acetic anhydride (3.5 ml). The mixture is stirred at room temperature for 15 hr, poured into ice–water, and extracted with ether. The organic layer is washed with 5% HCl saturated with NaCl, saturated aqueous $NaHCO_3$ solution, and then brine. Evaporation of the dried (Na_2SO_4) extract *in vacuo* gives the diacetate (13) (320 mg). [IR $\nu_{max}^{CHCl_3}$ cm^{-1}: 3600, 3420 (OH), 1930 (C=C=C), 1725 ($OCOCH_3$); UV λ_{max}^{EtOH} nm: 315 (sh), 329, 345, 364.] To a solution of the diacetate (13) (320 mg) in 2-propanol (6 ml) are added water (2 ml) and $PhSO_2Na \cdot 2H_2O$ (204 mg). The mixture is refluxed for 20 hr under N_2. After cooling, the reaction mixture is diluted with ether and washed with brine. Evaporation of the solvent *in vacuo* gives an oil that was purified by silica gel short column chromatography (acetone–hexane, 1:3) under reduced pressure and then by pHPLC [column, LiChrosorb Si-60 (5 μm), 1 × 30 cm; mobile phase, THF–hexane, 3:7; detection, 320 nm; flow rate, 2 ml/min] to provide a yellow oil (3) (272 mg, 63% from 11). IR $\nu_{max}^{CHCl_3}$ cm^{-1}: 3590, 3470 (OH), 1930 (C=C=C), 1728 ($OCOCH_3$), 1305, 1295 (split) (SO_2), 1140 (SO_2); UV λ_{max}^{EtOH} nm: 321 (sh), 336, 353, 372; ^{1}H NMR (500 MHz, $CDCl_3$) δ: 1.10 (3H, s, 1-methyl), 1.37, 1.41 (each 3H, s, 1- and 5-methyl), 1.81 (3H, s, 9-methyl), 1.92 (3H, s, 15'-methyl), 2.08 (3H, s, $OCOCH_3$), 3.84 (2H, s, 14'-H_2), 5.42 (1H, m, 3-H), 5.76 (1H, d, $J = 11$ Hz, 15-H), 6.06 (1H, s, 8-H), 6.08 (1H, d, $J = 12$ Hz, 10-H), 6.16 (1H, dd, $J = 14.5$, 12 Hz, 13-H), 6.30 (1H, dd, $J = 14.5$, 12 Hz, 12-H), 6.36 (1H, dd, $J = 14.5$, 11 Hz, 14-H), 6.53 (1H, dd, $J = 14.5$, 12 Hz, 11-H), 7.58 (2H, t, $J = 8$ Hz, ArH), 7.68 (1H, t, $J = 8$ Hz, ArH), 7.89 (2H, t, $J = 8$ Hz, ArH); high-resolution MS *m/z*: 510.241 (M$^+$, $C_{30}H_{38}O_5S$ requires 510.244).

In the same manner as described for the preparation of 3 from 11, the (11Z)-apocarotenal (12)[8] (135 mg) provides the (all-*E*)-C_{22}-allenic sulfone (3) (103 mg, 51%).

A mixture of the C_{22}-apocarotenals (11 and 12) (p. 19) prepared from the C_{15}-allenic aldehyde (10) (320 mg) is treated in a manner similar to that used for the preparation of 3 from 11 to provide the (all-*E*)-sulfone (3) (244 mg, 51% from 10).

Palladium-Catalyzed Olefination

Palladium-coupling reaction between a vinyl bromide and a Z-enynol was applied by Pattenden and Robson[9] to a total synthesis of the tetra-Z-lycopene. This is a typical example, showing that a carbon–carbon bond formation in polyene chains is done not at a double bond but at a single bond.

In preparing the C_{15}-epoxyformyl ester (18) (Scheme II), palladium-

---

[9] G. Pattenden and D. C. Robson, *Tetrahedron Lett.* **28**, 5751 (1987).

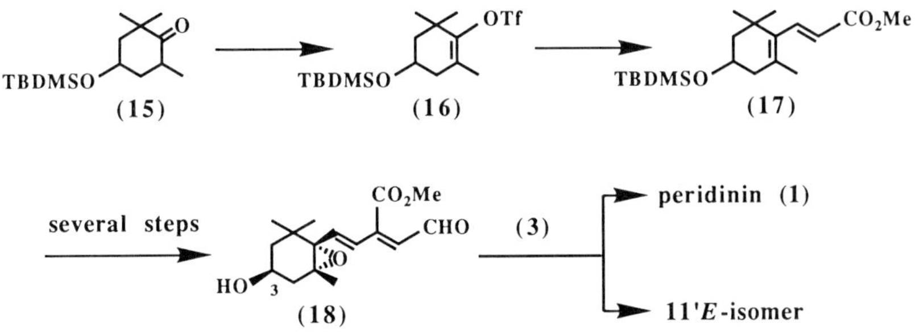

SCHEME II. Palladium-catalyzed olefination leading to the C_{15}-epoxyformyl ester (18), a key intermediate to peridinin (1).

catalyzed olefination[10] of the vinyl triflate (16) is used. This reaction gives a high yield and illustrates a new approach to polyene-chain formation.

Preparation of Vinyl Triflate

To a stirred solution of diisopropylamine (5.13 ml, 1.1 Eq) in dry THF (75 ml) is added a solution (23.1 ml, 1.1 Eq) of *n*-butyllithium (1.59 M in hexane) at $-78°$ under N_2 and the mixture is stirred for 30 min at $-78°$. To this LDA solution is added a solution of the silyloxyketone (15) (9.00 g) prepared from 4-hydroxy-2,2,6-trimethylcyclohexanone[11] and TBDMSCl, in dry THF (75 ml). After the addition is complete, the mixture is stirred for 1 hr at $-78°$, then a solution of Tf_2NPh[12] (12.50 g, 1.1 Eq) in dry THF (75 ml) is added at the same temperature. The mixture was stirred at about $0°$ for 5 hr. The reaction is quenched with saturated aqueous NH_4Cl solution. After evaporation of THF under reduced pressure, the residue is extracted with ether. The organic layer is washed with brine and dried (Na_2SO_4). Evaporation of the solvent *in vacuo* gives the residue, which is purified by silica gel column chromatography (ether–hexane, 4:96) to afford a colorless oil (16) (11.88 g, 89%). IR $v_{max}^{CHCl_3}$ cm^{-1}: 1398, 1130 (OSO_2); ^{1}H NMR (200 MHz, CDCl$_3$) δ: 0.08 (6H, s, SiMe × 2), 0.89 (9H, s, *tert*-butyl), 1.15, 1.21 (each 3H, s, *gem*-methyl), 1.75 (3H, s, 2-methyl), 2.16 (1H, ddd, $J = 17, 9, 1$ Hz, 5-H), 2.36 (1H, br dd, $J = 17, 6$ Hz, 5-H), 4.02 (1H, m, 4-H); ^{13}C NMR (50.3 MHz, CDCl$_3$) δ: 118.76 (q, $J = 318$ Hz, CF_3); high-resolution MS *m/z*: 402.151 (M$^+$, $C_{16}H_{29}O_4SSiF_3$ requires 402.151).

[10] W. J. Scott, M. R. Peña, K. Swärd, S. J. Stoessel, and J. K. Stille, *J. Org. Chem.* **50**, 2302 (1985).
[11] E. Widmer, *Pure Appl. Chem.* **57**, 741 (1985).
[12] J. E. McMurry and W. J. Scott, *Tetrahedron Lett.* **24**, 979 (1983).

Palladium-Coupling Reaction

To a solution of vinyl triflate (16) (6.49 g), methyl acrylate (5.73 ml, 4.0 Eq), and triethylamine (7.94 ml, 3.5 Eq) in dry DMF (45 ml) is added Pd(PPh$_3$)$_2$Cl$_2$[10] (330 mg, 3 mol%). The mixture is heated with stirring at 75° for 22 hr under N$_2$. After cooling, the reaction solution is diluted with ether and washed with 5% HCl saturated with NaCl, saturated aqueous NaHCO$_3$ solution, and then brine. Evaporation of the dried (Na$_2$SO$_4$) solvent *in vacuo* gives the residue, which is purified by silica gel short column chromatography (ether–hexane, 7:93) under reduced pressure to provide a colorless oil (17) (5.10 g, 93%). UV λ_{max}^{EtOH} nm: 278; IR $\nu_{max}^{CHCl_3}$ cm^{-1}: 1707 (conjugated CO$_2$CH$_3$); ^{1}H NMR (200 MHz, CDCl$_3$) δ: 0.08 (6H, s, SiMe × 2), 0.90 (9H, s, *tert*-butyl), 1.08, 1.10 (each 3H, s, *gem*-methyl), 1.48 (1H, t, $J = 12.5$ Hz, 2$_{ax}$-H), 1.66 (1H, ddd, $J = 12.5, 4, 1.5$ Hz, 2$_{eq}$-H), 1.76 (3H, s, 5-methyl), 2.08 (1H, br dd, $J = 17.5, 9$ Hz, 4$_{ax}$-H), 2.27 (1H, br dd, $J = 17.5, 6$ Hz, 4$_{eq}$-H), 3.76 (3H, s, CO$_2$CH$_3$), 3.94 (1H, m, 3-H), 5.82 (1H, d, $J = 16$ Hz, 8-H), 7.37 (1H, br d, $J = 16$ Hz, 7-H); high-resolution MS *m/z*: 338.228 (M$^+$, C$_{19}$H$_{34}$O$_3$Si requires 338.228).

As shown in Scheme II, condensation of an α-sulfonyl carbanion of 3 with 18, using the sulfone method, leads to peridinin (1).

[3] Structure and Characterization of Carotenoids from Various Habitats and Natural Sources

By TAKAO MATSUNO

Acetone is the most commonly used solvent for the extraction of carotenoid-containing tissues. Carotenoids are susceptible to light, heat, oxygen, and acids. The removal of solvents from the extract must be done carefully at low temperatures, preferably below 40°, with a stream of oxygen-free inert gas such as N$_2$, and under a low-intensity light. Saponification is also a common practice to remove irrelevant lipid components and to concentrate the carotenoid fraction. Before the saponification step, care must be taken to remove the last trace of acetone remaining in the sample, because it may cause aldol condensations during the process and from unwanted artifacts. Enzymatic hydrolysis,[1-3] instead of a traditional

---

[1] P. B. Jacobs, R. D. LeBoef, S. A. McCommams, and J. D. Tauber, *Comp. Biochem. Physiol. B* **72B**, 157 (1982).

[2] T. Matsuno, M. Ookubo, T. Nishizawa, and I. Shimizu, *Chem. Pharm. Bull.* **32**, 4309 (1984).

chemical saponification, is undertaken to isolate the unstable carotenoids in alkaline medium. Traces of HCl in the chloroform used as solvent must be removed before use, as it also may produce carotenoid artifacts. Column chromatography, preparative thin-layer chromatography (TLC), recrystallization, high-performance liquid chromatography (HPLC), and combinations thereof are techniques adopted in recent carotenoid chemistry. Identification of a purified carotenoid sample may be done through co-TLC or co-HPLC with an authentic specimen. Ultraviolet (UV) and visible (VIS) light absorption spectra and mass spectrometry (MS) spectra are basic means for identification. Structural determination may be based on data from derivative formation, mass spectra, infrared (IR) spectra, ^{1}H NMR, ^{13}C NMR, circular dichroic (CD) spectra, and so on. For carotenoids with highly complicated structures, study of the deliberately decomposed derivatives may be necessary. In some cases, either total synthesis or partial synthesis starting from a carotenoid with known structure may be necessary for a full elucidation of the structures of the carotenoid in question.

This is an example of extraction, isolation, structure determination, and characterization of carotenoids from various habitats and natural sources. Nineteen carotenoids isolated from the carapace of hermit crab *Paralithodes brevipes* (*Hanasakigani* in Japanese) will be described here. *Paralithodes brevipes,* which belongs to Anomura (order Decapoda), inhabits the Sea of Okhotsk and the Bering Sea, and is one of the important edible crabs in Hokkaido, Japan.

In general ketocarotenoids are dominant in the Arthropoda, and astaxanthin is widely distributed in free, esterified, and protein-complexed forms. They occur mostly in the exoskeleton, ectoderm, egg, and ovary.[4]

Extraction and Isolation of Carotenoids

Paralithodes brevipes (4300 g, five specimens) is used for this experiment. The carotenoids are extracted with acetone from the carapace (1500 g) three times. They are then transferred to *n*-hexane–diethyl ether (1:1) by addition of water. The upper layer is washed with water and dried over anhydrous sodium sulfate. The extracted solution is evaporated under reduced pressure in N_2 below 40°. The resulting red oil is chromatographed on silica gel (Kieselgel 60, 70–230 mesh; Merck, Rahway, NJ)

---

[3] T. Matsuno, M. Katsuyama, T. Hirono, T. Maoka, and T. Komori, *Nippon Suisan Gakkaishi* **52**, 115 (1986).

[4] T. Matsuno and S. Hirao, *in* "Marine Biogenic Lipids, Fats and Oils" (R. G. Ackman, ed.), Vol. 1, p. 251. CRC Press, Boca Raton, Florida, 1989.

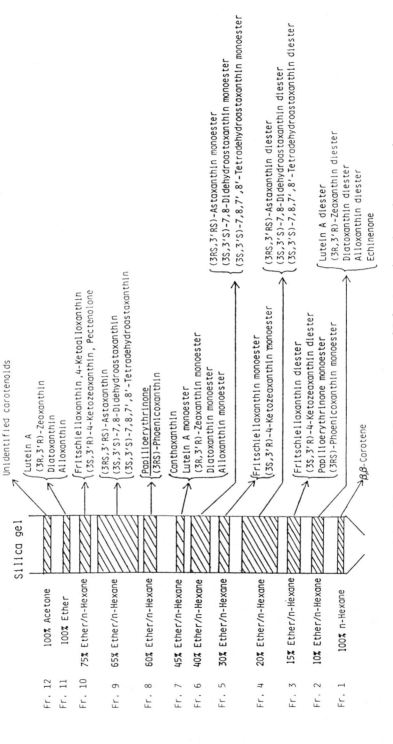

Fig. 1. Isolation of the carotenoids from the carapace of *P. brevipes* using column chromatography.

using an increasing percentage of ether in n-hexane. Twelve fractions are obtained, as shown in Fig. 1. Fractions 2–6 and 7 are esterified carotenoid fractions. Fractions 2 and 7 are hydrolyzed by the usual saponification and fractions 3–5 and 6 (alkaline-labile carotenoids) are hydrolyzed with lipase.

Enzymatic Hydrolysis of Esterified Carotenoids[2,3]

An acetone–0.05 M Tris-HCl buffer (1:99) solution (100 ml) of the carotenoid esters (0.5 mg) is incubated with lipase OF-360 (activity 360,000 units/g; Meito Sangyo Co., Ltd., Nagoya, Japan) (2 g) in N_2 with stirring at 37° for 24 hr. The reaction mixture is extracted three times with 200 ml of ether–n-hexane (1:1).

After hydrolysis each fraction is submitted to further purification by semipreparative high-performance liquid chromatography (HPLC).

HPLC column: Cosmosil 5SL (5 μm), 250 × 8 mm i.d. (Nakarai Chemicals, Kyoto, Japan); mobile phase, n-hexane–ethyl acetate–acetone (7:1:2); flow rate, 2.0 ml/min

Separation of each stereoisomer of carotenoids is carried out by HPLC with a chiral resolution column.[5]

HPLC column: Sumipax OA-2000 (5 μm), 300 × 4 mm i.d. (Sumitomo Chemical Co., Ltd., Osaka, Japan); mobile phase, n-hexane–CH_2Cl_2–ethanol (48:16:0.6); flow rate, 2.0 ml/min

Figures 2 and 3 show the separation of each stereoisomer of phoenicoxanthin and astaxanthin fractions by HPLC on Sumipax OA-2000.

Characterization of Carotenoids[6–9]

Characterization of carotenoids is carried out by means of visible light absorption spectra (VIS), infrared spectra (IR), mass spectra (MS), ^{1}H NMR spectra, circular dichroism spectra (CD), chemical derivatizations such as acetylation, trimethylsilylation, methylation of allylic hydroxy group, reduction with $NaBH_4$ in the case of ketocarotenoids, and co-TLC and co-HPLC with authentic samples.

[5] T. Maoka, K. Komori, and T. Matsuno, *J. Chromatogr.* **318**, 122 (1985).
[6] G. Britton and T. W. Goodwin, this series, Vol. 18, p. 654.
[7] S. Liaaen-Jensen and A. Jensen, this series, Vol. 23, p. 586.
[8] G. Britton, this series, Vol. 11, p. 113.
[9] T. Matsuno, M. Katsuyama, T. Maoka, T. Hirono, and T. Komori, *Comp. Biochem. Physiol. B.* **80B**, 779 (1985).

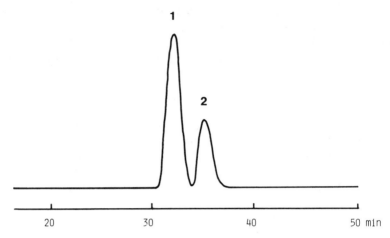

FIG. 2. HPLC separation of two stereoisomers of phoenicoxanthin obtained from *P. brevipes*. Column, Sumipax OA-2000 (5 μm), 300 × 4 mm i.d.; mobile phase, *n*-hexane–CH_2Cl_2–ethanol (48:16:0.6); flow rate, 2.0 ml/min; detection, 450 nm. Peak 1, (3*R*)-phoenicoxanthin; peak 2, (3*S*)-phoenicoxanthin.

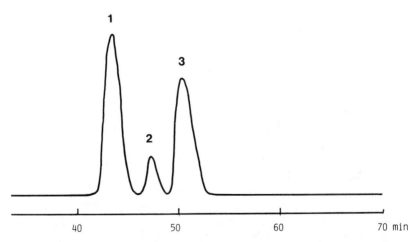

FIG. 3. HPLC separation of three stereoisomers of astaxanthin obtained from *P. brevipes*. Conditions as in Fig. 2. Peak 1, (3*R*,3'*R*)-astaxanthin; peak 2, (3*R*,3'*S*, *meso*)-astaxanthin; peak 3, (3*S*,3'*S*)-astaxanthin.

Instruments

Melting points are measured on a Yanagimoto (Kyoto, Japan) micromelting point apparatus. Visible and IR spectra are recorded on a Shimadzu (Kyoto, Japan) UV-240 spectrophotometer in ether and on a Shimadzu IR-27G spectrophotometer in KBr disks, respectively. MS spectra are obtained with a Hitachi (Tokyo, Japan) M-80 instrument with a direct inlet system at 70 eV, 190–210°. ^{1}H NMR (300 MHz) spectra are recorded with a Varian (Sunnyvale, CA) XL-300 instrument in CDCl$_3$ with tetramethylsilane as an internal standard. CD spectra are measured on a Jasco (Tokyo, Japan) 500-C spectropolarimeter in EPA (ether–isopentane–ethanol, 5:5:2) solvent at 20°. HPLC is carried out on a Shimadzu LC-6A instrument with a Shimadzu SPD-6VA UV–VIS spectrophotometric detector set at 450 nm.

TABLE I
CAROTENOIDS FROM CARAPACE OF *Paralithodes brevipes*

Carotenoid[a]	VIS (λ_{max} nm)[b]	MS (M$^+$)	R_f[c]	Composition (%)
β-Carotene (1)	(425),[d] 449, 475	536	0.82	0.8
Echinenone (2)	455	550	0.62	0.2
Canthaxanthin (3)	468	564	0.49	0.2
(3R)-Phoenicoxanthin (4)	470	580	0.45	2.1
(3S)-Phoenocoxanthin (5)	470	580	0.45	0.9
Lutein A (6)	419, 443, 473	568	0.38	0.3
(3R,3'R)-Zeaxanthin (7)	(425), 449, 475	568	0.38	0.5
Diatoxanthin (8)	(425), 451, 479	566	0.37	0.8
Alloxanthin (9)	(426), 452, 480	564	0.37	0.6
Papilioerythrinone (10)	450–470	580	0.43	5.6
Fritschiellaxanthin (11)	450–470	582	0.39	4.5
(3S,3'R)-4-Ketozeaxanthin (12)	462	582	0.39	1.5
Pectenolone (13)	454–473	580	0.39	1.8
4-Ketoalloxanthin (14)	454–473	578	0.39	1.0
(3R,3'R)-Astaxanthin (15)	471	596	0.40	19.0
(3R,3'S,*meso*)-Astaxanthin (16)	471	596	0.40	9.5
(3S,3'S)-Astaxanthin (17)	471	596	0.40	11.4
(3S,3'S)-7,8-Didehydroastaxanthin (18)	475	594	0.40	27.2
(3S,3'S)-7,8,7',8'-tetradehydroastaxanthin (19)	476	592	0.40	6.1
Unidentified carotenoids				6.0
Total carotenoids				6.1 mg/100 g

[a] Structures can be found in Fig. 4.
[b] In ether.
[c] R_f value, TLC on silica gel G, acetone–*n*-hexane (3:7).
[d] Shoulder.

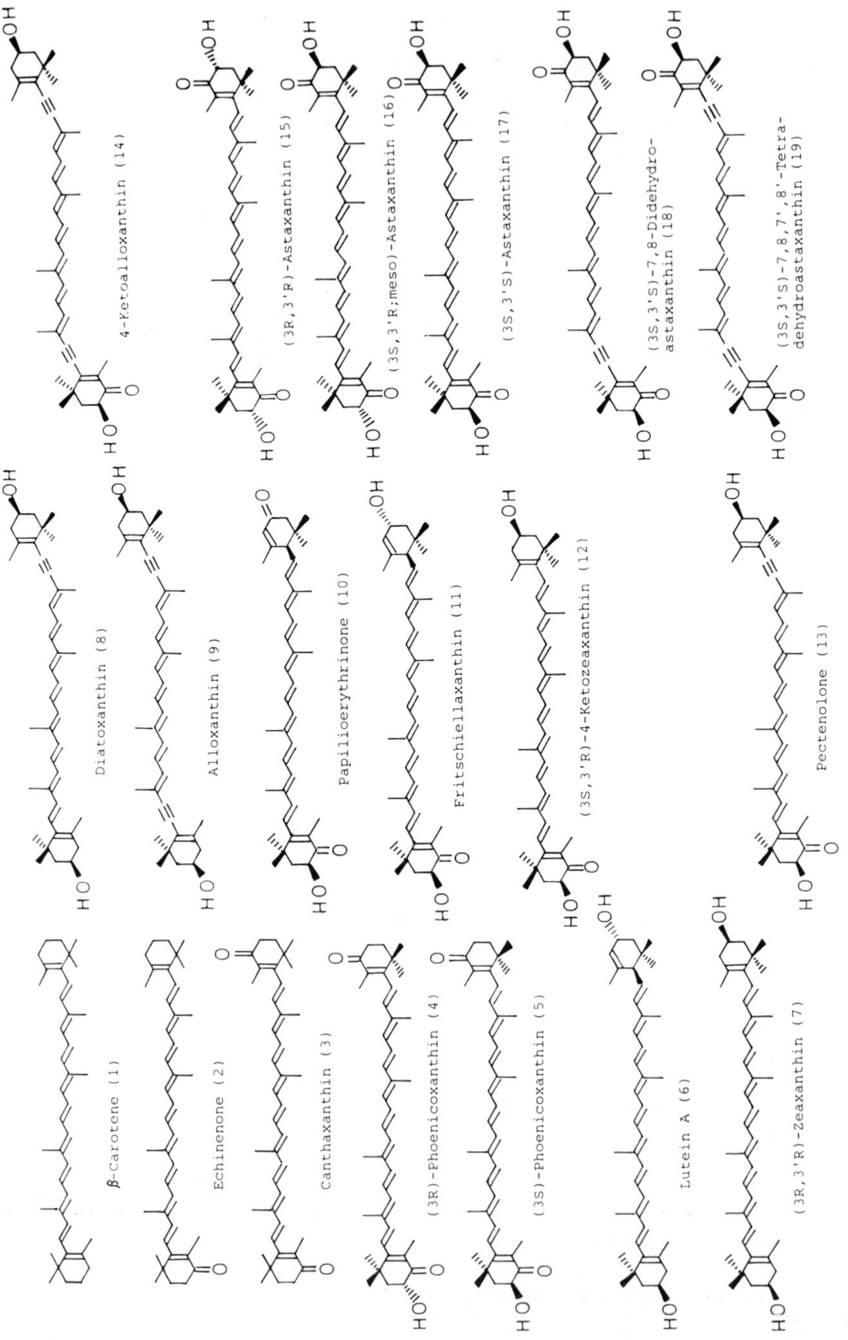

FIG. 4. Structures of the carotenoids obtained from *P. brevipes* (**1–19**).

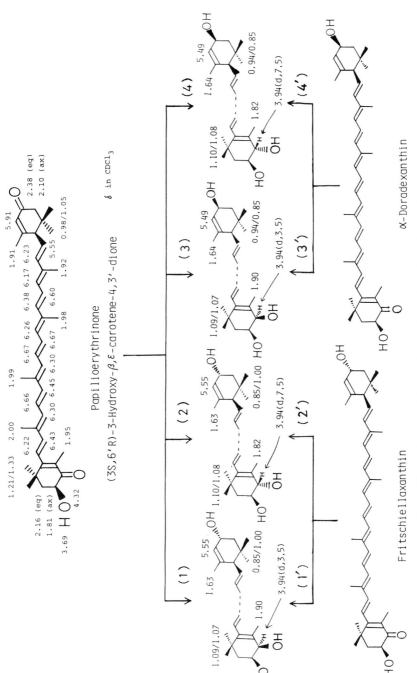

FIG. 5. Determination of the absolute configuration of papilioerythrinone.

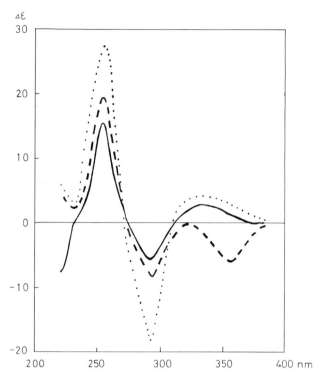

FIG. 6. CD spectra of papilioerythrinone (—), fritschiellaxanthin (----), and α-doradexanthin (····) in EPA at 20°.

Carotenoids in *Paralithodes brevipes*

Nineteen carotenoids have been identified. Table I shows the carotenoids from the carapace of *P. brevipes* and their VIS, MS, and R_f data and percentage compositions. Figure 4 shows the structures of the carotenoids obtained from *P. brevipes*.

Determination of Absolute Configuration of Papilioerythrinone

Papilioerythrinone was isolated as reddish needles with mp 155–157° (yield, 0.8 mg from 1500 g carapace). The validity of the planar structure of papilioerythrinone (3-hydroxy-β,ε-carotene-4,3′-dione) proposed by

Harashima et al.[10] has been confirmed by MS, IR, and ^{1}H NMR (Fig. 5) data.[11]

Papilioerythrinone possesses two chiral centers at C-3 and C-6'. The absolute configuration was determined by CD spectral data (Fig. 6) and comparison of NaBH$_4$ reduction products with those of fritschiellaxanthin [(3S, 3'R,6'R)-3,3'-dihydroxy-β,ε-caroten-4-one][12,13] and α-doradexanthin [(3S,3'S,6'R)-3,3'-dihydroxy-β,ε-caroten-4-one].[12] The CD spectrum of papilioerythrinone was closely similar to those of fritschiellaxanthin and α-doradexanthin (Fig. 6). Therefore it was assumed that papilioerythrinone possesses the same chiralities at C-3 and C-6' as those of fritschiellaxanthin and α-doradexanthin.

Reduction of papilioerythrinone (0.6 mg) with NaBH$_4$ (20 mg) in methanol (20 ml) for 10 min at 20° provided four stereoisomers of triols **(1)**, **(2)**, **(3)**, and **(4)** on Fig. 5, which were separated by HPLC on Sumipax OA-2000 with n-hexane–CH$_2$Cl$_2$–ethanol (48 : 16 : 1.5) as shown in Fig. 5. As was expected, triols **(1)** and **(2)** were completely identical to triols **(1')** and **(2')** derived from fritschiellaxanthin by NaBH$_4$ reduction, respectively, and triols **(3)** and **(4)** also completely coincided with triols **(3')** and **(4')** from α-doradexanthin (Fig. 5). Thus the absolute configuration of papilioerythrinone was determined to be (3S,6'R)-configuration by the accumulated evidence described above.

[10] K. Harashima, J. Nakahara, and G. Kato, *Agric. Biol. Chem.* **40,** 711 (1976).
[11] G. Englert, in "Carotenoid Chemistry and Biochemistry" (G. Britton and T. W. Goodwin, eds.), Proc. 6th Int. Symp. Carotenoids, p. 107. Pergamon, Oxford, 1982.
[12] R. Buchecker, C. H. Eugster, and A. Weber, *Helv. Chim. Acta* **61,** 1962 (1978).
[13] T. Matsuno and M. Ookubo, *Nippon Suisan Gakkaishi* **48,** 653 (1982).

[4] Surface-Enhanced Raman Scattering Spectroscopy of Photosynthetic Membranes and Complexes

By MICHAEL SEIBERT, RAFAEL PICOREL, JAE-HO KIM, and THERESE M. COTTON

Introduction

Surface-enhanced Raman scattering (SERS) results when ion complexes, molecules, or chromophores adsorbed onto or near roughened silver, gold, or copper substrates are excited with laser light. The effect was first reported for pyridine on anodized silver electrodes[1-3] where an en-

[1] M. Fleischmann, P. F. Hendra, and A. J. McQuillan, *Chem. Phys. Lett.* **26,** 123 (1974).

hancement of 10^6-fold over normal Raman scattering intensity was determined.[2,3] Both chemical and electromagnetic interactions between the adsorbate and the metal contribute to the enhancement phenomenon, depending on the metal, the adsorbate, and the surface structure.[4] Surface-enhanced resonance Raman scattering (SERRS) occurs when the laser excitation wavelength is in resonance with an electronic transition in the adsorbate and provides additional enhancement. The major advantages of SERS/SERRS include extreme analytical sensitivity, molecular selectivity, minimization of fluorescence, and distance sensitivity. Thus, sample concentrations in the range that is used for fluorescence spectroscopy are appropriate, and chromophores can be selected for or discriminated against, depending on the excitation wavelength. Since the Raman intensity is proportional to (a/r),[10] where a is the local radius of curvature at the metal surface and r is the distance of the adsorbate from the center of the local radius of curvature, only scattering centers located within tens of angstroms are detectable.[5] Interference by Rayleigh scattering generally is not a problem because Raman scattering is inelastic (scattering is shifted by vibrational frequencies), although SERS/SERRS is limited to the near-ultraviolet (UV) through the near-infrared (IR) region of the spectrum.

Although many small molecules have been studied by SERS/SERRS, the investigation of complex biomolecules has been limited. Still, spectra for amino acids, proteins, nucleic acids, polynucleotides, DNA, and lipid monolayers have been reported.[6-9] Valuable information about surface interactions, the types of chemical structures leading to enhancement, electron transfer, denaturation, redox effects, enzyme activity, selective chromophore excitation, spin states, distance effects, and frequency shifts has been obtained.[7-9] All of this work set the stage for the application of SERS/SERRS to membrane systems[10,11] and their components, which by virtue of their complexity are well suited to investigation by a sensitive, selective, distance-dependent technique.

---

[2] D. L. Jeanmaire and R. P. Van Duyne, *J. Electroanal. Chem.* **84**, 1 (1977).
[3] M. G. Albrecht and J. A. Creighton, *J. Am. Chem. Soc.* **99**, 5215 (1977).
[4] H. Metiu, *in* "Surface Enhanced Raman Scattering" (R. K. Chang and T. E. Furtak, eds.), p. 1. Plenum, New York, 1982.
[5] T. M. Cotton, R. A. Uphaus, and D. Möbius, *J. Phys. Chem.* **90**, 6071 (1986).
[6] I. R. Nabiev, S. D. Trakhanov, E. S. Efremov, V. V. Marinyuk, and R. M. Lasorenko-Manevich, *Bioorg. Khim.* **7**, 941 (1981).
[7] T. M. Cotton, *Adv. Spectrosc.* **15**, 91 (1988).
[8] E. Koglin and J.-M. Séquaris, *Top. Curr. Chem.* **134**, 1 (1986).
[9] I. R. Nabiev, R. G. Efremov, and G. D. Chumanov, *Sov. Phys.—Usp. (Engl. Transl.)* **31**, 241 (1988).
[10] M. Seibert and T. M. Cotton, *FEBS Lett.* **182**, 34 (1985).
[11] I. R. Nabiev, R. G. Efremov, and G. D. Chumanov, *J. Biosci.* **8**, 363 (1985).

Experimental Procedures

Raman Instrumentation

As in conventional Raman and resonance Raman (RR) spectroscopy, lasers are used to excite SERS/SERRS. In most instances, continuous-wave gas lasers, such as Ar^+ or Kr^+, have been used, although pulsed lasers may be used as well. The advantage of the latter is that time-resolved spectra are obtained, and these can provide mechanistic information. The two most common sample illumination geometries for solution Raman and RR samples seen in Fig. 1A and B involve collection of the scattered light at 90° relative to the incident angle or in the backscattering mode. In SERS/SERRS, the laser light is focused on the sample (adsorbed on a substrate, see below) at an incident angle of approximately 60°, as seen in Fig. 1C. The scattered radiation is collected in the backscattering mode by a lens with a high collection efficiency. The light is then collimated through a second lens with an f number that matches the spectrometer.

The choice of spectrometer and detector is particularly important when applying SERS/SERRS to biological molecules. This is especially true when electrodes are used as the SERS-active substrate. Irradiation of the sample can lead to rapid photodegradation of the biomolecule. Rotation of the electrode in the laser beam can help to alleviate this problem, but the use of multichannel detection schemes is another important approach. With array detectors or charge-coupled devices, a substantial part of the spectrum [e.g., 850 cm^{-1} with an 1200-g/mm grating (Triplemate 1877, Spex Industries, Edison, NJ) and 514.5-nm excitation] can be collected within seconds. Figure 2 illustrates a monochromator/spectrograph combination with an attached diode array detector. The monochromator stage functions as a filter to remove Rayleigh light and transmit a band of frequencies to the spectrograph stage. The latter disperses the light across the detector. A typical diode array contains 1024 pixels, each of which is 25 μm in width and 2.5 mm in height.

SERS Substrate Preparation

The three most common SERS substrates include electrodes, vacuum-deposited island films, and silver colloids. In all cases, some type of surface roughness is necessary to achieve maximal enhancement. The particulate nature of the surface provides enhancement of the electromagnetic field near the metal surface according to the electromagnetic theory of enhancement. The methods for preparing these substrates are summarized here.

Silver electrodes are prepared by sealing flattened silver wire or foil in a glass tube. An inert resin, such as Torr Seal (Varian, Palo Alto, CA) is used

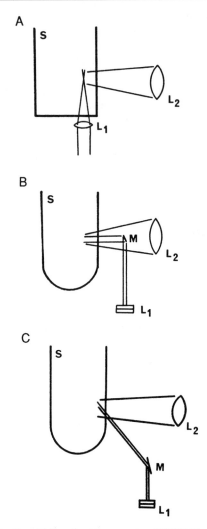

FIG. 1. Schematic for Raman illumination geometry. (A) 90° illumination: S is the sample (cuvette), L_1 is the focusing lens for the laser beam, and L_2 is the collection lens for the Raman scattered light. (B) 180° or backscattering illumination: S is the sample (test tube, cuvette, electrochemical cell, etc.), L_1 is a focusing lens (often a cylindrical lens is used to produce a line image at the spectrometer slit), L_2 is the collection lens, and M is a small front-surface mirror that directs the laser beam to the sample. (C) Same as (B), except the laser beam is directed into the sample at an angle of 60°.

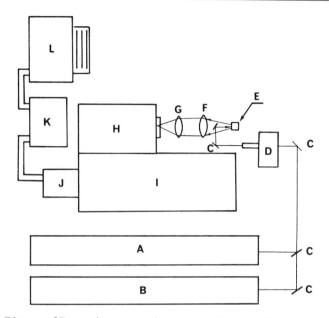

FIG. 2. Diagram of Raman instrumentation. A, Argon ion laser; B, krypton ion laser; C, mirrors; D, tunable premonochromater (Anaspec, Applied Photophysics, Hudson, MA); E, sample holder; F, collecting lens; G, focusing lens; H, monochromator stage of spectrometer (Spex Triplemate); I, spectrograph stage of spectrometer (Spex Triplemate); J, photodiode array detector (PARC 1420, Princeton Applied Research Corp., Princeton, NJ); K, detector controller (PARC 1418); L, PARC OMA II operating system and computer.

for this purpose. The surface is then polished with an alumina suspension in water, beginning with a coarse grade (5 μm) and ending with the finest grade (0.05 μm). A polishing wheel equipped with a nylon disk may be used for this purpose. At this point, the surface is mirror-like in appearance. The electrode is roughened by an oxidation–reduction cycle (ORC) in an electrochemical cell. Figure 3 shows a typical spectroelectrochemical cell containing the silver working electrode, a counter electrode (usually platinum), and a reference electrode, such as the standard calomel electrode (SCE). The potentials used for the ORC depend on the electrolyte and the size of the electrode. When 0.1 M Na_2SO_4 is used, a typical cycle involves stepping the potential from -0.6 to $+0.45$ V and allowing the silver to oxidize until approximately 25 mC/cm^2 of charge has passed. At this point, the electrode is stepped back to the starting potential, and the Ag^+ in the diffusion layer is reduced on the electrode surface. This produces a roughened surface containing particles approximately 250 Å in diameter.

Silver island films are prepared by vacuum ($<1 \times 10^{-6}$ torr) deposition of silver (utilizing resistive heating) onto a surface such as glass or quartz

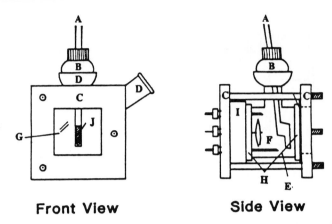

FIG. 3. Diagram of SERRS electrochemical cell. A, SERRS electrode submerged in cell; B, thermometer clamp; C, cell clamp assembly; D, cell body openings; E, nitrogen blanket inlet; F, platinum auxiliary electrode; G, front window of glass; H, O rings; I, rear plate; J, SERRS electrode surface.

slides. The rate of deposition, temperature of the glass substrate, and the thickness of the film are all important parameters with respect to the SERS activity of these surfaces. Typically, the films are prepared by depositing 50 Å (average mass) of silver at a rate of 0.02–5 Å/sec. Under these conditions, silver islands are formed with an average diameter of about 200–400 Å. The silver film is dipped into a solution of the molecule of interest for a sufficiently long period of time to adsorb the sample. The dipping time and solution concentration determine the surface coverage.

Silver colloids are prepared by reducing a solution of $AgNO_3$ with a reducing agent, such as sodium citrate or sodium borohydride. Although this is a relatively simple procedure, it is essential that the container be scrupulously clean. If not, the silver sol will aggregate and precipitate.

Photosynthetic Samples

A number of different photosynthetic protein–pigment complexes and membrane preparations have been examined by SERRS. These include isolated reaction center (RC) complexes,[12,13] chromatophores, and spheroplasts from photosynthetic bacteria and their mutants,[14,15] and photosystem II preparations.[10,13] Electrodes were used in each case. Following the

[12] T. M. Cotton and R. P. Van Duyne, *FEBS Lett.* **147**, 81 (1982).
[13] R. Picorel, R. E. Holt, R. Heald, T. M. Cotton, and M. Seibert, *J. Am. Chem. Soc.* **113**, 2839 (1991).
[14] R. Picorel, R. E. Holt, T. M. Cotton, and M. Seibert, *J. Biol. Chem.* **263**, 4374 (1988).
[15] R. Picorel, T. Lu, R. E. Holt, T. M. Cotton, and M. Seibert, *Biochemistry* **29**, 707 (1990).

roughening procedure, the samples were added to the electrochemical cell containing electrolyte buffer solution and the electrode, or the electrode was dipped into a solution containing sample. In the first case, spectra were obtained after about 15 min to allow for adsorption of the biological preparation onto the electrode surface. With the second procedure, the electrode was removed after a short period of time and then placed in the electrochemical cell containing electrolyte–buffer solution for spectral analysis. The potential of the electrode should be adjusted to the appropriate value for obtaining the spectrum of interest. In some cases, samples do not adhere well, and the electrode cannot be placed into an electrolyte solution without loss of sample from the surface. Low-temperature methods may then be used. In this case, the silver electrode can be dipped in sample solution, and excess liquid removed either by gently touching a tissue to the surface, allowing the liquid to wet the tissue by capillary action, or by shaking excess liquid off the electrode surface. The electrode can then be submerged in a Dewar flask containing liquid nitrogen for spectral analysis.

Methods for Obtaining Data on Photosynthetic Samples

The excitation wavelength and laser power are critical parameters for obtaining strong SERRS spectra. When samples contain multiple chromophores (e.g., chlorophylls *a* and *b*, carotenoids), it is possible to excite selectively only one class of molecules by choosing a wavelength that is in resonance with this class only. For example, in the case of photosynthetic bacterial preparations, excitation from 457.9 to 514.5 nm preferentially excites RR scattering from carotenoids. Laser power can have a dramatic effect if the sample is easily photodegraded. In all cases minimal power is used, and the integration time is adjusted to obtain good signal-to-noise ratios. Typically, the laser power is 20 mW or less.

Other important variables include the electrode potential and temperature. Adsorption depends on the surface charge of the sample. In the case of certain samples, the adsorption interaction may change with potential if different regions of the surface have different charged groups. For example, in the case of bacterial RC preparations, adsorption appears to occur closest to the bacteriochlorophyll (BChl) molecules at negative potentials and closest to bacteriopheophytin molecules at more positive potentials.[12] The temperature of the electrode can have a strong influence on the stability of photolabile samples. For some samples, it was determined that direct immersion of the electrode into liquid nitrogen produced the most stable spectra and preserved the native structure of the proteins.[16]

[16] T. M. Cotton, V. Schlegel, R. E. Holt, B. Swanson, and P. Ortiz de Montellano, *Proc. Int. Soc. Opt. Eng.* **1055**, 263 (1989).

The presence of extraneous chemicals (buffers, electrolyte) can also play a role in SERRS of photosynthetic samples. Buffers, such as Tris, can produce strong scattering and interference bands. Counterions, such as chloride, are important with respect to sample adsorption and the production of SERRS-active sites.[17] In most cases, the samples and the electrolyte solution are degassed to minimize reactions with oxygen.

Potential problems that may occur in SERRS studies include lack of adsorption of the sample to the metal surface, photoinstability (bleaching or photoreactivity), and disruption of the protein structure on the electrode surface with time. Adsorption can be affected by the potential of the metal surface. This can be varied by a potentiostat (electrode) or by adding extraneous small molecules that adsorb to the surface and change the surface potential (e.g., chloride ions or citrate on silver sols). Again, photoinstability and structural changes can be minimized by using low-temperature techniques. The use of a multichannel detector is also important because collection time can be drastically reduced in comparison to that required for single-channel detection. Finally, it is necessary that the samples and water used for the spectral analysis be as free from contaminants as possible. Often, impurities can produce strong SERS/SERRS spectra and/or prevent adsorption of the sample to the surface.

Examples of SERRS Results Obtained for Carotenoids

Bacterial Photosynthetic Systems

Location of Carotenoids in Chromatophore Membranes. Photosynthetic membranes isolated from *Rhodospirillum (Rs.) rubrum* have been examined by SERRS in liquid electrolyte suspensions. Chromatophore (cytoplasmic side out) and spheroplast (periplasmic side out) vesicles from wild-type (S1), carotenoidless mutant (G9), and reaction centerless mutant (F24) strains were used to locate spirilloxanthin (Spx) on the cytoplasmic side of the membrane within the B880 antenna complex.[14] The extreme distance-sensitive property of SERRS allowed observation of Spx peaks at 1508, 1151, and 1001 cm^{-1} with chromatophores but not spheroplasts. This was instrumental in drawing a model[14] for B880 in which the carotenoid partially spans the membrane, exposed on the cytoplasmic side but embedded near BChl molecules at some distance from the periplasmic side. Similar observations have been made with the carotenoids in the antenna complexes of *Rhodobacter (Rb.) sphaeroides* membranes.[15]

---

[17] P. Hildebrandt and M. Stockburger, *J. Phys. Chem.* **88**, 5935 (1984).

Orientation of Carotenoids in Chromatophore Membranes. In certain cases it is also possible to ascertain orientational information on asymmetric carotenoid molecules by comparing SERRS and RR frequency peaks.[15] This was demonstrated in liquid suspensions by comparing chromatophores isolated from photosynthetically grown *Rb. sphaeroides* and those grown aerobically in the dark. Spheroidene with a methoxy group on one end is synthesized in the former case, and spheroidenone with an oxo group on the methoxy end is synthesized in the latter. SERRS and RR peaks observed for spheroidene were the same (1519, 1153, and 1000 cm^{-1}), whereas the 1511 cm^{-1} RR peak observed for spheriodenone shifted to 1520 cm^{-1} in the case of SERRS (Fig. 4). The shift most likely results from interaction of the oxo group with the silver electrode in a manner that disrupts its conjugation with the polyene chain of the carotenoid and thus restricts delocalization of the π electrons. This demonstrates that the methoxy end of the carotenoids is oriented toward the cytoplasmic side of the chromatophore membrane, because the periplasmic side gives no SERRS signal.[15]

Reaction Center and Antenna Complexes. Low-temperature SERRS techniques have been used to examine spirilloxanthin in isolated RC and antenna complexes. Peaks at 1526, 1156, and 1000 cm^{-1} in RC complexes are characteristic of the cis isomer, whereas peaks at 1506–1510, 1149, and 1001 cm^{-1} in chromatophores and acetone-extracted spirilloxanthin are indicative of the all-trans conformation.[13] Carotenoid spectra in isolated antenna complexes are quite similar to those obtained with chromatophores (Picorel, Cotton, and Seibert, unpublished results). Large carotenoid signals obtained in RCs indicate that the molecule is located close to the surface, in agreement with X-ray crystallographic data.[18] It is noteworthy that the detergents [lauryl dimethylamine *N*-oxide (LDAO), Triton X-100, and sodium dodecyl sulfate (SDS)] used to isolate the pigment–protein complexes have little effect on the SERRS signals.

Green Plant Photosynthetic Systems

Location of Carotenoids in Thylakoid Membranes. Examination of stromal- and lumenal-side-out spinach thylakoid vesicles and (stromal side out) photosystem II (PSII) membrane fragments using liquid electrolyte SERRS techniques led to the conclusion that carotenoids are exposed on both sides of higher plant membranes.[19] Binding of antibodies (raised

---

[18] T. O. Yeates, H. Komiya, A. Chirino, D. C. Rees, J. P. Allen, and G. Feher, *Proc. Natl. Acad. Sci. U.S.A.* **85**, 7993 (1988).

[19] M. Bakhtiari, R. Picorel, T. M. Cotton, and M. Seibert, *Photochem. Photobiol.,* in press (1992).

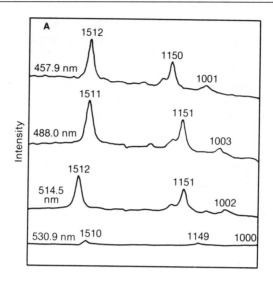
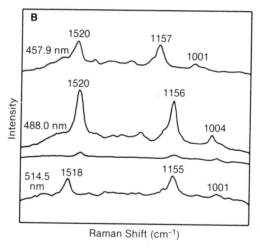

FIG. 4. Resonance Raman (A) and SERRS (B) spectra of chromatophores isolated from wild-type *Rb. sphaeroides* grown aerobically in the dark. Note the shift in peak frequency in the SERRS spectra, indicating that the methoxyl end of spheroidenone is located on the cytoplasmic side of the membrane.

against lumenal-side proteins) to PSII membrane fragments confirmed the distance sensitivity of SERRS and led to the conclusion that carotenoids are also exposed at the appressed membrane surfaces of thylakoids.[19]

Isolated Reaction Centers. Low-temperature SERRS studies revealed that β-carotene, the only carotenoid present in isolated D1/D2 PSII RC, was located near the surface of the complex and was in the all-trans configuration (peaks at 1527, 1158, and 1006 cm^{-1}), consistent with RR data. Again, the presence of detergents did not prevent detection of SERRS signals.

Do SERRS Techniques Denature Protein Complexes?

To observe SERRS signals, protein complexes must be adsorbed onto anodized metal surfaces. There has been a long-standing debate over whether physical and/or chemical interactions between the two disrupt protein structure (and hence affect conclusions obtained from biological SERRS studies). The fact that the cis conformation of the native bacterial RC carotenoid is maintained during SERRS experiments with isolated RCs and no new peaks are ascribed to heme denaturation of cytochrome b-559 in isolated PSII RC complex suggested that significant protein denaturation does not occur under the conditions used.[13] Although this is encouraging, a general rule cannot as yet be drawn, and every protein complex should be considered individually.

Other Applications of SERS/SERRS in Photosynthesis

Although emphasis in this chapter has been on carotenoids, similar SERRS techniques can be used to examine other photosynthetic chromophores. Cytochrome b-559 has been mentioned above, but other pigments, including other cytochromes, chlorophylls and other porphyrins, flavins, antenna pigments in addition to chlorophylls and carotenoids, quinones, and other electron-transport components, are under study.[13,20] SERS signals attributable indirectly to the presence of manganese, functional in the O_2-evolution process of photosynthesis, have also been observed.[10,21] Finally, SERS spectra of bacteriorhodopsin in purple membrane have been recorded.[9]

Conclusions and Future Developments

The results described above demonstrate that SERRS has considerable potential for analytical applications (determination of Raman spectra on very small and highly dilute samples), as well as for fundamental studies of

---

[20] R. Picorel, T. Lu, R. E. Holt, T. M. Cotton, and M. Seibert, in "Current Research in Photosynthesis" (M. Baltscheffsky, ed.), Vol. II, p. 907. Kluwer Academic, Dordrecht, The Netherlands, 1990.

[21] M. Seibert, T. M. Cotton, and J. G. Metz, *Biochim. Biophys. Acta* **934**, 235 (1988).

biomolecular structure on surfaces. The membrane studies suggest that SERRS can provide qualitative information about the location and orientation of chromophores within the membrane. Future developments will include the use of new detectors (e.g., charge-coupled devices), which have very low dark counts and can be used to integrate signals for long time periods. It should be possible to obtain unenhanced Raman spectra on any surface using this approach. Although no longer surface (distance) sensitive, this approach will provide complementary information to SERRS about the structure of macromolecules on surfaces. Also, the use of new lasers, such as the titanium:sapphire laser, will allow excitation throughout the red region of the electromagnetic spectrum (650–1000 nm). This will be valuable for studies of biological samples on gold substrates. Finally, new nonlinear techniques, such as hyper-Raman and surface-enhanced hyper-Raman scattering, will be possible using pulsed excitation.

Acknowledgment

This work was supported by the Divisions of Chemical Sciences (M.S. and T.M.C.) and Energy Biosciences (M.S.), Office of Basic Energy Sciences, U.S. Department of Energy; Consejo Superior de Investigaciones Cientificas, Spain (R.P.); and the National Institutes of Health (T.M.C., grant Number GM-35108). Ames Laboratory is operated for the U.S. Department of Energy by Iowa State University under Contract No. W-7405-Eng-82. SERI is a division of the Midwest Research Institute and operated for the U.S. Department of Energy under Contract No. DE-AC02-83CH-10093.

[5] Synthesis of Carotenoids Specifically Labeled with Isotopic Carbon and Tritium

By ARNOLD A. LIEBMAN, WALTER BURGER, SATISH C. CHOUDHRY, and JOSEPH CUPANO

Introduction

The chemistry used in the synthesis of carotenoids is quite extensive and has been periodically reviewed.[1] This literature reflects the expanding use of chiral reagents and other tools of modern organic synthesis and also the increasing use of isotopically labeled carotenoids in a variety of appli-

---

[1] K. Bernhard, *in* "Carotenoids: Chemistry and Biology" (N. I. Krinsky, M. M. Mathews-Roth, and R. F. Taylor, eds.), Proc. 8th Int. Symp. Carotenoids, p. 337. Plenum, New York, 1989.

cations. In general, carotenoids that have been specifically labeled at one or more particular positions have been prepared with stable isotope variants, while carotenoids labeled with a radioactive isotope have usually been prepared from a labeled precursor by biosynthesis. Notable exceptions are certain carotenoids labeled with tritium. Partial reduction of an alkyne precursor of these carotenoids with tritium gas produces these labeled materials. Carotenoids are not a class of chemically stable compounds and the use of specifically isotopically labeled variants facilitates their study. Methodology has now been developed to prepare these compounds specifically labeled with isotopic carbon as well as with tritium.

Labeling of Carotenoids

Specific labeling is necessary in areas such as mass spectral rationalizations of carotenoids, in which the use of deuterium enrichment is essential for meaningful interpretation. Considerable application of specific deuterium labeling to various carotenoids for this purpose appears in the literature and has been reviewed.[2] Specifically labeled carotenoids are also required in areas of study such as metabolism and pharmacokinetics. The introduction of carotenoids into functioning biological systems for this purpose requires an isotopic label that is not subject to exchange with various components of the system. Because isotopic hydrogen is likely to undergo significant exchange in a biological system, a carbon label is then necessary. Incorporating a labeled carbon atom (or atoms) into a specific position of the carotenoid skeleton yields a substrate that cannot undergo exchange and will also maintain the integrity of the position of the label throughout the study.

The element carbon consists of four isotopic variants, two of which are stable and two radioactive. Of the former, carbon-12 is considered to be the normotope while carbon-13, occurring with a natural abundance of about 1%, is the stable isotope of carbon used for specific labeling. Carbon-11 is a high-energy positron-emitting radioactive isotope with a radioactive half-life of 20 min and therefore has limited value in the preparation of labeled carotenoids at this time. Carbon-14 is the well-known and well-used radioactive isotope of carbon that possesses readily detected and measured soft β radiation and a very favorable half-life of more than 5000 years. It also has isotope effects that are negligible in biological applications.

To synthesize carotenoids with a radioactive label in a specific position, the molecule must be viewed as the sum of its parts, just as in an unlabeled

[2] H. Budzikiewicz, in "Carotenoid Chemistry and Biochemistry" (G. Britton and T. W. Goodwin, eds.), Proc. 6th Int. Symp. Carotenoids, p. 155. Pergamon, New York, 1982.

synthesis. One of the parts will become the radioisotope carrier and its synthesis is selected because of economy of manipulations, or because only a few radiochemical steps are required, coupled with a favorable overall yield. The various syntheses of nonlabeled carotenoids by assembly of appropriate parts have been reviewed.[3]

From a radiochemical point of view, the most economical synthesis of carotenoids is the tritium gas reduction of those materials that can be prepared with a suitable alkyne moiety as a precursor. There are a limited number of structures that fit this requirement because it necessitates synthesis of an unlabeled precursor that contains all of the carbon atoms of the final product. While thus limited, this method of radioisotopic labeling of carotenoids remains the only one that is carried out essentially in one radiochemical step.

It is always desirable to proceed with a radiochemical synthesis in the minimum number of steps with radioactive intermediates, but this cannot be considered to be a limiting factor. The intended use of the labeled material must be the first priority in considering the feasibility of a radiochemical synthesis.

Carotenoids with high specific radioactivity at metabolically useful positions can be prepared via the existing chemistry for the synthesis of unlabeled carotenoids. The C-15 phosphonium (PPh_3) salt (**I**) derived from β-ionone is particularly useful. Two equivalents of this molecule and one equivalent of the dialdehyde (**II**) combine under Wittig conditions to provide β-carotene (**III**) as described in the literature.[3]

[3] O. Isler, *Pure Appl. Chem.* **51**, 447 (1979).

The radiolabeled C-15 phosphonium salt is significantly more stable than either radiolabeled retinoids or carotenoids and can be prepared from any of several labeled carriers.[4,5]

Both carbon and hydrogen isotopes have been incorporated into various retinoids via the C-15 phosphonium salt[4] and the use of this reagent in the preparation of similarly labeled carotenoids is now reported. Reaction of β-ionone (**IV**) with acetylene (**V**) in the presence of *n*-butyllithium and tetrahydrofuran (THF) provides ethynyl β-ionol (**VI**). Both ^{13}C- and ^{14}C-labeled acetylene are available and their use in preparing

labeled **VI** is quite facile. Conversion of **VI** to the ^{14}C-labeled **I** is carried out as described[4] and, because the intrinsic radiation is not self-absorbed, better yields of this labeled phosphonium salt may be obtained by extending the reaction time up to 3 days. However, using tritium gas for the conversion of unlabeled **VI** to **VII** by partial reduction and then to **I(T$_2$)** (^{3}H-labeled **I**) yields a product with self-absorbed radiation that is much

[4] A. A. Liebman, W. Burger, D. H. Malarek, L. Serico, R. R. Muccino, C. W. Perry, and S. C. Choudhry, *J. Labelled Compd. Radiopharm.* **28**, 525 (1990).
[5] H. Kaegi, in "The Retinoids" (M. B. Sporn, A. B. Roberts, and D. S. Goodman, eds.), Vol. I, p. 147. Academic Press, New York, 1984.

less stable. If the tritium-labeled phosphonium salt is to be stored for any length of time, it must be diluted with unlabeled phosphonium salt to no more than 50% of full enrichment and stored as a dilute solution, preferably in toluene or a toluene–ethanol mixture. The ^{14}C-labeled phosphonium salt, on the other hand, remains reasonably stable in storage for periods up to 1 year.

The phosphonium salt (I) has also been prepared using ^{13}C-labeled acetylene as the isotope carrier and also in deuterated form by reduction of VI with deuterium gas. These materials exhibit the same stability characteristics as the unlabeled phosphonium salt and conditions for their subsequent incorporation into carotenoids may differ from those involving radioisotope-labeled phosphonium salt. Synthetic approaches to nonradioactively labeled materials can be carried out on any size scale and tend to be concerned with mass yields, while those involving radioactivity are highly restricted insofar as scale is concerned and radiochemical purity of products is a primary concern.

While it is certainly possible to prepare labeled phosphonium salts other than the one derived from β-ionone and consequently to directly prepare labeled carotenoids other than β-carotene, we have not done so. β-Carotene, however, can be oxidized to canthaxanthin and then to other carotenoids, which would constitute an indirect synthesis of these materials in labeled form. Our discussion here is limited to the preparation of β-carotene labeled with deuterium, tritium, carbon-13, and carbon-14.

Procedures

General

All solvents are distilled (tetrahydrofuran is distilled from sodium ribbon using benzophenone as indicator). Spectra are recorded on standard instruments by the staff of the Physical Chemistry Department of Hoffmann-La Roche, Inc. (Nutley, NJ). Radiochemical purity is determined on thin-layer chromatograms with an LB 2832 Berthold (Nashua, NH) linear analyzer system and radioactivity is measured by the liquid scintillation technique using a Beckman (Palo Alto, CA) model LS 7500 spectrometer. Reactions to prepare the various carotenoids are carried out under subdued lighting or under yellow lights. Preparative details for the deuterium, tritium, and ^{14}C-labeled C-15 phosphonium salts are described in the literature.[4]

[3-Methyl-5-(2,6,6-trimethylcyclohex-1-enyl)-2,4-pentadienyl][1,2-$^{13}C_2$]-triphenylphosphonium Bromide

By transfer at $-78°$, 0.35 mol of [1,2-$^{13}C_2$]acetylene (7.8 liters)[6] is added to a solution of 0.34 mol (47.6 ml) of diisopropylamine in 600 ml of dry tetrahydrofuran (THF). A solution of n-butyllithium, 2.5 N, in 132 ml of hexane is added, with stirring, over a 30-min period and then stirring is continued an additional 10 min. A solution of 63.36 g (0.33 mol) of β-ionone in 150 ml of dry THF is added and the resulting mixture is stirred for 2 hr, during which time the bath temperature rises to $-10°$. Ethyl ether (500 ml) is added and the resulting solution is extracted with two portions, 150 ml each, of saturated brine. The organic layer is dried over anhydrous magnesium sulfate, filtered, and concentrated under reduced pressure to a residue of 79.62 g of yellow oil, which is used as is in the next step.

All of the material obtained above is dissolved in 750 ml of hexane and 2 g of Lindlar vitamin A catalyst [5% (w/w) Pd/CaCO$_3$ + Pb] and 1.33 ml of quinoline is added. The mixture is placed under an atmosphere of hydrogen and stirred for 1 hr, when an additional 1-g portion of the catalyst is added. Stirring under the hydrogen atmosphere is continued for two more hours, at which time a thin-layer chromatogram [SiO$_2$; 20% (v/v) ethyl acetate in hexane] shows the absence of starting material. The catalyst is removed by filtration and the filtrate is concentrated under reduced pressure to a residue, which is distilled. The fraction boiling between 106 and 108°/0.9 mmHg is collected and weighs 68.25 g (0.307 mol).

A solution of 106.33 g (0.31 mol) of triphenylphosphine hydrobromide in 300 ml of methanol is chilled to 0° and the ^{13}C-labeled vinyl β-ionol, obtained above, dissolved in 75 ml of methanol, is added over a 30-min period with stirring. The resulting mixture is stirred at room temperature for 22 hr and then diluted with 100 ml of water. This mixture is washed successively with 400 ml, then 200 ml, of hexane, which are combined and extracted with 100 ml of 80% aqueous methanol. The combined methanol–water layer is concentrated under reduced pressure to remove most of the methanol and then 400 ml of dichloromethane is added. After separating the layers, the organic layer is washed with two 100-ml portions of 5% (w/v) aqueous sodium chloride solution that, in turn, are combined and extracted with 100 ml of dichloromethane. The combined dichloromethane layers are dried over anhydrous magnesium sulfate, filtered, and concentrated under reduced pressure to a syrup that is dissolved in 100 ml of dichloromethane. One liter of ethyl acetate is added and the mixture is

---

[6] Obtained in one cylinder from Cambridge Isotope Laboratories (Woburn, MA).

reduced to half of its volume under reduced pressure. The resulting solution is chilled overnight in the refrigerator to yield 85.24 g of the desired product as a white solid, mp 172–174°.

β,β-[10,10',11,11'-$^{13}C_4$]Carotene

All of the C-15 phosphonium salt obtained above (85.24 g, 0.156 mol) is dissolved in 500 ml of ethanol. A 12.46-g (0.076 mol) portion of the C-10 dialdehyde (II) and 69 ml (0.79 mol) of 1,2-epoxybutane are added and the resulting mixture is heated in an inert atmosphere under reflux for 24 hr, then cooled and the solid formed is filtered off. The solid is washed with cold ethanol then resuspended in 100 ml of ethanol and, under nitrogen, heated at reflux for 16 hr. The red solid that forms on cooling is filtered off and further purified two more times by this resuspension technique in refluxing ethanol, under nitrogen, for 4 hr each time. The product isolated after the final purification weighs 13.89 g (0.0257 mol). The ^{1}H NMR spectrum is compatible and the carbon-13 (proton decoupled) spectrum shows the $^{13}C=^{13}C$ group at 124.88 and 130.85 ppm each as doublets. The mass spectrum is compatible with a molecular ion at 540.5.

β,β-[10,10',11,11'-$^{14}C_4$]Carotene

On a scale of 0.58 mmol (317 mg, 44 mCi) of the correspondingly ^{14}C-labeled C-15 phosphonium salt, the material is dissolved in 4 ml of ethanol and treated with 0.29 mmol (47.5 mg) of dialdehyde (II). Again, 1,2-epoxybutane, 0.3 ml, is employed as a mild organic base (generated *in situ*) and the reaction is carried out under argon at reflux for 24 hr. After cooling, 50 mg of unlabeled β-carotene is added, the mixture concentrated, and the resulting crude product is chromatographed over 50 g of fine silica with hexane elution to yield 51.6 mg (5.57 mCi) of product, the radiochemical purity of which is greater than 98%. The radiochemical yield is 12.65%.

β,β-[10,10'-$^3H_2$]Carotene

The tritium-labeled C-15 phosphonium salt, 101 mg, obtained by partial reduction of VI with approximately 10 Ci of tritium gas followed by reaction with triphenylphosphine hydrobromide, is treated with 6 mg (0.037 mmol) of dialdehyde (II) in 2.5 ml of dichloromethane while stirring under an argon atmosphere for 30 min at room temperature. A 0.11-ml portion of 0.75 N sodium methoxide solution is added[7] and the mixture

---

[7] R. Marbet (Roche, Basel), unpublished observations (1976).

is stirred an additional 90 min at room temperature. After this time, 5 ml of 1 N sodium methoxide solution is added. The resulting mixture is extracted with two 15-ml portions of dichloromethane, which are combined and concentrated to a residue that is chromatographed over 10 g of fine silica using 15% (v/v) dichloromethane in hexane for elution. The carotene-containing fractions are pooled, concentrated, and dissolved in 5 ml of ethanol. Under an argon atmosphere, the solution is maintained at reflux for 16 hr, then chilled slowly to 10°, at which temperature the product crystallizes to provide 7 mg of the desired material with a specific activity of 91.4 mCi/mg (49.35 Ci/mmol) and radiochemical purity of 97%.

Acknowledgment

We thank Dr. J. Blount and the staff in the Hoffmann-La Roche Physical Chemistry Department, particularly Dr. W. Benz for mass spectra, Dr. F. Scheidl for microanalysis, Dr. V. Toome for UV spectra, and Dr. T. Williams for NMR spectra.

[6] Synthesis of Deuterated β-Carotene

By H. ROBERT BERGEN, III

Introduction

Routine studies of the metabolism of carotenoids in humans require carotenoids labeled with nonradioactive isotopes. Appropriately labeled compounds should be symmetrically labeled and contain sufficient nonlabile isotope such that the interfering ions from naturally occurring carbon isotopes do not appreciably interfere with the analysis. An appropriate deuterium (D)-labeled analog of β-carotene may be prepared by the following procedure, as outlined in Fig. 1.[1] This synthesis is a modification of a synthesis originally reported by Johansen and Liaaen-Jensen.[2]

---

[1] This synthesis originally appeared in (a) H. R. Bergen, H. C. Furr, and J. A. Olson, *J. Labelled Compd. Radiopharm.* **25**, 11 (1988) and (b) H. R. Bergen and J. A. Olson, *J. Labelled Compd. Radiopharm.* **27**, 783 (1989). Copyright © (a) 1988 and (b) 1989, John Wiley & Sons, Ltd. Used with permission of John Wiley & Sons, Ltd.

[2] J. E. Johansen and S. Liaaen-Jensen, *Acta Chem. Scand., Ser. B* **B28**, 349 (1974).

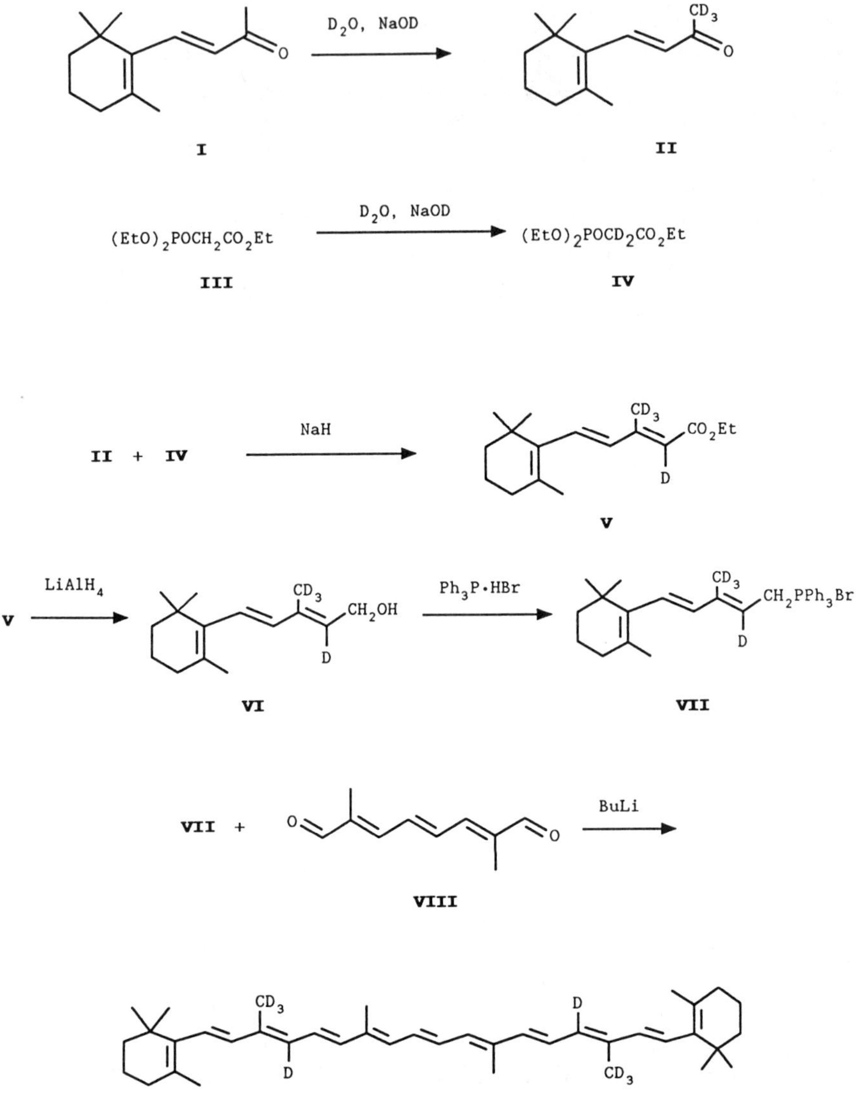

IX

FIG. 1. Scheme for synthesis of D_8-β-carotene.

Materials

Sodium hydride (dry powder), butyllithium, deuterium oxide (99.9 atom% deuterium), sodium deuteroxide (NaOD; 99.9 atom% deuterium), β-ionone, triethyl phosphonoacetate, and triphenylphosphine hydrobromide are all available from Aldrich Chemical Company (Milwaukee, WI). β-Ionone and triethyl phosphonoacetate can be purified by distillation under vacuum. Solvents are anhydrous reagent grade. Tetrahydrofuran (THF) is distilled from lithium aluminum hydride under an inert atmosphere prior to use. The C_{10}-dial was a gift from Hoffmann-LaRoche, Inc. (Nutley, NJ), or is readily synthesized.[3]

Experimental

D_3-β-Ionone (II)

D_3-β-Ionone (II) is prepared by stirring 50 g (0.26 mol) of freshly distilled β-ionone (I) with 156 g (7.8 mol) of deuterium oxide and sufficient anhydrous pyridine to disperse the emulsion formed. Fifty drops of a 40% (w/v) solution of sodium deuteroxide (NaOD) is added and the mixture stirred for 3 hr. The mixture is transferred to 0.5 liter of hexane and the aqueous layer is removed. The organic layer is washed six times with water (0.4 liter each time), dried using $MgSO_4$, and concentrated by rotary evaporation. The deuterium exchange is repeated to yield 45 g (90%) of 13,13,13-D_3-β-ionone (II) with >98% of the labile α-keto protons replaced with deuterium.

D_2-Triethyl Phosphonoacetate (IV)

D_2-Triethyl phosphonoacetate (IV) may be prepared by base-catalyzed exchange of enolizable hydrogens in deuterium oxide–sodium deuteroxide. To 175 g (0.78 mol) of freshly distilled triethyl phosphonoacetate (III) add 310 g of deuterium oxide and 5 ml of 40% NaOD and stir for 2 hr. The mixture is added to 0.5 liter of anhydrous ethyl ether and the aqueous layer removed and extracted with ether (five times, 300 ml each) to recover the phosphonate. The combined ether extracts are dried over $MgSO_4$, and concentrated by rotary evaporation. The exchange is repeated to yield after concentration 152 g (87%) of dideuterotriethyl phosphonoacetate (IV) with >99% of the enolizable protons replaced with deuterium.

---

[3] H. Mayer and O. Isler, in "Carotenoids" (O. Isler, ed.), p. 431. Birkhäuser, Basel, 1971.

D_4-Ethyl-β-ionylidine Acetate (V)

D_4-Ethyl-β-ionylidine acetate (V) is prepared via the Wittig–Horner reaction of D_3-β-ionone (II) and D_2-triethyl phosphonoacetate (IV). Dry NaH (10.56 g, 0.44 mol) is weighed under a dry, oxygen-free atmosphere and covered with 80 ml of anhydrous diethyl ether. D_2-Triethyl phosphonoacetate (100 g, 0.44 mol) dissolved in 100 ml of diethyl ether is added dropwise with stirring. This solution becomes amber after stirring for 2 hr. To this stirred solution, 28.6 g (0.15 mol) of D_3-β-ionone (II) in 100 ml of anhydrous diethyl ether is added dropwise and the mixture allowed to stir overnight. Thin-layer chromatography (TLC) on silica gel [20% (v/v) ethyl acetate in hexane] can be used to monitor the progress of the reaction. When all the ionone has reacted, water is added at a rate slow enough to prevent boiling. The mixture is poured into 0.5 liter of hexane and the organic layer is washed with water (several volumes), dried with $MgSO_4$, and concentrated by rotary evaporation. Yields of 9,9,9,10-D_4-ethyl-β-ionylidene acetate (V) of up to 92% can be obtained with >98% deuterium incorporation.[1a]

9,9,9,10-D_4-β-Ionylidene Ethanol (VI)

9,9,9,10-D_4-β-Ionylidine ethanol (VI) is prepared by reduction of V. D_4-Ethyl-β-ionylidene acetate (V) (80 g, 0.31 mol) in 0.3 liter of ethyl ether is added dropwise to a solution of lithium aluminum hydride (12.8 g, 0.34 mol) in 0.5 liter of anhydrous ethyl ether at −70° in a dry ice–acetone bath. After stirring for 1 hr at −70° the solution is gradually warmed to room temperature. The reduction is monitored by TLC and additional lithium aluminum hydride added at −70° if required. When the reduction is complete, the solution is cooled to −70° and 1 N H_2SO_4 added dropwise to deactivate remaining lithium aluminum hydride and to dissolve the aluminum hydroxide precipitate. The solution is poured into 0.5 liter of diethyl ether and the organic layer washed with 1 N H_2SO_4 and water. The dried ($MgSO_4$) organic layer is rotary evaporated to yield 60 g (90%) of D_4-β-ionylidene ethanol (VI) with only minor impurities detected.

9,9,9,10-D_4-β-Ionylidene Ethyltriphenylphosphonium Bromide (VII)

9,9,9,10-D_4-β-Ionylidene ethyltriphenylphosphonium bromide (VII) is prepared by adding 15.3 g (0.045 mol) of triphenylphosphonium hydrobromide to 9.86 g (0.045 mol) of 9,9,9,10-D_4-β-ionylidene ethanol dissolved in 100 ml of freshly distilled methanol. This solution is stirred for 1 hr at room temperature. When TLC (5% ethyl acetate in hexane) shows complete conversion of the alcohol to the Wittig salt, the solution is rotary evaporated at <30° and the remaining oil washed two times with 100 ml hexane. A crude crystalline solid can be recovered by removing excess solvent under vacuum (0.5 mmHg). The crude product can be crystallized from a minimum volume of THF to give white crystals of 9,9,9,10-D_4-β-ionylidene ethyltriphenylphosphonium bromide (VII), mp 133–136°.

D_8-β-Carotene (IX)

D_8-β-Carotene (IX) The recrystallized Wittig salt (VII) above (20.6 g, 37.8 mmol) is dissolved in 1 liter of THF and cooled in a dry ice–acetone bath. Butyllithium (37.4 mmol) in hexane is added dropwise. After the addition is complete, the solution is stirred for 2 hr and 2.95 g (18 mmol) of recrystallized (from ethyl acetate) 2,7-dimethyl-2,4,6-octatrienedial (VIII) in 0.025 liter of THF is added dropwise with stirring. The reaction is allowed to come to room temperature and stirred overnight. The product is poured into hexane and washed two times with water (0.5 liter), two times with 1 liter of methanol–water (95/5 v/v), and brine. After drying over $MgSO_4$ and removing the solvent, the residue is crystallized from hexane–methanol (2/1, v/v) to yield 10,10′,19,19,19,19′19′19′-D_8-β-carotene (IX) (2.5 g, 25%). Greater than 98% of the exchangeable protons can be replaced with deuterium in the final product.[1a] Routine synthesis yields β-carotene with >96% of the product containing six or more atoms of deuterium.[1b]

Comments

Thin-layer chromatography, to monitor reactions, can be performed on silica gel plates containing fluorescent indicators (Machery-Nagel, Polygram-Sil G/UV$_{254}$). Visualization is performed by illumination with an ultraviolet (UV) lamp. The dry NaH can be weighed out under N_2 with an inflatable chamber ("glove bag"; I²R, Cheltenham, PA).

[7] Formation of Volatile Compounds by Thermal Degradation of Carotenoids

By J. CROUZET and P. KANASAWUD

Several volatile compounds, such as β-ionone, dihydroactinidiolide, 2-methyl-2-hepten-6-one, and citral, present in plant products are considered degradation components of β-carotene[1-3] and lycopene[4] resulting from the action of heat, oxygen, or light on these precursors.

However, most of the research related to the production of volatile compounds by thermal degradation of carotenoids consists of reactions performed in organic solvent solution or in the pyrolysis of dry compounds at elevated temperature for a specific reaction time.[5-8]

In the present work aqueous suspensions of β-carotene and lycopene are treated at about 100° for different reaction times. The model system so realized is considered more representative of the conditions found during the treatment of vegetable products than the models previously studied.

Heat Treatment of Carotenoids

Reagents

β-Carotene (>97% pure; Merck, Darmstadt, Germany)

Lycopene (>90% pure; Sigma, St. Louis, MO); or extract from tomato puree and purify according to Karrer and Jucker.[9] Purity >90% as checked by thin-layer chromatography (TLC) and spectrophotometry.

---

[1] E. Demole and D. Berthet, *Helv. Chim. Acta* **55**, 1866 (1972).
[2] K. F. Murray, J. Shipton, and F. B. Whitfield, *Aust. J. Chem.* **25**, 1921 (1972).
[3] S. Isoe, S. B. Hyeon, and T. Sakan, *Tetrahedron Lett.* p. 279 (1969).
[4] E. R. Cole and N. S. Kapur, *J. Sci. Food Agric.* **8**, 360 (1957).
[5] W. C. Day and J. G. Erdman, *Science* **141**, 808 (1963).
[6] E. G. La Roe and P. A. Shipley, *J. Agric. Food Chem.* **18**, 174 (1970); G. Ohloff, *Flavour Ind.* **3**, 501 (1972).
[7] P. Schreier, F. Drawert, and S. Bhiwapurkar, *Chem. Mikrobiol. Technol. Lebensm.* **6**, 90 (1979).
[8] F. Drawert, P. Schreier, S. Bhiwapurkar, and I. Heindze, in "Flavour '81" (P. Schreier, ed.), p. 449. de Gruyter, Berlin, 1985.
[9] P. Karrer and E. Jucker, "Carotenoids." Elsevier, London, 1950.

Procedure

A 50-mg portion of β-carotene or 15 mg of lycopene is suspended by sonication in 50 ml of distilled water in a Kjeldahl flask. This flask is sealed, wrapped in aluminum foil, and heated at 97 ± 2° in an oil bath for different reaction times from 30 min to 4 hr or at 30 to 97° for 3 hr.

In some experiments, the suspension is saturated with oxygen before sealing the flask.

Isolation of Volatile Compounds

Two isolation procedures are used for the recovery of volatile compounds thermally induced during heat treatment of carotenoids.

Method 1 involves gas stripping of the most volatile compounds, trapping on Tenax GC,[10] thermal desorption, and cold trapping. The less volatile compounds are isolated by solvent extraction of the reactive medium.

Reagents for Method 1

Tenax CG (Chrompack, Middelburg, The Netherlands; 80–100 mesh)
Nitrogen purified through an 18 ft × ¼ in. column filled with Sil O Cel C 22 (Prolabo, Paris, France) and cooled to −70°

Reagents for Method 2

Dichloromethane (Merck)
Anhydrous sodium sulfate (Labosi, Paris, France)

Procedure for Method 1

The undissolved carotenoid compounds are separated by filtration of the reactive medium and the filtrate is poured into a two-neck 100-ml flask fitted with a refrigerant and gas inlet tubing. A glass trap, 150 × 2 mm filled with 80 mg of Tenax GC, is connected to the refrigerant (Fig. 1).

Before use, the trap is conditioned by heating at 220° for 1 hr under a purified nitrogen stream (30 ml/min).[11]

The volatile compounds are stripped by a purified nitrogen stream (45 ml/min) for 3 hr and the trap is disconnected from the stripping device and flashed by a stream of purified nitrogen (45 ml/min) at room temperature over 30 min to obtain water desorption. The nitrogen flow (30 ml/min) is

---

[10] R. Tressl and W. Jennings, *J. Agric. Food Chem.* **20**, 189 (1972).
[11] T. Tsugeta, T. Imai, Y. Doi, T. Kurata, and H. Kato, *Agric. Biol. Chem.* **43**, 1351 (1979).

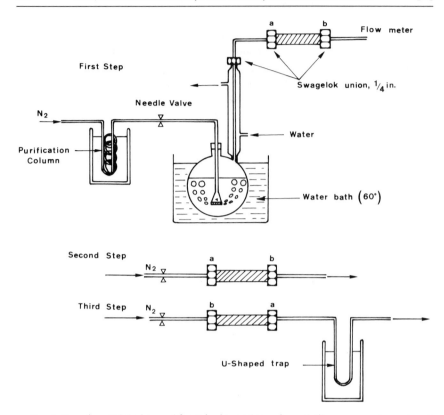

FIG. 1. Experimental device used for stripping and trapping volatile components produced during heat treatment of carotenoids in water suspension.

reversed and the volatile compounds eluted at 150° for 30 min. These compounds (about 0.5 μl) are trapped in U-shaped capillary tubing cooled at $-70°$ and used for gas–liquid chromatography (GLC) separation.

Procedure for Method 2

The filtrate obtained from the reactive medium is extracted three times with 15 or 50 ml of dichloromethane. The organic phase is dried over anhydrous sodium sulfate and concentrated to about 20 μl by solvent elimination under a smooth flow of nitrogen.

Separation of Volatile Compounds

A Varian (Sunnyvale, CA) 3700 gas chromatograph fitted with a flame ionization detector (FID), an on-column injector, and a 40 × 0.4 mm (i.d.) wall-coated open tubular (WCOT) glass capillary column coated with

Carbowax 20M (Varian) is used. The column is run isothermally at 50° for 10 min and then programmed to 170° at 4°/min with nitrogen (10 psi) as the carrier gas.

Identification of Volatile Compounds

Procedure

In order to increase the quantities of volatile compounds produced for identification purposes the heat treatment is performed under drastic conditions: 97° for 3 hr in the presence of oxygen at saturation level.

A Hewlett-Packard (Palo Alto, CA) 5996 quadrupole mass spectrometer is coupled to a gas chromatograph fitted with a fused silica capillary column DB5 (25 m × 0.256 mm; J&W Scientific, Folsom, CA). The source temperature is 150° and the ionization energy is 70 eV.

The volatile compounds are tentatively identified by automatic library search and the presence of compounds confirmed by Kovats index determination and, when possible, by comparison of retention times with those of authentic samples.

Results

The main volatile compounds identified after thermal degradation of β-carotene are 2,6,6-trimethylcyclohexanone (**I**), 2,6,6-trimethyl-2-cyclohexen-1-one (**II**), 2-hydroxy-2,6,6-trimethylcyclohexanone (**III**), β-cyclocitral (**IV**), 2-hydroxy-2,6,6-trimethylcyclohexane-1-carboxaldehyde (**V**), β-ionone (**VI**), 5,6-epoxy-β-ionone (**VII**) and dihydroactinidiolide (**VIII**).[12]

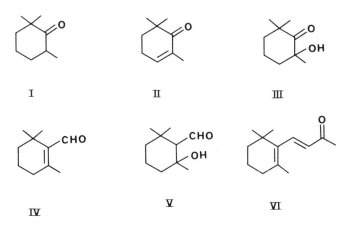

[12] P. Kanasawud and J. Crouzet, *J. Agric. Food Chem.* **38**, 237 (1990).

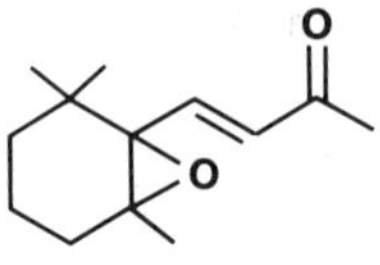

VII 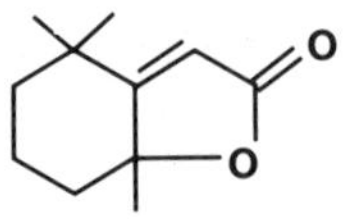   **VIII**

In the case of lycopene degradation, 2-methyl-2-hepten-6-one (**IX**), pseudo-ionone (**X**), 6-methyl-3,5-heptadien-2-one (**XI**), geranial (**XII**), and neral (**XIII**) are identified.[13]

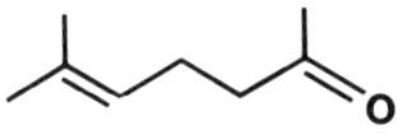

IX 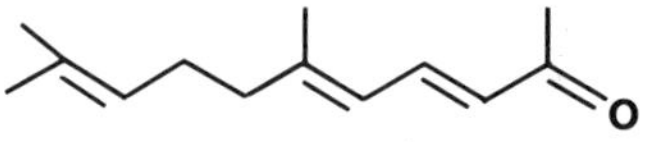   **X**

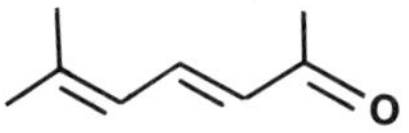

XI 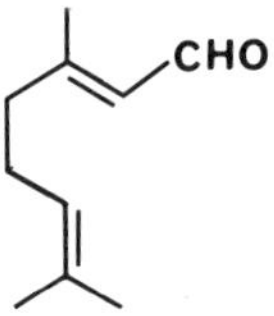   **XII** 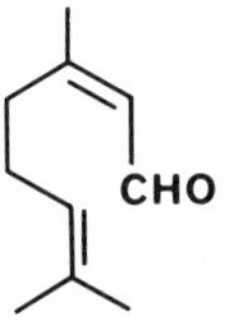   **XIII**

Kinetics of Formation of Volatile Compounds during Heat Treatment of Carotenoids

Procedure

β-Carotene and lycopene in water suspension are treated as indicated above after saturation with oxygen before sealing the flask. The volatile compounds produced are extracted using method 2. Linalool for β-carotene degradation and 1-heptanol for lycopene degradation are used as internal standards and are added to the reaction medium before the extraction process.

Quantitation of volatile compounds is made by gas chromatography, with the chromatograph coupled to a Shimadzu (Kyoto, Japan) CR 1B integrator. The response coefficients of **VI** (0.98), **IX** (1.17), **XII** (1.15) and

[13] P. Kanasawud and J. Crouzet, *J. Agric. Food Chem.* **38**, 1238 (1990).

XIII (1.11) are determined. For the other compounds the coefficients are assumed to be equal to 1.

Results

Dihydroactinidiolide (DHA) and 5,6-epoxy-β-ionone are the most important products detected in the reactive medium. However, if the concentration of DHA increases with reaction time the quantity of 5,6-epoxy-β-ionone produced reaches a maximum after 2 hr of reaction and then decreases (Fig. 2). This kinetic behavior shows clearly that DHA is an end product whereas 5,6-epoxy-β-ionone is a reactional intermediate.

The main compound produced during heat treatment of lycopene under the same conditions is 2-methyl-2-hepten-6-one, which may be considered a final product (Fig. 3). The kinetic comportment of geranial and pseudo-ionone are indicative of the transient nature of these com-

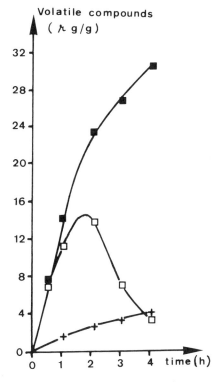

FIG. 2. Kinetics of formation of volatile compounds during heat treatment of β-carotene at 97° in the presence of oxygen: (■) dihydroactinidiolide; (□) epoxy-β-ionone; (+) β-ionone. (From Kanasawud and Crouzet.[12])

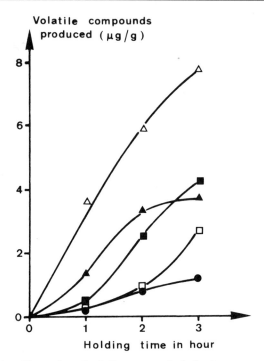

FIG. 3. Kinetics of formation of volatile compounds during heat treatment of lycopene at 97° in the presence of oxygen: (△) 2-methyl-2-hepten-6-one; (●) neral; (▲) geranial; (□) 6-methyl-3,5-heptadien-2-one; (■) pseudo-ionone. (From Kanasawud and Crouzet.[13])

pounds. trans–cis isomerization of geranial and cleavage of pseudo-ionone to 6-methyl-3,5-heptadien-2-one and 2-methylbutanal (not detected) probably occur during the process.

Isolation of Nonvolatile Compounds

Reagents

Aluminum oxide thin-layer chromatography plates (F 254, 20 × 20 cm, 1.5 cm thick; Merck)
Acetone (Merck)–petroleum ether, fraction 40° (Merck) (5:95, v/v)

Procedure

Nonvolatile compounds recovered by direct extraction of the reactive medium with dichloromethane are separated by TLC on aluminum oxide. The compounds separated are scraped and extracted with acetone. After filtration the solvent is evaporated under a nitrogen stream. These operations are performed in the dark.

Identification of Nonvolatile Compounds

Reagents

Hexane for spectroscopy (Merck)
Chlorohydric acid (Merck)

Procedures

Three identification procedures were used.

Procedure 1. The compounds isolated by TLC are rechromatographed on aluminum oxide; after solvent evaporation the plates were sprayed with concentrated chlorohydric acid.[9] The development of a blue color is indicative of the presence of an epoxide function.

Procedure 2. Adsorption spectra of unreacted β-carotene and lycopene and of degradation products are recorded on a Perkin-Elmer (Norwalk, CN) 550 spectrometer with hexane as solvent.

Procedure 3. The ^{1}H NMR spectra of the three most important β-carotene degradation compounds are recorded on a Bruker Spectrospin (Wissembourg, France) VM (360 MHZ). Deuteriochlorobenzene is the solvent and chemical shifts are given in parts per million (ppm) relative to tetramethylsilane (TMS), used as internal standard. 5,6-Epoxy derivatives and 5,8-furan oxides may be distinguished from the chemical shifts of methyl groups in positions 1 and 5,[13,14] respectively: 0.95 and 1.14 ppm and 1.1 and 1.16 ppm for the *gem*-dimethyl group in position 1 for epoxides and furan oxides and 1.24 and 1.42 ppm for the methyl group in position 5 for the previous compounds.[14,15]

Results

The identities of nonvolatile compounds produced during thermal treatment of β-carotene are given in Table I.

Kinetic studies show that mono- and diepoxy-β-carotene and mutatochrome are the first compounds detected, whereas aurochrome appears only after 2-hr treatment at 97°. The quantity of 5,6-epoxy-β-carotene is constant with time, whereas the quantity the other nonvolatile compounds increase.

[14] M. S. Barber, J. B. David, L. M. Jackman, and B. C. L. Weedon, *J. Chem. Soc.*, p. 2870 (1960).
[15] W. Vetter, G. Englert, N. Rigassi, and V. Schweeter, in "Carotenoids" (O. Isler, H. Gutmann, and V. Solms, eds.), p. 204. Birkhäuser, Basel, 1971.

TABLE I
IDENTIFICATION OF NONVOLATILE COMPOUNDS PRODUCED DURING
DEGRADATION OF β-CAROTENE[a]

Spot number (TLC)	Reaction (HCl)	λ_{max}^{hexane} (nm)			^{1}H NMR (ppm)	Identification
2	−	425	398	381		Aurochrome
3	−	446	428	403	1.10 (s) 1.14 (s) 1.48 (s)	Mutatochrome
5	+	453	426		0.98 (s) 1.15 (s)	5′,6,5′,6′-Diepoxy-β-carotene
6	+	473.5	444.5	422	0.93 (s) 1.14 (s)	5,6-Epoxy-β-carotene

[a] At 97° over 3 hr in the presence of oxygen.

In the case of lycopene four nonvolatile compounds are detected. Only the most important was tentatively identified as *cis,trans*-lycopene according to its visible spectra λ_{max}^{hexane} 442, 469, and 501 nm.

Kinetic studies indicate that this compound is probably an end product, the volatile compounds being produced from all-*trans*-lycopene.

According to the results obtained from either volatile or nonvolatile compound kinetics, mechanisms have been proposed for β-carotene and lycopene degradation.[12,13]

[8] Analysis of the Formation of Carotene Chromophore and Ionone Rings in *Narcissus pseudonarcissus* L. Chromoplast Membranes

By PETER BEYER and HANS KLEINIG

Carotene biosynthesis in higher plants is localized in plastids (for a review, see Ref. 1). Within these organelles this biosynthetic pathway is distributed between two sites. Phytoene formation from isopentenyl diphosphate is mediated by stromal proteins, whereas desaturation of phytoene leading to red-colored lycopene and cyclization of lycopene to α/β-carotene are catalyzed by membrane integral enzymes.[2,3] The membrane integral proteins can be solubilized and reconstituted into liposomes, while

---

[1] H. Kleinig, *Annu. Rev. Plant Physiol. Plant Mol. Biol.* **40**, 39 (1989).
[2] K. Kreuz, P. Beyer, and H. Kleinig, *Planta* **154**, 66 (1982).
[3] B. Camara, F. Bardat, and R. Monéger, *Eur. J. Biochem.* **127**, 255 (1982).

maintaining their enzymatic activities, by methods which we presented earlier in this series.[4] In this chapter we describe methods for working with chromoplast membranes, with a special view to obtaining substrates by *in vitro* biosynthesis or by extraction and purification, as well as analytical methods crucial in the investigation of the reactions leading from phytoene to α/β-carotene.

Biosynthesis of Radiolabeled Substrates

Because isopentenyl diphosphate is the only commercially available radioactive precursor, all other intermediates must be synthesized either biosynthetically or chemically. As an alternative, carotene substrates or carotene reference substances can also be extracted and purified from natural sources (see below). However, biosynthetically active carotene intermediates have a cis and poly-cis configuration,[5] whereas the corresponding species normally accumulating as end products have the all-trans configuration and are thus unsuitable as substrates. Indeed, with the exception of 15-*cis*-phytoene, which is astonishingly stable, all other *cis*- and poly-*cis*-carotenes are highly unstable in organic solution, especially at low concentrations and at a high degree of purity. This is the case with biosynthesized radiolabeled substrates or purified products from *in vitro* assays. The danger of nonenzymatic isomerization is minimal as long as these carotene substrates and products are in membranes or in crude extracts. It is essential to develop a rapid protocol for their analysis, in which a limited number of samples are processed from extraction through analysis, avoiding storage. Strong light and elevated temperatures must also be avoided.

Preparative Biosynthesis of Radiolabeled 15-cis-Phytoene

Chromoplasts (3.2–3.6 mg protein/ml in incubation buffer: Tris-HCl, 100 mM, pH 7.2; MgCl$_2$, 10 mM; DTE, 2 mM), isolated according to Liedvogel *et al.*,[6] are disintegrated in a French pressure cell at 3.5 MPa (5000 psi). Subsequent centrifugation at 150,000 g for 2 hr yields the chromoplast stromal fraction as the supernatant. Aliquots (1 ml) of this supernatant are incubated overnight at 25° in the dark in the presence of Mn^{2+} (1 mM) and ATP (3 mM), to suppress the formation of monoterpenes.[7] The addition of liposomes made from chromoplast lipids, to a

---

[4] P. Beyer, this series, Vol. 148, p. 392.
[5] P. Beyer, M. Mayer, and H. Kleinig, *Eur. J. Biochem.* **184**, 141 (1989).
[6] B. Liedvogel, P. Sitte, and H. Falk, *Cytobiologie* **12**, 155 (1976).
[7] U. Mettal, W. Boland, P. Beyer, and H. Kleinig, *Eur. J. Biochem.* **170**, 613 (1988).

concentration of about 0.2 mg/ml, greatly stimulates the synthesis of 15-*cis*-phytoene. The reaction is started by supplying the substrate [1-^{14}C]isopentenyl diphosphate (2.072 GBq/mmol) at 27.5 μM (55 kBq/ml).

Preparative Biosynthesis of 15-cis-ζ-[^{14}C]Carotene

With [1-^{14}C]isopentenyl diphosphate as substrate, the reaction is performed with chromoplast homogenates (instead of the stromal fraction and omitting liposomes) under otherwise identical incubation conditions. When radiolabeled 15-*cis*-phytoene is used as substrate, no other cofactors (except molecular oxygen) are required. This synthesis is carried out in chromoplast membranes obtained by resuspending the pellet of the above centrifugation in the original volume of incubation buffer. 15-*cis*-[^{14}C]Phytoene in acetone solution (1000 cpm/μl) is distributed to the samples in a maximum volume of 30 μl/ml membrane suspension. The addition is followed immediately by vigorous vortexing. Lipophilic effectors such as herbicides[8] or quinones[9] in varying amounts are predistributed from acetone stock solutions; then substrate is added, and the mixture dried and redissolved in a constant volume of acetone (maximally 30 μl/ml) before adding the chromoplast membrane suspension.

Preparative Biosynthesis of 7,9,9',7'-Tetra-cis-[^{14}C]lycopene (Prolycopene) and Cyclization Reaction

The incubation regime used for prolycopene formation employing 15-*trans*-ζ-[^{14}C]carotene as substrate is the same as described for the biosynthesis of 15-*cis*-ζ-[^{14}C]carotene. The purified (see below) 15-*cis*-ζ-[^{14}C]carotene is photoisomerized to trans by illumination (prolonged storage has the same effect). The illumination time required is only about 30 sec with the pure cis isomer; several minutes are necessary in the presence of higher amounts of other carotenes, e.g., β-carotene [ζ/β-carotene peak from high-performance liquid chromatography (HPLC) system I]. The degree of isomerization can be monitored using HPLC system II (see Fig. 1 and below). The prolycopene thus obtained is purified and used as substrate for cyclization reactions. As above, it is added as an acetone solution (up to 30 μl/ml membrane suspension). Lycopene cyclization additionally requires NADPH (1 mM) as an essential cofactor as well as strict anaerobic conditions. This is achieved with the aid of an enzymatic oxygen trap consisting

---

[8] M. P. Mayer, D. L. Bartlett, P. Beyer, and H. Kleinig, *Pestic. Biochem. Physiol.* **34**, 111 (1989).

[9] M. P. Mayer, P. Beyer, and H. Kleinig, *Eur. J. Biochem.* **191**, 359 (1990).

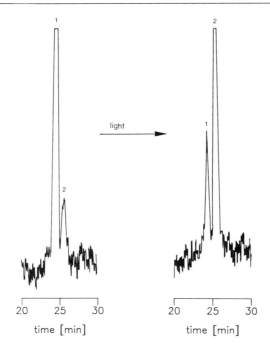

FIG. 1. Photoisomerization of ζ-carotene. On illumination 15-cis-ζ-carotene (1), biosynthetically formed, is converted into the trans isomer (2), which is used as substrate for further desaturation. (Analysis on HPLC system II.)

of (in the order of addition from stock solutions to the indicated final concentrations) catalase (45 μg/ml), glucose oxidase (120 μg/ml), and glucose (24 mM). Also, the assays are flushed with nitrogen and sealed. Incubations are at 25° for 3 hr in the dark.

For product analysis all incubation assays are extracted twice with an equal volume of $CHCl_3$–methanol (2:1). The pooled organic extracts are dried down under vacuum, immediately redissolved in $CHCl_3$, and processed (see below).

Reference cis-Carotenes and Unlabeled cis-Carotene Substrates

The daffodil chromoplast system is sufficiently active enzymatically to allow the use of unlabeled cis-carotene intermediates as substrates. Moreover, because chromoplast membranes contain amounts of carotene intermediates too small for spectral identification (especially neurosporene and lycopene), reference carotenes are used, e.g., as an internal standard. Unlabeled cis-carotene substrates as well as reference cis-carotenes are obtained from the 'Tangerine' variety of tomato fruit. 'Tangerine' tomato fruits are ground in a small volume of acetone. Acetone extraction is repeated several times until the material is almost colorless. The pooled extracts are transferred to a separatory funnel and partitioned against a one-third volume of a 1:1 mixture of petroleum benzene–diethyl ether. The upper phase is collected and evaporated to dryness at ambient temperature. For complete removal of water the extract is redissolved in a minimal volume of acetone; then petroleum benzene is added until phase separation occurs. The upper phase is dried and redissolved in $CHCl_3$ for the isolation of standard carotenes or carotene substrates.

To isolate the corresponding *trans*-carotenes a semiripe "normal" tomato fruit is processed analogously. Excess amounts of *trans*-lycopene are precipitated out of a concentrated petroleum benzene–diethyl ether (1:1) solution of the corresponding carotene fraction (see below) at $-20°$.

Larger amounts of 15-*cis*-phytoene are obtained from mustard (*Sinapis alba* L.), germinated under white light (about 30 W/m²) on three layers of filter paper soaked in H_2O with 10^{-5} M norflurazone, added to water from an acetone stock solution. The cotelydons are harvested, frozen in liquid nitrogen and pulverized. Repeated extraction with acetone and partitioning against petroleum benzene diethyl ether (2:1) yields phytoene and neutral lipids in the upper phase. After evaporation, a few microliters of $CHCl_3$ is added to take up the residue. Using ethanolic KOH, saponification takes place for 30 min at 50°. The voluminous precipitate is removed by centrifugation and the supernatant extracted with petroleum benzene–diethyl ether (1:1). The upper phase is washed thoroughly with water. For purification, HPLC systems I and II (see below) are used, revealing 15-*cis*-phytoene accompanied by small amounts of the trans isomer.

Preparation of *cis*-Carotene Fraction

A separation of carotenes from polar lipids and xanthophylls is necessary to permit isocratic HPLC. For this purpose $CHCl_3$ solutions of extracts are applied directly onto silica gel plates [thin-layer chromatography (TLC) system I]. The solvent is allowed to migrate only 5–10 cm; the sharp front, yellow from the presence of β-carotene, is scraped off and extracted with acetone. This procedure is a critical step as regards spontaneous isomerization and calls for rapidity. Once on the plate, drying or exposure to air must be minimized. Applying several samples on one plate is too time consuming and may be deleterious, so it is recommended that plates be cut into appropriate sizes for individual extracts. Scraping off and transfer into acetone for extraction is faster than drying the plate. Once in acetone, the silica gel is removed by centrifugation. After taking the supernatant to dryness under a stream of nitrogen, it is then dissolved in an appropriate volume of $CHCl_3$ (20–30 µl). The samples are now ready for further analysis on the isocratic HPLC system II or MgO/silica gel plates.

Alternatively, a carotene fraction is made on an HPLC gradient (HPLC system I). The hydrocarbon fraction (see Fig. 3) is collected and petroleum benzene as well as water are added until phase separation occurs. The upper phase is dried under a stream of nitrogen and immediately redissolved in a small volume of $CHCl_3$.

Purification of cis-Carotene Substrates/Reference Carotenes and Analysis of Incubation Assays

A combination of three analytical methods is needed to cover three specific requirements: (1) separation according to degree of desaturation, (2) state of cis–trans isomerism, and (3) number of cyclic end groups.

TLC System I. Silica gel 60 F_{254} plates (Merck, Darmstadt, Germany) are developed with the solvent system petroleum benzene–diethyl ether–acetone (40:10:5).

TLC System II: MgO/Silica Gel Plates. Silica gel 60 G and light MgO extra pure (both from Merck) are mixed 1:1 by volume to make a semifluid suspension in distilled water. After passing through a sieve and degassing, this suspension is spread on clean, fat-free glass plates and allowed to dry overnight. After activation (30 min, 140°) and cooling to ambient temperature, the plates are ready for use. Two solvent systems are recommended. Solvent A [petroleum benzene–toluene (40:5)] is used for the analysis of phytoene, phytofluene, and ζ-carotene. Note that this separation system, although more rapid and just as precise as HPLC system II, has the disadvantage of being very sensitive toward moisture in the air. On humid days the content of toluene must be reduced. Separation is good when β-carotene, as a reference, migrates at an R_f of about 0.3–0.4 (see Fig. 2). Phytofluene is visible under near-UV light. Solvent B [petroleum benzene–toluene–diethyl ether (40:10:6)] is used for the analysis of ζ-carotene, neurosporene, and lycopene; for analytical purposes, radioactive zones are scraped off and quantified by liquid scintillation counting.

HPLC System I

Column: ET 300/8/4 Nucleosil 10 C_{18} (Macherey-Nagel, Düren, Germany)
Polar solvent A: methanol–water (3:1)
Nonpolar solvent B: acetonitrile–tetrahydrofuran (1:1)

At a constant flow rate of 1.5 ml/min a gradient is run within 15 min from 100 to 30% solvent A (0 to 70% solvent B). The final conditions are

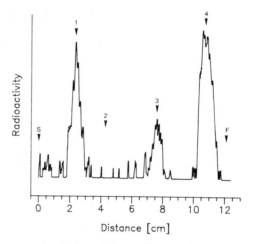

FIG. 2. A [^{14}C]phytoene incubation of chromoplast membranes, analyzed on TLC system II. The use of solvent A allows the separation of the products ζ-carotene (peak 1) and phytofluene (peak 3), as well as of nonconverted substrate, phytoene (peak 4). Nonlabeled β-carotene is present in membranes and migrates at position 2. S, Origin; F, solvent front.

maintained for 25 min, followed by reequilibration to the initial conditions (100% solvent A for 15 min) all at the same flow rate (Fig. 3).

HPLC System II

Column: 250/8/4 Nucleosil 5 C_{18} (Macherey-Nagel)
Eluent: acetonitrile; flow rate, 1–1.3 ml/min

Before analyzing unknowns, a carotene fraction from tomato is run through the column for standardization and the separation optimized to maximal, near baseline resolution of the two ζ-carotene isomers (see Fig. 4). The system is necessarily a compromise between separation according to the number of double bonds and the number of cyclic end groups. Carotenes elute with an increasing degree of desaturation as a homologous series, with phytoene as the most nonpolar and lycopene as the most polar. Because cyclization of the latter is accompanied by a gain of lipophilicity the separation of the mono- and bicyclic products may interfere by overlapping with the desaturation intermediates. This is further complicated by the formation of α- as well as β-ionone rings and by the variable state of

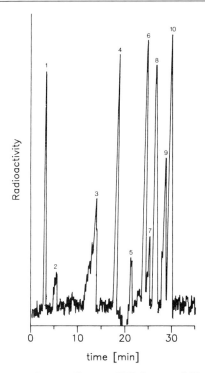

FIG. 3. Analysis of ^{14}C-labeled intermediates on HPLC system I. To show the capabilities of this gradient system, the extracts of three different incubations were mixed. Stroma fraction with [1-^{14}C]isopentenyl diphosphate as substrate, yielding isopentenol-dimethylallyl alcohol (peak 1), geraniol (peak 2), geranylgeraniol (peak 4), and phytoene (peak 10). Farnesol (peak 3) is normally not present and was externally added. Membranes with [^{14}C]phytoene as substrate, yielding phytofluene (peak 9) and ζ-carotene (peak 8) as products. Membranes with ζ-[^{14}C]carotene as substrate, yielding neurosporene (peak 7) and lycopene (peak 6). For a carotene fraction, peaks 6 through 10 are collected.

cis/trans isomerism of the desaturation intermediates. Thus, overlaps of relevant peaks sometimes cannot be avoided and reexamination on TLC system II is required.

HPLC System III. The same column is used as in HPLC system II, with methanol as the only eluent at a flow rate of 0.8 ml/min. Although HPLC system II exhibits some stereoselectivity, a significantly better separation of stereoisomers is achieved on system III. System III, however, does not separate desaturation intermediates. Hence individual peaks collected on HPLC system II are reexamined here. An example is shown in Fig. 5.

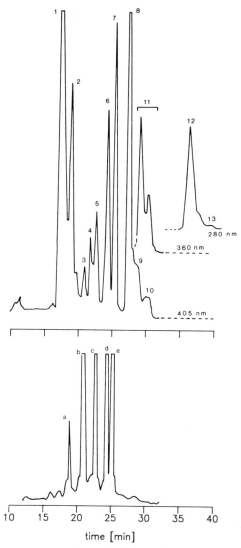

FIG. 4. Calibration of HPLC system II with carotenes from tomato. Upper scan: separation of a carotene fraction from semiripe "normal" tomato fruit. The system is optimized (flow, addition of 1–3% of water) to maximal resolution between 15-*cis*-ζ-carotene (peak 6) and the corresponding trans isomer (peak 8). The other compounds are as follow: peak 1, lycopene (essentially trans configured); peak 2, unknown *cis*-lycopene; peak 3, neurosporene, overlapping (if present) with δ-carotene; peak 4, γ-carotene; peak 5, 7,9,9'-tri-*cis*-neurosporene (proneurosporene); peak 8, *trans*-β-carotene; peaks 9 and 10, cis isomers; peak 11, phytofluene, probably 15-cis (larger peak) and trans (smaller peak), usually overlapping with the cis isomers of β-carotene; peak 12, 15-*cis*-phytoene; peak 13, *trans*-phytoene. α-Carotene, when present, elutes between peaks 7 and 8. Lower scan: separation of a carotene fraction from the 'Tangerine' variety of tomato fruit. Peak a, unknown *cis*-lycopene; peak b, 7,9,9',7'-tetra-*cis*-lycopene (prolycopene); peak c, 7,9,9'-tri-*cis*-neurosporene (proneurosporene); peaks d and e, the two isomers of ζ-carotene, as above.

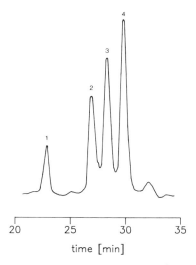

FIG. 5. Separation of cis-lycopene standards using HPLC system III. Peak 1, 7,9,9',7'-Tetra-cis-lycopene (prolycopene); peak 2, trans-lycopene; peak 3, 5,5'-di-cis-lycopene; peak 4, 7,7'-di-cis-lycopene.

Purification of Individual Carotene Species

Phytoene

Radioactive phytoene (see above) is purified from the carotene fraction on HPLC system II. The all-trans and 15-cis isomers become clearly separated and are identifiable by their distinct spectral properties (see Fig. 4). For purification from lipid-rich (liposomes) bulk samples a prepurification of the concentrated $CHCl_3$ extract, using HPLC system I, is recommended. Saponified extracts from herbicide-treated mustard cotyledons are first run through the gradient system (HPLC system I) to remove residual amounts of neutral glycerolipids. Although there is a substantial baseline shift at 280 nm when stabilized tetrahydrofuran is used, the massive phytoene peak can clearly be detected. The fraction is collected, partitioned against petroleum benzene–diethyl ether (1:1), and repurified with HPLC system II. Phytoene is stored in acetone solution at $-20°$ for several weeks.

Phytofluene and ζ-Carotene

Incubations with $[1-{}^{14}C]$ isopentenyl diphosphate using chromoplast homogenates or with phytoene (labeled or nonlabeled) using chromoplast membranes yield lipid-rich extracts. For analytical purposes, carotene

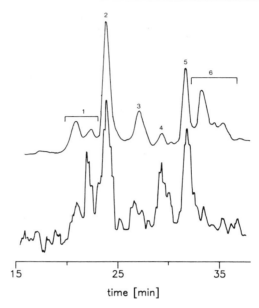

FIG. 6. Separation of substrate and products from a cyclization assay using HPLC system II. Upper scan: UV/VIS detection at 436 nm. Lower scan: Radioactivity detection. The assay was incubated as described in text in the presence of [^{14}C]prolycopene, 30,000 cpm (0.03 μM) isotope diluted with [^{12}C]prolycopene (1 μM). The control, incubated in parallel in the presence of denaturing amounts of CHCl$_3$/methanol (2/1), showed only the substrate peak 2, prolycopene. On incubation the following products were enzymatically formed. Peak 1, Lycopene isomers other than in peak 2; peak 3, revealed on reexamination on TLC system II small amounts of δ-carotene, γ-carotene, and proneurosporene; peak 4, α-carotene; peak 5, *trans-β*-carotene. Cis isomers of β-carotene (peak 6) were not enzymatically formed.

fractions from 1-ml assays (maximal volume) are made and subjected directly to HPLC system II or TLC system II using solvent A. When the state of cis–trans isomerism of ζ-carotene is investigated—a crucial point in the pathway—contact with adsorptive surfaces must be avoided. Using HPLC system I, a ζ-carotene/β-carotene fraction is obtained. After partition against petroleum benzene–diethyl ether (in the dark), it is analyzed immediately on HPLC system II. As the biosynthetic product, the 15-cis isomer of ζ-carotene will be detected almost exclusively.

For preparative purposes, however, such care must not be taken, because only the corresponding trans isomer will be further desaturated in the pathway. Thus carotene fractions may be applied directly to HPLC system II, where the two isomers are separated. Illumination enhances the yield of the trans isomer. Nonlabeled ζ-carotene isomers are obtained in the same way.

trans-ζ-Carotene can be stored at $-20°$ in acetone solution for several weeks. The pure 15-cis isomer cannot be stored. Smaller amounts of 15-*cis*-phytofluene are also obtained from the eluate of HPLC system I.

7,9,9'-Tri-cis-neurosporene (Proneurosporene) and 7,9,9',7'-Tetra-cis-lycopene (Prolycopene)

Carotene fractions from incubations with *trans*-ζ-[^{14}C]carotene as substrate yield proneurosporene and prolycopene. They are best analyzed on HPLC system II. Again, the critical step is the extraction from the silica gel. For preparative purposes the same system is used as described above for ζ-carotene. Purification is also possible using TLC system II with solvent B, but it may sometimes yield nonspecific artifacts at the origin (especially on hot, humid days). It should not be used for the preparation of radiolabeled substrates, but this system is well suited for the preparation of unlabeled prolycopene for cyclization experiments as well as for the preparation of proneurosporene. Carotene fractions from the 'Tangerine' tomato, rich in prolycopene, exhibit some lycopene isomerization while migrating on the plate. On TLC system II the isomers separate as a set with R_f values between 0 and about 0.4, prolycopene being the most nonpolar compound. Proneurosporene migrates just beyond prolycopene, whereas phytoene, phytofluene, and ζ-carotene are found close to the solvent front. Prolycopene is scraped off rapidly and repurified on HPLC system II to remove residual isomers and any possible proneurosporene contamination. It can be stored in acetone solution at $-20°$ for up to 2 weeks.

Analysis of Cyclic Carotenes

Carotene fractions are analyzed directly on HPLC system II. Figure 6 shows a typical elution profile. The cyclization of prolycopene includes an NADPH-dependent isomerization of this substrate.[10] Preparation of carotene fractions is thus a critical step. Blanks incubated in the presence of 1 ml $CHCl_3$–methanol are used to correct for any nonenzymatic isomerization. Interestingly, the reaction may also involve a reverse step, the reduction of prolycopene to proneurosporene, which interferes in the separation of γ-carotene; because both separate well on TLC system II with solvent B, they are reexamined here after collecting from HPLC. This is generally also necessary with the two monocyclic compounds δ- and γ-carotene, because optimization for resolution of the monocyclic compounds, in our experience, is accompanied by a loss of isomer resolution.

---

[10] P. Beyer, U. Kröncke, and V. Nievelstein, *J. Biol. Chem.* **226**, 17072 (1991).

Spectral Identification and Quantitation

The use of a photodiode array detector is necessary for these investigations. This device should be coupled to a radioactivity detector, equipped with a flow-through glass scintillation cell. The identification of proneurosporene and prolycopene employs both coelution with authentic standards and the comparison of the characteristic absorption spectra (see Ref. 5) to give unequivocal structural assignments. This is also the case with 15-*cis*- and all-*trans*-ζ-carotene. However, the state of isomerism of the 9 and 9′ double bonds is still tentative with these compounds and their precursors. It can only be finally elucidated by NMR analysis.

The quantitation of incubation assays using nonlabeled carotene substrates is done in analytical separations (HPLC system II) by integrating the peaks of carotene substrates and products at their respective λ_{max}. These values are normalized to a single coefficient, e.g., that of β-carotene. As correction factors the ratios of the respective molar extinction coefficients are as follow: β-carotene, 1; phytoene, 2.04; phytofluene, 1.89; ζ-carotene, 1.12; proneurosporene, 1.15; prolycopene, 1.38, δ-carotene, 0.79; and γ-carotene, 0.94. A second calculation to correct for nonspecific loss of material (e.g., volume errors on injection) is made by normalizing to a constant sum of peak areas in the control [carotene substrate, incubated in parallel in the presence of the same volume of $CHCl_3$–methanol (2:1)] and the enzymatically active assay. Subtraction of the individual peak areas from the control then allows the determination of mole% product formation and substrate consumption. Such calculations are not necessary for radioactive samples, but when absolute quantities are required it must be kept in mind that the biosynthetically obtained substrates are isotope diluted with their respective nonradioactive counterparts, present in chromoplast membranes and analogously converted on incubation.

Acknowledgments

We thank Dr. Randall Cassada for correcting the English version of the manuscript. This work was supported by the Deutsche Forschungsgemeinschaft.

[9] C_{50} Bicyclic Carotenoids: Sarcinaxanthin Synthesis

By J. P. FÉRÉZOU

Introduction

The bacterial bicyclic C_{50} carotenoids exhibit two extra C_5 isoprene units substituting the C-2 and C-2' centers of the cyclohexyl rings. Three bicyclic C_{50} carotenoids are known (Fig. 1): decaprenoxanthin (**1**) from *Flavobacterium dehydrogenans* with two substituted ε-end groups (**A**),[1] C.p.450 (**2**) from *Corynebacterium poinsettiae* with two substituted β-end groups (**B**),[2] and sarcinaxanthin (**3**) from *Sarcina lutea* with two substituted γ-end groups (**C**).[3]

From a biosynthetic point of view, lycopene is thought to be the immediate precursor of these carotenoids as for C_{40} carotenoids: the cyclization reaction is believed to be biogenetically initiated by an electrophilic attack of a C_5 unit at the C-2, C-2' centers of the acyclic C_{40} precursor.[4]

In agreement with such a postulated prenylation-cyclization process and prompted by progress in laboratory acid-promoted alkylation reactions,[5] we developed a "biomimetic" approach to the synthesis of these three C_{50} bicyclic carotenoids. These efforts led to the synthesis of racemic decaprenoxanthin (**1**) and C.p.450 (**2**), starting from pseudo-ionone[6] and, more recently, of racemic sarcinaxanthin (**3**), starting from geranyl acetate (**4**).[7] These works parallel other reports concerning synthesis of racemic decaprenoxanthin[8] or one of its diastereoisomers[9] as well as optically active C.p.450[10] and, more recently, decaprenoxanthin.[11]

---

[1] S. Liaaen-Jensen, S. Hertzberg, O. B. Weeks, and U. Schwieter, *Acta Chem. Scand.* **22**, 1171 (1968); U. Schwieter and S. Liaaen-Jensen, *Acta Chem. Scand.* **23**, 1057 (1969); A. G. Andrewes, S. Liaaen-Jensen, and O. B. Weeks, *Acta Chem. Scand., Ser. B* **29**, 884 (1975).
[2] A. G. Andrewes and S. Liaaen-Jensen, *Tetrahedron Lett.* **25**, 1191 (1984); G. Britton, A. P. Mundy, and G. Englert, *J. Chem. Soc., Perkin Trans. 1* p. 601 (1985).
[3] S. Hertzberg and S. Liaaen-Jensen, *Acta Chem. Scand., Ser. B* **31**, 215 (1977).
[4] I. E. Swift and B. V. Milborrow, *J. Biol. Chem.* **256**, 11607 (1981); G. Britton, *Pure Appl. Chem.* **47**, 223 (1976); G. Britton, *Pure Appl. Chem.* **57**, 701 (1985).
[5] M. Julia and C. Schmitz, *Tetrahedron* **42**, 2491 (1986).
[6] J. P. Férézou and M. Julia, *Tetrahedron* **41**, 1277 (1985).
[7] J. P. Férézou and M. Julia, *Tetrahedron* **46**, 475 (1990).
[8] A. K. Chopra, B. P. S. Khambay, H. Madden, G. P. Moss, and B. C. L. Weedon, *J. Chem. Soc., Chem. Commun.* p. 357 (1977).
[9] A. K. Chopra, B. P. S. Khambay, H. Madden, G. P. Moss, and B. C. L. Weedon, *J. Chem. Res., Synop.* p. 208 (1989).
[10] H. Wolleb and H. Pfander, *Helv. Chim. Acta* **69**, 646 (1986).
[11] M. Gerspacher and H. Pfander, *Helv. Chim. Acta* **72**, 151 (1989).

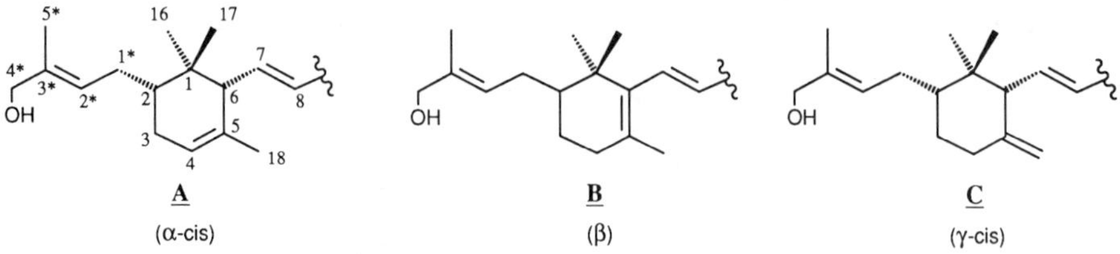

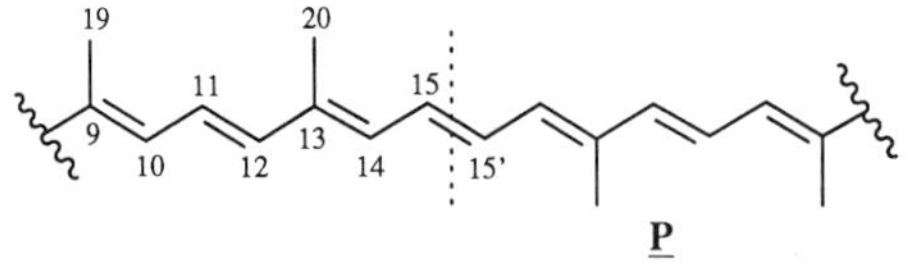

Decaprenoxanthin **1**: A - P - A

C.p.450 **2** : B - P - B

Sarcinaxanthin **3** : C - P - C

FIG. 1. Structures of the three known C_{50} bicyclic carotenoids (**1–3**).

A multistep synthesis of racemic sarcinaxanthin **(3)** is described here (Fig. 2) using as a key step an acid-promoted electrophilic alkylation–cyclization of geranyl acetate **(4)** with either isoprene epoxide **(5)** or with 2-methyl-3-buten-2-ol **(6)**. Careful attention has been paid to this crucial reaction that generates, in one step, the rather complex terminal cyclic C_{15} building blocks **(8)**. This efficiency, particularly in the case of isoprene epoxide **(5)**, highly counterbalanced the modest yield of the operation.

Discussion

The first step of sarcinaxanthin synthesis (Fig. 2) involves the acid-promoted electrophilic alkylation of geranyl acetate **(4)**: two different routes are used for the construction of the required C_{15} cyclic diol acetate intermediate **(8)**.

Reaction of isoprene epoxide **(5)** with geranyl acetate **(4)** under $ZnCl_2$ conditions leads to a complex mixture from which **8** is isolated through the careful silica gel column chromatography (6.3% yield based on isoprene epoxide, 18% on consumed geranyl acetate). Its structure has been assigned from exhaustive ^{1}H and ^{13}C nuclear magnetic resonance (NMR) spectro-

FIG. 2. Synthesis of sarcinaxanthin **(3)**. All depicted stereochemistries refer to relative configurations. (a) $ZnCl_2$, CH_3NO_2; (b) CF_3COOH, CH_3NO_2; (c) SeO_2, *t*-BuOOH; (d) $(CH_3)_3(C_6H_2)COCl$, pyridine; (e) $POCl_3$, pyridine; (f) NaOH, ethanol, then SiO_2 chromatography; (g) CH_3SO_2Cl, $(C_2H_5)_3N$; (h) C_6H_5SH, KH, ethanol; (i) H_2O_2, ammonium molybdate; (j) LDA, THF, then **15**; (k) *t*-BuOK, THF, then citric acid, methanol; (l) CBr_4, $P(C_6H_5)_3$, THF;(m) $P(C_6H_5)_3$, THF; (n) **18,** 40% aqueous KOH/CH_2Cl_2; (o) $LiAlH_4$, $(C_2H_5)_2O$.

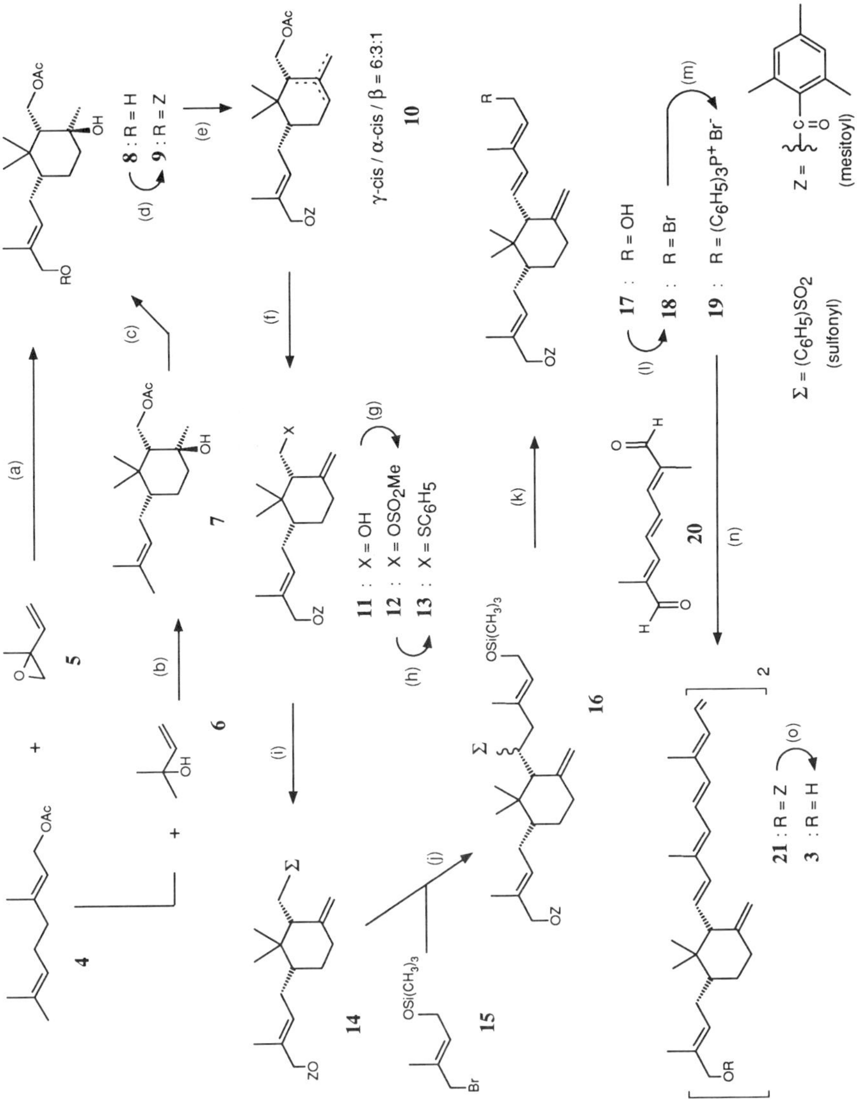

FIG. 1. *Continued*

metric analysis: this isomer exhibits an E-stereochemistry of the allylic double bond as well as the 2,6-*cis* relative stereochemistry required for sarcinaxanthin synthesis.

The second route involves a two-step approach to **8**: alkylation of **4** with 2-methyl-3-buten-2-ol **(6)** affords a complex mixture from which the alkylated product **7** is isolated in 14% yield (30% from consumed geranyl acetate). This yield is improved to 21% after further prenylation of recovered geranyl acetate **(4)** with **6**. Subsequent allylic oxidation of **7** with SeO_2/t-BuOOH[12] gives the diol acetate **(8)** (50% yield), identical to the same product obtained from isoprene epoxide.

Selective production of the primary allylic hydroxyl group of **8** as the mesitoate **9**[13] followed by $POCl_3$-promoted dehydration of the tertiary hydroxyl group gives a favorable 6:3:1 mixture of γ-cis, α-cis, and β isomer **(10)** (95% yield). After hydrolysis of the acetoxy function, silica gel chromatography affords the pure, expected γ-*cis*-C_{15} alcohol **(11)**.

Conversion of **11** into the sulfone **(14)** is straightforward: the corresponding mesylate **(12)** affords the sulfide **(13)** using potassium thiophenate as nucleophile (75% overall yield).[14] Subsequent oxidation of the sulfide function with H_2O_2–ammonium molybdate[15] gives the sulfone **(14)** in 76% yield (mp, 101–102°).

Having secured the preparation of the nucleophilic γ-*cis*-C_{15} sulfone **(14)** we next turned to the elaboration of the sarcinaxanthin C_{50} skeleton. We opted for a $C_{20} + C_{10} + C_{20}$ approach already developed for the synthesis of decaprenoxanthin **(1)** and C.p.450 **(2)**.[6] This route involves a C_5 homologation of **14** to the corresponding C_{20} phosphonium salt **(19)** followed by a final condensation onto the known central C_{10}-dialdehyde **(20)**.[16]

The C_{15}-protected sulfone **(14)** is first converted into the C_{20} alcohol **(17)** through condensation with the C_5-trimethylsilyloxybromobutene **(15)**, which is readily obtained by silylation of the corresponding known bromo alcohol.[17] Alkylation with **15** takes place smoothly to give a mixture of the diastereomers **(16)**, which are directly submitted to sulfonyl group elimi-

---

[12] M. A. Umbreit and K. B. Sharpless, *J. Am. Chem. Soc.* **99**, 5526 (1977).
[13] E. J. Corey, K. Achiwa, and J. A. Katzenellenbogen, *J. Am. Chem. Soc.* **91**, 4318 (1969).
[14] R. J. Amstrong and L. Weiler, *Can. J. Chem.* **61**, 2530 (1983).
[15] M. J. Parrott and D. I. Davies, *J. Chem. Soc., Perkin Trans. 1* p. 2205 (1973).
[16] We are grateful to Dr. H. Gutmann (Hoffman-La Roche, Basel, Switzerland) for a generous supply of C_{10}-dialdehyde **20**.
[17] V. L. Heasley, C. L. Frye, R. T. Gore, Jr., and P. S. Wilday, *J. Org. Chem.* **33**, 2342 (1968); J. B. Babler and W. J. Buttner, *Tetrahedron Lett.* p. 239 (1976); D. Uguen, Ph.D. thesis. Université Pierre et Marie Curie, Paris, 1977.

nation using potassium *tert*-butoxide.[18] Subsequent aqueous acidic workup affords, after purification, the pure all-E-C_{20} alcohol (**17**) in 45% yield from **14**. Conversion of **17** into the bromide (**18**) followed by triphenylphosphine treatment gives the required C_{20} phosphonium salt (**19**) in 85% yield.

Final Wittig condensation of **19** with the C_{10}-dialdehyde (**20**) under heterogeneous conditions[19] [40% (w/v) aqueous KOH/CH_2Cl_2] gives dimesitoylsarcinaxanthin (**21**), which is then deprotected to the (±)-*meso-threo*-diol (**3**) by $LiAlH_4$ reduction. The yield was 95% for these two steps. Repeated cystallizations to a constant melting point gave a pure compound whose spectroscopic data are in full agreement with those reported for the natural product.[3]

Experimental Section

General Procedures

Melting points were determined on a Büchi apparatus and are not corrected. Infrared (IR) spectra were measured on a Perkin-Elmer (Norwalk, CN) 599 spectrophotometer (in $CHCl_3$) and ultra violet (UV) spectra on a Varian Superscan 3 spectrophotometer. Mass spectra (MS) were obtained by direct introduction on a Nermag R10-10 apparatus, using either electron impact (EI) or chemical ionization (CI, NH_3) modes. Thin-layer chromatography (TLC) analyses were carried out using Merck (Darmstadt, Germany) silica gel 60 F_{254} plastic sheet (No. 5735). Merck silica gel 60H (No. 7736) or 60 (No. 7734) was used for column chromatography. Medium-pressure liquid chromatography (MPLC) was performed with a prepacked silica Lobar 440-37 column (No. 10402; Merck). ^1H NMR spectra were recorded at 250 MHz on a Cameca 250 apparatus and ^{13}C spectra on a Bruker (Wissembourg, France) WP 90 (22.63 MHz), in $CDCl_3$ using tetramethylsilane as internal standard. Only signals for diagnostic value are reported. ^1H and ^{13}C assignments are made using carbon numbering of carotenoids. Gas–liquid chromatography (GLC) analyses were carried out on a Girdel 30 (Delsi, France) chromatograph using a 5°/min temperature gradient from 120 to 290° on a 10% OV 101/Chromosorb W HP 100/120 glass column (3 m × 1/8-in. i.d.).

Commercial geranyl acetate (**4**) (Fluka AG, Buchs, Switzerland) was distilled before use. Isoprene epoxide (**5**) was synthesized from isoprene via the corresponding bromohydrin using established procedure.[20] Its purity

---

[18] C. Hervé Du Penhoat and M. Julia, *Tetrahedron* **42**, 4807 (1986).
[19] S. Hünig and I. Stemmler, *Tetrahedron Lett.* p. 3151 (1974); H. Mayer and H. Ruttimann, *Helv. Chim. Acta* **63**, 1451 (1980).
[20] E. J. Reist, I. C. Junga, and B. R. Baker, *J. Org. Chem.* **25**, 1673 (1960).

was checked by ¹H NMR spectrometry. Anhydrous $ZnCl_2$ was prepared on a 50-g scale by melting twice under vacuum and, after solidification, was crushed to powder under a dry atmosphere and stocked under nitrogen. All anhydrous solvents were freshly distilled according to standard procedures.[21]

Alkylation of Geranyl Acetate with Isoprene Epoxide

An oven-dried, three-necked, round-bottomed flask (1 liter) is equipped with a 500-ml pressure-equalizing dropping funnel with a rubber septum, a mechanical stirrer, and an alcohol thermometer and is maintained under a slight positive pressure of nitrogen throughout the reaction. The flask is charged with 375 ml of dry nitromethane and 137 g of anhydrous zinc chloride and the dropping funnel with a mixture of 125 ml of nitromethane, 98 g of geranyl acetate (0.5 mol) and 26 ml (0.25 mol) of isoprene epoxide. The $ZnCl_2$ suspension is cooled down to $-20°$ and the funnel contents added dropwise over a 30-min period under vigorous mechanical stirring. After stirring a further 3 hr at $-20°$ the reaction mixture is poured into 1 liter of ice-cold saturated aqueous sodium bicarbonate. The zinc salts are eliminated by filtration through a Celite pad and washed three times with diethyl ether (250 ml each). The organic phase is separated, the aqueous layer reextracted twice with 250 ml of diethyl ether, and the combined organic phases are washed with saturated brine (300 ml) and dried over $MgSO_4$. The solvent is removed under reduced pressure and the oily residue analyzed by GLC and TLC.

Flash chromatography of this crude product on silica gel [600 g of silica gel (Merck No. 7734) on a 8.5-cm i.d. column] using a stepwise ethyl acetate–petroleum ether gradient (0:10, 1:9, 2:8, 3:7, . . . , 10:0, 500 ml each) gives in order of elution unchanged geranyl acetate (81 g), an intermediate fraction that is not further analyzed (11.4 g), and a more polar fraction (12.6 g) containing **8** and its epimer at the tertiary hydroxyl group. Chromatography of this latter fraction by MPLC (ethyl acetate–petroleum ether, from 1:1 to 4:1, 400 ml each, 30-ml fractions) affords the colorless pure oily acetoxydiol (**8**) as the most polar product (6.3 g, 8.4% from isoprene epoxide, 24% calculated on consumed geranyl acetate).

v_{max}: 3650 (2), 3580 (m), 3430 (m), 2980 (s), 2920 (s), 2860 (m), 1720 (s), 1445 (m), 1370 (s), 1205–1250 (s), 1150 (m), 1030 (m), and 915 (m) cm^{-1}. ¹H NMR, δ: 0.76 and 1.05 (6H, 2s, 16,17-CH_3), 1.17 (3H, s, 18-CH_3), 1.65 (3H, s, 5*-CH_3), 2.07 (3H, s, CH_3CO), 3.98 (2H, s, 4*-H_2),

---

[21] D. D. Perrin and W. L. F. Armarego, "Purification of Laboratory Chemicals." Pergamon, Oxford, 1988.

4.33 (2H, AB part of ABX system, $J_{AB} = 11$, $J_{AX} = J_{BX} = 5$ Hz, 7-H$_2$), and 5.36 (1H, t, $J = 7$ Hz, 2*-H). ^{13}C NMR, δ: 13.6; 16.6; 21.3; 23.5; 25.4; 28.0 (2C); 37.6; 42.4; 48.5; 56.4; 63.1; 68.3; 70.3; 124.9; 135.3; and 170.7. MS (CI), m/z: 299 (3%, M$^+$ + 1), 281 (100%, M$^+$ + 1 − H$_2$O), 221 (9%, M$^+$ − H$_2$O − CH$_3$COOH), and 203 (42%, M$^+$ + 1 − 2H$_2$O − CH$_3$COOH).

*Alternative Route: Preparation of **8** from 2-Methyl-3-butten-2-ol*

A dry, four-necked, round-bottomed reactor (4 liter) is equipped with a 500-ml dropping funnel, an alcohol thermometer, a mechanical stirrer, and a calcium chloride drying tube. The flask is loaded with nitromethane (2 liters), geranyl acetate (196 g, 1 mol), and 2-methyl-3-butten-2-ol (86 g, 1 mol). This solution is cooled to 0° on stirring, then added dropwise over 45 min with a mixture of trifluoroacetic acid (100 ml, 1.3 mol) and nitromethane (300 ml), stirred at 0° for a further 3 hr, and poured into 2 liters of an ice-cold saturated aqueous NaHCO$_3$ (130 g) solution. The organic phase is separated and the aqueous phase reextracted three times with diethyl ether (1 liter). The combined organic phases are washed twice with water (1 liter), dried over MgSO$_4$, filtered, and evaporated under reduced pressure to give an oily residue from which unreacted geranyl acetate (105 g, 0.54 mol) is distilled under reduced pressure (3 × 10^{-2} mmHg) and submitted to a second alkylation reaction under the above conditions (using 47 g, 0.55 mol of **6** and 55 ml, 0.7 mol of trifluoroacetic acid). After reaction workup and elimination of unreacted geranyl acetate (53 g) by distillation as above, this second oily residue is combined to the first distillation residue and chromatographed on silica gel (1 kg, 100-mm i.d. column; Merck No. 7734). Elution is performed with a stepwise increasing gradient of diethyl ether in petroleum ether at 40–60° (3:7, 4:6, 5:5, . . . , up to 9:1, 2 liters each; then 10:0, 4 liters). After TLC control (ethyl acetate–petroleum ether, 2:3) the most polar fractions are combined to give 59 g of the cyclic acetate (**7**) (21% from **4**), which is shown from GLC analysis to be 85% pure. ^{1}H NMR, ^{13}C NMR, MS and IR spectra are reported elsewhere.[5]

Oxidation at the allylic distal position of **7** is carried out as follows according to Ref. 12: under a ventilated hood, at 25°, a 500-ml, three-necked, round-bottomed flask fitted with a 100-ml dropping funnel and stoppered is successively loaded with a magnetic stirring bar, 20 ml of CH$_2$Cl$_2$ containing 10 mol% of salicylic acid (2 g), 8.3 g of freshly sublimed SeO$_2$ (75 mmol), and, with stirring, 175 ml of a 3 N solution of *tert*-butylhydroperoxide (0.52 mol) in CH$_2$Cl$_2$. After stirring 15 min, a solution of hydroxyacetate (**7**) (43 g, 0.152 mol) in 20 ml of CH$_2$Cl$_2$ is added dropwise to the reaction mixture (15 min), which is then stirred at room temperature for 30 hr, washed with saturated NaHCO₃ (100 ml) and twice with saturated brine (100 ml), then dried over MgSO₄. The solvent and the excess of *t*-BuOOH are carefully removed under vacuum with a rotary evaporator (behind a safety shield) at 25° to give 50 g of crude product, which is dissolved in dry methanol (70 ml), transferred to a 250-ml round-bottomed flask, and NaBH₄ (4 g) was added portionwise at 0°. The reaction mixture is stirred for 30 min at 0° and then acidified with 100 ml of an aqueous solution of 1% acetic acid (dropwise addition). Subsequent extraction with ethyl acetate affords after usual workup 30 g of an oil that is subjected to SiO_2 column chromatography. Elution with an increasing ethyl acetate–petroleum ether gradient (from 0:1 to 1:0, by 10% steps, 500 ml each) gives 22.5 g (50% yield) of the pure hydroxylated synthon (**8**) (colorless oil) as the most polar fraction. This product is identical to the same product obtained by direct hydroxyprenylation of geranyl acetate with isoprene epoxide (see above).

Protection of Primary Allylic Hydroxyl Group of **8**

A dry, two-necked, round-bottomed flask (250 ml) equipped with a magnetic stirring bar, a rubber septum, and a gas inlet is successively loaded under a slight positive pressure of nitrogen at 0° with dry CH_2Cl_2 (100 ml), diol acetate (**6**) (22.3 g, 75 mmol), pyridine (13 ml), and 2,4,6-trimethylbenzoyl chloride (17.1 g, 94 mmol). The reaction mixture is allowed to stir at 0° for 24 hr and then poured into ice-cold aqueous 1 N HCl (200 ml). The aqueous phase is decanted and reextracted twice with CH_2Cl_2 (75 ml). The combined organic phases are washed with water, diluted aqueous NaHCO₃, and saturated brine and dried over MgSO₄. After filtration and evaporation of the solvent, the resulting oil is purified by flash chromatography using an ethyl acetate–petroleum ether gradient to obtain pure mesitoate (**9**) (30 g, 90%) as an oil.

Dehydration of **7**: *Obtention of Exocyclic Olefin* **11**

To an oven-dried, one-necked, round-bottomed flask (250 ml) conditioned under an in-line nitrogen atmosphere and equipped with a magnetic stirring bar and a rubber septum are added, via a syringe, 100 ml of a solution of mesitoate ester (**9**) (29 g, 65 mmol) in dry pyridine and, slowly, after cooling to 0°, freshly distilled POCl₃ (59 ml, 1.5 Eq). After stirring overnight at room temperature the reaction mixture is poured into 1.5 liters of ice-cold aqueous 1 N HCl and extracted three times with CH_2Cl_2 (400 ml). The combined organic phases are washed with water, dilute NaHCO₃, and brine, dried over MgSO₄, and concentrated to give the mixture of dehydrated products (**10**) (26.5 g, 96%). GLC and ¹H NMR analysis of the mixture showed 60% γ-cis, 30% α-cis, and 10% β isomers to be present.

This crude reaction product (25 g, 59 mmol) is transferred to a 500-ml one-necked round-bottomed flask and added at 0° with 100 ml of an ethanolic 1 N NaOH solution. After stirring for 4 hr the reaction mixture is diluted with water (250 ml) and extracted twice with CH_2Cl_2 (200 ml). The combined organic phases are washed with saturated brine, dried over $MgSO_4$, and concentrated to give 24 g of an oily residue that was submitted to column chromatography (silica, 75 × 7 cm i.d.; Merck No. 7734). The elution with diethyl ether–petroleum ether fractions (2:8, 25:75, 3:7, 35:65, up to 50:50, 2 liters of each) gives the expected γ-cis synthon (**11**) (12.3 g, 55%) as a pale yellow oil.

ν_{max}: 3580 (m), 3450 (w), 3070 (w), 3030 (w), 3005 (m), 2970 (s), 2940 (s), 2860 (m), 1715 (s), 1645 (w), 1615 (m), 1575 (w), 1455–1440 (m), 1380 (w), 1370 (w), 1275 (s), 1175 (s), 1090 (s), 1030 (w), 1010 (w), 960 (w), 900 (m), and 860 (m) cm^{-1}. ^{1}H NMR, δ: 0.63 and 1.09 (6H, 2s, 16,17-CH_3), 1.70 (3H, s, 5*-CH_3), 2.28 and 2.29 (3H + 6H, 2s, 3 mesitoyl-CH_3) centr. 3.85 (2H, AB part of ABX system, $J_{AB} = 11$, $J_{AX} = 9.5$ and $J_{BX} = 4$ Hz, 7-H_2), 4.68 and 4.97 (2H, d, $J = 1$ Hz, 18-H_2), 4.70 (2H, s, 4*-H_2), 5.57 (1H, t, $J = 7$ Hz, 2*-H) and 6.86 (2H, s, mesitoyl-H_2). ^{13}C NMR, δ: 14.4, 15.6, 19.6, 20.9, 26.3, 28.3, 29.6, 36.6, 38.2, 47.9, 55.9, 59.1, 70.4, 106.5, 127.8, 129.9, 130.0, 130.7, 134.4, 138.5, 147.0, 169.3. MS (CI), m/z: 402 (20%, M$^+$ + 1 + NH$_3$), 385 (3%, M$^+$ + 1), 367 (14%, M$^+$ + 1 − H$_2$O), and 221 (100%, M$^+$ + 1 − mesitoic acid).

Obtention of Sulfone **14**

Transformation of the primary alcohol (**11**) into the crucial sulfone is straightforward, as depicted in Fig. 2, and involves three steps according to well-established procedures: mesylation to **12**,[22] nucleophilic displacement of the mesylate group of **12** to the sulfide (**13**),[14] and oxidation to the sulfone (**14**).[7] This latter sulfone is obtained in 57% yield from **11** and is recrystallized from petroleum ether–isopropyl ether mixtures to give pure **14** as white crystals (mp 101–102°).

ν_{max}: 3070 (w), 3020 (w), 2930 (s), 2860 (m), 1715 (s), 1645 (w), 1610 (m), 1445 (s), 1390 (w), 1370 (w), 1315 (m), 1305 (m), 1270 (s), 1210 (m), 1170 (s), 1140 (m), 1090 (s), 890 (m), 855 (m) and 690 (m) cm^{-1}. ^{1}H NMR, δ: 0.53 and 0.95 (6H, 2s, 16,17-CH_3), 1.67 (3H, s, 5*-Me), 2.28 and 2.29 (3H + 6H, 2s, 3 mesitoyl-CH_3), 2.43 (1H, d, $J = 9$ Hz, 6-H), 3.36 (2H, AB part of an ABX system, $J_{AB} = 15$, $J_{AX} = 1.5$, $J_{BX} = 9$ Hz, 7-H_2), 4.50 and 4.78 (2H, 2 br s, 18-H_2), 4.69 (2H, s, 4*-H_2), 5.53 (1H, t, $J = 7$ Hz, 2*-H), 6.86 (2H, s, mesitoyl-H_2), and 7.50–7.92 (5H, m, $C_6H_5SO_2$). ^{13}C NMR, δ:

---

[22] T. Rosen, M. T. Taschner, T. A. Thomas, and C. H. Heathcock, *J. Org. Chem.* **50**, 1190 (1985).

14.4; 15.2; 19.6 (2); 21.0; 26.0; 29.1; 29.4; 36.6; 39.6; 47.8; (2); 53.1; 70.3; 107.6; 127.6 (2); 127.9 (2); 128.6 (2); 129.5; 130.2; 130.7; 133.0; 134.5 (2); 138.6; 139.3; 144.9; and 169.2. MS (CI), m/z: 526 (100%, M$^+$ + 1 + NH$_3$), 201 (13%), and 190 (24%).

1-Trimethysilyl-3-methyl-4-bromo-2-butene

To a dry, one-necked, round-bottomed flask (50 ml) equipped with a magnetic stirrer and a rubber septum fitted with a nitrogen inlet are successively added, at 0° and by means of a syringe, a solution of 3-methyl-4-bromo-2-buten-1-ol[19] (1.65 g, 10 mmol) in pentane (2 ml), hexamethyldisilazan (0.83 ml, 4 mmol), and trimethylchlorosilane (0.44 ml, 3.5 mmol). After stirring 30 min at 0° and a further 1 hr at room temperature, the reaction mixture is diluted with pentane (25 ml) and directly poured onto a short KOH-treated silica gel column (3 × 2 cm i.d.). Elution with 20 ml of pentane–ether fractions (9:1, 8:2, and 7:3) gives, after TLC monitoring (pentane–ether, 1:1) and solvent evaporation, 2.2 g (92%) of the pure silyloxy derivative **(15)** as a colorless oil.

^{1}H NMR: the product was shown to be 87% E, δ: 0.14 (9H, s, SiMe$_3$), 1.79 (2.6H, s, 3-E-CH$_3$), 1.87 (0.4H, s, 3-Z-CH$_3$), 3.94 (2H, s, 4-H$_2$), 4.18 (2H, d, J = 6.5 Hz, 2-H$_2$), 5.55 (0.13H, t, J = 6 Hz, 2-Z-H), and 5.73 (0.87H, t, J = 6 Hz, 2-E-H). MS (EI) m/z: 237 (100%).

Preparation of C_{20} Alcohol **17**

An oven-dried, one-necked, round-bottomed flask (50 ml) equipped with a magnetic sirring bar, charged with sulfone **(14)** (1.52 g, 3 mmol) and stoppered with a rubber septum, is evacuated and filled with nitrogen (four cycles) and maintained under a slight positive pressure of nitrogen. Tetrahydrofuran (THF, 15 ml) is transferred to the flask via a syringe and then, at −78°, a solution of lithium diisopropylamide (LDA) prepared from diisopropylamine (0.5 ml, 3.6 mmol) and n-butyllithium (1.45 N, 2.50 ml) in THF (10 ml) is added. After 15 min at −78° the bromide **(15)** (1.6 g, 6.6 mmol) is added in one portion by means of a microsyringe and the reaction mixture is stirred overnight while allowing it to warm slowly to room temperature. It is then poured into an ice-cold saturated NH$_4$Cl solution (25 ml) and diluted with diethyl ether (20 ml). The organic layer is separated and the aqueous phase reextracted twice with diethyl ether (20 ml). The combined organic phases are washed with saturated brine, dried over MgSO$_4$, and concentrated to give 2.58 g of an oily residue containing the protected sulfone **(16)** as a mixture of diastereomers [TLC, ethyl acetate–petroleum ether (2:3), and ^{1}H NMR analysis].

This crude product is then transferred into a 50-ml, three-necked, round-bottomed flask equipped with a magnetic stirring bar, an alcohol thermometer, a rubber septum, and a gas inlet, then maintained under a slight positive pressure of nitrogen and dissolved in THF (15 ml added via a syringe). This solution is cooled to $-30°$ and freshly sublimed potassium *tert*-butoxide (625 mg) is quickly added in one portion. The reaction mixture is allowed to warm to room temperature and is stirred for 2 hr. After cooling to $0°$ a solution of citric acid (1 g, 5 mmol) in methanol (15 ml) is added to the reaction mixture, which is stirred for a further 1 hr at $0°$ before dilution with diethyl ether (15 ml) and neutralization with dilute aqueous sodium bicarbonate (20 ml). After decantation, the aqueous layer is reextracted twice with diethyl ether (20 ml) and the combined organic fractions washed with saturated aqueous $NaHCO_3$ and saturated brine, dried over $MgSO_4$, and the solvent evaporated. The crude product is purified by silica gel column chromatography (15 × 3 cm i.d.; Merck No. 9385) using a stepwise diethyl ether–pentane gradient from 0:100 to 60:40, 5 by 5%, 100 ml each; 25-ml fraction are collected to give the pure all-*E*-C_{20} synthon (17) (605 mg, 45% from 14) as a colorless oil.

v_{max}: 3600 (w), 3530 (w), 3070 (w), 3020 (w), 2960 (m), 2920 (s), 2860 (m), 1715 (s), 1645 (w), 1610 (m), 1455 (m), 1390 (w), 1380 (w), 1370 (w), 1270 (s), 1175 (s), 1090 (s), 895 (m), 860 (w), and 825 (m) cm^{-1}. ^1H NMR, δ: 0.71 and 0.94 (6H, 2s, 16,17-CH_3), 1.71 (3H, s, 5*-CH_3), 1.84 (3H, s, 19-CH_3), 2.28 and 2.30 (3H + 6H, 2s, 3-mesitoyl-CH_3), 2.46 (1H, d, $J = 10$ Hz, 6-H), 4.26 (2H, d, $J = 7$ Hz, 11-H_2), 4.52 and 4.77 (2H, 2 br s, 18-H_2), 4.71 (2H, s, 4*-H_2), 5.59 (2H, 2 superposed t, $J = 7$ Hz, 10-H, 2*-H), 5.81 (1H, q, $J = 16$ and 10 Hz, 7-H), 6.07 (1H, d, $J = 16$ Hz, 8-H) and 6.86 (2H, s, mesitoyl-H_2). ^{13}C NMR, δ: 14.6; 15.1; 19.8(2); 21.1; 27.6; 28.7; 28.9; 36.3; 39.0; 48.2 (2); 58.1; 59.1; 70.8; 107.9; 127.9 (2); 128.0; 128.4 (2); 130.0; 130.3; 130.8; 134.7; 135.6; 136.4; 138.8; 149.6; and 169.6. MS (CI), m/z: 468 (12%, $M^+ + 1 + NH_3$), 433 (100%, $M^+ + 1 - H_2O$), 269 (42%, $M^+ + 1 + H_2O$ − mesitoic acid), and 182 (23%).

Transformation of Alcohol **17** *into Phosphonium Salt* **19**

The C_{20} alcohol (17) (470 mg, 1.05 mmol) is converted into the C_{20} phosphonium salt via the corresponding bromide using carbon tetrabromide and triphenylphosphine according to a described procedure.[23] The crude product of these two steps is submitted to silica gel chromatography (elution with methanol–dichloromethane mixtures from 1 to 8%) to give

---

[23] H. Hayashi, K. Nakanishi, C. Brandon, and J. Marmur, *J. Am. Chem. Soc.* **95**, 8749 (1973).

680 mg of pure phosphonium bromide (19) as an amorphous white solid (84% from 17).
MS [fast atom bombardment (FAB)], m/z: 695 (100%, M$^+$ − Br).

(±)-all-E,cis-Sarcinaxanthin

This reaction is carried out under reported heterogeneous conditions.[19]

To a 25-ml, one-necked, round-bottomed flask equipped with a magnetic stirrer and containing a mixture of 40% aqueous KOH solution (3 ml) and the C_{10} dialdehyde (20) (33 mg, 0.2 mmol) in CH_2Cl_2 (1 ml) is added dropwise at −10° under N_2 a solution of the phosphonium bromide (19) (465 mg, 0.6 mmol) in CH_2Cl_2 (5 ml). The reaction mixture is stirred for 1 hr at −10°, then for 30 min at room temperature. After dilution with water and diethyl ether (10 ml each) the decanted aqueous phase is reextracted twice with diethyl ether (10 ml) and the combined organic phases washed three times with saturated brine (5 ml), dried over $MgSO_4$, and concentrated to give a deep red oily residue. This is purified by silica gel column chromatography (10 × 2.5 cm i.d.; elution with diethyl ether–pentane fractions 2:8, 3:7, 4:6, 5:5, then 1:0, 50 ml each) to give the oily red protected carotenoid (21) [198 mg, quantitative yield from the dialdehyde (20)] as the less polar product (TLC control with pentane–diethyl ether, 1:1). This racemic carotenoid is probably a mixture of meso-threo isomers.

This carotenoid fraction (198 mg, 0.2 mmol) is transferred into a dry, 50-ml, one-necked, round-bottomed flask equipped with a magnetic stirrer and a rubber septum, then conditioned under nitrogen. Dry THF (20 ml) and, at 0° after dissolution, a decanted 1 M solution of $LiAlH_4$ in dry diethyl ether (2 ml) are added via syringes. The reaction mixture is warmed to room temperature, stirred for 2 hr, and 20 ml of water is added at 0°, carefully at the beginning. The hydrolyzed reaction mixture is then extracted three times with diethyl ether (20 ml) and the combined organic fractions are washed with saturated brine, dried over $MgSO_4$, and concentrated. The red oily residue is purified by silica gel chromatography (14 × 2 cm i.d.; diethyl ether–pentane fraction 2:8, 3:7, . . . , 10:0, 50 ml each) to give pure sarcinaxanthin (3) (127 mg, 90%), which is recrystallized from acetone until a constant melting point (mp 208–210°) is obtained.

γ_{max} (acetone): 397 (shoulder), 416, 442, and 470 nm (ε: 5.10^4; 9.710^4; 1.47 10^5; and 1.46 10^5). These data as well as NMR and MS spectra are in agreement with those reported for the natural product.[3,7]

Acknowledgments

We are grateful to the CNRS (UA 1110) and to the Université Pierre et Marie Curie (Paris, 6) for financial help.

[10] Synthesis of Carotenoporphyrin Models for Photosynthetic Energy and Electron Transfer

By DEVENS GUST, THOMAS A. MOORE, ANA L. MOORE, and PAUL A. LIDDELL

Introduction

Molecular dyads consisting of carotenoid polyenes covalently linked to porphyrin or chlorophyll derivatives serve as components of useful model systems for the study of photosynthetic energy and electron transfer.[1] The carotenoporphyrins mimic three important processes that occur in photosynthetic antennae and/or reaction centers. One of these is carotenoid antenna function, wherein carotenes absorb light in regions of the spectrum where chlorophylls have small extinction coefficients and transfer the resulting excitation to chlorophyll via a singlet-singlet energy transfer process. Second, carotenoids quench the triplet states of chlorophylls by triplet-triplet energy transfer, and thereby help prevent chlorophyll-sensitized production of singlet oxygen. In this way, they provide photoprotection from the deleterious effects of this highly reactive oxygen species. Finally, carotenoporphyrins have been incorporated into a number of more complex molecular devices that mimic the photoinitiated electron transfer cascades that lie at the heart of the photosynthetic conversion of light energy into useful chemical potential in the form of long-lived charge separation. Carotenoporphyrins are also potentially useful for a variety of other studies in which it is desirable to sensitize the production of carotenoid triplet states, control the lifetime of porphyrin or chlorophyll triplet states, control singlet oxygen sensitization, or sensitize the production of porphyrin or chlorophyll singlet states with light of wavelengths not readily absorbed by these chromophores.

---

[1] D. Gust, T. A. Moore, P. A. Liddell, G. A. Nemeth, L. R. Makings, A. L. Moore, D. Barrett, P. J. Pessiki, R. V. Bensasson, M. Rougée, C. Chachaty, F. C. De Schryver, M. Van der Auweraer, A. R. Holzwarth, and J. S. Connolly, *J. Am. Chem. Soc.* **109**, 846 (1989); D. Gust, T. A. Moore, A. L. Moore, G. Seely, P. A. Liddell, D. Barrett, L. O. Harding, X. C. Ma, S.-J. Lee, and F. Gao, *Tetrahedron* **45**, 4867 (1989); D. Gust, T. A. Moore, A. L. Moore, F. Gao, D. Luttrull, J. M. DeGraziano, X. C. Ma, L. R. Makings, S.-J. Lee, T. T. Trier, E. Bittersmann, G. R. Seely, S. Woodward, R. V. Bensasson, M. Rougée, F. C. De Schryver, and M. Van der Auweraer, *J. Am. Chem. Soc.* **113**, 3638 (1991). Also, for recent reviews, see Refs. 2-4.

[2] D. Gust and T. A. Moore, *Science* **244**, 35 (1989).

[3] D. Gust and T. A. Moore, *Adv. Photochem.* **16**, 1 (1991).

[4] D. Gust and T. A. Moore, *Top. Curr. Chem.* **159**, 103 (1991).

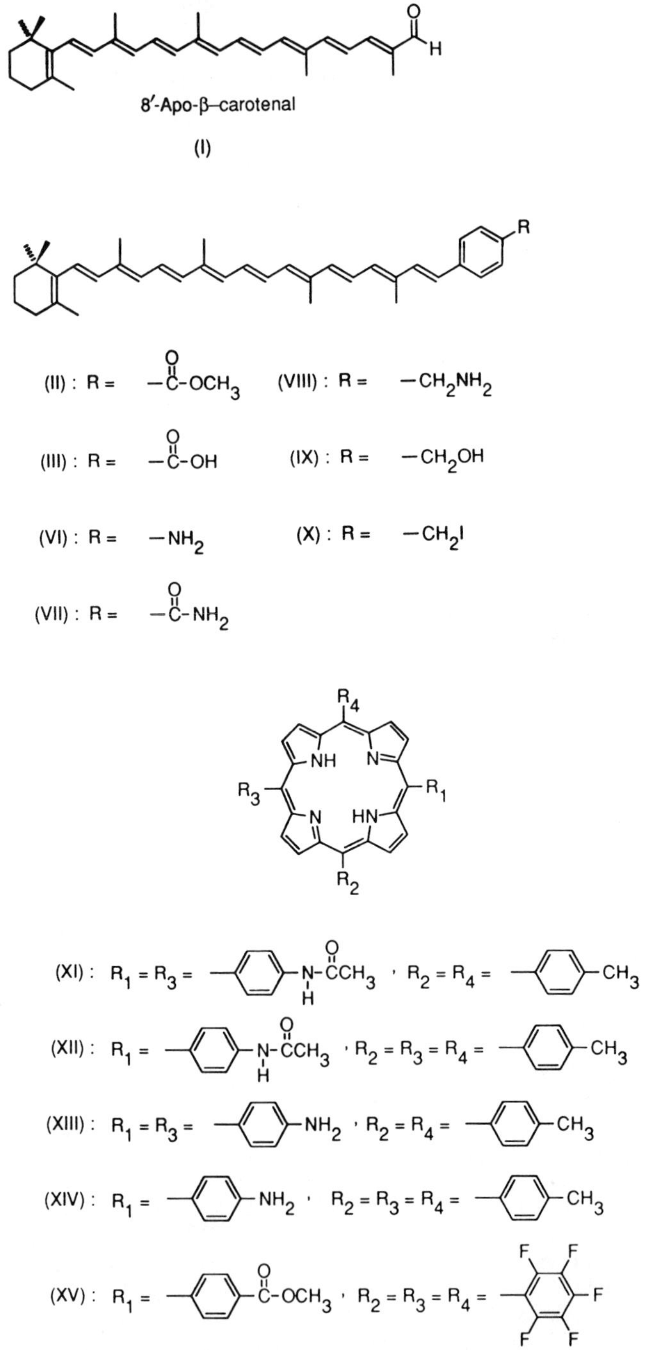

FIG. 1. Structure of carotenoids (**I–III, VI–X**), porphyrin and chlorophyll derivatives (**XI–XVI**), and carotenoporphyrins (**XVII–XXI**).

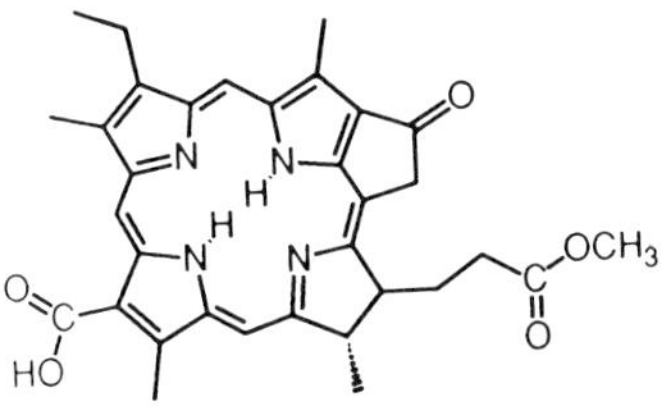

(XVI)

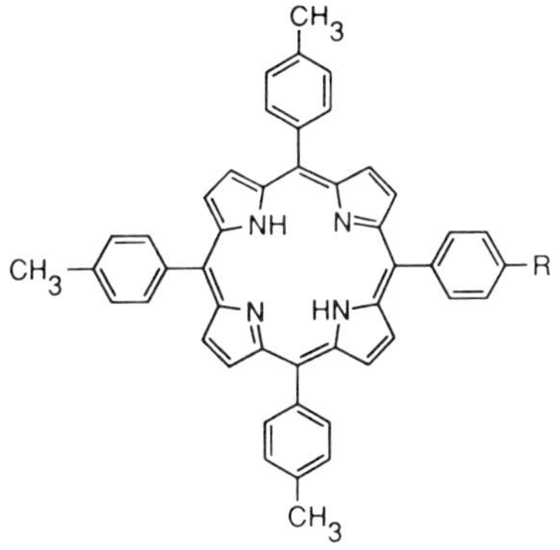

(XVII) : R= —NH-C(=O)-⟨benzene⟩-carotenoid chain

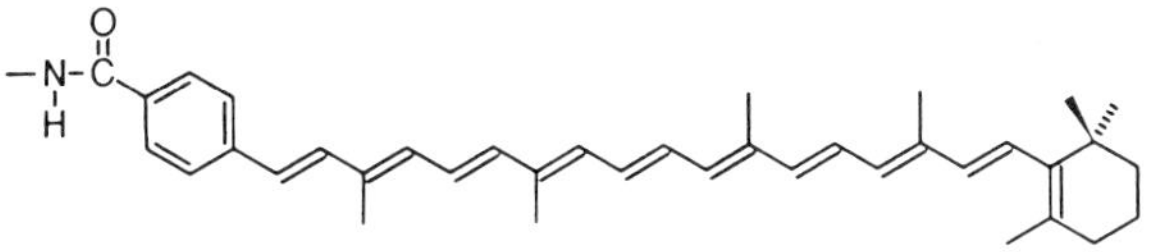

(XVIII) : R= —C(=O)-NH-CH₂-⟨benzene⟩-carotenoid chain

(XIX) : R= —O-CH₂-⟨benzene⟩-carotenoid chain

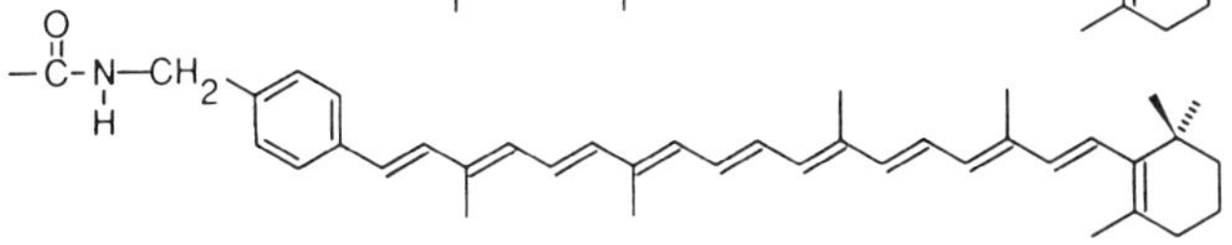

(XX) : R= —C(=O)-O-CH₂-⟨benzene⟩-carotenoid chain

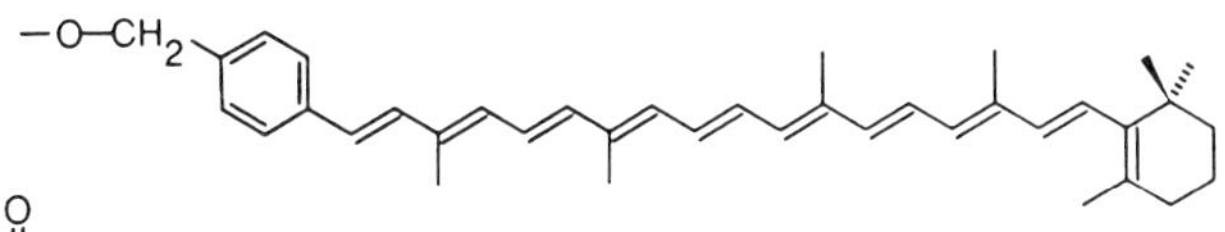

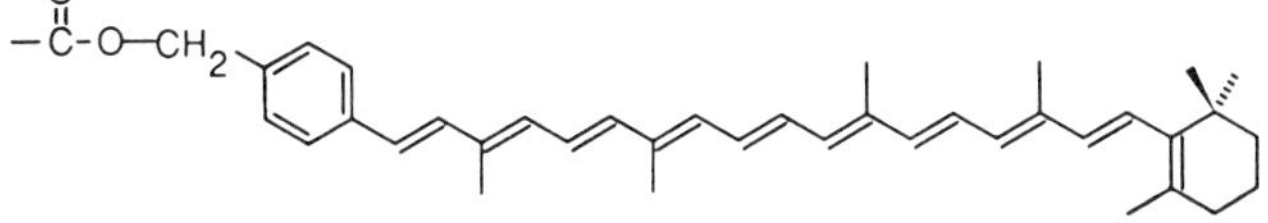

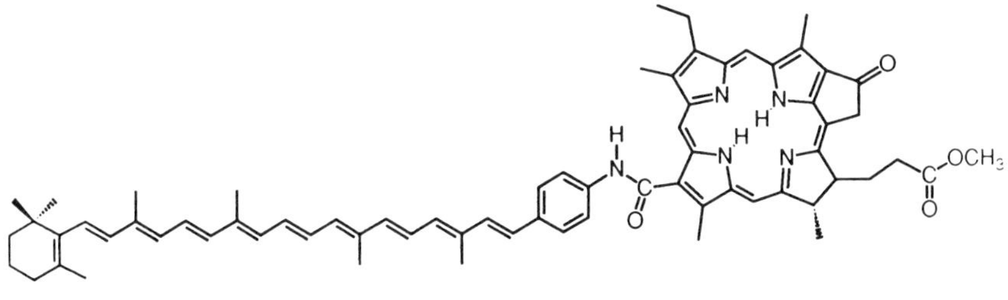

(XXI)

The design of a carotenoporphyrin suitable for the study of one or more of the energy and electron-transfer processes described above requires the proper choice of spectral and electrochemical properties for the two chromophores. Moreover, the selection of a suitable covalent linkage between the chromophores is just as crucial. This is the case because the rates of singlet energy transfer, triplet energy transfer, and electron transfer all depend very strongly on donor-acceptor separation and orientation, and on the involvement of the linkage bonds in the transfer process. In this chapter, we present synthetic methods for the preparation of seven carotenoid polyenes and six porphyrins and chlorophyll derivatives. Finally, the preparation of carotenoporphyrin dyads having five different covalent linkages is described. Figure 1 shows some of the basic structures for carotenoids (**I-III, VI-X**), porphrin and chlorophyll derivatives (**XI-XVI**), and carotenoporphyrins (**XVII-XXI**). The synthetic methods have been chosen to illustrate a variety of linkage strategies. With minor modifications, these preparative methods may be employed to prepare a host of other carotenoporphyrins for various purposes.

Because multistep synthesis requires reasonable amounts of starting materials, the precursors of the various synthetic carotenoids should be easily obtainable. With this in mind, we have chosen 8'-apo-β-carotenal (**I**, Fig. 1) as the basis for the carotenoids described here. This compound features a synthetically versatile aldehyde group, and because it is used as a food colorant, it is readily available commercially. As shown in Scheme I, this compound may be easily converted to 7'-apo-7'-(4-carbomethoxyphenyl)-β-carotene (**II**) via a Wittig reaction. Ester (**II**) is the starting material for the preparation of a variety of other carotenoids, as shown in Scheme I. Alternatively, as illustrated in Scheme II, the Wittig reaction may be used to convert 8'-apo-β-carotenal (**I**) to aminocarotenoid (**VI**).

The tetraarylporphyrins described in this chapter may be prepared either by the classic method developed by Rothemund[5] and Adler *et al.*,[6] or via the newer Lindsey variation.[7] Examples of both procedures are given below. One of the porphyrins described (**XV**) bears three pentafluorophenyl groups, which make this porphyrin moiety an especially good electron acceptor. The synthesis of a chlorophyll derivative based on the readily available and relatively stable methylpyropheophorbide a[8] is also presented.

---

[5] P. Rothemund, *J. Am. Chem. Soc.* **61**, 2912 (1939).
[6] A. D. Adler, F. R. Longo, F. D. Finarelli, J. Goldmacher, J. Assour, and L. Korsakoff, *J. Org. Chem.* **32**, 476 (1967).
[7] J. S. Lindsey, H. C. Hsu, and I. C. Schreiman, *Tetrahedron Lett.* **27**, 4969 (1986).
[8] G. W. Kenner, S. W. McCombie, and K. M. Smith, *J. Chem. Soc., Perkin Trans. 1* p. 2517 (1973).

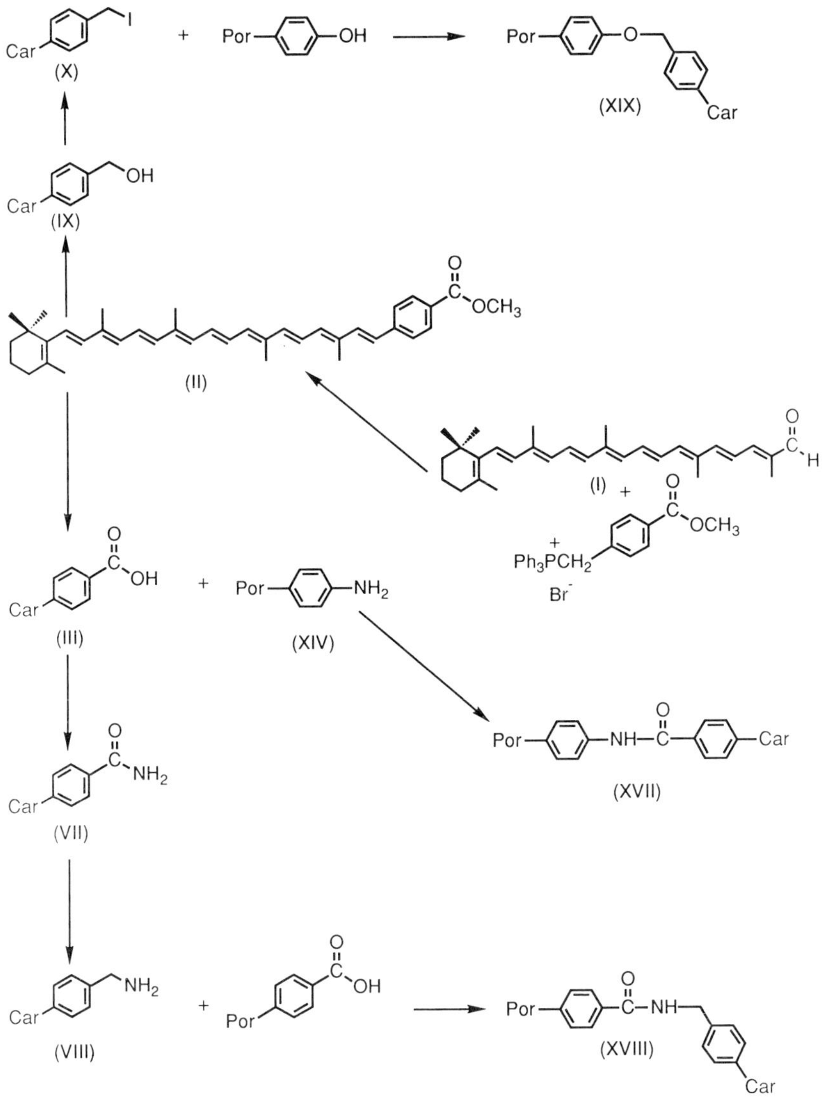

SCHEME I

Finally, the coupling of carotenoids and cyclic tetrapyrroles by a variety of amide, ester, and ether linkages is described. Only ultraviolet–visible (UV–VIS) spectral data are given below in order to save space. For ^{1}H nuclear magnetic resonance (NMR) data, the original reports should be consulted.[1]

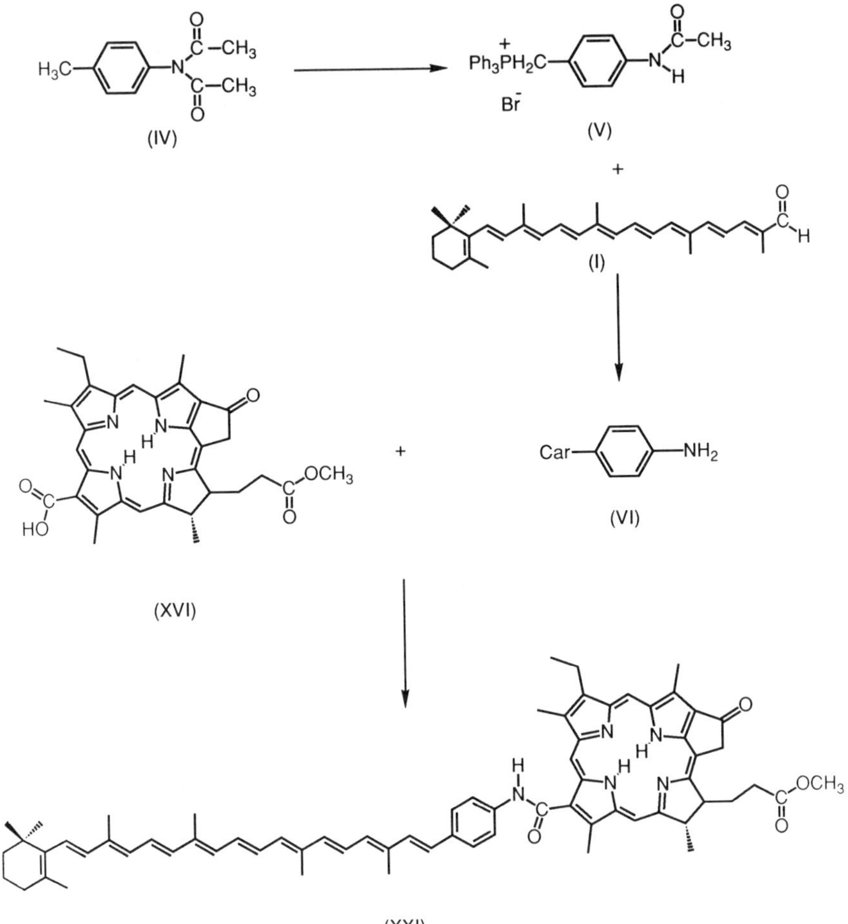

SCHEME II

Synthetic Methods

Carotenoids

Carotenoic Acid

4-Carbomethoxybenzyltriphenylphosphonium bromide. To a 150-ml flask equipped with a condenser, magnetic stirring bar, and nitrogen purge line are added 1.50 g (6.55 mmol) of methyl-α-bromo-*p*-toluate, 1.72 g (6.55 mmol) of triphenylphosphine, and 50 ml of toluene. The solution is refluxed for 2 hr and filtered, and the residue is washed with dry toluene. The white solid is dried under vacuum to give 3.0 g (93%) of the phosphonium salt.

7'-Apo-7'-(4-carbomethoxyphenyl)-β-carotene (II). Into a 200-ml flask outfitted with a magnetic stirring bar, a condenser, and a gas inlet tube are placed 1.0 g (2.4 mmol) of 8'-apo-β-carotenal (I), 50 ml of dimethyl sulfoxide, 1.4 g (2.9 mmol) of 4-carbomethoxybenzyltriphenylphosphonium bromide, and 0.17 g (3.1 mmol) of sodium methoxide. The suspension is heated to 80° and stirred under an argon atmosphere. After 16 hr a supplemental amount of both the phosphonium bromide (1.18 g, 2.4 mmol) and sodium methoxide (0.13 g, 2.4 mmol) is added and the reaction mixture is stirred for an additional 16 hr. The reaction mixture is then poured into ethyl ether (800 ml) and the organic solution is washed six times with 150-ml portions of water to remove all traces of dimethyl sulfoxide. The ether layer is dried over anhydrous magnesium sulfate and filtered, the solvent is evaporated, and the residue is recrystallized from dichloromethane–methanol to afford 1.12 g (85% yield) of II. [UV–VIS absorption (toluene) λ_{max} (nm) 302, 376, 458, 482, and 514.]

7'-Apo-7'-(4-carboxyphenyl)-β-carotene (III). A 110-mg (0.201 mmol) portion of II is dissolved in 16 ml of a mixture of tetrahydrofuran and methanol (3:1). To this solution is added 2 ml of 10% (w/v) aqueous potassium hydroxide and the mixture is stirred under an argon atmosphere for 18 hr. The reaction mixture is then partitioned between chloroform and water (pH 1–2) and the aqueous layer is washed with chloroform until all the carotene is extracted. The combined chloroform extracts are dried over anhydrous sodium sulfate and filtered, and the solvent is evaporated to yield 98 mg (91%) of the pure carotenoic acid. [UV–VIS absorption (dichloromethane) λ_{max} (nm) 302, 376, 458, 482, and 514.]

Aminocarotenoids

1-(N,N-Diacetylamino)-4-methylbenzene (IV). To a 500-ml flask equipped with a condenser are added 40 g (0.37 mol) of *p*-toluidine and 200 ml of acetic anhydride. On mixing the two reagents, a vigorous reaction takes place that gives rise to long prismatic crystals. These crystals redissolve as the mixture is heated. After refluxing for 22 hr the excess acetic anhydride is distilled under vacuum. The last traces are removed by an azeotropic distillation with toluene. As the solvent is removed, crystals begin to precipitate. NMR analysis indicates that this material is the monoacetylated compound. The remaining material is distilled under vacuum, and the fraction distilling between 109 and 111° (0.5 mmHg) is collected as a colorless liquid. A total of 48.9 g of the desired product is obtained, which corresponds to a 69% yield.

4-(N-Acetylamino)benzyltriphenylphosphonium bromide (V). To a flask equipped with a condenser are added 12 g (63 mmol) of **IV**, 11.1 g (62.4 mmol) of *N*-bromosuccinimide, and 150 ml of carbon tetrachloride. The suspension is brought to reflux and exposed to light from a 100-W tungsten lamp. After 2 hr the reaction mixture is cooled and filtered, and the solvent is evaporated. The ^{1}H NMR spectrum indicates that approximately one-half of the starting material is still present. The crude product is dissolved in a solution containing 180 ml of benzene and 16 g (61 mmol) of triphenylphosphine and is heated at 70° with stirring. After 22 hr the thick paste is filtered, and the residue is recrystallized from dichloromethane–benzene to yield 14.9 g (52%) of pure **V** as a white powder.

7'-Apo-7'-(4-aminophenyl)-β-carotene (VI). To a 100-ml flask are added 0.50 g (1.2 mmol) of 8'-apo-β-carotenal (**I**), 80 ml of dimethyl sulfoxide, 1.1 g (2.4 mmol) of **V**, and 0.20 g (3.7 mmol) of sodium methoxide. The mixture is stirred for 5 hr under an argon atmosphere at 60–70° and then quenched by pouring the dark orange solution into 500 ml of ethyl ether. This is followed by repeated washings with water in an attempt to remove all the dimethyl sulfoxide. The organic layer is dried over anhydrous magnesium sulfate and filtered, and the solvent is evaporated under reduced pressure. Thin-layer chromatography (TLC; silica gel with dichloromethane) indicates that all the carotenal has been consumed and that partial deprotection of the amide has taken place. The crude product is dissolved in 30 ml of tetrahydrofuran (freshly distilled from lithium aluminum hydride) to which 75 ml of saturated methanolic potassium hydroxide solution is added. This solution is heated to 63°, stirred under an argon atmosphere for 5.5 hr, and then poured into 500 ml of ethyl ether and washed six times with 150-ml portions of water. The organic layer is dried over anhydrous magnesium sulfate and filtered, and the solvent is evaporated. The residue is chromatographed with chloroform on a dry-packed silica gel column to give 319 mg (53% yield) of the pure amimocarotenoid (**VI**). [UV–VIS absorption (dichloromethane) λ_{max} (nm) 299, 376, 478, and 506.]

7'-Apo-7'-(4-carbamylphenyl)-β-carotene (VII). To a 500-ml flask that has been flushed with argon are added 735 mg (1.40 mmol) of 7'-apo-7'-(4-carboxyphenyl)-β-carotene (**III**), 50 ml of benzene, and 664 mg (8.40 mmol) of pyridine. With the addition of pyridine, the mixture immediately becomes homogeneous. Next, 500 mg (4.20 mmol) of thionyl chloride is added and the reaction is stirred for 20 min at room temperature. The solvent is removed and the residual pyridine and thionyl chloride are removed via azeotropic distillation with 100 ml of benzene. A 100-ml portion of benzene that has been saturated with ammonia gas is added immediately and the reaction mixture stirred vigorously for 15 min and then poured into a separatory funnel containing 300 ml of water. The aqueous phase is extracted three times with 200-ml portions of dichloromethane and the combined organic extracts are concentrated to an orange-red product. Residual pyridine and water are removed by azeotropic distillation using 400 ml of toluene. The product is recrystallized from methanol–chloroform to afford 555 mg (76% yield) of pure **VII**. The mother liquor from the recrystallization is concentrated and subjected to flash column chromatography on silica gel with chloroform containing 1% (v/v) methanol to yield 89 mg of pure amide, bringing the total yield to 88%. [UV-VIS absorption (dichloromethane) λ_{max} (nm) 301, 373, 480, and 506.]

7'-Apo-7'-(4-aminomethylphenyl)-β-carotene **(VIII)**. Amide **VII** (89 mg, 0.17 mmol) and 10 ml of tetrahydrofuran are added to a 20-ml round-bottomed flask. The solution is cooled to 0° and an excess of lithium aluminum hydride is added with vigorous stirring. After 1 hr at 0° no reaction has occurred. The reaction mixture is warmed to room temperature and after 12 hr, TLC with chloroform–5% methanol indicates that the reaction is complete. The reaction mixture is poured into a separatory funnel containing 200 ml of water and the aqueous phase is extracted with three 100-ml portions of dichloromethane. The combined organic extracts are concentrated to an orange-red residue. The remaining water is removed via azeotropic distillation with 200 ml of toluene. The residue is then subjected to flash column chromatography on silica gel with chloroform–2% methanol to afford 39 mg (44% yield) of pure compound **VIII**.

Carotene Alcohol and Iodide

7'-Apo-7'-(4-hydroxymethylphenyl)-β-carotene **(IX)**. To a 100-ml flask equipped with a nitrogen purge line are added 0.31 g (0.57 mmol) of **II**, 60 ml of tetrahydrofuran, and 0.022 g (0.580 mmol) of lithium aluminum hydride. After stirring for 15 min, the reaction mixture is poured into 150 ml of water and the resulting mixture washed three times with 75-ml portions of dichloromethane. The combined extracts are dried over sodium sulfate and filtered, and the solvent is evaporated. The residue is recrystallized from dichloromethane–methanol to give 0.27 g of the pure carotenoid alcohol (93% yield). [UV–VIS absorption (toluene) λ_{max} (nm) 361, 474, 506, and 656.]

7'-Apo-7'-(4-iodomethylphenyl)-β-carotene **(X)**. To a 30-ml flask equipped with an efficient condenser, an argon inlet tube and a magnetic stirring bar are added 0.036 g (0.064 mmol) of diphosphorus tetraiodide and 10 ml of carbon disulfide. The flask is cooled to 0°, 0.10 g (0.19 mmol) of **IX** is added, and the reaction mixture is stirred for 30 min. After this period the contents of the flask are poured into saturated potassium carbonate and the resulting mixture extracted several times with ether. The organic layer is then filtered through celite, dried over anhydrous magnesium sulfate, and filtered. Evaporation of the solvent gives 0.10 g (83% yield) of the desired product.

Porphyrin Derivatives

5,15-Bis(4-acetamidophenyl)-10,20-bis(4-methylphenyl)porphyrin **(XI)** *and 5-(4-Acetamidophenyl)-10,15,20-tris(4-methylphenyl)porphyrin* **(XII)**. To 70.2 g (0.430 mol) of 4-acetamidobenzaldehyde and 44.5 g (0.370 mol) of *p*-tolualdehyde in 3 liters of propionic acid at 90° is added 53.7 g (0.800 mol) of pyrrole. This solution is refluxed for 0.5 hr and then poured hot into 12 liters of water, followed by the addition of 500 g of sodium chloride. The material that precipitates as a green mass is collected by vacuum filtration. This precipitate is dried in a vacuum oven at 160° at aspirator pressure for 24 hr and then for an additional 24 hr at 0.5 mmHg to remove all traces of propionic acid. During the drying the color changes to purple. The resulting material is pulverized with a mortar and pestle to a fine powder. The porphyrins are extracted from the crude product using a Soxhlet extractor and chloroform. After extracting for 24 hr the remaining solid material is removed, pulverized again, and then extracted for an additional 24 hr. Methanol (100 ml) and 3 g of potassium carbonate are added to the extracts for every 500 ml of chloroform. This solution is refluxed for 1 hr and filtered hot. Methanol (300 ml) is added to the filtrate and the chloroform is removed under vacuum to precipitate the porphyrins as a dull purple solid that still contains some tarry by-products. This material is refluxed for 0.5 hr in 400 ml of methanol and 100 ml of dichloromethane, followed by filtration to collect the porphyrins. This procedure is repeated to give 15.5 g of tar-free porphyrin mixture. The chlorin impurities are oxidized to the corresponding porphyrins by refluxing for 45 min with 2,3-dichloro-5,6-dicyano-1,4-benzoquinone (DDQ) in 2 liters of chloroform plus 100 ml of toluene. The complex between the porphyrins and the DDQ is broken by adding 100 ml of methanol and 500 ml of acetone and refluxing for an additional 15 min. The excess DDQ is removed by chromatography on a short bed of alumina with chloroform–5% methanol. Removal of the solvent gives the chlorin-free porphyrins. Column chromatography on silica gel with chloroform and a methanol gradient gives 1.72 g (1.14% yield) of **XI** plus 3.17 g of 5-(4-acetamidophenyl)-10,15,20-tris(4-methylphenyl)porphyrin **(XII)**, 1.46 g of 5,10-bis(4-acetamidophenyl)-15,20-bis(4-methylphenyl)porphyrin, and 0.75 g of 5,10,15-tris(4-acetamidophenyl)-20-(4-methylphenyl)porphyrin. [UV-VIS absorption (dichloromethane) of **XI** λ_{max} (nm) 242, 418, 516, 552, 592, and 648.]

5,15-Bis(4-aminophenyl)-10,20-bis(4-methylphenyl)-porphyrin **(XIII)**. Amide **XI** (1.72 g, 2.27 mmol) is dissolved in 100 ml of 6 N HCl and kept at refluxing temperature for 1 hr. After cooling, the green protonated porphyrin is collected by vacuum filtration, dissolved in chloroform, and washed with aqueous sodium carbonate. The organic layer is dried over magnesium sulfate and filtered. Methanol is added and slow evaporation of the solvent under vacuum results in the crystallization of the pure porphyrin **(XIII)**, which is collected by filtration and dried under vacuum to give 1.25 g (82% yield). [UV-VIS absorption (dichloromethane) λ_{max} (nm) 243, 301, 418, 516, 552, 592, and 648.]

5-(4-Aminophenyl)-10,15,20-tris(4-methylphenyl)porphyrin **(XIV)**. **XIV** is prepared from **XII** by a method similar to that outlined for **XIII**.

5-(4-Carbomethoxyphenyl)-10,15,20-tris(pentafluorophenyl)porphyrin **(XV)**. **XV** is prepared using a modification of the porphyrin synthesis developed by Lindsey and co-workers.[7] In a 2-liter round-bottomed flask are placed 1.20 g (17.9 mmol) of pyrrole, 2.65 g (13.5 mmol) of pentafluorobenzaldehyde, and 0.74 g (4.5 mmol) of methyl-4-formylbenzoate together with 1.4 ml of trifluoroacetic acid and 1.8 liters of chloroform. After refluxing the mixture for 4.5 hr and cooling to room temperature, 4.09 g of DDQ is added and the mixture is left to stir overnight. The reaction mixture is neutralized with aqueous sodium bicarbonate and extracted with four 100-ml portions of dichloromethane. The organic layer is separated and filtered through a short silica gel column to remove tar. The porphyrins are finally separated by chromatography on silica gel with toluene elution. Recrystallization from dichloromethane and methanol gives 61 mg of pure **XV** (1.49% yield). [UV-VIS absorption (dichloromethane) λ_{max} (nm) 414, 510, 540, 586, and 640.]

2-Desvinyl-2-carboxy-methylpyropheophorbide a **(XVI)**. To a 500-ml flask are added 1.0 g (1.8 mmol) of methylpyropheophorbide a[8] and 300 ml of acetone. As the solution is stirred, 3% (w/v) aqueous potassium permanganate (60 ml) is added over a period of 1 hr. The reaction mixture is diluted with acetone (200 ml) and filtered through Celite. The filtrate is first reduced in volume and diluted with chloroform (400 ml) and then washed with two 300-ml portions of acidified brine solution. The solvent is evaporated and the residue is chromatographed on silica gel with chloroform-1% (v/v) methanol as the solvent to give 316 mg (30% yield) of the acid **(XVI)**. [UV-VIS absorption (dichloromethane) λ_{max} (nm) 234, 282, 325, 384, 414, 478, 516, 548, 626, and 686.]

Dyads

Carotenoporphyrin XVII. To a 50-ml flask are added 70 mg (0.10 mmol) 7'-apo-7'-(4-carboxyphenyl)-β-carotene (**III**), 20 ml of dry benzene, 29 μl (0.40 mmol) of thionyl chloride, and 80 μl (0.99 mmol) of dry pyridine. The initial orange suspension is rapidly converted into the acid chloride, as indicated by the dark red color. After stirring the solution for 30 min under argon, the solvent is distilled under vacuum. Benzene (40 ml) is added and evaporated to dryness, under vacuum, to remove the excess thionyl chloride. The residue that remains is dissolved in 30 ml of dry dichloromethane and added to a solution of 133 mg (0.198 mmol) of 5-(4-aminophenyl)-10,15,20-tris(4-methylphenyl)porphyrin (**XIV**), which is dissolved in 60 ml of dry dichloromethane and 0.2 ml of dry pyridine. This solution is stirred under argon for 60 min and then partitioned between dichloromethane and water. The organic layer is washed twice and 70-ml portions of water, the solvent is evaporated, and the residue is dried under vacuum. Chromatography on silica gel with toluene–0.5% (v/v) ethyl acetate as the solvent and subsequent recrystallization from dichloromethane–methanol give 82 mg (53% yield) of the carotenoporphyrin **XVII**. [UV–VIS absorption (dichloromethane) λ_{max} (nm) 305, 376, 418, 480, 512, 590, and 648.]

Carotenoporphyrin XVIII. To a 50-ml flask equipped with a condenser and a nitrogen gas line are added 120 mg (0.171 mmol) of 5-(4-carboxyphenyl)-10,15,20-tris(4-methylphenyl)porphyrin,[9] 30 ml of dry dichloromethane, and 3 ml of oxalyl chloride. The dark green solution is refluxed under a nitrogen atmosphere for 1 hr and cooled, and the solvent is removed under vacuum. To remove all traces of oxalyl chloride, two consecutive 25-ml portions of toluene are added and evaporated. The residue is dissolved in a mixture of 50 ml of dichloromethane and 1 ml of pyridine. The resulting solution is added to 70 mg (0.14 mmol) of **VIII** in 50 ml of dichloromethane and stirred under an argon atmosphere. After 1 hr the reaction mixture is poured into 180 ml of dichloromethane and washed twice with 100-ml portions of water. The organic layer is separated and the solvent is evaporated. Remaining traces of water and pyridine are removed by azeotropic distillation with toluene. The residue is chromatographed on silica gel with dichloromethane and the product is recrystallized from dichloromethane–methanol to afford 76 mg (46% yield) of the pure carotenoporphyrin **XVIII**. [UV–VIS absorption (dichloromethane) λ_{max} (nm) 304, 373, 418, 476, 510, 592, and 648.]

Carotenoporphyrin XIX. To a 50-ml flask equipped with a drying tube,

---

[9] J. A. Anton and P. A. Loach, *J. Heterocycl. Chem.* **12,** 573 (1975).

a magnetic stirring bar, and an argon inlet tube are added 0.064 g (0.095 mmol) of chlorin-free 5-(4-hydroxyphenyl)-10,15,20-tris(4-methylphenyl) porphyrin,[10] 4 ml of dry dimethyl sulfoxide, and 1 ml of a 0.095 M solution of sodium methoxide in dimethyl sulfoxide. The reaction is stirred for 15 min at room temperature and then 0.030 g (0.048 mmol) of 7'-apo-7'-(4-iodomethylphenyl)-β-carotene (X) dissolved in 20 ml of dry toluene is added. After stirring for 30 min, the reaction is quenched by adding 30 ml of water. The aqueous suspension is extracted four times with 100-ml portions of ether. The combined extracts are dried over anhydrous magnesium sulfate and filtered, and the solvent is evaporated under reduced pressure to yield a dark red crude product. Chromatography on silica gel with deaerated toluene gives 0.053 g (93% yield) of the pure carotenoporphyrin XIX. [UV–VIS absorption (toluene) λ_{max} (nm) 367, 420, 472, 506, 550, 592, and 650.]

Carotenoporphyrin XX. A 0.143-g (0.204 mmol) portion of 5-(4-carboxyphenyl)-10,15,20-tris(4-methylphenyl)porphyrin,[9] 10 ml of dry benzene, and 1 ml of oxalyl chloride are combined in a 25-ml flask equipped with a reflux condenser, magnetic stirring bar, and nitrogen line. The mixture is heated for 3 hr at reflux. The solvent is then removed by evaporation at reduced pressure and 5 ml of dry benzene added. The benzene is again removed at reduced pressure to eliminate traces of unreacted oxalyl chloride. Benzene (7 ml) is added to the green residue. A solution of 96.3 mg (0.185 mmol) of 7'-apo-7'-(4-hydroxymethylphenyl)-β-carotene (IX), 3 ml of benzene, and 0.5 ml of pyridine is added, and the mixture is stirred at room temperature for 43 hr. Water (1 ml) is then added and the stirring is continued for 15 min. The reaction mixture is diluted with 50 ml of dichloromethane and washed with two 20-ml portions of water. The combined aqueous extracts are back extracted with 20 ml of dichloromethane and the organic layers are combined and dried over anhydrous sodium sulfate, filtered, and evaporated to dryness. The residue is chromatographed on silica gel with toluene. The solvent is evaporated and the residue recrystallized from dichloromethane–methanol to yield 54 mg (24% yield) of the pure carotenoporphyrin XX. [UV-VIS absorption (toluene) λ_{max} (nm) 367, 420, 453, 474, 507, 550, 592, and 651.]

Carotenopyropheophorbide XXI. To a 50-ml flask are added 100 mg (0.176 mmol) of 2-desvinyl-2-carboxy-methylpyropheophorbibe *a* (XVI), 20 ml of dry benzene, 142 μl of dry pyridine, and 51 μl (0.70 mmol) of thionyl chloride. The mixture turns from brownish green to a characteristic reddish brown on the addition of the thionyl chloride. Stirring is continued

---

[10] R. G. Little, J. A. Anton, P. A. Loach, and J. A. Ibers, *J. Heterocycl. Chem.* **12**, 343 (1975).

under an argon atmosphere for 30 min. The solvent is then removed under vacuum, followed by the successive addition and removal of two 20-ml portions of benzene to the reaction mixture to eliminate excess thionyl chloride. The residue is dissolved in 25 ml of dichloromethane containing 0.1 ml of pyridine and this mixture is added to 89 mg (0.18 mmol) of aminocarotenoid **VI**. After stirring for 1 hr, TLC (silica gel with methylene chloride–3% methanol) indicates that the reaction is complete. The contents of the flask are diluted with 120 ml of dichloromethane and washed successively with aqueous citric acid, aqueous sodium bicarbonate, and brine. The organic layer is then dried over anhydrous sodium sulfate and filtered, and the solvent is evaporated under vacuum. The residue is purified by flash chromatography on silica gel with chloroform–0.3% (v/v) methanol to afford 154 mg (81% yield) of carotenopyropheophorbide **XXI**. [UV-VIS absorption (dichloromethane) λ_{max} (nm) 305, 412, 476, 508, 616, and 674.]

Acknowledgments

This work was supported by the National Science Foundation (CHE-8903216), the U.S. Department of Energy (DE-FG02-87ER13791), and the American Chemical Society (PRF#23911-AC4). This is Publication 71 from the Arizona State University Center for the Study of Early Events in Photosynthesis. The Center is funded by U.S. Department of Energy Grant DE-FG02-88ER13969 as part of the U.S. Department of Agriculture–Department of Energy–National Science Foundation Plant Science Center Program.

[11] Use of Borohydride Reduction Methods to Probe Carotenoid–Protein Binding

By R. Gómez and J. C. G. Milicua

Since the beginning of the twentieth century, several scientists have devoted much effort to the study of carotenoid–protein complexes, and various reviews have been published.[1-6] In spite of this, much is still not

---

[1] W. L. Lee, "Carotenoids in Animal Coloration." Dowdens, Hutchinson and Ross, Stroudsburg, 1966.
[2] D. F. Cheesman, W. L. Lee, and P. F. Zagalsky, *Biol. Rev. Cambridge Philos. Soc.* **42**, 131 (1967).
[3] P. F. Zagalsky, *Pure Appl. Chem.* **47**, 103 (1976).
[4] G. Britton, G. M. Armitt, S. Y. M. Lau, A. K. Patel, and C. C. Shone, "Carotenoid Chemistry and Biochemistry" (G. Britton and T. W. Goodwin, eds.), Proc. 6th Int. Symp. Carotenoids, pp. 237–251. Pergamon, Oxford, 1982.
[5] P. F. Zagalsky, *Oceanis* **9**, 43 (1983).

understood about the physiological and ecological significance of these complexes: is their function simply to afford protective coloration, a means for storage of nutrients during development, or something entirely different? Of particular biochemical interest is the mechanism of the interactions between carotenoid and protein, which give rise to the large bathochromic shift in the absorption spectrum. The elucidation of this mechanism will provide information that may be highly relevant in connection with studies on the visual pigments[7] and enzyme–substrate interactions.[8]

According to Lee,[1] the carotenoid–protein complexes may be classified in two groups: true carotenoproteins, i.e., those complexes not containing lipids other than carotenoids and showing carotenoid–protein stoichiometry, and carotenolipoproteins, which contain lipids and are not necessarily stoichiometric. If the carotenoprotein spectral characteristics are considered, however, a more detailed classification may be devised because the color of the complexes are due to different interactions.

The first class of carotenoproteins contains those that have strong carotenoid–protein interaction (Fig. 1a). These proteins show a large bathochromic shift from the red color of the astaxanthin to the blue or purple color of the complex, due to carotenoid–protein interactions, and may be included in the first group of the classification according to Lee. Representative examples are the α-crustacyanin from *Homarus americanus*,[9,10] the α-crustacyanin from *Homarus gammarus*,[11,12] and the blue carotenoproteins of several other species, including *Procambarus clarkii*,[13] *Carcinus maenas*,[14] *Upogebia pusilla*,[15] and *Astacus leptodactylus*.[16]

---

[6] J. B. C. Findlay, D. J. C. Pappin, M. Brett, and P. F. Zagalsky, "Carotenoids: Chemistry and Biology" (N. I. Krinsky, M. M. Mathews-Roth, and R. F. Taylor, eds.), Proc. 8th Int. Symp. Carotenoids. Plenum, New York, 1989.
[7] R. Callender and B. Honig, *Annu. Rev. Biophys. Bioeng.* **6**, 273 (1977).
[8] J. G. Belasco and J. R. Knowles, *Biochemistry* **22**, 122 (1983).
[9] M. Buchwald and W. P. Jenks, *Biochemistry* **7**, 834 (1968).
[10] R. Quarmby, D. A. Norden, P. F. Zagalsky, M. J. Ceccaldi, and R. Daumas, *Comp. Biochem. Physiol. B* **55B**, 55 (1977).
[11] D. F. Cheesman, P. F. Zagalsky, and M. J. Ceccaldi, *Proc. R. Soc. London, B* **164**, 130 (1966).
[12] P. F. Zagalsky and M. L. Tidmarsh, *Comp. Biochem. Physiol. B* **80B**, 599 (1985).
[13] A. M. Gárate, J. C. G. Milicua, R. Gómez, J. M. Macarulla, and G. Britton, *Biochim. Biophys. Acta* **881**, 446 (1986).
[14] A. M. Gárate, E. Urréchaga, J. C. G. Milicua, R. Gómez, and G. Britton, *Comp. Biochem. Physiol.* **77B**, 605 (1984).
[15] A. Villarroel, A. M. Gárate, R. Gómez, and J. C. G. Milicua, *Comp. Biochem. Physiol. B* **81B**, 547 (1985).
[16] R. Gómez, J. de las Rivas, A. M. Gárate, P. G. Barbón, and J. C. G. Milicua, *Comp. Biochem. Physiol. B* **83B**, 855 (1986).

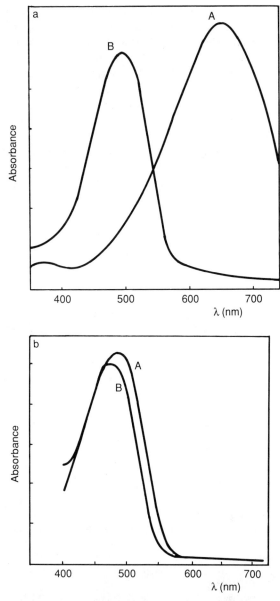

FIG. 1. (a) Carotenoproteins with carotenoid–protein interactions. The carotenoprotein (A) generally shows a blue color, λ_{max} 590–640 nm, in contrast to the orange color of the free carotenoid (mainly astaxanthin) (B), λ_{max} 474 nm in ether. (b) Carotenoproteins with Carotenoid–lipid interactions. The carotenolipoprotein complexes (A), λ_{max} 482 nm, have spectra similar to those of the free carotenoids (mainly astaxanthin and astaxanthin esters)

[11] CAROTENOPROTEIN REDUCTION 103

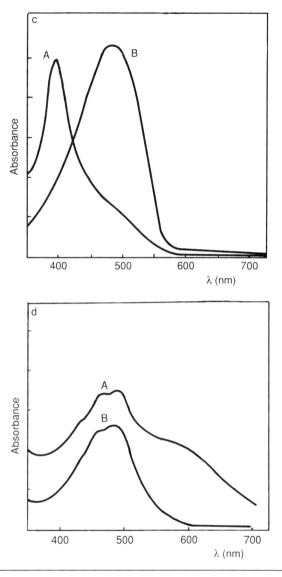

(B), $\lambda_{max} = 474$ nm in ether. (c) Carotenoproteins with carotenoid–carotenoid interactions. Complexes of this kind (A) are yellow in color, $\lambda_{max} = 390$ nm, compared with the orange-red of the free carotenoid (B), $\lambda_{max} = 474$ nm in ether. (d) Carotenoproteins with mixed interactions. Some carotenoproteins exhibit more than one interaction pattern, e.g., the spectrum of the native purple carotenoprotein from eggs of *Pa. marmoratus* (A) shows a double peak in the region of 450–550 nm, and a shoulder in the wavelengths between 600 and 650 nm. When the purple carotenoprotein is denatured (i.e., with SDS), it becomes red and this shoulder disappears, giving rise to spectrum (B).

The second group includes the carotenoproteins with carotenoid-lipid interaction. These carotenoproteins show spectra very similar to those of the free carotenoids (Fig. 1b). These carotenolipoproteins include the red pigments from *Macropipus puber*[17] and *P. clarkii*.[18]

In the third group are carotenoproteins showing a strong hypsochromic shift from the carotenoid to the carotenoprotein spectra, like the yellow carotenoprotein from the lobster[11,19] or *A. leptodactylus*[20] (Fig. 1c). The change in color of this third group is due to carotenoid-carotenoid interaction.[21]

Finally, there are carotenoproteins with two different kinds of interactions; these can be considered as mixed carotenoproteins, such as the purple carotenoprotein from the eggs of the crab *Pachygrapsus (Pa.) marmoratus*,[22] which shows both carotenoid-protein and carotenoid-lipid interactions (Fig. 1d).

Protein Reduction

For the borohydride test, the carotenoprotein must be solubilized and also pure to avoid unwanted reactions. First, it is necessary to extract the carotenoprotein from the tissues. Two different methods can be used to solubilize the proteins, depending on their solubility. The first, the third, and the fourth groups of carotenoproteins are water-soluble and may be extracted with aqueous buffers.[2,9,13-16,19,20] The carotenoproteins belonging to the second group are not soluble in water and need to be solubilized with detergents such as Triton X-100.[17,18]

Distinct carotenoproteins occur in the calcified layer of the crustacean carapace and may be extracted following decalcification with ethylenediamine tetraacetic acid (EDTA). Examples of this are the extraction and purification of the blue water-soluble carotenoprotein[13] and the red insoluble proteins from *P. clarkii*.[18]

[17] R. Gómez, J. C. G. Milicua, and A. M. Gárate, *Rev. Esp. Fisiol.* **40,** 319 (1984).
[18] J. C. G. Milicua, R. Gómez, A. M. Gárate, and J. M. Macarulla, *Comp. Biochem. Physiol. B* **81B,** 1023 (1985).
[19] V. R. Salares, N. M. Young, H. J. Bernstein, and P. R. Carey, *Biochemistry* **16,** 4751 (1985).
[20] J. C. G. Milicua, I. Arberas, P. G. Barbón, A. M. Gárate, and R. Gómez, *Comp. Biochem. Physiol. B* **85B,** 615 (1986).
[21] P. F. Zagalsky, *Comp. Biochem. Physiol. B* **71B,** 243 (1982).
[22] P. G. Barbón, Tesina de Licenciatura. Univ. Basque Country, Bilbao, Spain, 1983.

Extraction of Water-Soluble Blue Carotenoprotein from *Procambarus clarkii*

Carapaces are dissected, washed free of adhering hypodermic tissues under cold running water, rinsed in 0.3 M boric acid adjusted to pH 6.8 with solid Tris, spread on paper to dry, and left overnight in a cold room. They are then ground to fine powder in a coffee grinder, care being taken to avoid heating, and the powder is stirred overnight (magnetic stirrer) with 8 vol of 0.2 M sodium phosphate buffer/0.08 M EDTA (pH 7.5) in the cold and either in the dark or in diffuse light. The homogenate is then centrifuged at 9500 g for 30 min at 2°. The pellet is reextracted until no more solubilized blue carotenoprotein is detected in the supernatant. Six extractions are usually sufficient to achieve complete solubilization of the carotenoprotein, depending on the state of molt of the animals and the size of the particles after grinding. The combined supernatants are then filtered and stored at $-20°$ in the dark.

For extraction of the water-insoluble red carotenoprotein from *P. clarkii*, the same procedure is used, but Triton X-100 (0.1–5%) is included in the extracting buffer.

Purification of Carotenoproteins

The carotenoproteins are purified to avoid unwanted reactions, such as the reaction of borohydride with β-denatured forms.

Purification of Blue Carotenoprotein from Procambarus clarkii[13] *(First Group)*

All steps in the isolation and purification procedures are carried out at 2–4°.

The extracted blue carotenoprotein is subjected to ammonium sulfate treatment, and the fraction precipitating between 30 and 45% saturation is recovered and dissolved in the minimal volume of 0.02 M sodium phosphate buffer (pH 7.5) and dialyzed overnight against this buffer in the dark. After centrifugation at 9500 g for 30 min at 2° to remove the red (denatured?) material from the dialysate, the blue chromoprotein is adsorbed onto a 12 × 2.2 cm column of DEAE-cellulose (DE-52; Whatman, Clifton, NJ) equilibrated with 0.02 M phosphate buffer (pH 7.5). Elution is carried out with a pulsed linear gradient of sodium chloride (2 × 100 ml, 0–1.0 M) in the same buffer, i.e., as soon as a colored band begins to move, the gradient is discontinued and salt-free buffer is added until elution is achieved: the gradient is then resumed. Two main fractions are separated by this procedure: a purple fraction (β form) is eluted at zero

NaCl concentration, and the main blue protein (α form) is eluted at an NaCl concentration of 0.15–0.20 M.

The blue carotenoprotein (α form) is purified further by chromatography on a column (87 × 2.2 cm) of Sephacryl S-300 SF (Pharmacia, Piscataway, NJ) equilibrated in and eluted with 0.02 M KCl, 0.05 M sodium phosphate buffer (pH 7.5) with a constant flow rate of 3.6 ml/(cm$^2 \cdot$ hr). Finally, the blue carotenoprotein is adsorbed on a second DEAE-cellulose column, under the same conditions as before, but eluted with a shallower gradient of NaCl concentration (0–0.3 M).

This purification procedure is sometimes modified by repeating one or more of the chromatographic steps whenever the carotenoprotein appears not to be pure after the second ion-exchange column.

Yellow Carotenoproteins (Third Group)

A modification of the method described above may be used for the yellow carotenoproteins (λ_{max} 385), by changing the percentages of saturation of ammonium sulfate and the concentration of NaCl used in the ion-exchange chromatography on DE-52. For the yellow carotenoprotein from *A. leptodactylus*,[20] 0–30% ammonium sulfate saturation is used for precipitation and a range of NaCl (0.24–0.50 M) for elution from DE-52. The yellow carotenoprotein is purified further by chromatography on a column of Sephacryl S-300 as described above.

Carotenoproteins with Mixed Interactions (Fourth Group)

The carotenoproteins with mixed interactions are also soluble in aqueous buffers. We have found that the optimum pH to solubilize and purify the carotenoprotein from the eggs of *P. marmoratus* is pH 5.[23] Extraction is achieved with 0.5 M NaCl, 5 mM EDTA, pH 5, followed by precipitation with ammonium sulfate in the range of 50–60%. The ion-exchange step is made after dialysis against 0.1 M sodium acetate buffer, pH 5, and the protein elutes at 0.6–1 M NaCl. Further purification of this purple carotenoprotein is performed by chromatography on a column of Sephacryl S-300 with 0.5 M NaCl, 5 mM EDTA buffer.

Red or Orange Carotenoproteins (Second Group)

All the steps in the purification of carotenoproteins are performed in subdued light in the cold (4°).

For the water-insoluble red and orange carotenoproteins of this kind, some of which are found in the carapace of the crayfish *P. clarkii,* the use

---

[23] J. A. Fresno, Tesina de Licenciatura. Univ. Basque Country, Bilbao, Spain, 1990.

of detergents such as Triton X-100 (0.1-5% in 50 mM sodium phosphate buffer, pH 7.0) is essential for solubilization.[17,18] The purification steps are similar to those described above for the blue or yellow carotenoproteins.

Solid ammonium sulfate is added to the total material to give 17% saturation. After 24 hr, the extract is centrifuged at 10,000 g for 15 min, and the red upper layer is removed and dissolved in 0.05 M sodium phosphate buffer, pH 8.0. The carotenoid-protein complex is chromatographed through a 2.2 × 90 cm column of Sephadex G-200 (Pharmacia), equilibrated with 0.05 M phosphate buffer, pH 8.0. The column is eluted with the same buffer and the fraction having maximal absorption at 482 nm is collected, dialyzed against 0.02 M Tris-HCl, pH 8.0 buffer, and then passed through a 17 × 3.5 cm DE-52 column. The red carotenoprotein is eluted immediately. Excess Triton X-100 is removed by a gel-filtration step on a Sephadex G-200 column.

Borohydride Reduction of Carotenoproteins

Each group of carotenoproteins has a different interaction between the carotenoid and the protein, and these properties may be revealed by reduction with borohydride. If the interaction between carotenoid and protein is strong the highest concentrations of borohydride are needed to reduce astaxanthin to crustaxanthin.

Blue Carotenoproteins

The blue carotenoproteins belonging to the first group have the strongest interaction between carotenoid and protein and a large amount of borohydride must be used to give the reaction in a reasonable time (3 hr).

An example is the reaction of borohydride with the blue carotenoprotein from *P. clarkii*.[24] The amount of pure carotenoprotein recommended is about 1.2 mg/ml in samples of 4 ml in 0.05 phosphate buffer, pH 7.5. (This is suitable both for the detection of spectral shifts and for carotenoid extraction.)

The samples are reduced by adding 40 mg (0.18 M) of potassium borohydride (KBH$_4$) at a constant temperature of 20° in the dark. Absorption spectra are recorded every 5 min on a spectrophotometer. Carotenoids are extracted from the incubated carotenoprotein by the acetone-ether method of Powls and Britton.[25] The carotenoid prosthetic group of the carotenoprotein is extracted by addition of an equal volume of acetone to break the protein-carotenoid complex, and then diethyl ether (2 vol) is

---

[24] J. C. G. Milicua, A. M. Gárate, P. G. Barbón, and R. Gómez, *Comp. Biochem. Physiol. B* **95B**, 119 (1990).

[25] R. Powls and G. Britton, *Biochim. Biophys. Acta* **453**, 270 (1976).

added to separate the carotenoid into the upper ethereal phase. The ether is then evaporated with nitrogen and stored under nitrogen, in the dark at $-18°$, until further analysis.

The carotenoid samples are chromatographed by thin-layer chromatography (TLC) on silica gel G with hexane–acetone (75:25, v/v) and the carotenoids are analyzed quantitatively on the TLC plates by means of a scanner at a wavelength of 450 nm.

The rate (r) of reduction may be followed by monitoring loss of native complex (Cp) (indicated by absorbance at 635 nm) vs elapsed time (t), measured every 5 min after KBH_4 addition. In this particular case it is a first-order reaction (Fig. 2). A plot of $\ln[Cp]_0/[Cp]$ versus time is almost strictly linear, according to the following equation:

$$\ln[Cp]_0/[Cp] = 0.0952t - 0.2009 \quad (r = 0.995)$$

A rate constant value ($k = 9.5 \times 10^{-2} M$) for the reduction is obtained from this equation.

The accessibility of astaxanthin in the blue carotenoprotein of *P. clarkii* to KBH_4 may be compared with that of the most extensively studied carotenoprotein complex, α-crustacyanin, the blue carotenoprotein from the carapace of the lobster *H. americanus*[9] for which a first-order reduction rate constant of 5.8×10^{-2}/min has been determined, indicating that astaxanthin is slightly more accessible inside the *P. clarkii* carotenoprotein.

In a parallel experiment the carotenoids are extracted from the carotenoprotein, at each elapsed time following addition of KBH_4, by using acetone and then transferring into ether. No significant amount of carot-

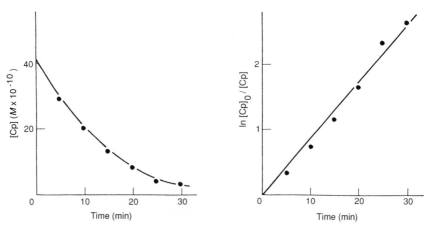

FIG. 2. KBH_4 reduction of the blue carotenoprotein from *P. clarkii* following a first-order reaction.

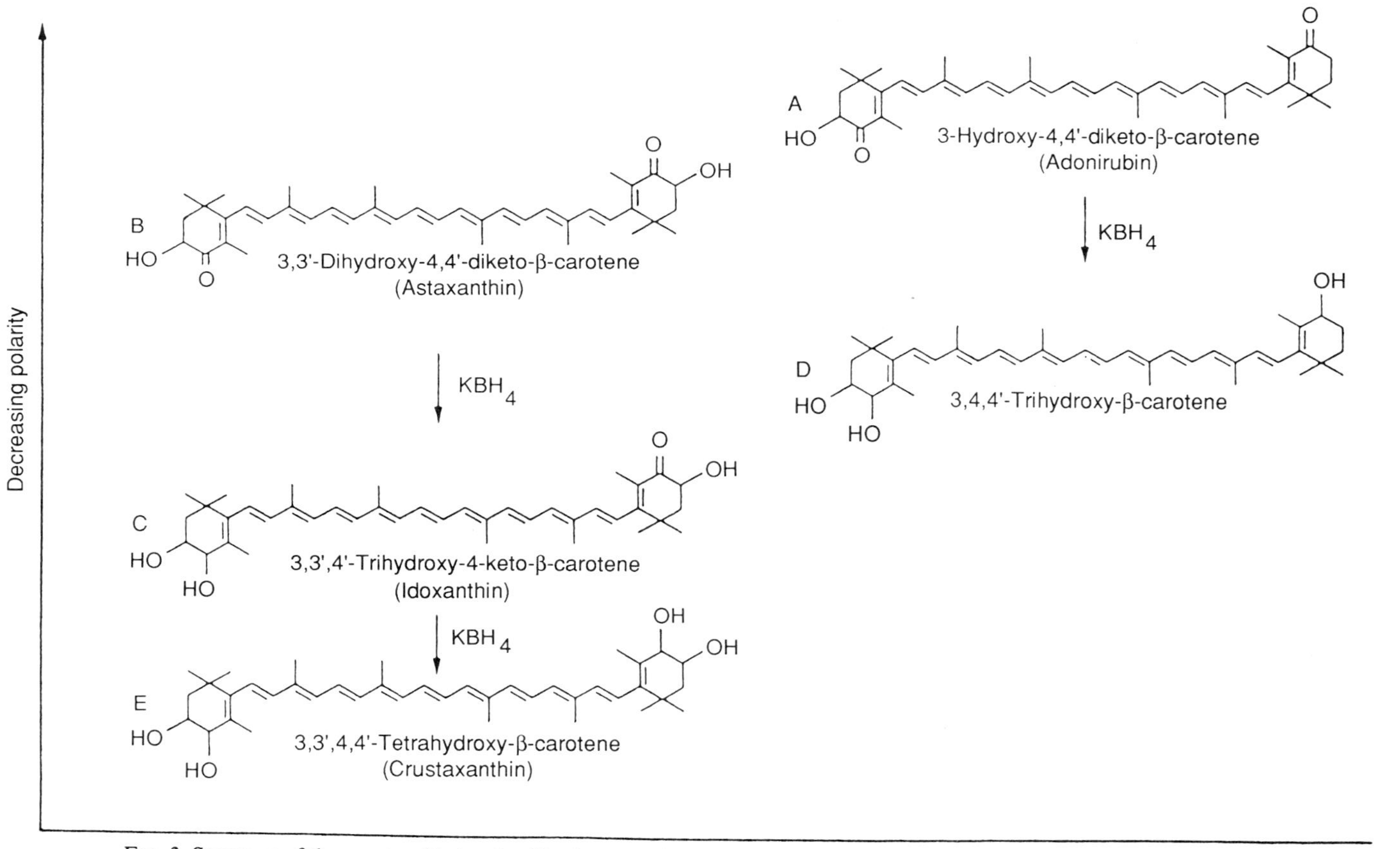

FIG. 3. Structure of the carotenoids involved in the reduction process of *P. clarkii* blue carotenoprotein reduction.

enoid migrates at the upper ethereal phase, when the reduced carotenoprotein is shaken solely with ether, without the prior addition of acetone, suggesting that the carotenoid remains bound to the protein after borohydride reduction. The reduction of the carotenoids astaxanthin and adonirubin, which comprise the prosthetic group of the blue carotenoprotein from *P. clarkii,* is shown in Fig. 3.

Carotenolipoproteins (Second Group)

The red and orange carotenoproteins belonging to the second group have only a weak interaction between the carotenoid and the protein, and this is reflected in the rapid borohydride reduction of the carotenoids (1 min) under the conditions described for the blue carotenoproteins. To extend the reduction of these lipoproteins to a reasonable time, use a small amount of borohydride: 1 mg/ml and a concentration of 1.2 mg/ml of carotenoprotein.

In order to calculate the rate of reduction it is necessary to calculate the percentages of each carotenoid in the reduced and the unreduced forms present at different times and separated by TLC. Extraction and TLC are performed as described above for the blue carotenoproteins.

Carotenoproteins Showing Hypsochromic Shift (Third Group)

The yellow carotenoproteins belonging to the third group are those that show a hypsochromic shift due to binding of the carotenoids to the carotenoprotein. Only one of these carotenoproteins, namely the yellow one of *H. americanus,* has been studied with respect to borohydride reduction.[9] In this case the reduction is made as follows: 0.1 M sodium borohydride is added to the yellow carotenoprotein in 0.01 M sodium borate buffer (pH 9.0). The reduction is complete in 10 min. This reaction cannot easily be followed kinetically because the shift in the adsorption maximum is small, from 410 to 392 nm, but the rapid rate of reduction indicates that the carbonyl groups are not so strongly protected in the yellow pigments as they are in the blue carotenoproteins.

Carotenoproteins with Mixed Interactions (Fourth Group)

Borohydride reduction of the carotenoproteins with mixed interactions has not yet been investigated to determine if differences can be detected between the two different interactions, namely that with the protein (protected pigments) and that with the lipids (unprotected pigments).

Acknowledgments

We are indebted to Dr. G. Britton, Department of Biochemistry, University of Liverpool, for helpful and invaluable advice.

[12] Gas-Phase Reaction Mass Spectrometric Analysis of Carotenoids

By WILLIAM R. LUSBY, FREDERICK KHACHIK, GARY R. BEECHER, and JAMES LAU

Introduction

Advances in gas-phase reaction mass spectrometry (MS) have provided a range of new tools to monitor progress in the isolation of carotenoids and to probe the structure of native and synthetically elaborated carotenoids.

Classical methods of mass spectrometric analysis that use gradual heating of the sample from a direct insertion probe coupled with conventional electron ionization[1] (EI) by means of 70-eV electrons have both produced an abundance of reference data[2,3] and contributed to the understanding[2,3] of mechanisms by which structurally significant ions are produced. Among the types of ions produced by this method of analysis are those resulting from in-chain losses of toluene, xylene, and dimethyldihydronaphthalene moieties, those diagnostic for end groups, and ions indicative of specific functional groups,[4] such as aldehydes, alcohols, and ethers. These fragment ions result from either a redistribution of excess kinetic energy deposited in the molecular ion during formation[5] or from the ionization of thermally generated neutral species.[2] While for many carotenoids these mechanisms produce ions providing sufficient information for confirmation of structure, frequently the formation of ions with less internal energy is desirable. Reduction of the energy of electrons in the ionizing beam from the usual 70 to 12 eV impedes high-energy processes, thereby decreasing[6] the abundance of less massive fragments and, sometimes, accentuating otherwise obscured ions. There is, however, a concomitant attenuation of the absolute abundance of all ions. An alternative procedure for increasing the relative abundance of ions at the upper end of the spectrum is to place the sample on a metallic surface, as opposed to glass or ceramic, and in close proximity to the beam of ionizing electrons.[2]

---

[1] Electron ionization was formerly referred to as electron impact.
[2] W. Vetter, G. Englert, N. Rigassi, and U. Schwieter, *in* "Carotenoids" (O. Isler, ed.), p. 243. Birkhäuser, Basel, 1971.
[3] C. R. Enzell and I. Wahlberg, *in* "Biochemical Applications of Mass Spectrometry" (G. R. Waller and O. C. Dermer, eds.), p. 407. Wiley, New York, 1980.
[4] C. R. Enzell, G. W. Francis, and S. Liaaen-Jensen, *Acta Chem. Scand.* **23,** 727 (1969).
[5] K. Bieman, "Mass Spectrometry." McGraw-Hill, New York, 1962.
[6] U. Schwieter, G. Englert, N. Rigassi, and W. Vetter, *Pure Appl. Chem.* **20,** 365 (1969).

Gas-phase reactions within the ionizing chamber of the mass spectrometer not only have the ability to produce initial ions with low energy content but also provide a variety of means to probe the structure of the analyte. Here, the expression "gas-phase reactions" relates to the production of ions by the processes of chemical ionization (CI) or electron capture.

Chemical ionization[7] is the production of analyte ions by gas-phase reaction of the sample with ions derived from a reagent gas that is continuously added to the ion source chamber. These ion–molecule reactions often produce analyte ions with considerably less energy content relative to EI, and as a consequence fragmentation is reduced. The nature of reagent gas ions produced for reaction with the sample is a function of the gas employed, gas pressure, and temperature of the ionizing chamber. A variety of reagent gas ions can be produced by varying these three parameters. Analyte ions in CI spectra may be rationalized by the application of three general rules: random attack by the reagent gas ions, localized reaction of the attacking ion, and elimination of neutral molecules. Chemical ionization spectra are characterized by ions reflecting the molecular mass and the loss of structural moieties.

Electron-capture negative-ion (ECNI) production[8,9] of analyte ions requires gas-phase reaction between sample molecules and electrons possessing thermal levels of energy. A copious supply of thermal energy-level electrons is produced when a buffer gas, such as methane, is continuously supplied to the ion source chamber. Molecular anions produced by ECNI[10] are less energetic than ions produced by either EI or CI and, therefore, are less likely to fragment.[11]

Conventionally, carotenoid samples are introduced to the ionizing chamber via a direct insertion probe. The carotenoid is placed, as a solution, onto a glass or ceramic surface, the solvent removed by evaporation, and the probe inserted through vacuum locks and positioned against an aperture of the source block. The glass or ceramic surface is then gradually heated to effect either a distillation or sublimation of the sample. The sample drifts, by molecular flow, across the chamber and into the path of ionizing electrons.[12] Under these conditions, thermally labile compounds suffer extensive degradation with concomitant reduction in the amount of

---

[7] M. B. S. Munson and F. H. Field, *J. Am. Chem. Soc.* **88**, 2621 (1966).
[8] D. F. Hunt and F. W. Crow, *Anal. Chem.* **50**, 1781 (1978).
[9] D. F. Hunt and S. K. Sethi, *ACS Symp. Ser.* **70**, 150 (1978).
[10] The ECNI process is not the same as negative-ion chemical ionization (NICI). For ECNI, the ionizing species are electrons, whereas for NICI, the ionizing species are negative ions.
[11] E. A. Stemmler and R. A. Hites, *Biomed. Environ. Mass Spectrom.* **17**, 311 (1988).
[12] J. R. Chapman, "Practical Organic Mass Spectrometry." Wiley, New York, 1985.

sample transferred, intact, into the gas phase. Among sample introduction methods[13] that minimize decomposition is the use of probes that penetrate into the interior of the ionizing chamber, thereby reducing the distance from probe tip to the source of ionization. Additionally, the sample is deposited directly onto a wire surface to which a rapidly increasing electrical current is applied. This type of probe is commonly referred to as a direct exposure probe (DEP) to differentiate it from a conventional probe. By use of a DEP, the time of sample distillation/sublimation is reduced from minutes[6] to seconds.[14,15]

Another alternative means of sample introduction to the mass spectrometer is by particle beam interface between the ionizing chamber and a high-performance liquid chromatograph (HPLC) equipped with a diode array detector. This mode of sample introduction provides not only a method for separation of sample mixtures but also characteristic chromatographic data and ultraviolet/visible (UV/VIS) spectra immediately prior to mass spectrometric analysis. The approach taken in this chapter is to discuss the application of these techniques to various chemical classes of carotenoids, and to illustrate each technique by presenting data for a representative compound of each class.

Methods

Sample Handling

To avoid artifacts and degradation, carotenoid samples should be analyzed as soon as possible after isolation. Samples that cannot be immediately examined should be stored dry in sealed, evacuated, ampules at $-80°$. The use of lighting containing a minimum of UV radiation, such as incandescent or special fluorescent,[16] for all manipulations of samples reduces photoinduced degradation. Exposure of samples to air, heat, acids, and bases should be avoided as much as possible. Factors that determine the amount of carotenoid required for analysis are the method of ion production, rate of heating of the probe, range of mass scanned, and geometry of the sample probe tip.

---

[13] R. J. Cotter, *Anal. Chem.* **52**, 1589A (1980).
[14] W. R. Lusby, G. R. Beecher, and F. Khachik, *Abstr. Int. Symp. Carotenoids, 8th* p. 13 (1989).
[15] W. R. Lusby, F. Khachik, and G. R. Beecher, *Proc. Am. Soc. Mass Spectrom. Conf. Mass Spectrom. Allied Top., 37th* p. 772 (1989).
[16] A suitable fluorescent lamp is the F40GO model, manufactured by General Electric.

Source of Samples

Canthaxanthin (98%; Fluka, Ronkonkoma, NY), β-apo-8'-carotenal (>99%; Fluka), and β-carotene (type IV, Sigma Chemical Co., St. Louis, MO) were used without additional purification. Lutein, violaxanthin, and lutein epoxide were isolated from kale (*Brassica oleracea*, var. 'Acephala') according to published procedures.[17] Isozeaxanthin, isozeaxanthin bispelargonate, and β-apo-8'-carotenol were prepared by partial synthesis according to published procedures.[18] Lutein bisdecanoate was prepared from lutein and decanoyl chloride according to general procedures.[19] β-Carotene 5,6-epoxide and β-carotene 5,6,5',6'-diepoxide were donated by Hoffmann-La Roche (Nutley, NJ). Samples of these carotenoid epoxides were also prepared by epoxidation of β-carotene with m-chloroperbenzoic acid followed by chromatographic isolation and purification. Bixin powder was generously contributed by D. M. Scheffel (Miles, Inc., Madison, WI). Bixin was converted to bixindiol by reduction with lithium aluminum hydride and purified by thin-layer chromatography on silica gel.

Instrumentation

Mass spectra were obtained from a Finnigan-MAT (San Jose, CA) model 4500 spectrometer equipped with a direct exposure probe (DEP) and 5-kV conversion dynodes for the detection of positive and negative ions. Under yellow fluorescent light [General Electric (Schenectady, NY) F40GO], benzene solutions of samples were deposited onto the DEP probe, and solvent removed within seconds by a gentle stream of room-temperature air. Insertion of sample into the instrument was followed by heating the probe with a current rate of 50 mA/sec (except where noted) from 0 to 1000 mA. All carotenoids yielded spectra between 250 and 350 mA of applied heating current. Electron ionization spectra were obtained at indicated source temperatures of 60 and 150°, and at 70, 40, 30, 20, and 14 eV. Ammonia (electronic grade; Matheson Gas Products, Elk Grove Village, IL) (0.6 torr), deuteroammonia (Cambridge Isotopes Laboratory, Woburn, MA) (0.6 torr), isobutane (99.5%; Matheson Gas Products) (0.4 torr), and methane (99.97%; Matheson Gas Products) (0.3 to 0.35 torr) positive-ion CI spectra were obtained at 60°. ECNI mass spectra were collected at a source temperature of 60° and the following indicated pressures (measured at calibration gas inlet): methane and argon (99.99%; Matheson Gas Products), 0.3 torr; ammonia and deuteroammonia, 0.25

---

[17] F. Khachik, G. R. Beecher, and N. F. Whittaker, *J. Agric. Food Chem.* **34**, 603 (1986).
[18] F. Khachik and G. R. Beecher, *J. Agric. Food Chem.* **36**, 929 (1988).
[19] F. Khachik and G. R. Beecher, *J. Chromatogr.* **449**, 119 (1988).

torr. Mass spectra were collected beginning at 50 Da for EI-MS and ECNI-MS, 100 Da for positive-ion methane CI-MS, and 130 Da for positive-ion ammonia and deuteroammonia CI-MS. The upper limit of the scan range was limited to the molecular mass of the analyte plus 50 Da. Except where noted, the mass range was scanned in 1-sec cycles.

On-line coupled LC-UV-MS analyses were performed on a combination of Hewlett-Packard (Palo Alto, CA) instruments. A model 1090L liquid chromatograph equipped with a diode array detector was connected by means of a particle beam interface to a model 5989A mass spectrometer. Details of LC conditions are given in Chapter [18] of this volume. Following exit from the LC diode array detector, the eluate was split 2.9:1, with the lesser amount being delivered to the particle beam interface operated at a desolvation temperature of 45°. ECNI spectra were produced by methane, 0.85 torr measured at the gas chromatography (GC) transfer line, at a source temperature of 320°. Spectra were collected from 150 to 700 Da, with a scan cycle time of 1.5 sec.

Techniques

Hydrocarbons

β-Carotene

Electron Ionization. The EI spectrum of β-carotene[6] (Fig. 1) obtained by conventional probe is characterized by a molecular ion ($M^{\ddot{+}}$) with a relative intensity of 40%, with the base peak occurring at *m/z* 69. In contrast, analysis of β-carotene with direct exposure probe (DEP) introduction and a total sample heating time of less than 10 sec yielded a spectrum (Fig. 2a) with the molecular ion as the base peak and the ion at *m/z* 69 reduced to a relative abundance of 50%. In either case, the fragment ion at *m/z* 444, resulting from the loss of toluene from $M^{\ddot{+}}$, remains at approximately 20%. A 10- to 20-fold increase in absolute sensitivity is realized by use of the DEP mode of sample introduction.

A classical technique to increase the relative abundance of $M^{\ddot{+}}$ is to reduce the energy level of the incident ionizing electrons from the normal value of 70 eV. For example, analysis of β-carotene at 12 eV[6] using a conventional probe yielded a spectrum almost devoid of low mass-to-charge ions, with most of the ion beam current being carried by $M^{\ddot{+}}$ and the fragment ion at *m/z* 444. A series of analyses that employed both a reduction of electron energy level together with introduction by a DEP revealed several features. First, absolute signal intensity, and therefore sensitivity, passed through a maximum at 40 eV (Fig. 2b) with the intensity at 40 eV about twice that obtained at 70 eV. This effect was observed

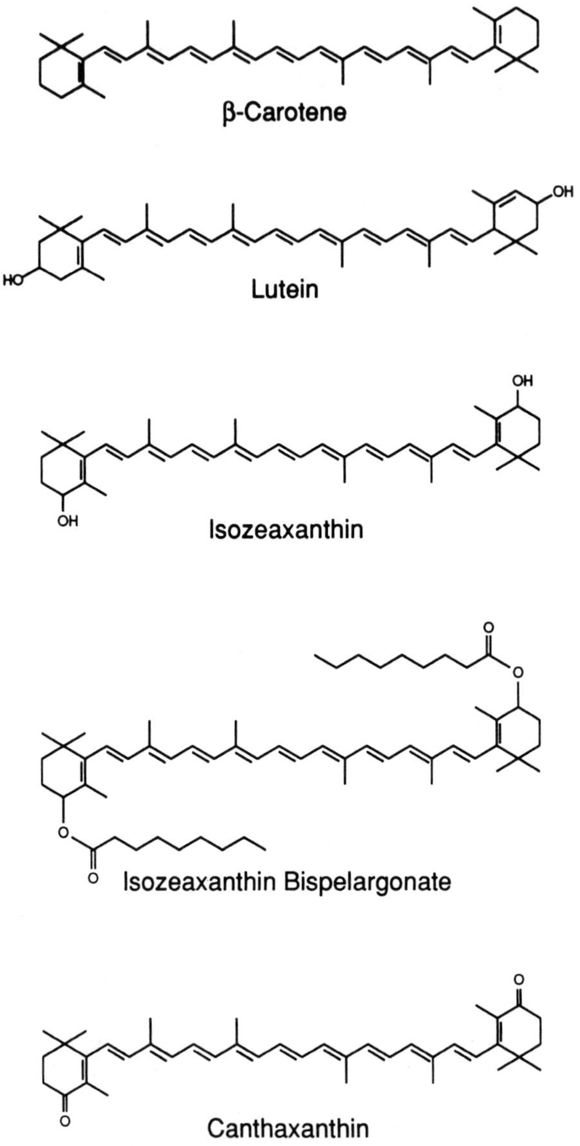

FIG. 1. Structures of selected carotenoids.

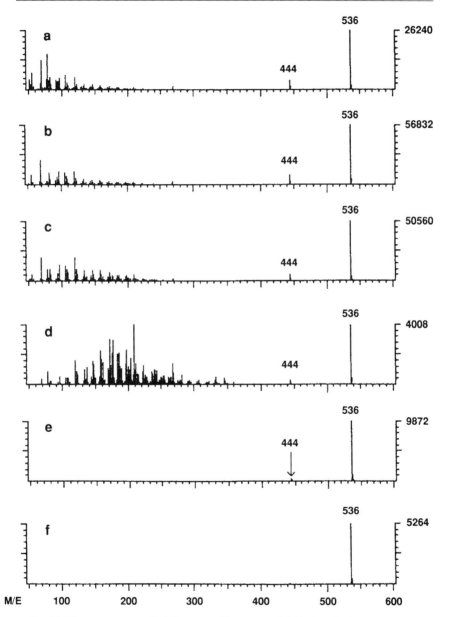

FIG. 2. Direct exposure probe EI spectra of β-carotene: (a) 70 eV, (b) 40 eV, (c) 30 eV, (d) 20 eV, (e) 18 eV, (f) 14 eV.

for most carotenoids irrespective of structural class. Qualitatively the spectra obtained at 70, 40, and 30 eV (Fig. 2a–c) are nearly identical. However, at energy levels of 20 eV (Fig. 2d) and lower, a reduction in the abundance of lower m/z ions was observed together with a gradual progressive appearance of high-mass fragments that culminated at 14 eV (Fig. 2f) with a spectrum composed of two ions, the base peak at $M^{\ddot{+}}$ and an extremely weak intensity at m/z 444 for the $(M - \text{toluene})^{\ddot{+}}$ ion. The above series of spectra were obtained at an ionizing source temperature of 150°. A parallel series of experiments at a reduced source temperature of 60° caused a reduction in both relative abundances of lower mass-to-charge fragments as well as an attenuation of absolute signal intensity for all ions.

Chemical Ionization. Methane CI analysis of β-carotene at an indicated reagent gas pressure of 0.3 torr and a source temperature of 60° yielded a spectrum containing the protonated molecular species $(M + H)^+$ at m/z 537 as the base peak, with much less abundant ions at m/z 565, the ethyl adduct $(M + C_2H_5)^+$, and at m/z 521 an ion that corresponds to protonation followed by the loss of a neutral molecule of methane $(M + H - CH_4)^+$. The sensitivity of this method, as measured by the ratio of $(M + H)^+$ for methane CI to $M^{\ddot{+}}$ for electron ionization, was 15 times greater (Table I). The absolute number of $(M + H)^+$ ions detected was directly

TABLE I
SENSITIVITY[a] FOR DETERMINATION OF MOLECULAR WEIGHT

Compound	Molecular weight	EI[b]	CI		
			NH_3	CH_4[c]	ECNI[d]
β-Carotene	536	1	3.2[e]	15.0	290
α-Carotene	536	1	5.1[e]	18.0	280
Lutein	568	1	13.0[f]	30.0	254
Isozeaxanthin	568	1	38.0[f]	27.0	110
Isozeaxanthin bispelargonate	848	—[g]	1.0[f]	1.0	660[h]
Canthaxanthin	564	1	3.0[f]	30.0	490
β-Apo-8′-carotenal	416	1	1.3[e]	1.1	25
Lutein epoxide	584	1	24.0[e]	67.0	210
Violaxanthin	600	1	35.0[e]	10.0	540

[a] Relative to electron ionization.
[b] $M^{\ddot{+}}$ observed; 70 eV at source temperature of 150°.
[c] $(M + H)^+$ observed; methane, 0.3 torr at source temperature of 60°.
[d] M^- observed; methane, 0.25 torr at source temperature of 60°.
[e] $(M + H)^+$ observed; ammonia, 0.6 torr at source temperature of 60°.
[f] $(M + NH_4)^+$ observed; ammonia, 0.6 torr at source temperature of 60°.
[g] —, No $M^{\ddot{+}}$ ions were observed.
[h] Sensitivity calculated relative to methane-positive ion.

related to the rate at which current was applied to the DEP. A series of experiments analyzing 20 ng of β-carotene at heating rates that ranged from 1 to 200 mA/sec is presented in Table II. While the greatest total yield of $(M + H)^+$ ions occurred at a probe-heating rate of 200 mA/sec, the period of ion production (about 2 sec) was too brief for optimum data collection with a 1-sec scan cycle time.

As previously observed,[6] prolonged thermal exposure of carotenoids in the solid state should be avoided. For example, methane CI analysis of β-carotene at a source temperature of 60° yielded 12-fold the number of $(M + H)^+$ ions produced at a source temperature of 190°. However, the loss in signal at 190° was only threefold if the residence time in the ion chamber was reduced by both rapid insertion and immediate application of heating current. This effect was also noted for other classes of carotenoid compounds.

Compared to EI at 70 eV, ammonia CI of β-carotene provided an approximate threefold increase (Table I) in abundance of ions diagnostic for the molecular weight. The actual analyte ions observed were a function of the reagent gas ions and the source temperature. For example, at 60° and an indicated source pressure of 0.6 torr, the reagent gas ions produced were NH_4^+, $(NH_3)_2^+$, $(NH_3)_3^+$, and $(NH_3)_4^+$ at m/z 18, 35, 52, and 69 and with a ratio of 45:35:100:70. At 190° and 0.6 torr, the reagent gas ions observed were NH_4^+, $(NH_3)_2^+$, and $(NH_3)_3^+$ with a ratio of 100:65:14. The ability to modify the composition of the reagent gas ions that participate in these gas-phase reactions provided a useful tool with which to probe carotenoids. For β-carotene, at 190°, the bulk of the beam current was carried by $(M + H)^+$ at m/z 537, but at 60° both the ammonium adduct ion

TABLE II
RATE OF HEATING VS PRODUCTION OF $(M + H)^+$ IONS FOR β-CAROTENE

Heating rate (mA/sec)	Total number of $(M + H)^+$ detected[a]
1	122
5	382
10	991
50	4,260
100	14,549
200	15,214

[a] β-Carotene, 20 ng; $(M + H)^+$ at 537, methane at 0.35 torr.

$(M + NH_4)^+$ at m/z 554 (18%) and the ammonium–ammonia adduct ion $[M + (NH_3)_2H]^+$ at m/z 571 (31%) are observed in addition to the base peak at 537 $(M + H)^+$.

Isobutane at either 60 or 190° provided predominantly the *tert*-butyl ion for participation in ion–molecule CI reactions. Major analyte ions produced were $(M + H)^+$ at m/z 537 and $(M + C_4H_9)^+$ at m/z 593, an observation that is in agreement with other studies.[20] The major disadvantages of isobutane were the presence of higher molecular weight hydrocarbons, which limited the low-mass range of the scan cycle, and the propensity of isobutane to contaminate the source and reduce sensitivity.

Electron Capture Negative Ionization. Methane at 0.25 torr provided abundant thermal energy-level electrons for capture by carotenoids. For β-carotene, ECNI coupled with sample entry via a DEP was approximately 300 times as sensitive (Table I) as 70-eV electron ionization for the detection of ions from which the molecular weight can be determined. The spectrum consisted of the molecular anion (M^-) at m/z 536 as the base peak and associated isotopic ions at m/z 537 and 538 due to contributions from the 1.1% natural abundance of ^{13}C. Perhaps of greater importance than the very high sensitivity is the ability of ECNI to detect carotenoids in the presence of substantial amounts of non-electron-capturing contaminants. This provides a powerful technique for monitoring progress during isolation procedures. Argon functioned equally well as a buffer gas for generation of thermal energy-level electrons and produced a similarly clean spectrum. Curiously, there was the consistent presence of a weak ion (0.1%) at m/z 444. A likely explanation would be the capture of an electron by the thermally generated neutral species resulting from the loss of toluene from β-carotene. Care must be exercised to exclude oxygen and water from the buffer gas and the ionization chamber. Presence of these contaminants will cause appreciable levels of OH^- and other negative-ion reagent gas reactants that participate in negative-ion CI rather than in the desired electron capture process. The presence of unwanted negative reagent gas ions can be determined by scanning, at reduced electron multiplier voltage, a suitable low-mass range in the negative-ion detection mode.

Hydroxycarotenoids

Lutein

Electron Ionization. Lutein (Fig. 1), when introduced by a conventional slow heating probe and subjected to 70-eV electron ionization,

[20] J. Carnevale, E. R. Cole, D. Nelson, and J. S. Shannon, *Biomed. Mass Spectrom.* **5**, 641 (1978).

yielded a spectrum[21] the upper end of which was characterized by ions at m/z 550 (31%, M $-$ H$_2$O), 476 (7%, M $-$ toluene), 462 (5%, M $-$ xylene), 458 (11%, M $-$ H$_2$O $-$ toluene), 444 (8%, M $-$ H$_2$O $-$ xylene), and 392 (29%, M $-$ H$_2$O $-$ C$_{12}$H$_{14}$), where C$_{12}$H$_{14}$ is a dimethyldihydronaphthalene moiety.[3] In contrast, when lutein was admitted to the source via a DEP and ionized by 70-eV electron ionization (Fig. 3a) the base peak was the molecular ion M^{+} at m/z 568, and the ion at m/z 550, reflecting the loss of water, was extremely weak (1%) as were other ions that are formed by dehydration. Other ions present in the DEP-EI spectrum above m/z 400 were as follows: m/z 476 (9%, M $-$ toluene), m/z 430 (9%, M $-$ 139), and m/z 338 (10%, M $-$ toluene $-$ 138) where 139 corresponds to either the ε or β end group of lutein. These data suggest that sample entry via a direct exposure probe under conditions of electron ionization promoted in-chain losses and suppressed the elimination of water. Reduction of the ionizing voltage to 20 eV yielded a spectrum (Fig. 3b) with decreased relative abundances for low-mass fragments and increased abundances of ions resulting from elimination of in-chain units and end groups. The above DEP-introduced lutein spectra were obtained at a source temperature of 60°. Elevation of the source temperature to 150° reduced the relative abundance of higher m/z ions and did not provide additional diagnostic information.

Chemical Ionization. Analysis by methane CI (Fig. 3c) at a source temperature of 60° provided a relative abundance of 97% for the protonated molecular species (M + H)$^{+}$, with the base peak at m/z 551 reflecting a loss of water (M + H $-$ H$_2$O)$^{+}$, and ions at m/z 597 (8%) and 609 (2%), respectively, for the adduct ions (M + C$_2$H$_5$)$^{+}$ and (M + C$_3$H$_5$)$^{+}$. Relative to DEP-EI, methane CI was 30 times more sensitive (Table I). The spectrum obtained by ammonia CI (Fig. 3d) was characterized by a base peak at m/z 586 corresponding to an ammonium adduct ion, (M + NH$_4$)$^{+}$, which provided a 13-fold increase in sensitivity for the determination of molecular weight. The ammonium–ammonia adduct ion [M + (NH$_3$)$_2$H]$^{+}$ at m/z 603 (4%) provided additional confirmation of the molecular weight. The use of deuteroammonia, N^{2}H$_3$, as a reagent gas permitted the facile determination[22] of the number of exchangeable hydrogens on minute amounts of sample. For lutein, the base peak of the N^{2}H$_3$-CI spectrum (Fig. 3e) occurred at m/z 592 and corresponds to M + N^{2}H$_4$ + H$_x$ where H$_x$ equals the number of exchangeable hydrogens. Subtracting from 592 the molecular mass of 568, typically determined by NH$_3$ $-$ CI,

---

[21] S. R. Heller and G. W. A. Milne, "EPA/NIH Mass Spectral Data Base." U.S. Government Printing Office, Washington, D.C., 1978.

[22] D. F. Hunt, C. N. McEwen, and R. A. Upham, *Tetrahedron Lett.* p. 4539 (1971).

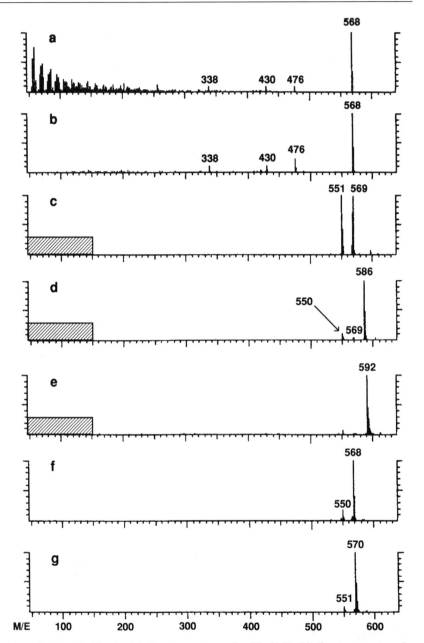

FIG. 3. Spectra of lutein: (a) direct exposure probe, 70-eV EI; (b) direct exposure probe, 20-eV EI; (c) methane CI; (d) ammonia CI; (e) deuteroammonia CI; (f) methane ECNI; (g) deuteroammonia ECNI. Cross-hatched regions were not scanned because of presence of reagent gas ions.

and 22, for the adducting species N^2H_4, leaves 2, which is the number of exchangeable hydrogens. In some cases, adduction by the ammonium ion is followed by the numerically equivalent loss of a molecule of water. These ambiguities can be resolved by employing $^{15}NH_3$ as a CI reagent gas to determine the presence or absence of a CI adducted species.

Electron Capture Negative Ionization. The electron-capture process using methane (Fig. 3f) yielded an increase in sensitivity (Table I), compared to DEP-EI, of greater than 250 with the base peak corresponding to the molecular anion M^- at m/z 568 and with lesser peaks at m/z 550 (19%) for the $(M - H_2O)^-$ ion, and m/z 532 (1%) for $(M - 2H_2O)^-$. The ECNI spectrum of the isomeric compound isozeaxanthin (Fig. 1) contained the same major ions; however, the order of abundance was reversed, with the base peak at m/z 532 corresponding to a double elimination of water, ions at m/z 550 (77%) for a single elimination of water, and the molecular anion M^- at m/z 568 (32%). Differences in relative abundances between lutein and isozeaxanthin can be rationalized by recognizing that the loss of water from the two allylic hydroxyl groups in isozeaxanthin is driven by extension of the conjugated system, whereas in lutein loss of water from the ε end group does not increase the extent of conjugation. Approximately the same increase in sensitivity, compared to methane, was realized when argon was used for the generation of thermal-level electrons for capture. Employment of deuteroammonia as an electron capture buffer gas permitted simultaneous electron capture and deuterium exchange (Fig. 3g). The result provided a determination of the number of exchangeable hydrogens at very high sensitivity and without background ions from non-electron-capturing contaminants.

Carotenol Fatty Acyl Bisesters

Isozeaxanthin Bispelargonate

Electron Ionization. Fatty acyl bisesters of hydroxycarotenoids, present in fruits[23] and vegetables,[18,24] are particularly challenging compounds for mass spectrometric analysis. For example, DEP-EI analysis of a synthetic sample of isozeaxanthin bispelargonate (Fig. 1) at a heating current rate of 50 mA/sec and a source temperature of 150° yielded a spectrum (Fig. 4a) dominated by nondiagnostic ions below m/z 250. Very weak abundances are observed at m/z 690 for $(M - C_9H_{18}O_2, 0.2\%)$, where $C_9H_{18}O_2$ corresponds to pelargonic acid, and at m/z 532 for $(M - 2C_9H_{18}O_2, 0.3\%)$. Reducing the ionizing voltage to 20 eV and the source temperature to 60°

---

[23] F. Khachik, G. R. Beecher, and W. R. Lusby, *J. Agric. Food Chem.* **37**, 1465 (1989).
[24] F. Khachik, G. R. Beecher, and W. R. Lusby, *J. Agric. Food Chem.* **36**, 938 (1988).

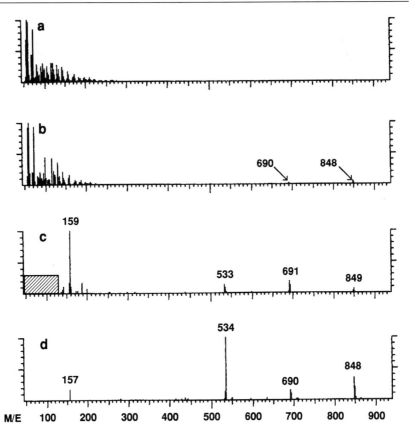

FIG. 4. Spectra of isozeaxanthin bispelargonate: (a) direct exposure probe, 70-eV EI and source temperature of 150°; (b) direct exposure probe, 20-eV EI and source temperature of 60°; (c) methane CI; (d) methane ECNI. Cross-hatched region was not scanned because of presence of reagent gas ions.

provided a spectrum (Fig. 4b) with M$^+$ at m/z 848 (5%) and (M − C$_9$H$_{18}$O$_2$)$^+$ (0.2%).

Chemical Ionization. Methane CI provided (Fig. 4c) a limited number of diagnostically significant ions. The protonated molecular species (M + H)$^+$ occurred at m/z 849 (8%) and was accompanied by ions at m/z 691 (19%), reflecting the loss of a molecule of pelargonic acid and protonation, m/z 533 (12%), corresponding to the loss of two molecules of pelargonic acid and protonation, and m/z 159 (100%), corresponding to protonated pelargonic acid.

Electron Capture Negative Ionization. The use of methane for electron-capture ionization produced an equally simple spectrum (Fig. 4d) with the molecular anion M^- at m/z 848 (37%), $(M - C_9H_{18}O_2)^-$ at m/z 690 (15%), $(M - 2C_9H_{18}O_2 + 2H)^-$ at m/z 534 (100%), and the pelargonate anion $(C_9H_{17}O_2)^-$ at m/z 157 (18%). The production of higher mass ions was directly related to the rate of heating of the direct exposure probe (Table III). While m/z 534 remained the base peak, a progressive increase in the relative abundance of the molecular anion M^- at m/z 848 was observed. Again, the advantage of operating in the electron-capture mode of ionization is greater sensitivity, a 600-fold increase when compared to either ammonia or methane CI (Table I), and the invisibility of non-electron-capturing contaminants.

Ketocarotenoids

Canthaxanthin

Electron Ionization. Analysis of canthaxanthin (Fig. 1) by DEP-EI at 70-eV ionizing voltage produced a spectrum (Fig. 5a) with a molecular ion $M^{\ddot+}$ intensity of 32% at m/z 564. Reduction of the ionizing voltage to 30 eV increased the abundance (Fig. 5b) of the molecular ion to 100%, with the next most abundant ion at m/z 133 (40%). Further reduction of the ionizing voltage to 20 eV (Fig. 5c) provided a spectrum dominated by $M^{\ddot+}$.

Chemical Ionization. Employment of methane CI yielded an increase in sensitivity (Table I) of 30-fold when comparing $(M + H)^+$ from methane CI to $M^{\ddot+}$ from DEP-EI at 70 eV. In addition, the methane CI spectrum (Fig. 5d) contained a confirmatory, albeit weak, ion $(M + C_2H_5)^+$ at m/z 593 (6%). In contrast, ammonia CI provided only three times the sensitiv-

TABLE III
RELATIVE ABUNDANCE OF IONS VS RATE OF PROBE HEATING CURRENT FOR ISOZEAXANTHIN DIPELARGONATE

Rate (mA/sec)	m/z^a		
	534	690	848
10	100	12	2
20	100	41	13
50	100	43	35
100	100	15	58

^{a} ECNI, methane at 0.3 torr and source temperature of 60°.

ity and generated a spectrum characterized by a base peak, $(M + NH_4)^+$, at m/z 582, and ions at m/z 565, 63% $(M + H)^+$.

Electron Capture Negative Ionization. The use of either methane (Fig. 5e) or argon (Fig. 5f) for the formation of the molecular anion M^- by ECNI resulted in an approximate 500-fold increase in sensitivity for the determination of molecular weight. The spectra were uni-ionic and non-electron-capturing contaminants were not observed.

Other Classes of Compounds

We have observed results similar to those detailed above for carotenoids with various additional functional groups. These are carotenoid epoxides (namely, β-carotene 5,6-epoxide, β-carotene 5,6,5',6'-diepoxide, lutein epoxide, and violaxanthin), an apocarotenal (β-apo-8'-carotenal), an ether (3'-methoxy-β,ε-carotene-3-ol), and a carotenol fatty acyl monoester (lutein monopalmitate).

Sample Introduction of Carotenoids by HPLC/Photodiode Array Detector/Particle Beam Mass Spectrometer Interface

Introduction of samples via a particle beam interface between an HPLC system equipped with a photodiode array detector and mass spectrometer provides several advantages. First, there is a reduction in both the amount of handling during the transfer of the sample from vial to probe tip and concomitant degradation of the sample from exposure to air. Another advantage is that the continuous chromatographic introduction permits deconvolution of closely eluting components and exclusion of spurious ions from spectra.

The ECNI spectrum for β-carotene obtained by this mode of introduction was identical to that observed by direct exposure probe introduction. The use of high temperature in the source region during particle beam introduction can lead to slightly different spectra for those compounds where the loss of easily generated neutrals is possible. For example, in the case of the negative-ion electron-capture spectrum of lutein, introduction by a particle beam interface yields the same three ions observed with introduction by direct exposure probe; however, the abundance of ions reflecting elimination of water is increased. Comparing direct exposure probe to particle beam, the abundance for m/z 568 M^- decreases from 100 to 44% while m/z 550 $(M - H_2O)^-$ increases from 19 to 100%, and m/z 532 $(M - 2H_2O)^-$ increases from 1 to 29%. It is reasonable to assume these shifts in relative intensities reflect the higher source temperature employed during particle beam introduction (320°) compared to direct exposure

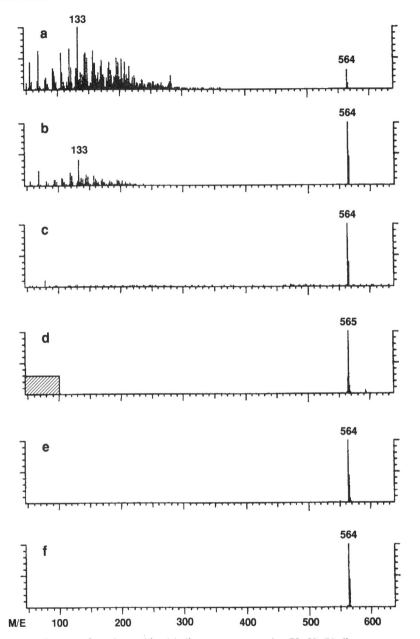

FIG. 5. Spectra of canthaxanthin: (a) direct exposure probe, 70 eV; (b) direct exposure probe, 30 eV; (c) direct exposure probe, 20 eV; (d) methane CI; (e) methane ECNI; (f) argon ECNI. Cross-hatched region was not scanned because of the presence of reagent gas ions.

probe introduction (60°). Finally, in-line consecutive determination of UV–VIS spectra and mass spectra coupled with chromatographic retention data provides a powerful combination of complementary information for structural elucidation.

Conclusions

The coupling of sample introduction via a direct exposure probe together with gas-phase reaction chemistry within the mass spectrometer ionizing chamber provides several powerful techniques for examining the structure of carotenoids and determination of their molecular weights. During isolation procedures, electron-capture negative-ion mass spectrometry provides a useful method for detection of carotenoid compounds in the presence of non-electron-capturing contaminants. Increased sensitivity for molecular weight determinations is realized with the application of chemical ionization techniques, and electron-capture negative ion spectrometry provides even greater sensitivity. ECNI sensitivity is usually in excess of 100-fold that of conventional electron ionization using a direct exposure probe. The use of deutero-containing reagent and buffer gases allows the determination of the number of exchangeable hydrogens by either chemical ionization or the more sensitive electron-capture negative-ion process. Ions indicative of the acyl groups of fatty acyl esters of carotenols are readily determined by chemical ionization methods and with even greater sensitivity by electron-capture negative-ion techniques. These gas-phase reaction techniques do not replace conventional EI techniques but rather provide complementary and confirmatory evidence of structure. The coupling of an HPLC equipped with a diode array detector to a mass spectrometer provides chromatographic retention data and UV/VIS spectra as well as mass spectra, and, in many cases, is sufficient for on-line identification of unknown carotenoids.

Acknowledgments

We are grateful to D. J. Harrison, K. R. Wilzer, and M. B. Goli for skillful technical assistance.

[13] Formation of Nonvolatile Compounds by Thermal Degradation of β-Carotene: Protection by Antioxidants

By CLAUDETTE BERSET and CLAIRE MARTY

Introduction

A large number of publications dealing with the structural modifications of carotenoid pigments brought about by heating at high temperatures have been published since 1960. Most research has dealt with the identification of the compounds in the volatile fraction, judiciously using the new possibilities offered by gas chromatography (GC) coupled with mass spectrometry (MS). The nonvolatile fraction, on the other hand, has not been extensively studied. This heavy fraction merits study, however, because it includes derivatives that may retain coloring properties and sometimes vitamin or antioxidant activities. They are often unstable reaction intermediates, however, present in small quantities, whose purification and identification present numerous problems.

The most thorough structural data concern the behavior of all-*trans*-β-carotene in extrusion cooking. A systematic study of the degradation products that absorb in the visible region has been undertaken and the results compared to those obtained after prolonged heating.

Study of Degradation of all-*trans*-β-Carotene in Extrusion Cooking

Extrusion cooking is a process for shaping starting materials that over a short period of time, combines the action of high temperatures (150 to 220°), high pressures, and intense shearing. At the extruder output, a paste (or starch) is forced through dies, the water evaporates, and the product acquires an expanded, brittle, porous texture. β-carotene incorporated into the starch before extrusion cooking is then extracted from the extrudates for analysis.

Purification of β-Carotene

Pure *trans*-β-carotene is an unstable molecule. This is why the commercial product must be purified just before use in order to eliminate the oxidation and isomerization products it contains.

The method we have developed[1] involves two successive chromatographies through an open column. Synthetic β-carotene (100 mg, purum-

---

[1] C. Marty and C. Berset, *Riv. Ital. Sostanze Grasse* **53**(3), 163 (1986).

grade) (Fluka, Ronkonkoma, NY) mixed with 400 mg of alumina II–III (0.063–0.200 nm; Merck, Rahway, NJ) is placed on top of a column of alumina II–III, 20 cm high and 3 cm i.d. The deposit is covered with a 1-cm-thick layer of sand. Elution is with hexane–diethyl ether (90:10, v/v). All operations are carried out at 4° in the dark. The eluent recovered is then rechromatographed on a column of alumina I of the same size. The cis isomers of β-carotene are eluted with hexane–diethyl ether (80:20, v/v). *trans*-β-carotene is eluted with hexane–diethyl ether (70:30, v/v). After eliminating the solvent under a stream of nitrogen, *trans*-β-carotene is taken up in dichloromethane (Carlo Erba RPE) and allowed to crystallize at room temperature.

Conditions of Extrusion Cooking

trans-β-carotene (800 mg) is dissolved in 400 ml of chloroform and mixed with 500 g of corn starch. The mixture is stirred for 1 hr, dried under vacuum, and then under a stream of nitrogen. It is incorporated into 10 kg of corn starch in a mixer.

The extruder cooker used is a Clextral BC 45 with two copenetrating and corotating screws. Assembly is as shown in Table I.

The temperature of the barrel in the compression region near the dies is held constant at 170 to 185°, depending on the test. Screw rotation speed is set at 150 rpm and material flow rate is 25 kg/hr. The water content of the paste is adjusted to 20–22% by the addition of water at screw entry, at a flow rate of 2.4 liters/hr.

After stabilizing temperature, current, and pressure, the machine is run for about 15 min before samples are collected in polyethylene bags, which are closed by heat welding and stored at −30°. A rotation knife attached in front of the dies cuts spherical extrudates about 1 cm in diameter, whose mean water content after cooling is 8%.

TABLE I
ASSEMBLY OF SCREWS IN CLEXTRAL EXTRUDER COOKER

Screw length (mm)	Pitch length (mm)
200	50
100	35
50	25
100	15
50	−15[a]
100	15

[a] Reverse flight.

Pigment Extraction

The following are added to 50 g of extrudates ground in a mill: 1 liter of distilled water, 100 ml of a solution of 2 g of hydroquinone per liter of ethanol, 250 ml of 0.05 M phosphate buffer, pH 7.0, and 50 ml of an aqueous solution of α-amylase at 10 g/liter.

Enzymatic digestion of the starch is carried out by mixing for 2 hr at room temperature in the dark. Pigments are then extracted with two 1-liter aliquots of hexane–acetone (60:40, v/v). The organic phase is washed twice with 500 ml of distilled water, dried over anhydrous sodium sulfate, and concentrated under vacuum at 30°. The residue is stored under nitrogen at −30° until analysis.

The fractionation diagram of *trans*-β-carotene derivatives obtained during extrusion cooking is shown in Fig. 1.

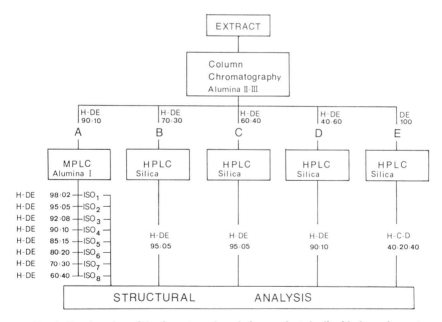

FIG. 1. Fractionation of the β-carotene degradation products by liquid-phase chromatography. MPLC, Medium-pressure liquid chromatography; HPLC, high-performance liquid chromatography; H, *n*-hexane; DE, diethyl ether; C, chloroform; D, dichloromethane. [From Marty and Berset.[4] Reprinted with permission of *Journal of Food Science,* 1988, **53**(6): 1880–1886. Copyright © 1988 Institute of Food Technologists.]

Analysis of cis Isomers of β-Carotene

The main chain of β-carotene contains nine conjugated double bonds that can give rise to cis–trans isomerism. Although this can theoretically yield 272 potential isomers, in reality steric constraints limit this number considerably.[2]

Vecchi et al.[3] identified eight cis isomers of β-carotene produced by photoisomerization. We used their spectral characteristics (Table II) and elution order from alumina as references.

After extrusion cooking, seven geometric isomers of *trans*-β-carotene in fraction A are separated by medium-pressure chromatography (Fig. 1). The system is composed of a 1-cm i.d., 30 cm long thermostatted column, a 302 pump, and a Gilson 8020 manometric module. The stationary phase is composed of alumina I (0.063–0.200 mesh; Merck). Pressure is 1.5 ± 0.5 atm and the temperature is $\pm 2°$. The solvent is *n*-hexane containing increasing proportions of diethyl ether, as indicated in Fig. 1, and the flow rate is 5 ml/min. Chemical desorption/ionization mass spectrometry [CI(NH$_3$)] gives a base peak at *m/z* 537 for all the compounds studied, corresponding to the protonated molecule $C_{40}H_{57}^+$.[4]

The shape of the spectra, absorption maxima of the compounds, and their chromatographic behavior enabled the following structures to be established, in comparison to those described by Vecchi et al.[3] ISO$_1$, 13,13'-di-*cis*-β-carotene; ISO$_3$, 9,13'-di-*cis*-β-carotene; ISO$_4$, 15-*cis*-β-carotene; ISO$_5$, 13-*cis*-β-carotene (major compound); ISO$_6$, 9,9'-di-*cis*-β-carotene; ISO$_7$, *trans*-β-carotene. ISO$_2$, whose absorption maxima are at 467.5, 438.6, and 335 nm, was not identified.

Analysis of Epoxy Compounds

In an aliphatic chain forming a conjugated system, the electron density for the double bonds decreases from the extremities toward the median double bond.[5]

In these conditions, it is consistent that the epoxidation reaction, requiring a high electron density, occurs preferentially at the 5,6- and/or 5',6'-carbon double bond. As a result of the internal stress due to the epoxy ring formed, however, a subsequent rearrangement leads to the formation of a 5,8- (or 5',8')-furanoid oxide ring. This reaction sequence is identified

---

[2] L. Pauling, *Forstschr. Chem. Org. Naturst.* **3**, 203 (1983).
[3] M. Vecchi, G. Englert, R. Maurer, and V. Meduna, *Helv. Chim. Acta* **64**, 2746 (1981).
[4] C. Marty and C. Berset, *J. Food Sci.* **53**, 1880 (1988).
[5] A. H. El-Tinay and C. O. Chichester, *J. Org. Chem.* **35**(7), 2290 (1970).

TABLE II
SPECTRAL CHARACTERISTICS OF CIS ISOMERS OF
β-CAROTENE IN n-HEXANE[a]

Isomer	Absorption maxima in n-hexane			
9,13,13'-tri-cis	454	429	341	272
13,13'-di-cis	459	443	339	279
9,13-di-cis	461	435	334	276
9,13'-di-cis	461	435	334	281
9,9'-di-cis	465	438	?	265
15-cis	472	446	334	275
13-cis	467	441	335	271
9-cis	474	446	340	255

[a] From Vecchi et al.[3]

in fraction B of the extrudates colored by *trans*-β-carotene. Compounds are identified using chemically synthesized standards.

Epoxidation Reaction. M-Chloroperbenzoic acid (0.3 mol; Aldrich, Milwaukee, WI) is added to 1 mol of *trans*-β-carotene in solution in dichloromethane. The mixture is added to a round-bottomed flask containing a $CaCl_2$ guard and immersed in a 0° bath with magnetic stirring for 1 hr. Epoxy compounds are then fractionated on a plate of alumina I (60 F 254 aluminum oxide, 0.2 mm thick; Merck) using 10% (v/v) diethyl ether in hexane as solvent. Each colored band is taken up in dichloromethane and rechromatographed

Characterization of Epoxy and Furanoid Oxide Functions. Aluminum chloride in ethanol reacts with 5,6- or 5,8-epoxy groups, forming a blue compound absorbing between 650 and 700 nm.[6] In addition, epoxy functions are rapidly isomerized into furanoid oxides in the presence of traces of acid.

This structural change causes a hypsochromic shift of the absorption maximum in ethanol of about 20 nm for a monoepoxy and about 40 nm for a diepoxy molecule.[7] To verify the presence of one or two epoxy function(s), one drop of 0.1 M HCl is added directly to the spectrophotometer cuvette containing the derivative in question and the spectrum is compared to that obtained before adding the acid.

Fractionation and Identification of Epoxy Derivatives of trans-β-

---

[6] B. Monties and C. Costes, *C. R. Hebd. Seances Acad. Sci., Ser. C* **226,** 481 (1968).

[7] T. W. Goodwin, "The Biochemistry of the Carotenoids, 2nd Ed., Vol. 1. Chapman and Hall, London, 1980.

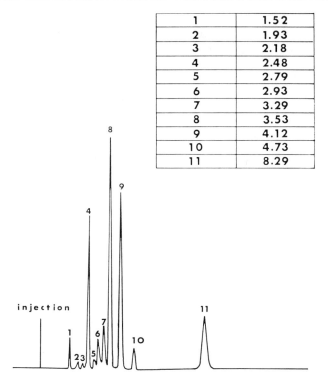

FIG. 2. HPLC chromatogram of β-carotene epoxides (fraction B) formed during extrusion cooking. Stationary phase, Lichrosorb SI 60; pressure, 32 atm; flow rate, 1.0 ml/min; temperature, 22°; detection, 420 nm; eluent, 5% diethyl ether in n-hexane. [From Marty and Berset.[8] Reprinted with permission of *Journal of Food Science,* 1986, **51**(3): 698–702. Copyright © 1986 Institute of Food Technologists.]

Carotene.[8] Four mono- or diepoxy compounds are present in fraction B of the extrudates (Fig. 2), 5,8 : 5′,8′-β-carotene eluting with fraction C. When each derivative in fraction B is subjected to high-performance liquid chromatography (HPLC) on Lichrosorb Si 60 (Merck) as stationary phase, several cis and trans isomers are detected.

Each peak is identified with (1) ultraviolet (UV)–visible absorption spectrophotometry by comparing the spectra to those of the standards and by the hypsochromic shift, (2) infrared spectrophotometry by the detection of epoxy functions, and (3) electron impact (EIMS) and chemical ionization mass spectrometry (CIMS). M = 552 for monoepoxides and M = 568 for diepoxides. Peak 1 is the all-*trans*-β-carotene; peaks 2 and 3 are cis

[8] C. Marty and C. Berset, *J. Food Sci.* **51,** 698 (1986).

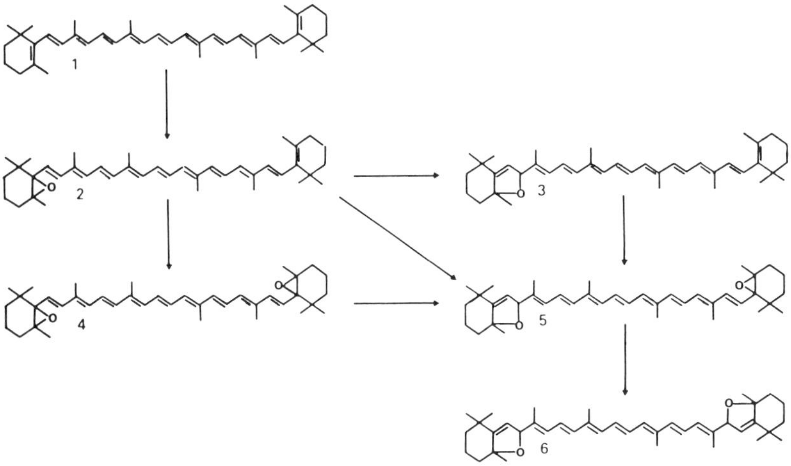

FIG. 3. Reaction sequence for the formation of β-carotene epoxides during extrusion cooking or heating. **1**, β-Carotene; **2**, β-carotene 5,6-epoxide; **3**, β-carotene 5,8-epoxide; **4**, β-carotene 5,6,5′,6′-diepoxide; **5**, β-carotene 5,6,5′,8′-diepoxide; **6**, β-carotene 5,8,5′,8′-diepoxide. [From Marty and Berset.[8] Reprinted with permission of *Journal of Food Science*, 1986, **51**(3): 698–702. Copyright © 1986 Institute of Food Technologists.]

isomers of β-carotene 5,6-epoxide; peaks 5, 6, 7, and 9 are cis isomers of β-carotene 5,8-epoxide; peaks 8, 10, and 11 are, respectively, the all-trans isomers of β-carotene 5,8-epoxide, 5,6,5′,6′-diepoxide, and 5,8,5′,8′-diepoxide

The possible sequential mechanism for the appearance of the five derivatives identified is shown in Fig. 3.

Analysis of Aldehyde or Ketone Derivatives

Fractions C and D (Fig. 1) are chromatographed on a column of Lichrosorb Si 60 (Merck) at 22° (column: 125 × 4 mm, 5 μm). Peaks are detected at 420 and 430 nm, respectively, and are collected automatically with a Gilson 201 fraction collector during repeated injections. Identification is accomplished with the techniques described above: visible and infrared spectrophotometry, EIMS and CIMS by ammonia, and by the chemical reactivity of carbonyl groups[9]: (1) when carotenoid pigments bearing aldehyde or ketone functions are transferred from *n*-hexane to alcohol, the fine structure of the absorption spectrum disappears, presenting only a single broad band; (2) the reduction of an aldehyde group by

[9] P. Critchley, J. Friend, and T. Swain, *Chem. Ind. (Berlin)* **18**, 596 (1958).

TABLE III
SPECTROMETRIC AND CHEMICAL CHARACTERIZATION OF
OXIDIZED DERIVATIVES OF FRACTIONS C AND D

Compound	Molecular formula	m/z	Absorption maxima in n-hexane
β-Apo-15-carotenal	$C_{20}H_{28}OH^+$	285	385.5
β-Apo-14'-carotenal	$C_{22}H_{30}OH^+$	311	398.4
β-Apo-12'-carotenal	$C_{25}H_{34}OH^+$	351	414.0
β-Apo-10'-carotenal	$C_{27}H_{36}OH^+$	377	434.9
β-Apo-8'-carotenal	$C_{30}H_{40}OH^+$	417	454.9–481
β-Caroten-4-one	$C_{40}H_{54}OH^+$	551	456.4

sodium borohydride results in a 20- to 30-nm shift of the absorption maximum; and (3) hydroxylamine hydrochloride reacts specifically with aldehyde groups, yielding an oxime.

The use of these methods led to the identification of five apocarotenals, as well as β-caroten-4-one.[4] The characteristics of these derivatives are summarized in Table III. Extrusion cooking caused the oxidative break of all double bonds in the main chain of β-carotene (Fig. 4).

According to Pullman,[10] the 7,8-carbon double bond in the polyene chain of β-carotene has the highest mobile index, 0.731, calculated by the method of molecular orbitals. This electron structure favors the preferential oxidative break of this double bond. This hypothesis is supported by examining the quantities of the five apocarotenals detected in fraction D, whose order of appearance was (1) β-apo-8'-carotenal, (2) β-apo-10'-carotenal, (3) β-apo-12'-carotenal, (4) β-apo-14'-carotenal, and (5) β-apo-15-carotenal. Extrusion cooking thus causes a sequential shortening of the molecule, two carbon atoms at a time. β-Apo-15-carotenal, however, can form directly by rupture of the central double bond.[11]

Formation of Hydroxyl Derivatives

The chromatogram in Fig. 5 was obtained by eluting fraction E from a column of Lichrosorb Si 60 under the same conditions as described above. Elution was with hexane–chloroform–dichloromethane (40:20:40, v/v/v). This fraction contained a large number of compounds in small

---

[10] A. Pullman, *C. R. Hebd. Seances Acad. Sci, Ser. D* p. 1430 (1960).
[11] C. Marty and C. Berset, *J. Agric. Food Chem.* **38**, 1063 (1990).

quantities. Only the two main peaks, DP13 and DP14, were identified after collecting at column emergence.

The molecular mass of DP13 was 584, corresponding to the empirical formula $C_{40}H_{56}O_3$, and it presented the reactions characteristic of furanoid oxide functions: formation of a blue compound absorbing at 671 nm in the presence of $AlCl_3$ in ethanol, and hypsochromic shift of the maximum by 46.8 nm with respect to *trans*-β-carotene. Furthermore, the addition of 10 µl/ml of 0.1 N hydrochloric acid to a chloroform solution of the compound led to its dehydration and a structural rearrangement, demonstrating the existence of a hydroxyl group.[7] These observations are consistent with DP13 being a 5,8,5',8'-β-carotene diepoxide with an OH group on C-3 or C-4.

The molecular mass of DP14 was 568, corresponding to the empirical formula $C_{40}H_{56}O_2$. The mass of its HCl dehydration product corresponded to the loss of two molecules of water. DP14 is thus a dihydroxyl derivative. Its absorption spectrum was that of *trans*-β-carotene-3,3'-diol, or zeaxanthin, naturally present in the fruits of *Capsicum annuum*.

In summary, after extrusion cooking, the *trans*-β-carotene incorporated in corn starch accounted for only 10 to 20% of the colored molecules present in extrudates. The degradation compounds in the nonvolatile fraction belonged to six different families: six mono- or poly-cis stereoisomers of β-carotene, five mono- or diepoxy compounds, five apocarotenals, one polyene ketone, one diepoxymonohydroxyl derivative, and one dihydroxyl derivative.

We have shown elsewhere[11] that residence of less than 1 min at 180° in the extruder cooker led to far more degradation of *trans*-β-carotene than heating in a sealed ampoule at the same temperature for 2 hr. This is explained by the different conditions of the medium in which the heat treatment is applied. The presence of starch and water, as well as the incorporation of oxygen, play an important role in the oxidative degradation of the pigment.

Protective Effect of Antioxidants

The protective effect of various synthetic or natural antioxidants on all-*trans*-β-carotene may be evaluated at the end of the extrusion cooking and during the storage of the extrudates by the assay of residual all-*trans*-β-carotene and its breakdown products with liquid chromatography.[12]

---

[12] C. Berset and C. Marty, *Proc. Eur. Food Chem., V Int. Symp.* **2**, 427 (1989).

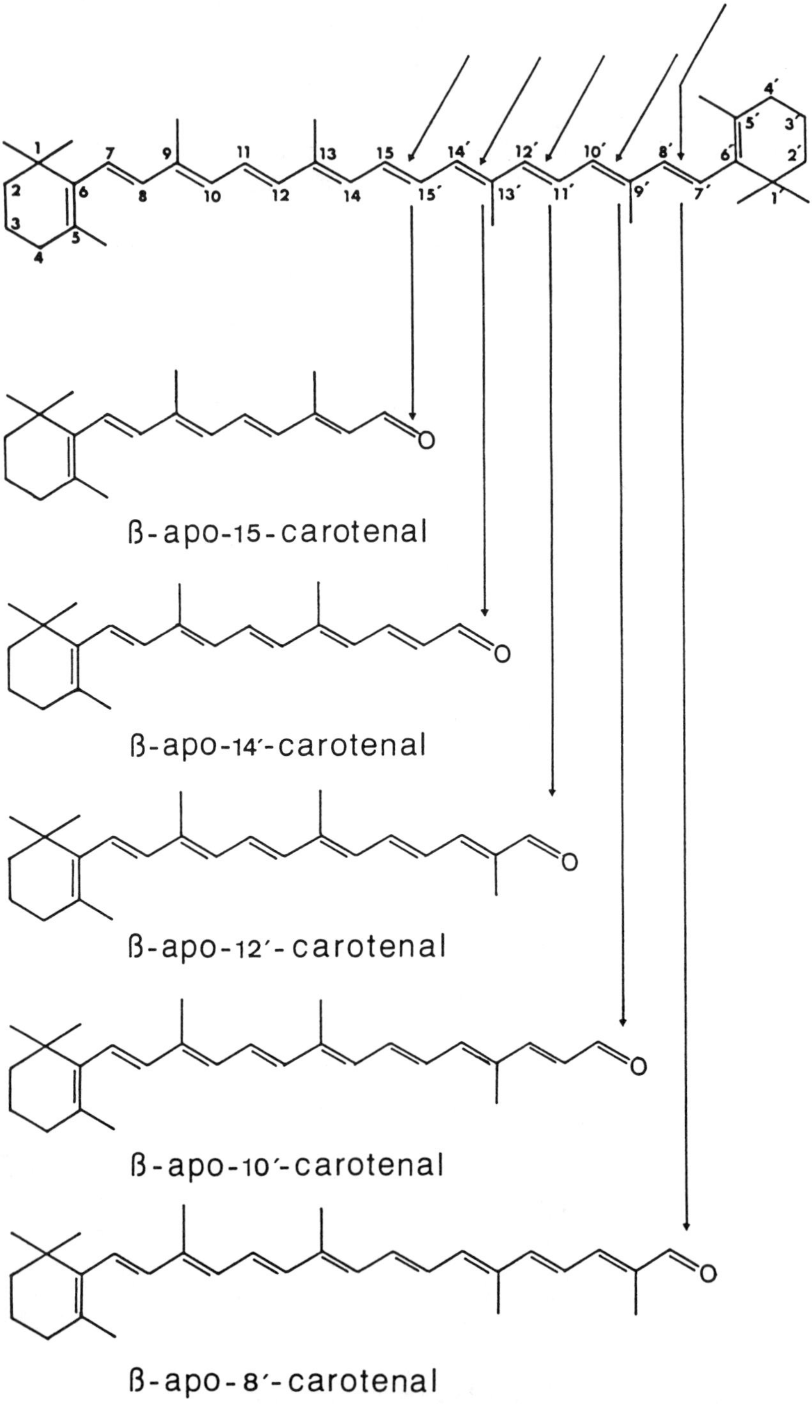

β-apo-15-carotenal

β-apo-14′-carotenal

β-apo-12′-carotenal

β-apo-10′-carotenal

β-apo-8′-carotenal

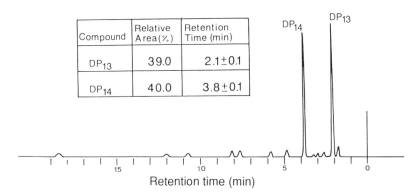

FIG. 5. HPLC chromatogram of fraction E. Stationary phase, Lichrosorb SI 60; flow rate, 1.0 ml/min; temperature, 22° ± 1°; detection, 450 nm; eluent, n-hexane–chloroform–dichloromethane, 0:20:40 (v/v/v). [From Marty and Berset.[4] Reprinted with permission of *Journal of Food Science*, 1988, **53**(6): 1880–1886. Copyright © 1988 Institute of Food Technologists.]

Methods

trans-β-Carotene (18 g, type I; Sigma, St. Louis, MO) is mixed with 60 kg of native corn starch. The mixture is divided into six batches of 10 kg each, to which the following antioxidants are added: (1) control without antioxidant; (2) 100 ppm BHA (=BHA 100); (3) 250 ppm deodorized extract of rosemary (=ROM 250); (4) 500 ppm deodorized extract of rosemary (=ROM 500); (5) 500 ppm *dl*-α-tocopherol (=TOCO 500); and (6) 50 ppm BHT (=BHT 50).

The six tests are carried out in the extruder, following the procedure described. The temperature of the product just before the dies is between 181 and 186°. Total residual pigments are extracted from the extrudates and evaluated in *n*-hexane as β-carotene equivalents, using absorbance at 450 nm ($E^{1\%} = 2597$).

β-Carotene 5,6-epoxide, β-carotene 5,8-epoxide, and β-carotene 5,6,5',6'-diepoxide are eluted with 1.5% *n*-hexane in diethyl ether as solvent on a Lichrosorb Si 60 stationary phase. β-Carotene 5,8,5',8'-diepoxide and β-apo-8'-carotenal are fractionated using the same HPLC column with *n*-hexane–diethyl ether (95:5, v/v). The flow rate is 1 ml/min and the

FIG. 4. Positions of oxygen attack on the polyene chain of β-carotene corresponding to the formation of five apocarotenals isolated after extrusion cooking. [From Marty and Berset.[11] Reprinted with permission of *Journal of Agricultural and Food Chemistry*, 1990, **38**(4): 1063–1067. Copyright © 1990 American Chemical Society.]

TABLE IV
PERCENTAGE OF LOSS OF trans-β-CAROTENE
DURING EXTRUSION COOKING

Sample	Losses (%)
Control	72
BHA 100	68
ROM 250	55
ROM 500	56
TOCO500	50
BHT 50	51

temperature is $20 \pm 0.5°$. cis and trans isomers are separated on a Brownlee C_{18} polymeric phase.[13] The chromatographic conditions are as follow: column, encapped ODS R18 (250 × 4.6 mm) equipped with a precolumn of RP18; eluent, a ternary nonaqueous phase of acetonitrile–methanol–dichloromethane (63:27:10, v/v/v); flow rate, 1 ml/min; and temperature of the column, constant $22 \pm 0.5°$.

The colorimetric parameters of the ground extrudates are calculated in the Cielab 1976 system (L*a*b*).[14] The bleaching of the extrudates during storage is given by the formula $(\Delta E^*/\Delta E^*_{max}) \times 100$, where ΔE^* is the difference in color between the extrudate samples at the outlet of the extruder and at a later storage time

$$\Delta E^* = (\Delta L^{*2} + \Delta a^{*2} + \Delta b^{*2})^{1/2}$$

ΔE^*_{max} is total bleaching expressed as the difference in color between the extrudates at the outlet of the extruder and colorless samples

$$\Delta E^*_{max} = (\Delta L'^{*2} + \Delta a'^{*2} + \Delta b'^{*2})^{1/2}$$

Results

Table IV shows the protective effect of each antioxidant on β-carotene during the extrusion cooking. All the antioxidants tested had a similar action, except for BHA 100, whose effectiveness was much lower. This latter result is surprising because BHA is known to be a powerful antioxidant of lipids.

As noted before, residual *trans*-β-carotene is not the only colored species in the extrudates, and other molecules improved the color. Figure 6

---

[13] C. Marty, E. Lesellier, P. Saint Martin, J. P. Moissonnier, and C. Berset, *Analusis* **18**, 78 (1990).
[14] C. Berset, J. Trouiller, and C. Marty, *Lebensm.-Wiss. Technol.* **22**, 1 (1989).

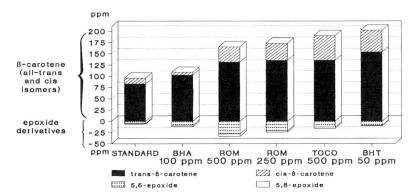

FIG. 6. Residual pigments after extrusion cooking: β-carotene and epoxide derivatives. The sum of the amounts of β-carotene 5,6,5′,6′-diepoxide and β-carotene 5,8,5′,8′-diepoxide is lower than 3 ppm. [From Berset and Marty.[12] Reprinted from *Pro. Eur. Food Chem., V Int. Symp.*, 1989, **2**: 427–432, with permission of Institut National de la Recherche Agronomique.]

shows that there were fewer cis isomers in the control and BHA samples than in the other batches. Moreover, a higher proportion of epoxy compounds was present in the samples protected by the rosemary extracts. Thus, in the absence of an effective antioxidant, pigment breakdown leads

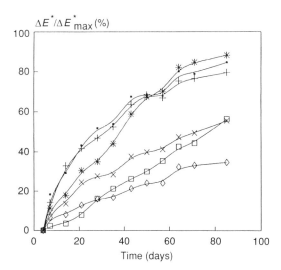

FIG. 7. Bleaching kinetics of the extrudates. ■, Standard; +, BHA (100 ppm); *, ROM (250 ppm); □, ROM (500 ppm); ×, TOCO (500 ppm); ◊, BHT (50 ppm). [From Berset and Marty.[12] Reprinted from *Proc. Eur. Food Chem., V Int. Symp.*, 1989, **2**: 427–432, with permission of Institut National de la Recherche Agronomique.]

mainly to the formation of colorless low molecular weight derivatives, while antioxidants slow the process by blocking some of the steps in the oxidative process.

During storage (Fig. 7), BHT was found to be the best long-term protective agent for β-carotene; 500 ppm α-tocopherol or deodorized rosemary oleoresin gave similar moderate effects. A dose of 250 ppm of these two is not sufficient in the drastic conditions of extrusion cooking.

Conclusion

Studies of the behavior of purified all-*trans*-β-carotene in extrusion cooking have shown that this type of treatment leads to extensive degradation of the molecule. The fraction of residual nonvolatile compounds at extruder output accounts for only 10 to 20% of the initial pigment content. It is to be noted that in commercial preparations of β-carotene for industrial use in which the colorant is coated in a matrix, losses are much lower.

Several types of chemical reactions are revealed, including multiple isomerizations on certain trans double bonds leading to modifications of the spatial conformation of the molecule and undoubtedly modifying its stability. In addition, oxidations of the β-carotene molecule occur (epoxy formation, hydroxylation, ketone function). Finally, oxidative rupture progressively shortens the unsaturated chain.

The rates of these degradation phenomena can be decreased by the presence of antioxidants, provided the compounds themselves are heat stable.

[14] Natural Sources of Carotenoids from Plants and Oils

By AUGUSTINE S. H. ONG and E. S. TEE

Introduction

A large number of pigments in living organisms—carotenoids, chlorophylls, anthocyanins, and porphyrins—provide the rich variety of color in nature. The carotenoids, believed to have derived their name from the fact that they constitute the major pigment in the carrot root, *Daucus carota*, are undoubtedly among the most widespread and important. This group of pigments is found throughout the plant kingdom (although their presence is often masked by chlorophyll) and in insects, birds, and other animals. These pigments provide a whole range of light yellow to dark red colorings; and, when complexed with proteins, they produce green and blue colorations. Thus, a wide variety of foods and feeds, e.g., yellow vegetables, tomatoes, apricots, oranges, egg yolk, chicken, butter, shrimp, lobster, salmon, trout, and yellow corn, owe their color principally to carotenoids, as do certain food color extracts from natural sources such as palm oil, paprika, annatto, and saffron.[1-3]

Aside from providing aesthetic qualities as colorants in the plant and the animal kingdoms, carotenoids, synthesized exclusively by photosynthetic microorganisms and by members of the plant kingdom, play a fundamental role in metabolism. Most importantly, the carotenoids serve the animal kingdom as sources of vitamin A activity.

The main source of vitamin A in the diet of the rural communities in developing countries is carotenoids, the precursors of vitamin A, from vegetables and fruits. Preformed vitamin A found in meat, liver, and eggs is frequently out of reach of the economically deprived. Many of these countries have abundant vegetation, which is a rich source of carotenoids, so there should indeed be no vitamin A deficiency problem. However, vegetable consumption among young children is known to be low and whatever small amounts of carotenoids are ingested, together with a diet poor in fat and protein, makes absorption of these precursors of vitamin A poor. At the same time, it is also not clear which fruits and vegetables are richest in β-carotene. These are obviously important areas for studies and intervention.[4]

A new frontier in carotenoid research has been the examination of a possible association between carotenoids and the development and prevention of cancer. In addition to their role as vitamin A precursors, carotenoids are known to be able to act as antioxidants, and to inactivate (quench) highly reactive chemical species such as singlet oxygen, triplet photochemical sensitizers, and free radicals, which would otherwise induce potentially harmful processes (e.g., lipid peroxidation). Through these actions, it has been suggested that carotenoids may play important roles in cancer causation and prevention.[5-9]

---

[1] B. Borenstein and R. H. Bunnell, *Adv. Food Res.* **15,** 195 (1966).
[2] B. C. L. Weedon, *in* "Carotenoids" (O. Isler, H. Gutmann, and U. Solms, eds.), pp. 29–59. Birkhäuser, Basel, 1971.
[3] J. C. Bauernfeind, *J. Agric. Food Chem.* **20,** 456 (1972).
[4] E. S. Tee, *CRC Crit. Rev. Food Sci. Nutr.* **31,** 103 (1992).
[5] R. Peto, R. Doll, J. D. Buckley, and M. B. Sporn, *Nature (London)* **290,** 201 (1981).
[6] J. A. Olson, *J. Nutr.* **116,** 1127 (1986).
[7] N. J. Temple and T. K. Basu, *Nutr. Res.* **8,** 685 (1988).
[8] N. I. Krinsky, *Clin. Nutr.* **7,** 107 (1988).
[9] G. W. Burton, *J. Nutr.* **119,** 109 (1989).

Several lines of evidence have suggested that important relations exist involving carotenoids and cancer.[8,10-12] These include evidence of the effects of carotenoids in experimental cancer, epidemiological evidence suggesting associations between reduced risk of cancer and carotene status, and findings from cancer treatment and prevention trials. The possibility that carotenoids may play a role in the etiology and prevention of cancer has added a new, exciting dimension to the study of these nutrients in human nutrition.

Analysis of Carotenoids in Plant Sources

Particular emphasis has been placed on understanding the types and concentrations of carotenoids in foods. It is now apparent that previously reported values of vitamin A activity in food composition tables may be unreliable because the methodologies used were not sufficiently discriminative and thus had included carotenoids that do not possess vitamin A activity. Advances in studies into the structures and properties of various carotenoids have shown that only a handful of the hundreds of carotenoids occurring in nature possess vitamin A activity. Some of these may occur in higher concentrations than β-carotene, the most potent precursor of vitamin A. The identification and the quantitation of such carotenoids devoid of vitamin A activity are also important, as it has been suggested that, like β-carotene, they may also be associated with lower cancer risk. It is now evident that epidemiological studies associating cancers with diet would require more accurate data on carotenoid content of foods. Currently, association is on a broad basis, linking vegetable consumption or "carotene" content of foods to risk of developing cancer. Improved analytical methods would greatly help to identify the compounds of interest, not only in foods but in blood serum as well.

In cognizance of the need for more accurate analysis of carotenoids, the International Vitamin A Consultative Group (IVACG) has called on national laboratories to develop techniques for determining accurately the provitamin A content of native fruits and vegetables in order to support vitamin A deficiency prevention and control programs.[13] The group also has emphasized that because carotenoids may be protective against some

---

[10] M. M. Mathews-Roth, *Pure Appl. Chem.* **57**, 717 (1985).
[11] R. G. Ziegler, *J. Nutr.* **119**, 116 (1989).
[12] C. M. Williams and J. W. Dickerson, *Nutr. Res. Rev.* **3**, 75 (1990).
[13] K. L. Simpson, S. T. C. Tsou, and C. O. Chichester, "Biochemical Methodology for the Assessment of Carotenes." International Vitamin A Consultative Group, Washington, D.C., 1987.

forms of cancer, the quantitation of these compounds in foods assumes even broader importance. The lack of information on the amount and form of carotenoids in existing food composition databases has also been highlighted.[14] Although high-pressure liquid chromatography (HPLC) has been proposed as an efficient tool for the separation and quantitation of carotenoids, the IVACG recognizes that there are as yet limited data on the carotenoid content of foods using this technique.

The analysis of carotenoids and retinoids is complicated and beset with various problems because of the large number of naturally occurring carotenoids (as many as 600 have been reported). The occurrence of cis and trans isomers of carotenoids further complicates the analysis. The composition and content of carotenoids in various plant materials vary widely. Not all of the naturally occurring carotenoids are precursors of vitamin A, and for those with provitamin A activity, the biological activity varies widely.

It is, therefore, rather difficult to obtain accurate data on carotenoid content and composition. The literature abounds with attempts at improving the analytical techniques. The procedure should be able to remove interfering compounds effectively, to separate the carotenoids, and to quantitate them accurately. There has been much work on method development and improvement in this field.[15] A wide variety of separation, detection, and quantitation procedures have been used in studies of carotenoids and retinoids. An early technique used for the separation of carotenoids in plant materials, mainly in the 1960s, was countercurrent distribution. A few early studies using paper chromatography were also reported. In the 1970s, studies of carotenoids and retinoids using gas–liquid chromatography and gel-permeation chromatography were introduced and a number of studies employed thin-layer chromatography (TLC) as a single technique for the separation of carotenoids in plant materials. Thin-layer chromatography was also used by several investigators in combination with other separation techniques. The procedure was less widely used in studies of retinoids. Adsorption open-column chromatography, utilizing primarily descending, gravity-flow columns, was widely used for the study of carotenoids and retinoids in foods, even in the 1960s. This procedure remained very much in use in the 1970s and 1980s. Since the late 1970s, however, HPLC has become a widely used procedure for the separation of

---

[14] B. A. Underwood, M. Chavez, J. Hankin, J. A. Kusin, A. Omololu, F. Ronchi-Proja, R. Butrum and S. Ohata, "Guidelines for the Development of a Simplified Dietary Assessment to Identify Groups at Risk for Inadequate Intake of Vitamin A." International Vitamin A Consultative Group, Washington, D.C., 1989.

[15] E. S. Tee and C. L. Lim, *Food Chem.* **41**, 147 (1991).

carotenoids, mainly because the technique effects rapid separation, it is nondestructive and, more importantly, better resolution is achieved. The ability of HPLC to separate rapidly and to quantitate various carotenoids, at least in standard preparations, has been demonstrated. Its application to the analysis of foods, however, is still being developed and improved.

Many investigators have shown that the use of other techniques in combination with HPLC will greatly enhance the usefulness of the procedure. The combined use of open-column chromatography, TLC, and ultraviolet–visible (UV–VIS) absorption spectra, for example assists in the identification and confirmation of carotenoids.

Natural Sources of Carotenoids from Plants and Oils

The ability of organisms to produce carotenoids seems to have developed at an early stage in evolution. Photosynthetic bacteria, the algae, spore-bearing vascular plants, and the higher plants preserve this capability. Animals, certainly those of the higher orders, are not capable *do novo* carotenogenesis and are dependent for their carotenoids on those preexisting in their diet.[2] This chapter deals mainly with those carotenoids present in higher plants, which are the more important food sources.

Between 1933 and 1948, the number of known naturally occurring carotenoids increased from 15 to about 80, and rose sharply to about 300 some 20 years later.[16] In 1986 Goodwin[17] reported that over 500 of them were known. Most of these, however, are xanthophylls, which are oxygenated carotenoids. The common name and the semisystematic equivalent, together with the structure of about 300 natural carotenoids, have been listed in the 800-reference publication of Straub,[18] who updated the list to 400 compounds in a 1600-reference monograph.[19] Bauernfeind[3] tabulated the occurrence of some 40 common carotenoids in various foodstuffs.

It has been estimated that nature produces about 100 million tons of carotenoid pigments per year.[20] Most of this output is in the form of four major carotenoids, namely, fucoxanthin, the characteristic pigment of many marine algae and the most abundant natural carotenoid, and lutein, violaxanthin, and neoxanthin,[2] the three main carotenoids in green leaves. All other carotenoids are produced in relatively small amounts. However,

---

[16] O. Isler, "Carotenoids" (O. Isler, H. Gutmann, and U. Solms, eds.), pp. 11–27. Birkhäuser, Basel, 1971.
[17] T. W. Goodwin, *Annu. Rev. Nutr.* **6**, 273 (1986).
[18] O. Straub, *in* "Carotenoids" (O. Isler, H. Gutmann, and U. Solms, eds.), pp. 771–850. Birkhäuser, Basel, 1971.
[19] O. Straub, "Key to Carotenoids: Lists of Natural Carotenoids." Birkhäuser, Basel, 1976.
[20] O. Isler, R. Ruegg, and U. Schwieter, *Pure Appl. Chem.* **14**, 245 (1965).

some, like β-carotene and zeaxanthin, occur very widely and others, such as lycopene, capsanthin, bixin, and spirilloxanthin, constitute the principal pigments in particular organisms.

The overall carotenoid pattern may vary from relatively simple mixtures to extremely complex ones. The simplest mixtures may be found in foods of animal origin, owing to the limited ability of the animal to absorb, modify, and deposit carotenoids. The other extreme is the formidable array of carotenoids encountered in citrus products, dehydrated alfalfa meal, and paprika.[1,3]

The concentrations of carotenoids vary enormously from one source to another. The highest concentration of carotenes has been found in the red fringe of the corona of the pheasant's-eye narcissus. *Narcissus majalis.* Here β-carotene can constitute up to 16% of the dry weight. Furthermore, the daily rate of β-carotene formation reaches 70 μg/mg dry weight, which is more than 10,000 times the rate observed in carrot roots.[21]

The distribution of carotenoids in various plant materials has been reviewed in various publications.[1,3,22-24] Most of the data presented include those reported some years ago (generally pre-1970) using open-column chromatography and thin-layer chromatography. These methods generally do not permit adequate separation of the carotenoids, so results were usually reported as carotene, total carotene, or β-carotene. There have been rapid advances in the development of methodologies for more accurate quantitation of various carotenoids in foods in recent years, and HPLC has become widely used for these analyses.[15] This chapter includes mainly data obtained using HPLC and some aspects of the analytical conditions used by the investigators will also be presented.

Leaves and Other Vegetables

One of the earlier studies on plant materials using HPLC made use of a mixture of calcium and magnesium hydroxide for the separation of carotene stereoisomers in several vegetables, including spinach, broccoli, kale, carrot, and sweet potatoes.[25] Rather large-diameter (15-mm i.d.) columns were used, especially designed to withstand pressures up to 20 psi. The

---

[21] V. H. Booth, *Biochem. J.* **87**, 238 (1963).
[22] T. W. Goodwin, in "Chemistry and Biochemistry of Plant Pigments" (T. W. Goodwin, ed.), 2nd Ed., Vol. 1, pp. 225–261. Academic Press, New York, 1976.
[23] T. W. Goodwin, in "The Biochemistry of the Carotenoids," 2nd Ed., Vol. 1, pp. 96–142. Chapman and Hall, New York, 1980.
[24] T. W. Goodwin, in "The Biochemistry of the Carotenoids," 2nd Ed., Vol. 1, pp. 143–203. Chapman and Hall, New York, 1980.
[25] J. P. Sweeney and A. C. Marsh, *J. Assoc. Off. Anal. Chem.* **53**, 937 (1970).

isomers were eluted using p-methylanisole in petroleum ether and acetone in petroleum ether. Fractions were concentrated and studied in a spectrophotometer. Only the percentages of the various isomers were reported.

The ease with which isomers are formed has been recognized as a problem in carotenoid analysis. Various procedures have been known to induce isomerization, including refluxing in an organic solvent, photochemical reactions, contact with active surfaces, and irradiation in the presence of iodine. Thus, inadvertent isomerization of carotenoids can take place during sample handling and analysis, e.g., during prolonged heating, exposure of solutions at room temperature to sunlight or artificial light, or even during contact with various active surfaces such as column chromatography adsorbents. The formation of cis isomers of carotenoids during the analysis of vegetables has been observed by several investigators.[25-29]

Carotenoids in leafy vegetables have been reported by several other investigators. The following studies are all carried out using reversed-phase HPLC C_{18} columns from various manufacturers. Using a mixture of methanol, acetonitrile, chloroform, and water (200:250:90:11, v/v), β-carotene and total carotenoid concentrations of 20 Thai vegetables were reported and vitamin A activity calculated based on β-carotene.[30] In the green leafy vegetables studied, β-carotene concentration ranged from 600 to 7800 μg/100 g of sample and made up 3 to 30% of the total carotenoid concentration. In a subsequent study,[31] a larger series of vegetables (50) were analyzed, and β-carotene and total carotenoid concentrations were reported. The effect of various food-processing procedures on the levels of carotenoids was also studied. β-Carotene concentrations of the leafy vegetables ranged from low levels of less than 2000 μg/100 g, e.g., in pickled mustard leaves and Asian pennywort, to high levels of about 20,000 μg/100 g for spinach and coriander, to very high concentrations of over 30,000 μg/100 g in common fennel and bitter melon leaves (Table I). The percentage of β-carotene was reported to range from 2 to 20% of total carotenoids in the vegetables.

Tuberous vegetables and beans were found to have considerably lower carotenoid concentrations. β-Carotene content was reported to range from 40 μg/100 g (cauliflower) to about 2000 μg/100 g of onion plant and garden peas.

[26] I. Stewart and T. A. Wheaton, *J. Chromatogr.* **55**, 325 (1971).
[27] S. J. Schwartz and M. Patroni-Killam, *J. Agric. Food Chem.* **33**, 1160 (1985).
[28] L. A. Chandler and S. J. Schwartz, *J. Food Sci.* **52**, 669 (1987).
[29] P. W. Simon and X. Y. Wolff, *J. Agric. Food Chem.* **35**, 1017 (1987).
[30] A. J. Speek, S. Speek-Saichua, and W. H. P. Schreurs, *Food Chem.* **27**, 245 (1988).
[31] A. J. Speek, C. R. Temalilwa, and J. Schrijver, *Food Chem.* **19**, 65 (1986).

TABLE I
LUTEIN AND β-CAROTENE CONCENTRATIONS[a] IN GREEN VEGETABLES

Vegetable	Range/mean carotenoid concentration		Ref.
	Lutein	β-Carotene	
Green, leafy (4 types)	—	330–5,030	34
Green, nonleafy (6 types)	—	217–763	
"Cruciferous" vegetables (5 types)	280–34,200	80–14,600	32
Leafy vegetables (32 types)	—	1,000–44,400	31
Tuberous vegetables and beans (16 types)	—	40–1,700	
Green, leafy (7 types)	250–10,200	1,000–5,600	35
Other vegetables (19 types)	trace–440	11–430	
Green, leafy (27 types)	73–29,900	97–13,600	36
Green, nonleafy (8 types)	142–460	74–569	

[a] Data given in micrograms per 100 g fresh weight.

In a study of five vegetables, mainly of the genus *Brassica,* three solvent systems were employed to effect the separation of several carotenoids and chlorophylls.[32] A combination of isocratic and gradient chromatography and various mixtures of methanol, acetonitrile, dichloromethane, and hexane were used. Three classes of compounds were separated: (1) xanthophylls, (2) chlorophylls and their derivatives, and (3) the hydrocarbon carotenoids. Lutein was the most abundant xanthophyll, and was accompanied by three minor cis isomers. The other two major xanthophylls were neoxanthin and violaxanthin. The only hydrocarbon carotenoids present were all-*trans*-β-carotene and its 15,15′-cis isomer. The lutein and β-carotene (the latter includes the cis isomer) concentrations are tabulated in Table I.

Other investigators had also focused on the analysis of the α- and β-carotene content of vegetables and fruits. A study of nine vegetables and fruits from different supermarkets was carried out using a mobile phase of acetonitrile, methanol, and tetrahydrofuran (40:56:4, v/v).[33] In a study of carotenoids in 22 fruits and vegetables,[34] an aqueous mobile phase of acetonitrile, tetrahydrofuran, and water (85:12.5:2.5, v/v) was used. Vitamin A activity of the foods was calculated based on the α- and β-carotene and cryptoxanthin concentrations. With the exception of carrot, β-caro-

[32] F. Khachik, G. R. Beecher, and N. F. Whittaker, *J. Agric. Food Chem.* **34**, 603 (1986).
[33] R. J. Bushway, *J. Agric. Food Chem.* **34**, 409 (1986).
[34] J. L. Bureau and R. J. Bushway, *J. Food Sci.* **51**, 128 (1986).

tene contributed to 85% of the total provitamin A activity of the vegetables. β-Carotene concentration in green leafy vegetables is clearly higher than in nonleafy varieties (Table I). α-Carotene was found in low concentrations in these vegetables, whereas cryptoxanthin was not detected in any of them. Carotenoids in carrot and the fruits studied will be discussed later in this chapter. Of the foods studied, no significant differences were reported based on either locations or months of analyses.

Carotenoids in 69 types of Finnish vegetables, fruits, berries, mushrooms, and their respective products have been reported.[35] Good resolution was said to have been obtained using a mobile phase of acetonitrile, dichloromethane, and methanol (70:20:10, v/v). In addition, a combination of isocratic and gradient chromatography was also employed for the separation of carotenoids in most extracts. Lutein and β-carotene were the predominant carotenoids in the vegetables studied, and both carotenoids occurred in higher concentrations in green leafy vegetables (Table I). In most of the vegetables studied, however, α- and γ-carotenes, cryptoxanthin and lycopene were not detected.

A study of the carotenoid composition and content of 40 tropical vegetables and 14 fruits emphasized the major carotenoids that occur in sufficient quantities to contribute significantly to dietary intake.[36] A ternary mixture of acetonitrile, methanol, and ethyl acetate (88:10:2, v/v) was used to separate the carotenoids isocratically in a C_{18} column. The method gave satisfactory separation and quantitation for lutein, cryptoxanthin, lycopene, and γ-, α-, and β-carotenes. For most of the green leafy vegetables, carotenoid compositions were rather consistent. In most cases, only β-carotene and lutein were obtained. In all the vegetables studied, β-carotene was the major carotenoid found and in 20 of the green leafy vegetables studied it made up over 50% of the sum of all carotenoids quantitated. For the remaining seven samples, at least 20% of the carotenoids was β-carotene. Lutein was also detected in all the vegetables in fairly high proportions. Except for five samples, lutein made up over 25% of all the carotenoids in these vegetables. Other carotenoids were encountered infrequently.

The carotenoid compositions of the green nonleafy vegetables were similar to those for the green leafy varieties. Seven of the eight vegetables in the former group were found to have over 25% lutein and over 30% β-carotene.

In contrast to the green vegetables, the carotenoid composition of the

---

[35] M. I. Heinonen, V. Ollilainen, E. K. Linkola, P. T. Varo, and P. E. Koivistoinen, *J. Agric. Food Chem.* **37**, 655 (1989).

[36] E. S. Tee and C. L. Lim, *Food Chem.* **41**, 309 (1991).

fruit and root vegetables was rather different. Although β-carotene and lutein were found in all these vegetables, several other carotenoids were also determined. For example, α-carotene was detected in carrot and pumpkin, while cryptoxanthin was found in red chilli. However, lycopene was detected only in tomato, and made up about 60% of the carotenoids quantitated.

β-Carotene concentration in the 27 green leafy vegetables studied ranged from about 100 to 14,000 μg/100-g sample (Table I). Twenty-one of the vegetables had β-carotene concentrations of over 3000 μg/100 g, or 500 μg retinol equivalent (RE)/100 g, close to the Food and Agricultural Organization/World Health Organization (FAO/WHO) recommended safe level of vitamin A intake of 600 μg RE/day.[37] Lutein concentrations varied widely, ranging from about 70 to 30,000 μg/100 g of vegetable. The eight types of green nonleafy vegetables studied had much lower β-carotene concentrations (Table I), with none of them exceeding 1000 μg/100 g of vegetable.

In summary, various studies have shown that leafy vegetables contain very high carotenoid concentrations, particularly the green leafy varieties. β-Carotene and lutein are the major carotenoids, and together can account for over 80% of all carotenoids, whereas α- and γ-carotenes, cryptoxanthin, and lycopene are infrequently encountered or occur in small concentrations.

Most of the studies of green leafy and nonleafy vegetables cited above have emphasized the content of a few major carotenoids in a variety of plant materials. Several other studies carried out on a few vegetables, have reported the analysis of a large number of carotenoids. These studies frequently require the use of a number of separation techniques and more sophisticated instrumentation. In one study employing reversed-phase HPLC, chlorophylls and eight carotenoids were separated from extracts of spinach and green algae using methanol–acetonitrile–water in a linear gradient system.[38] The study cited above of several green vegetables of the genus *Brassica* is another example of the complexities encountered and the need for various analytical techniques for the identification and quantitation of these pigments.[32] Eighteen components, belonging to 3 classes of compounds (xanthophylls, chlorophylls and their derivatives, and the hydrocarbon carotenoids) were separated from the vegetable extracts. The major constituents were also separated by semipreparative TLC and HPLC and were identified by such tools as mass spectrometry (MS), nuclear

---

[37] FAO/WHO, "Requirements of Vitamin A, Iron, Folate and Vitamin B_{12}," FAO Food Nutr. Ser. 23. Food and Agriculture Organization, Rome, 1988.
[38] T. Braumann and L. H. Grimme, *Biochim. Biophys. Acta* **637**, 8 (1981).

magnetic resonance (NMR), and UV-visible spectroscopy. Detailed studies of squash using a combination of isocratic and gradient reversed phase HPLC systems to effect the separation of some 25 carotenoids have been reported.[39,40] A study of several carotenoids and carotenol fatty acids in various dried and canned fruits was made using two C_{18} reversed-phase HPLC columns, each with a different adsorbent.[41] Various eluants composed of mixtures of acetonitrile, methanol, dichloromethane, and hexane were used for isocratic and gradient elution, and various techniques were used in the identification of carotenoids.

The separation of carotenoids from paprika (red bell pepper) using gradient elution from a C_{18} column has been reported.[42,43] In the latter study 38 carotenoids and esters, including several unidentified components, were separated. TLC was used to assist in the separation and identification of the pigments whereas a combination of TLC and reversed-phase HPLC[44] was used in another study of several carotenoids found in paprika.

Carrots (*D. carota* L.) have traditionally been an important dietary source of carotenoids and have been studied by several investigators. The following studies all made use of a C_{18} column for reversed-phase HPLC. In a study of the separation of α- and β-carotene from raw, canned, and frozen carrots,[45] a solvent mixture of acetonitrile, tetrahydrofuran, and water (85:12.5:2.5, v/v) was used (Table II). Similar results were reported in a subsequent study, using the same mobile phase (Table II), and examining carrot from various sources obtained at three different times of the year.[34]

The separation of α- and β-carotene was also effected on a Microsorb C_{18} column using an isocratic system of methanol, acetonitrile, and dichloromethane-hexane delivered from three pumps.[46] However, only total carotene concentrations were reported for carrots. In a subsequent study of various yellow/orange vegetables, the separation of several carotenoids from raw, cooked, and canned carrot was effected using a mixture of methanol, acetonitrile, and dichloromethane (22:55:23, v/v).[47] In addition to α- and β-carotenes, ζ-carotene was also reported at a concentration of about 5.5-8.5% of the total carotenoids detected (Table II).

[39] F. Khachik and G. R. Beecher, *J. Agric. Food Chem.* **36,** 929 (1988).
[40] F. Khachik, G. R. Beecher, and W. R. Lusby, *J. Agric. Food Chem.* **36,** 938 (1988).
[41] F. Khachik, G. R. Beecher, and W. R. Lusby, *J. Agric. Food Chem.* **37,** 1465 (1989).
[42] C. Fisher and J. A. Kocis, *J. Agric. Food Chem.* **35,** 55 (1987).
[43] G. K. Gregory, T. S. Chen, and T. Philip, *J. Food Sci.* **52,** 1071 (1987).
[44] P. A. Biacs, H. G. Daood, A. Pavisa, and F. Hajdu, *J. Agric. Food Chem.* **37,** 350 (1989).
[45] R. J. Bushway and A. M. Wilson, *Can. Inst. Food Sci. Technol. J.* **15,** 165 (1982).
[46] F. Khachik and G. R. Beecher, *J. Chromatogr.* **346,** 237 (1985).
[47] F. Khachik and G. R. Beecher, *J. Agric. Food Chem.* **35,** 732 (1987).

TABLE II
α- AND β-CAROTENE CONCENTRATIONS[a] IN CARROTS

Carrot	Range/mean carotenoid concentration		Ref.
	α-Carotene	β-Carotene	
Raw	2,000–5,000	4,600–12,500	45
Canned	3,200–4,800	7,000–11,000	
Frozen	8,400–8,800	26,000–28,100	
Raw	3,790	7,600	34
Raw, A$^+$ hybrid	10,650	18,350	47
Freshly cooked, A$^+$ hybrid	15,000	25,650	
Canned	2,800	4,760	
Line B6273			
Lyophilized	3,400	6,000	29
Raw	3,200	5,200	
Frozen	3,100	5,100	
Line B9692			
Lyophilized	6,100	13,800	29
Raw	6,600	11,700	
Frozen	6,600	11,600	
HCM line			
Lyophilized	20,300	28,200	29
Raw	20,600	25,100	
Frozen	20,400	25,500	
Raw, 19 cultivars	2,200–4,900	4,600–10,300	48
Raw	3,410	6,770	36

[a] Data given in micrograms per 100 g fresh weight.

A more detailed study of carotenoids in several American carrot lines was reported using a mobile phase composed of acetonitrile, dichloromethane, and methanol for reversed-phase HPLC.[29] The carotenes separated were identified using column chromatography and thin-layer chromatography. Six carotenes, namely α-, β-, γ-, and ζ-carotenes, β-zeacarotene, and lycopene, were detected. The predominant carotenoid in all samples was β-carotene, accounting for 44–79% of all carotenes quantitated. Approximately 92–93% of the total carotenes was accounted for by β- and α-carotene, while ζ-carotene constituted about 2–4% of the total carotenes. The remaining 3–6% was made up of β-zeacarotene, γ-carotene, and lycopene. The α- and β-carotene content obtained for three of the carrot lines is tabulated in Table II. The report also included studies of changes in carotene content due to different treatment procedures.

An extensive study of 19 cultivars of carrot found in Finland has been reported.[48] The carotenoids were eluted isocratically using a mixture of acetonitrile, dichloromethane, and methanol (70:20:10, v/v) and α- and β-carotenes were the major carotenoids, making up an average of 28 and 57%, respectively, of all carotenes. Another 12% of the carotenes was γ-carotene, while lutein made up the remaining 2%. The ranges of α- and β-carotene content obtained are tabulated in Table II.

In a recent study of tropical vegetables and fruits, α- and β-carotenes were reported as the major carotenoids in carrot, composing over 90% of all carotenoids.[36] The ratio of α- to β-carotene was approximately 1:2.

Numerous studies on carotenoids in carrot have shown that α- and β-carotenes are the major carotenoids present in this root vegetable. Most of the data have shown quite consistently that the concentration of β-carotene is approximately twice that of α-carotene. The characteristic feature of carrot is its high concentration of α-carotene. Other minor carotenoids reported to be present, making up approximately 10% of total carotenoids, and γ- and ζ-carotene, β-zeacarotene, lutein, and lycopene.

Tomato, another orange-colored vegetable, has also been studied. Six samples of red-ripe Massachusetts greenhouse tomatoes were examined using several reversed-phase columns.[49] The best separation was obtained using a 5-μm particle size Partisil column (Whatman, Chifton, NJ) and a mobile phase consisting of 8% (v/v) chloroform in acetonitrile to elute the pigments isocratically. The carotenoids were detected at 470 nm. Lycopene and β-carotene occurred in a ratio of approximately 9:1 and their concentrations are tabulated in Table III.

A combination of TLC and HPLC was used in another study of carotenoids in tomatoes.[50] Two mobile phases were used for HPLC, namely, acetone–water (9:1, v/v) and acetonitrile–2-propanol–water (200:288:13, v/v). In overripe tomatoes, lycopene, β-carotene, and lutein constituted about 76, 12.4, and 3.5%, respectively, of all the pigments, with the remaining 3–5% consisting of several minor components. These minor components, particularly neurosporene and ζ-carotene, were found to occur in much higher concentrations in ripe tomatoes. The decrease in the concentrations of neurosporene and ζ-carotene in the overripe fruit was reported to be due to the conversion of these intermediates to lycopene and β-carotene.

Lycopene and β-carotene have also been reported by other investigators

---

[48] M. I. Heinonen, *J. Agric. Food Chem.* **38**, 609 (1990).
[49] M. Zakaria, K. Simpson, P. R. Brown, and A. Krstulovic, *J. Chromatogr.* **176**, 109 (1979).
[50] H. G. Daood, P. A. Biacs, A. Hoschke, M. Harkay-Vinkler, and F. Hajdu, *Acta Aliment.* **16**, 339 (1987).

TABLE III
β-CAROTENE AND LYCOPENE CONCENTRATIONS[a] IN TOMATO

Tomato	Range/mean carotenoid concentration		Ref.
	β-Carotene	Lycopene	
Raw	80.5–127	384–1,072	49
	660	3,100	35
	365	723	36
Paste	8,000	25,400	51

[a] Data given in micrograms per 100 g fresh weight.

as the major carotenoids present in tomatoes[35,36] (Table III). The concentrations of the two carotenoids, present in a ratio of approximately 2:1, made up close to 90% of total carotenoids quantitated. Lutein was detected as a minor component, comprising about 10% of the total carotenoids.[36] Lutein was also reported to comprise about 2.5% of all carotenoids, and another 4% was contributed by γ-carotene.[35]

More complicated chromatography procedures[51] have been used to detect the presence of other carotenoids in tomatoes. The separation of carotenoids from tomato paste using a combination of normal-phase open-column chromatography and HPLC was reported. Pigments were first eluted using a stepwise gradient of diethyl ether in petroleum ether and further analyzed by reversed-phase HPLC. For HPLC a photodiode array detector was used to monitor the carotenoids. The quantities of the four major carotenoids (lycopene, phytoene, β-carotene, and phytofluene) were 254, 167, 80, and 49 ppm, respectively (Table III). cis isomers of these carotenoids were also detected, and minor carotenoid components included α-carotene, lycoxanthin, and cis-mutatoxanthin.

Pumpkin is a common food item in many Asian communities, although not many studies of carotenoids in this vegetable have been reported. A group of Japanese investigators studied the composition and vitamin A value of the carotenoids in two varieties of pumpkin having different colors[52]: Cucurbita moschata is yellow fleshed, whereas Cucurbita maxima is of the orange-fleshed variety. The pigments were first separated by phase separation using liquid–liquid partition between hexane and 90%

[51] B. Tan, J. Food Sci. 53, 954 (1988).
[52] T. Hidaka, T. Anno, and S. Nakatsu, J. Food Biochem. 11, 59 (1987).

methanol. The fractions obtained were analyzed by open-column chromatography using a mixture of MgO–Hyflo Super Cel (1:1, w/w). Pigments were eluted from the column using different proportions of acetone in hexane. The three major carotenoids detected were lutein, β-carotene, and luteoxanthin, together making up about 75% of all the carotenoids, with concentrations of 922, 505, and 370 μg/100 g, respectively. Other carotenoids detected were, in descending order, taraxanthin, zeaxanthin, auroxanthin, β-cryptoxanthin, ζ-carotene, β-carotene 5,6-epoxide, and α-carotene. The vitamin A activity of the yellow variety was found to be higher than that of the orange variety, which had many oxygenated carotenoids.

Two other studies have also reported on the carotenoid composition of a pumpkin sample, determined by reversed-phase HPLC.[31,36] In one study,[36] total carotenoid concentration was less than 2300 μg/100-g sample, and the concentrations of lutein and α- and β-carotenes were 940, 756, and 578 μg/100 g, respectively. Cryptoxanthin, lycopene, and γ-carotene were not detected. The other study[31] reported a similar total carotenoid concentration, but the β-carotene content was much lower, at 49 μg/100-g sample.

Contrary to popular belief, the highest total carotenoid content or vitamin A activity is not found in orange-colored vegetables and fruits.[36] Instead, green leafy vegetables, including several local varieties used as ulam (vegetables consumed raw), have been found to be the richest sources of total carotenoids as well as provitamin A carotenes. The chlorophyll present in these leaves masks the carotenoids present. Several brightly colored vegetables and fruits, traditionally regarded as having "high carotene value," were found to have vitamin A activity values that were much lower than expected based on the color of the foods, because a large proportion of the carotenoid present in these foods was lycopene, a carotenoid with no vitamin A activity.

Fruits

Numerous investigators have studied the carotenoids in fruits, particularly the yellow- and orange-colored varieties. Data cited in this section were generated using reversed-phase HPLC and particular attention has been given to the few reports dealing with several fruits. Carotenoid compositions of fruits are generally more complex than those for green vegetables. Because of the numerous types of fruits, with different characteristics, there are large variations in carotenoid composition and content in these plant sources. Data for selected fruits are summarized in Table IV.

β-Carotene is found in most fruits, but the concentration varies widely. Low β-carotene concentrations of less than 100 μg/100 g fresh weight were

TABLE IV
CAROTENOID CONCENTRATIONS[a] IN FRUITS

Fruit	Range/mean carotenoid concentration				
	Lutein	Cryptoxanthin	Lycopene	α-Carotene	β-Carotene
Banana	20–40	0	0	60–160	40–100
Berries, grapes, black currant	20–200	0	0	0–60	6–150
Mango	—	—	—	—	63–615
Orange, mandarin	20–30	7–300	—	20	25–80
Papaya, watermelon	0	450–1500	2000–5300	0	228–324
Starfruit	60	1070	0	0	28

[a] Data given in micrograms per 100 g fresh weight. Compiled from Refs. 31, and 34–36.

reported for various types of berries, grapes, black and red currant, and several citrus fruits.[34,35] Watermelon and plum were found to have the highest β-carotene concentrations, approximately 200–400 μg/100 g of fruit.

Oranges were found to have a low β-carotene concentration of less than 50 μg/100 g.[34–36] Orange juice concentrates have been reported to have a higher β-carotene content, ranging from 80 to 260 μg/100 g of juice.[53]

In the study of 14 tropical fruits, β-carotene concentration was found to vary widely.[36] Fruits with low levels (less than 100 μg/100 g fruit) of this carotenoid included bananas, jackfruit *(Artocarpus heterophyllus),* musk lime *(Citrus microcarpa),* and starfruit *(Averrhoa carambola).* Watermelon, papaya, papaya exotica, tree tomato, and a Thai variety of mango have β-carotene concentrations ranging from 200 to 600 μg/100 g. The mango had the highest concentration of β-carotene in this study[36] and almost all the carotenoids detected were β-carotene. Ripe mangoes were found to have five times more carotenoids than the unripe fruits, and β-carotene content was reported to be 63 μg/100 g of fruit.[31]

α-Carotene has been reported to occur in several fruits, including banana, orange, mandarin, berries, peaches, and cantaloupe.[34–36] The level of α-carotene is generally low with less than 100 μg/100 g of fruit. Cryptoxanthin is also found in several fruits, at low concentrations. However, in several other tropical fruits, e.g., papaya, starfruit, and tree tomato, over 1000 μg cryptoxanthin/100 g fruit has been reported.[36]

Lycopene occurs in high concentrations (2000 to 4500 μg/100 g in

[53] T. Philip and T. S. Chen, *J. Food Sci.* **53,** 1703 (1988).

deep-orange or reddish-colored fruits such as papaya, watermelon, and pink grapefruit.[35,36,41] Lutein occurs more regularly in many fruits, at concentrations ranging from 12 to 450 μg/100 g of fruit.[35,36]

Besides the major carotenoids in fruits, the occurrence of various other carotenoids in apricots, peaches, cantaloupe, and pink grapefruit has also been reported.[41] A combination of isocratic and gradient reversed-phase HPLC was used to separate the three classes of carotenoids (xanthophylls, hydrocarbon carotenoids, and carotenol fatty acid esters). The carotenoids were monitored at six wavelengths using a photodiode array detector. The xanthophylls were identified as zeaxanthin and β-cryptoxanthin. The hydrocarbon carotenoids were identified as lycopene, γ-, ζ-, and β-carotene, phytofluene, and phytoene, as well as several of their cis isomers. The carotenol fatty acid esters were found to be present only in peaches.

The vitamin A activity of fruits is generally low. However, fruits contain more complex carotenoids, and several of these carotenoids may be of nutritional significance, in addition to being provitamin A compounds.

Roots and Tubers

Compared to vegetables and fruits, carotenoids in roots and tubers have gained less attention. Although generally low in carotenoids, roots and tubers could be significant sources of provitamin A carotenoids when consumed in sufficient quantities.

Sweet potato has been studied more frequently than other roots and tubers. Recent data, generated using HPLC, are tabulated in Table V. The large variation in carotenoid concentrations reported by various investigators[34,54,55] may depend on the variety of sweet potato studied, as these potatoes are known to vary considerably in color.

Few studies on carotenoids in cassava have been reported. The edible portion of the tuber is generally white, and would be expected to have a low carotenoid concentration. A study has shown that the total carotenoid concentration is less than 50 μg/100 g edible portion.[55] HPLC showed that the β-carotene content was only 20 μg/100 g, while lutein, cryptoxanthin, and lycopene were also detected in small quantities (Table V) and γ- and α-carotenes were not detected at all. In screening 654 clones in the cassava germplasm collection, it was found that several clones have different intensities of yellow color.[56] The β-carotene content of 21 of these clones was studied using the open-column chromatography method of the Association

[54] U. Singh and J. H. Bradbury, *J. Sci. Food Agric.* **45,** 87 (1988).
[55] E. S. Tee, unpublished observation (1991).
[56] S. N. Moorthy, J. S. Jos, R. B. Nair, and M. T. Sreekumari, *Food Chem.* **36,** 233 (1990).

TABLE V
β-CAROTENE AND LYCOPENE CONCENTRATIONS IN SELECTED ROOTS AND TUBERS

Root or tuber	Range/mean carotenoid concentration				Ref.
	Lutein	Cryptoxanthin	Lycopene	β-Carotene	
Sweet potato					
Different varieties	—	0	—	5–551	34
	—	—	—	1–4	54
Yellow variety	25	0	42	19	55
Orange variety	7	27	147	1140	55
Cassava					
White variety	2	3	1	20	55
Yellowish	—	—	—	40–790	56
Potato	13–60	Trace	Trace	3–40	34–36
Taro	3–31	1	1–3	2–16	55

^a Data given in micrograms per 100 g fresh weight.

of Official Analytical Chemists (AOAC). β-Carotene content was indeed higher, and was found to range from 40 μg/100 g for the faint yellow clones to 790 μg/100 g for yellow tubers. Potatoes and taro are other root crops with low carotenoid concentrations (Table V).

Vegetable Oils

Crude oil of yellow maize (corn) is a fairly good source of the provitamin A carotenoids.[57] The carotenoid pigments in peanut oil have been studied during maturation.[58,59] Carotenoid esters in soybean and rapeseed oils have also been studied.[60] The concentration of carotenoids contained in raw rapeseed and linseed oils is about 30–40 ppm.[61] The major carotenes found in olive oil are β-carotene and lutein, which constitute about 30 to 31 ppm, respectively.[62,63] Other vegetable oils such as barley oil, sunflower seed oil, and cottonseed oil also show the presence of carot-

[57] S. S. Mikhitaryan, A. P. Nechaev, Y. I. Denisenko, and V. T. Lyubushkin. *Izv. Vyssh, Uchebn. Zaved., Pischch — Teknol.* **6,** 29 (1966).
[58] H. E. Patte and A. E. Purcell, *J. Am. Oil Chem. Soc.* **46,** 629 (1969).
[59] E. P. Harold and E. P. Albert, *J. Am. Oil Chem. Soc.* **44,** 328 (1967).
[60] P. E. Froehling, G. Van Den Bosch, and H. A. Boekenoogen, *Lipids* **7,** 447 (1972).
[61] J. Peredi, *Elelmiszervizsgalati Kozl.* **22,** 255 (1976).
[62] B. Stancher, F. Zonta, and P. Bogoni, *J. Micronutr. Anal.* **2,** 97 (1987).
[63] M. J. Minguez-Mosquera, and J. Garrido-Fernandez, *Grasas Aceites (Seville)* **37,** 337 (1986).

TABLE VI
TOTAL CAROTENOIDS FROM VARIOUS OIL PALM SPECIES[a]

Oil palm species[b]	Total carotenoids[c] (ppm)
E.o.	4347
E.o. × E.g. (D)	1846
E.o. × E.g. (P)	1289
E.o. × E.g. (D) × E.g.(P)	864
E.g. (P)	380
E.g. (D)	948
E.g. (T)	610

[a] From Ref. 74.
[b] E.o., Elaeis oleifera; E.g., Elaeis guineensis; D, Dura; P, Pisifera; T, Tenera.
[c] Total carotenoids estimated at 446 nm.

enoids.[61,64,65] However, the concentration of carotenoids in vegetable oils is generally low. Of the vegetable oils that are widely consumed, palm oil contains the highest known concentration of agro-derived carotenoids.[66,67]

Red palm oil is obtained from the fruits of the palm tree *Elaeis guineensis,* of which a large number of subspecies are known.[68,69] Palm forests grow in West Africa, but the tree is cultivated in East Africa, Java, Malaysia, and South America.[69,70] Palm oil from the Far East and from the Belgian Congo contains 500–800 ppm of carotenes, whereas that from the Ivory Coast (especially Dahomey) contains 1000–1600 ppm, but the oil yield is less.[69,71] Palm oil derived from Malaysian-planted *Tenera (T)* species has a carotenoid content of about 500–700 ppm.[72]

It was also reported that other oil palm species such as *Elaeis oleifera (Melanococca)* from South America have a total carotenoid content of

---

[64] A. I. Demchenko, *Izv. Vyssh. Uscheb. Zaved. Pishch. Teknol.* **5,** 18 (1969).
[65] A. L. Markmen and A. U. Umarav, *Uzb. Khim. Zh.* **6,** 61 (1961).
[66] K. E. Ben and L. Brixius, *Fette Seifen* **67,** 65 (1965).
[67] B. Tan, *J. Am. Oil Chem. Soc.* **66,** 770 (1989).
[68] A. Beinayme, "Etudes sur de carotene de l'hulle de palme," Ser. Sci. 8. Inst. Rech. Thiles and Oleagineux, Paris, 1955.
[69] C. W. S. Hartley, *in* "The Oil Palm," 2nd Ed., Chs. 2 and 5. Longman, London, 1977.
[70] P. Blaizot and P. Cuvier, *J. Am. Oil Chem. Soc.* **30,** 586 (1953).
[71] J. C. Bauernfeind (ed.), "Carotenoids as Colorants and Vitamin A Precursors: Technical and Nutritional Applications." Academic Press, New York, 1981.
[72] S. H. Goh, Y. M. Choo, and A. S. H. Ong, *J. Am. Oil Chem. Soc.* **62,** 237 (1985).

TABLE VII
CAROTENOID CONTENTS OF PALM OIL EXTRACTS[a]

Palm oil extract	Carotenoid content[b] (ppm)
Crude palm oil	630–700
Crude palm olein	680–760
Crude palm stearin	380–540
Second pressed oil	1800–2400
Residual oil from fiber	4000–6000

[a] From Ref. 74.
[b] Total carotenoids estimated at 446 nm.

about 4000 ppm. The hybrid palms, a cross between the African oil palm *E. guineensis (E.g.)* and the South American oil palm *E. oleifera (E.o.)*, i.e., *E.o.* × *E.g.* hybrids, have been shown to provide several advantages including a more unsaturated oil, a lower height increment in trunk growth, and resistance to certain diseases.[69,73] They have been reported to have carotenoid content intermediate between that of *E. oleifera* and *E. guineensis*,[69,74,75] as shown in Table VI. The *Elaeis oleifera* species has a higher carotenoid concentration, being seven times higher than the present commercially planted species, *Tenera*, followed by the hybrids of *E.o.* × *E.g.* *(Dura)*.

A carotenoid-rich oil can also be obtained from the palm pressed fiber,[74,76] a palm oil by-product which is presently burnt as a fuel in palm oil mills. The pressed fiber was found to contain about 5–6% of residual oil with a carotenoid concentration of 4000–6000 ppm (Table VII), and the carotenoid content of the residual oil in the pressed fiber of hybrid oil palm fruits was even higher, at 6000–7000 ppm.[74,75]

Carotenoids from the commercial crude palm oil are concentrated during the extraction and fractionation process. A system of palm oil extraction based on a double-pressing technique has been implemented by several mills in Malaysia.[77] The expected advantages of double pressing over the conventional single-stage pressing are lower oil loss in fiber, higher

[73] J. Meunier, G. Valleja, and D. Boutin, *Oleagineux* **31**, 519 (1976).
[74] Y. M. Choo, S. C. Yap, A. S. H. Ong, C. K. Ooi, and S. H. Goh, *in* "Proceedings of the AOCS World Conference of Edible Fats and Oils: Basic Principle and Modern Practices" (D.R. Ericson, ed.), pp. 436–440.
[75] S. C. Yap, Y. M. Choo, C. K. Ooi, A. S. H. Ong, and S. H. Goh, *Elaeis* **3**, 369 (1991).
[76] Y. M. Choo, A. S. H. Ong, C. K. Ooi, and S. C. Yap, U.K. Patent Appl. 2212806 (1991).
[77] B. Sidek, and H. T. Lim, paper presented at the National Oil Palm/Palm Oil Conference —Current Development, October 11–15, 1988, Kuala Lumpur, Malaysia.

kernel extraction rate, less wear on the screw worm and cages, and reduction of contamination of the kernel oil in crude palm oil. More interestingly, the oil produced from the second pressing has a higher concentration of carotenoids (Table VII). This could be due to the fact that the first extraction (first pressing) in a double-pressing process is carried out at lower pressure (to avoid cracking of the nuts) and oil relatively higher in carotenoids was extracted. After removal of nuts the fiber is then subjected to higher pressure and more carotenoids are extracted out of the mesocarp, together with some residual oil from the first pressing.

Carotenoids are also being concentrated in an industrial process called fractionation.[78,79] Palm oil is a semisolid fat at ordinary room temperature; its semisolid nature is due to the presence of the solid, fully saturated triglycerides and the high melting point monooleo-glycerides and monolinoleoglycerides dispersed through the liquid dioleinglycerides and other more unsaturated glycerides. Fractionation is carried out to extend the uses of palm oil and the products obtained are the liquid oil (olein, 70–80%) and the solid fat (stearin, 20–30%). The liquid oil is used as cooking oil and the solid fraction can be used as a component of the harder frying fats, for the production of margarine, vanaspati component, or cocoa butter substitute. The carotenoid content in the palm olein (lower melting) fraction is enriched by 10–20%, as shown in Table VII.

Various analytical methods have been used for the determination of the carotenoid profile of crude palm oil. Column chromatography has been used in the earlier studies using different adsorbents[79–82] and reverse phase HPLC has been shown to have several advantages for the separation of carotenoids in the oil.[83,84] The major carotenoids present in crude palm oil are α- and β-carotenes, which constitute about 80–90% of the total carotenoid content. Other carotenoids are present as minor components, including ζ-carotene, phytofluene, phytoene, lycopene, neurosporene, γ-carotene, and δ-carotene, as well as xanthophylls such as zeaxanthin, α-carotene 5,8-epoxide, and β-carotene 5,6-epoxide[75,80,84] (see Fig. 1). The carotenoid composition of various species of palm oil is given in Table VIII.

The carotenoid profile for the oil extracted from fiber is slightly different in terms of chemical composition as compared with the carotenoid

[78] J. W. E. Coenen, *Rev. Fr. Corps Gras* **21**, 343 (1974).
[79] T. D. Tjang and J. J. Olie, *Planter* **48**, 201 (1972).
[80] A. T. Q. Jose, B. R. A. Delva, W. Esteves, and F. P. Gerhard, *Fat Sci. Technol.* **92**, 222 (1990).
[81] M. Agroud, *Oleagineux* **13**, 249 (1958).
[82] B. Tan, C. M. Grady, and A. M. Gawienowski, *J. Am. Oil Chem. Soc.* **63**, 1175 (1986).
[83] H. J. C. F. Nelis and A. P. De Leenheer, *Anal. Chem.* **55**, 270 (1983).
[84] J. H. Ng and B. Tan, *J. Chromatogr. Sci.* **26**, 463 (1988).

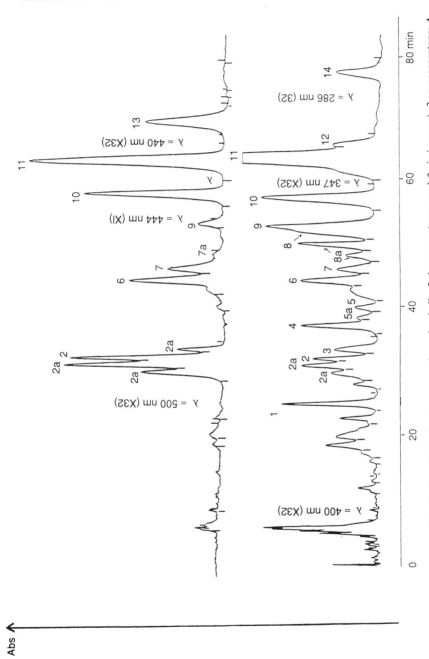

FIG. 1. HPLC of carotenoids of palm oil. Peaks: 1, presumed xanthophylls; 2, lycopene (trans and 3-cis isomers); 3, α-zeacarotene; 4, β-zeacarotene; 5, neurosporene (trans and cis isomers); 6, δ-carotene; 7, γ-carotene (trans and cis isomers); 8, ζ-carotene; 9, cis-α-carotene; 10, α-carotene; 11, β-carotene; 12, phytofluene; 13, cis-β-carotene; 14, phytoene. a, cis isomer.

TABLE VIII
PERCENT COMPOSITION OF CAROTENOIDS IN PALM OIL[a]

Carotenoids	Elaeis guineensis (E.g.)			Elaeis oleifera or Melanococca (E.o.)	Hybrids (E.o. × E.g.)		Palm	
	Tenera	Pisifera (P)	Dura (D)		E.o. × E.g. (P)	E.o. × E.g. (D)	Pressed fiber oil	Second pressed oil
Phytoene	1.27	1.68	2.49	1.12	1.83	2.45	11.87	6.50
Phytofluene	0.06	0.90	1.24	Trace	Trace	0.15	0.40	1.63
cis-β-Carotene	0.68	0.10	0.15	0.48	0.38	0.55	0.49	0.28
β-Carotene	56.02	54.39	56.02	54.08	60.53	56.42	30.95	31.10
α-Carotene	35.06	33.11	24.35	40.38	32.78	36.40	19.45	20.68
cis-α-Carotene	2.49	1.64	0.86	2.30	1.37	1.38	1.77	1.70
ζ-Carotene[b]	0.69	1.12	2.31	0.36	1.13	0.70	7.56	4.62
δ-Carotene	0.83	0.27	2.00	0.09	0.24	0.22	6.94	2.13
γ-Carotene	0.33	0.48	1.16	0.08	0.23	0.26	2.70	2.48
Neurosporene[c]	0.29	0.63	0.77	0.04	0.23	0.08	3.38	1.88
β-Zeacarotene	0.74	0.97	0.56	0.57	1.03	0.96	0.37	0.58
α-Zeacarotene	0.23	0.21	0.30	0.43	0.35	0.40	Trace	0.15
Lycopene[d]	1.30	4.50	7.81	0.07	0.05	0.04	14.13	26.45
Total (ppm)	673	428	997	4592	1430	2324	5162	2510

[a] From Ref. 74.
[b] Together with two cis isomers.
[c] Together with a cis isomer.
[d] Together with three cis isomers.

TABLE IX
CAROTENOID CONCENTRATES FROM VARIOUS METHODS[a]

Method	Carotenoid content[b] (ppm)	Recovery (%)
Vacuum distillation[c]	>20,784	<46
Molecular distillation[c]	>80,000	>80
Adsorption[c]		
C_{18}	8,000–9,000	>90
Carbon	5,000–7,000	<50
Moleculat distillation of crude palm oil	1,290–1,990	
Adsorption from crude palm oil (activated carbon)	3,700–5,600	<80

[a] From Ref. 74.
[b] Total carotenoids estimated at 440 nm.
[c] Through methyl ester route.

profile from crude palm oil. The major carotenoids are still α- and β-carotenes, but these constitute only about 50% of the total carotenoids present. Phytoene, lycopene, ζ-carotenes, and δ-carotenes are present at higher concentrations.[74] It is interesting to note that the carotenoid profile of the second pressed oil is similar to that of the fiber oil (Table VIII).

Because carotenoids are likely to grow in importance and value, the recovery of carotenoids from palm oil and palm oil by-products is important. In addition, some studies are also being carried out to recover and concentrate carotenoids from various palm oil sources in order to obtain an oil with high concentrations of palm-based carotenoids for pharmaceutical uses. Numerous extraction methods have been developed to recover the carotenoids from crude palm oil; these include the saponification method,[85,86] urea process,[87] adsorption,[88–91] selective solvent extraction,[91,92] molecular distillation,[93] and transesterification followed by distillation of esters.[94–97]

[85] J. M. Tabor, H. F. Seibert, and P. R. Frohring, U.S. Patent 2440029 (1948).
[86] P. P. Blaizot, U.S. Patent 2652433 (1953).
[87] G. Knafo, *Bull. Mens. Inst. Technol. Etudes Recherches Gras* **6**, 323 (1952).
[88] A. S. H. Ong and P. L. Boey, British Patent 1562794 (1980).
[89] Unilever, Ltd., British Patent 691924 (1953).
[90] H. Mamuro, Y. Kubota, and H. Shiina, Japanese Patent 61282357 (1986).
[91] Y. Tanaka, I. Hama, A. Oishida, and A. Okabe, British Patent 2160874 A (1986).
[92] H. J. Passino, U.S. Patent 2615927 (1952).
[93] T. L. Ooi, A. S. H. Ong, Y. Mamuro, W. Kubota, H. Shinna, and S. J. Nakasato, *J. Jpn. Oil Chem. Soc.* **35**, 543 (1986).
[94] Lion Fat and Oil Company, British Patent 1515238 (1976).

Most of the reported methods of recovering carotenoids directly from palm oil are difficult, inefficient, or costly. Volatile palm oil methyl esters have been prepared on a large scale as an oleochemical or diesel substitute.[98,99]

A mild reaction converts palm oil (triglycerides) to volatile methyl esters, leaving the valuable minor components unchanged[100] and allowing the recovery of carotenoids in palm oil. The carotenes are concentrated or recovered from the volatile ester by adsorption,[101] solvent–solvent extraction,[96] polymer membrane,[102] and distillation.[103] The method involves selective adsorption of the carotenoids, using reversed-phase adsorption material[101]; the esters with higher polarity are eluted first from the column and then the carotenoids are recovered. The carotenoid concentrations recovered range from 8000 to 9000 ppm (Table IX). A recovery greater than 90% can be obtained using this method and the column can be regenerated and reused more than 30 times without any loss of activity. The carotenoid concentration obtained through the carbon adsorption of the crude palm methyl ester is also shown in Table IX; the recovery, as well as the carotenoid concentration, is low compared to the C_{18} reversed-phase method. Data on carotenoid content reported using the activated carbon adsorption as well as the molecular distillation of crude palm oil are also included in Table IX.

The second method involves the distillation of the volatile alkyl ester using normal vacuum distillation or molecular distillation techniques.[103] Residual concentrates of 2.0% carotenoids (Table IX) content can be achieved by normal vacuum distillation with a recovery of about 46%. This residual carotenoid can be further concentrated to 8.4% by normal-phase column chromatography and at the same time other separated minor components were also concentrated.[74] Total tocopherol and tocotrienol contents are increased to 37% and sterols are concentrated to 32%, with a recovery of 83 and 81%, respectively, based on the crude methyl ester.[100] An oil with a final carotenoid concentration of 80,000 ppm has been

---

[95] E. W. Eckey, U.S. Patent 2460796 (1949).
[96] N. Hara, I. Hama, H. Izumimoto, and A. Nakamura, Japanese Patent 6305074 (1988).
[97] I. Hama, Y. Tanaka, Y. Yogo, and T. Okabe, Japanese Patent 61109764 (1986).
[98] N. O. V. Sonntag, *J. Am. Oil. Chem. Soc.* **61**, 229 (1984).
[99] Y. M. Choo, A. S. H. Ong, K. Y. Cheah, and A. Baker, Australian Patent PJ 1105/88 (1988).
[100] Y. M. Choo and Ab. Gapor, *Top. Palm Oil Dev.* **14** (special issue), 39 (1990).
[101] Y. M. Choo, S. H. Goh, A. S. H. Ong, and T. S. Kam, British Patent 2218989 (1991).
[102] K. Yamada, M. Egawa, Y. Endo, and I. Hoshiga, Japan Patent 76147533 (1976).
[103] C. K. Ooi, Y. M. Choo, and A. S. H. Ong, U.S. Patent No. 5019668 (1991).

achieved through molecular distillation.[104] A carotene with a concentration of 72% has been obtained through column adsorption.[105]

A process has also been developed to produce deacidified and deodorized red palm oil from degummed palm oil with >80% of the original carotenes still intact.[106] This red palm oil is of similar excellent quality to the refined, bleached, and deodorized (rbd) palm oil that is normally traded, but contains no carotenes.

Acknowledgment

We thank Yuen-May Choo and Soon-Chee Yap of the Palm Oil Research Institute of Malaysia for assistance in the preparation of this manuscript.

[104] Y. M. Choo, personal communication (1992).
[105] I. Hama, N. Hara, Y. Tanaka, and M. Nakamura, Japanese Patent Appl. 86/86,333 (1986).
[106] C. K. Ooi, Y. M. Choo, and A. S. H. Ong, Australian Patent PI 7267/88 (1988).

[15] Distribution of Carotenoids

By T. W. GOODWIN

Introduction

Carotenoids are probably the most widely distributed naturally occurring pigments, being found in almost every phylum in the plant and animal kingdoms. The extensive distribution of the pigments allied with some 500 different structures makes a full assessment of the biological significance of the patterns observed impossible in the space available, but a general picture will be presented. For full details specialist volumes[1-3] and reviews[4,5] must be consulted. Similarly no structures will be presented but the full chemical name of each pigment will be given.[1,3]

[1] T. W. Goodwin, "The Biochemistry of the Carotenoids," 2nd Ed., Vol. 1. Chapman and Hall, London, 1980.
[2] T. W. Goodwin, "The Biochemistry of the Carotenoids," 2nd Ed., Vol. 2. Chapman and Hall, London, 1984.
[3] T. W. Goodwin, and G. Britton, in "Plant Pigments" (T. W. Goodwin, ed.), pp. 61-132. Academic Press, San Diego, 1988.
[4] G. Britton, Terpenoids Steroids (a) **6**, 144 (1976); (b) **7**, 155 (1977); (c) **8**, 181 (1978); (d) **9**, 218 (1979); (e) **10**, 164 (1981); (f) **11**, 133 (1982); (g) **12**, 235 (1983).
[5] G. Britton, Nat. Prod. Rep. (a) **1**, 67 (1984); (b) **2**, 349 (1985); (c) **3**, 591 (1986).

TABLE I
MAIN CAROTENOIDS IN CHLOROPLASTS OF
HIGHER PLANTS[a]

Trivial name	Systematic name
α-Carotene	β,ε-Carotene
β-Carotene	β,β-Carotene
β-Cryptoxanthin	β,β-Caroten-3-ol
Lutein	β,ε-Carotene-3,3′-diol
Zeaxanthin	β,β-Carotene-3,3′-diol
Violaxanthin	5,6,5′,6′-Diepoxy-5,6,5′,6′-tetrahydro-β,β-carotene-3,5,3′-diol
Neoxanthin	5′,6′-Epoxy-6,7-didehydro-5,6,5′,6′-tetrahydro-β,β-carotene-3,5,3′-triol

[a] Relative amounts of the pigments can vary considerably from species to species.

Despite their wide distribution carotenoids are synthesized *de novo* only in higher plants and protists. The source of animal carotenoids can always be traced to the presence of higher plants and/or protists somewhere along the food chain. Carotenoids characteristic of animals are formed by metabolic transformations of these food carotenoids.[2,6]

It must be emphasized that carotenoids are present in all photosynthetic tissues where they play an essential role as protectors against photodynamic sensitization in the light-harvesting process. In the absence of carotenoids aerobic photosynthesis cannot take place.[7]

Higher Plants

Chloroplasts

Carotenoids occur in all chloroplasts, generally in a simple mixture (Table I). They exist as complexes in noncovalent binding with proteins.[7] Unusual carotenoids are noted in the leaves of some gymnosperms, in which they accumulate in extraplastidic oil droplets[3]; typical pigments are rhodoxanthin (4′,5′-didehydro-4,5′-*retro*-β,β-carotene-3,3′-dione) in red winter leaves of some members of the Cupressaceae and Taxaceae, and

---

[6] T. W. Goodwin, *Annu. Rev. Nutr.* **6,** 273 (1986).
[7] R. Cogdell, *in* "Plant Pigments" (T. W. Goodwin, ed.), pp. 183–230. Academic Press, San Diego, 1988.

semi-β-carotenone (5,6-*seco*-β,β-carotene-5,6-dione) in young red leaves of two cycads.[3]

Reproductive Tissues

Not all reproductive tissues contain carotenoids but those that do accumulate complex mixtures with some pigments being species specific, e.g., lilixanthin (3,4,3'-trihydroxy-$\beta\kappa$-caroten-6'-one) in *Lilium* species, and crocetin (8,8'-diapocarotene-8,8'-dioic acid) in *Crocus* species.[3] Well over 40 pigments unique to flower petals have been reported and many are cis (Z)-isomers.[1,3] Carotenogenic flowers can be roughly divided into three main groups: those synthesizing (1) highly oxygenated carotenoids, frequently 5,8-epoxides, (2) mainly carotenes, and (3) highly species-specific pigments, for example eschscholtzanthin (4,5'-didehydro-4,5'-retro-β,β-carotene-3,3'-diol) from the California poppy.

Fruits are even more prodigal in their synthetic ability than are flowers and over 70 characteristic fruit carotenoids have been described.[1,3] Table II indicates the eight general groups into which fruits can be placed according to their carotenoid-synthesizing proclivities.

Algae

The carotenoids present in chloroplasts of the different algal classes represent, with one exception, complex mixtures that tend to be characteristic of each class. This is in marked contrast to the chloroplast carotenoids of higher plants, which are always similar, rather simple mixtures (Table I). The exception is the class Chlorophyta, members of which tend to accu-

TABLE II
GROUPING OF FRUITS ACCORDING TO CAROTENOID COMPOSITION

Group	Pigment characteristic
I	Insignificant amounts
II	Small amounts of chloroplastidic carotenoids
III	Relatively large amounts of lycopene (ψ,ψ-carotene) and its hydroxy derivatives
IV	Relatively large amounts of β-carotene and its hydroxy derivatives
V	Large amounts of epoxides, particularly 5,8-epoxides
VI	Unusual carotenoids, e.g., capsanthin 3,3'-dihydroxy-β,κ-caroten-6'-one (peppers)
VII	Poly-*cis(Z)*-carotenes, e.g., prolycopene 7,9,7',9'-tetra-*cis*-ψ,ψ-carotene (tomato mutants)
VIII	Apocarotenoids, e.g., persicaxanthin (5,6,-dihydro-12'-apo-β-carotene-3,12-diol) (peaches)

mulate the same pigments as the higher plants but with some individuality, e.g., siphonaxanthin (3,19,3'-trihydroxy-7,8-dihydro-β,ε-caroten-8-one) occurs in many *Caulerpales*.[1,3] A brief indication of the carotenoid distribution in algal classes is outlined in Table III.[1,3,8]

In some stress situations, e.g., nitrogen deficiency, certain algae accumulate large amounts of β-carotene in oily droplets or in the cell wall. *Dunaliella bardawil* can synthesize up to 9% dry weight of β-carotene under conditions of high light intensity and high salt concentration.[9]

Bacteria

Photosynthetic Bacteria

The purple photosynthetic bacteria are characterized by their ability to synthesize acyclic carotenoids having tertiary hydroxyl groups, frequently methylated at C-1 and/or C-1', additional double bonds at C-3,4 and/or C-3',4' and, less frequently, keto groups at C-2 and/or C-2'; a typical pigment is spirilloxanthin (1,1'-dimethoxy-3,4,3',4'-tetradehydro-1,2,1',2'-tetradehydro-ψ,ψ-carotene). The pigments are located in chromatophores that are part of the membrane system of the bacterial cells.

A characteristic of the green photosynthetic bacteria is the appearance of carotenoids such as isorenieratene (ϕ,ϕ-carotene), which is also found, originally named leprotene, in certain mycobacteria (nonphotosynthetic). Members of the class Cyanophyta, until recently classed as blue-green algae, contain cyclic carotenoids characterized by keto groups at C-4 (echinenone, β,β-caroten-3-one) and by being glycosylated at C-2' [myxoxanthophyll, 2'-(β-L-rhamnopyranosyloxy)-3',4'-didehydro-1',2'-dihydro-β,ψ-carotene-3,1'-diol].

Nonphotosynthetic Bacteria

The sporadic appearance of carotenoids in nonphotosynthetic bacteria is matched by the appearance of unique C_{30} carotenoids (some staphylococci) and C_{45} and C_{50} carotenoids (many flavobacteria), as well as glycosidic C_{40} pigments (mycobacteria). Of the conventional C_{40} carotenoids xanthophylls, including the epoxides characteristic of higher plants, are rarely found.

[8] T. Bjørnland, in "Carotenoids: Chemistry and Biology" (N. I. Krinsky, M. M. Mathews-Roth, and R. F. Taylor, eds.), Proc. 8th Int. Symp. Carotenoids, p. 21. Plenum, New York, 1989.

[9] K. Schiedt, in "Carotenoids: Chemistry and Biology" (N. I. Krinsky, M. M. Mathews-Roth, and R. F. Taylor, eds.), Proc. 8th Int. Symp. Carotenoids, p. 247. Plenum, New York, 1989.

TABLE III
CAROTENOID CHARACTERISTICS OF ALGAL CLASSES[a]

Classes	Characteristic pigments
Chlorophyta	Frequently same as higher plants, but a number of exceptions
Rhodophyta	Simple mixture, generally of α- and β- carotenes and lutein and zeaxanthin. Epoxides rare
Pyrrophyta	Characteristic pigment, peridinin[b]
Chrysophyta	Epoxy, allenic, and acetylenic carotenoids, e.g., fucoxanthin[c], diadinoxanthin[d]
Euglenophyta	Acetylenic carotenoids, e.g., diatoxanthin[e]
Chloromonadophyta (Raphidophyta)	Acetylenic, allenic, and epoxycarotenoids, e.g., heteroxanthin[f]
Xanthophyta	Same as in Chloromonadophyta
Cryptophyta	Acetylenic carotenoids, e.g., alloxanthin[g]
Phaeophyta	Fucoxanthin[c]

[a] From Refs. 1, 3, and 8.
[b] $5',6'$-Epoxy-$3,3,3'$-trihydroxy-$6,7$-didehydro-$5,6,5',6'$-tetrahydro-$10,11,20$-trinor-β,β-caroten-$10',11'$-olide 3-acetate.
[c] $5,6$-Epoxy-$3,3',5'$-trihydroxy-$6',7'$-didehydro-$5,6,7,8,5',6'$-hexahydro-β,β-caroten-8-one $3'$-acetate.
[d] $5,6$-Epoxy-$7',8'$-didehydro-$5,6$-dihydro-β,β-carotene-$3,3'$-diol.
[e] $7,8$-Didehydro-β,β-carotene-$3,3'$-diol.
[f] $7',8'$-Didehydro-$5,6$-dihydro-β,β-carotene-$3,5,6,3'$-tetraol.
[g] $7,8,7',8'$-Tetradehydro-β,β-carotene-$3,3'$-diol.

Fungi

As in the nonphotosynthetic bacteria carotenoid distribution in fungi is apparently capricious, with phycomycetes characterized by the synthesis of predominantly β-carotene. Higher plant xanthophylls are rare but unique pigments, such as torulene ($3',4'$-didehydro-β,ψ-carotene), appear in the red yeasts; one such organism uniquely produces $3R,3'R$-astaxanthin ($3,3'$-dihydroxy-β,β-carotene-$4,4'$-dione), whereas astaxanthin from algae *(Haematococcus)* and aquatic animals has the opposite chirality.[3]

Animals

All animal carotenoids arise from plant carotenoids ingested at some point in the food chain. They exist either (1) unchanged, (2) after oxidation, or (3) after reduction. Until recently pathway 2, which accounted for the widespread distribution of oxygenated carotenoids in marine animals, was considered the only pathway available to animals. However, in fish and in chickens astaxanthin and canthaxanthin (β,β-carotene-$4,4'$-dione) can be reduced via hydroxy derivatives to β-carotene.[9] This discovery explains the source of vitamin A in marine animals, which accumulate large amounts of keto carotenoids but no β-carotene as such.[6] Considerable current activity continues to reveal many new carotenoids in marine animals.[10]

Many bright colors of marine animals, e.g., the dark blue of lobster carapace (crustacyanin) and the red of *Pomacea* eggs, are the result of binding between carotenoids and proteins. The green pigment of the exoskeleton of the Turkish crayfish is the result of the coexistence of blue, green, and yellow carotenoproteins. The structure of these complexes is still being actively pursued.[11]

[10] T. Matsumo, *in* "Carotenoids: Chemistry and Biology" (N. I. Krinsky, M. M. Mathews-Roth, and R. F. Taylor, eds.), Proc. 8th Int. Symp. Carotenoids, p. 59. Plenum, New York, 1989.

[11] J. C. B. Findlay, D. J. C. Pappin, M. Brett, and P. F. Zagalsky, *in* "Carotenoids: Chemistry and Biology" (N. I. Krinsky, M. M. Mathews-Roth, and R. F. Taylor, eds.), Proc. 8th Int. Symp. Carotenoids, p. 75. Plenum, New York, 1989.

Section II

Separation and Quantitation

[16] Stability of β-Carotene under Different Laboratory Conditions

By GIORGIO SCITA

Carotenoids are a well-characterized class of compounds present in microorganisms, algae, higher plants, animals, and humans.[1,2] More than 600 carotenoids occur in nature. About 38 are precursors of vitamin A. Of these, only a few occur in sufficient concentration to play a significant role in the human diet. The most plentiful carotenoid is β-carotene (BC), which also has the highest vitamin A activity. Multiple biological functions beside the provitamin A activity have been ascribed to this compound, for instance, the activity as singlet oxygen or free-radical scavenger,[3] as a stimulant of the immuno response,[4] and the action as anticarcinogenic agent.[5] In consequence of this variety of functions and in correlation with the nontoxicity of BC, its biological role and metabolic fate have been the subject of emphasis in recent research. Nevertheless, little has been done to test the stability of BC under conditions other than food storage or processing. Simpson and Chichester,[6] for instance, reviewed the metabolic degradation of BC in senescent tissues, showing also how a pure solution of this pigment was destroyed or altered by acid and, in some cases by alkalies. El-Tinay and Chichester[7] studied the oxidation of BC in toluene at various temperatures, finding that the rate of BC loss indicated a zero-order reaction in the presence of excess oxygen. Carnevale *et al.*[8] showed that fluorescent light catalyzed the autooxidation of BC dissolved in fatty acid. A free-radical destruction of BC was shown to occur during the aerobic conversion of sulfite, a chemical widely used in foods, to sulfate, a process known to generate radicals.[9]

The purpose of this chapter is to define the conditions to be used in

---

[1] T. W. Goodwin, "The Comparative Biochemistry of Carotenoids." Chapman and Hall, London, 1952.
[2] T. W. Goodwin, "The Biochemistry of the Carotenoids," 2nd Ed., Vol. 1. Chapman and Hall, London, 1980.
[3] G. W. Burton and K. U. Ingold, *Science* **224,** 569 (1984).
[4] A. Bendich, *Clin. Nutr.* **7,** 113 (1988).
[5] N. I. Krinsky, *Clin. Nutr.* **7,** 107 (1988).
[6] K. L. Simpson and C. O. Chichester, *Annu. Rev. Nutr.* **1,** 351 (1982).
[7] A. H. El-Tinay and C. O. Chichester, *J. Org. Chem.* **35,** 2290 (1970).
[8] J. Carnevale, E. R. Cole, and G. Crank, *J. Agric. Food Chem.* **27,** 462 (1979).
[9] G. D. Peiser and S. F. Yang, *J. Agric. Food Chem.* **27,** 446 (1979).

laboratory research on β-carotene that could prevent its degradation. To do this, (1) BC stability was examined under conditions usually used in an experimental setting to investigate its metabolic or functional relevance both *in vitro* and in cell culture systems; (2) the effect of some antioxidants in preventing this degradation was tested; and (3) its storage stability under controlled conditions with regard to the type of solvent, light, temperature, and atmosphere was tested.

Reagents

β-[^{14}C]carotene (BC) was a gift from Hoffmann-La Roche (Nutley, NJ). Unlabeled BC was purchased from Fluka Biochemica (Buchs, Switzerland). α-Tocopherol, butylated hydroxytoluene (BHT), and taurodeoxycholic acid (TDCA) were from Sigma (St. Louis, MO). All the organic solvents were Fischer-brand (Santa Clara, CA) high-performance liquid chromatography (HPLC) grade. The HPLC column was an Ultrasphere ODS, 5 μm, 4.6 × 150 mm, from Beckman (Fullerton, CA). Separation was performed on a Beckman/Altex HPLC system. Citoscynt, the scintillation liquid, was from ICN (Irvine, CA).

High-Performance Liquid Chromatography

The extracts from each experiment (see below) were analyzed by HPLC in the following way: after evaporating to dryness under a stream of N_2, the samples were reconstituted in a mixture of toluene–methanol (1:3) and injected into the HPLC column. The optimized HPLC analysis was performed at a flow rate of 1.5 ml/min, using a linear gradient elution system from 100% solvent A (methanol plus 0.5% ammonium acetate) to 80% solvent A plus 20% toluene, developed in 12 min. The retention time of BC was 12 min. The effluent was monitored at 452 nm.

Stability of BC under Simulated Incubation Conditions

Two different micellar solutions of BC were prepared. The first was a modification of the method of Bertram[10] obtained by rapidly mixing 20 μl of a 2 m*M* solution of BC in tetrahydrofuran–dimethylsulfoxide (THF–

---

[10] R. V. Cooney and J. S. Bertram, this series, Vol. 214 [6].

DMSO) (1 : 1, v/v) plus 5 μl of a 0.5 mM [^{14}C]BC solution in THF–DMSO (1 : 1, v/v) into 5 ml of 90% Iscove's modified Dulbecco's medium (IMDM) with 10% bovine fetal serum and stirring it for 30 min. The second was prepared according to a modified procedure by Westergaard and Dietschy.[11] Crystalline BC was dissolved in THF to provide a final concentration of 8 mM. This solution (400 μl) was evaporated to dryness under N_2 and the resulting precipitated crystals were resuspended in 50 μl of THF and added to 55 ml of 40 mM taurodeoxycholic acid (TDCA) in 40 mM Tris, pH 7.4. The mixture was gently stirred overnight in the dark at room temperature, then diluted 1 : 1 with the same buffer without TCDA and sonicated for 15 min using a Branasonic (Branson Sonic Power, Danbury, CN) 220 sonicator bath. The BC recovered in micellar solution was 60%. Both types of micellar solutions were incubated at 37° in Corning (Corning, NY) dishes (60-mm diameter) in the dark, in presence of a 5% CO_2 in air atmosphere. At each time point a volume of ethanol plus 0.025% BHT was added to aliquots of each solution and, after saturating with NaCl, extracted twice with 6 vol of hexane. The samples were then evaporated to dryness and reconstituted either in toluene–methanol (1 : 3, v/v) when analyzed by HPLC, or in benzene when the BC concentration was determined by spectrophotometer at 452 nm. When [^{14}C]BC was used, each fraction from the HPLC elution was collected, mixed with scintillation liquid, and counted by a liquid scintillation counter.

Exposure of BC to Light

Crystalline BC was dissolved in a solution of toluene with or without 0.025% BHT or α-tocopherol to provide a final concentration of 4 μM. This solution (50 ml) was put into a 200-ml beaker to produce a layer 1.5 cm deep and a surface area of 45.3 cm^2. Exposure was to a fluorescent cold lamp (40 W) giving an intensity of 6500 lx at the surface of the sample from a distance of 10 cm. A UV lamp at 325 nm was used for UV treatment, giving an intensity of 750 pW/cm^2 with the sample at a 30-cm distance. The BC concentration was tested at different times by monitoring the absorbance at 463 nm. The experiment with a smaller air exchange was carried out under the same conditions except that a test tube with an exchange surface of 0.785 cm^2 was used.

---

[11] H. Westergaard and J. M. Dietschy, *J. Clin. Invest.* **58**, 97 (1976).

Stability of BC under Simulated Incubation Conditions

This study was carried out to test the stability of two different micellar BC solutions at different times, under dark conditions at 37°. Detection limits of less than 1 pmol were obtained by the use of liquid chromatographic techniques together with the utilization of radioactive BC.

Furthermore, as a double check, a spectrophotometric monitoring system was used. As shown in Fig. 1 the BC appeared to be fairly stable over 24 hr of incubation (less than 4%) with no significant differences between the two different methods used to prepare micellar solutions. Even after 48 hr of incubation, the loss of the pigment did not exceed 10 and 15%, respectively, for the TCDA and the THF micellar solutions. No known putative metabolic products were detected in the breakdown products of β-carotene. To have shown the BC is quite stable under conditions of relatively high temperature and for a long time is relevant in relation to the widespread use of cell-free and cell culture systems, which require similar incubation conditions for the study of BC metabolism (uptake and cleavage) and for the investigation of a direct anticarcinogenic action of this pigment.

Light Degradation and Antioxidant Effect

To explore the factors that could cause BC degradation, the effects of different kinds of light exposure were investigated. As shown in Figs. 2 and 3, the exposure of BC both to UV and fluorescent light, concomitant with experimental conditions that allowed a good atmospheric oxygen exchange, was highly damaging. The loss of the pigment was complete after 48 hr, resulting in total bleaching under both conditions: UV light was three times more effective (50% loss occurs after 8 hr) than fluorescent light (50% loss occurs after 24 hr). In this respect the present results are in accord with those previously obtained by Carnevale et al.[8] for BC autooxidation catalyzed by fluorescent light. Nevertheless, it is not easy to compare their rates of BC loss to the present data because the composition of lipids used as solvent system by these authors highly affected this rate (50% loss occurred after 15 hr in laurate solution, while no loss was observed in oleate solution), showing that a protective effect had been introduced with greater efficiency as unsaturation was increased.

Having established the degradation rate induced by light–atmospheric air exposure on BC, the protective effect of two common antioxidants, butylated hydroxytoluene (BHT) and α-tocopherol, were further investigated. BHT, at the usual concentration used in biological assays (1 mM), accounted for a loss reduction that ranged from more than 210% (50% loss

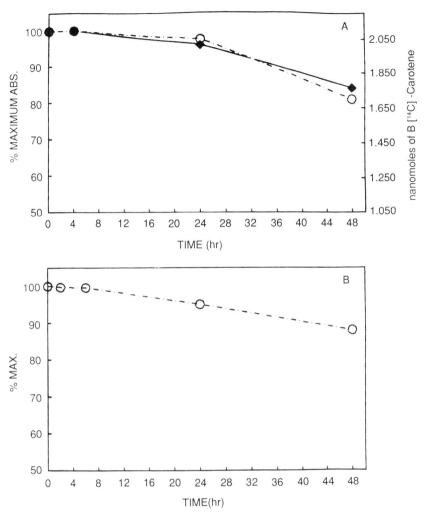

FIG. 1. Rate of degradation of β-carotene from a micellar solution under simulated incubation conditions. (A) a micellar solution in THF–DMSO (1:1, v/v) consisting of unlabeled (2 mM) and labeled BC (0.5 mM), was mixed with cell culture medium and incubated at 37° in the dark under a 5% CO_2:95% air atmosphere. An aliquot of each sample was extracted at different times and analyzed both for the unlabeled BC content (○), by measuring the absorbance at 452 nm in benzene, and for the labeled BC content (◆), by a monitoring system described in the text. (B) Micellar solutions of β-carotene (17 μM) in taurodeoxycholic acid were incubated and analyzed as described above. Each value is the mean of at least two determinations, which differed by no more than 5% of the mean.

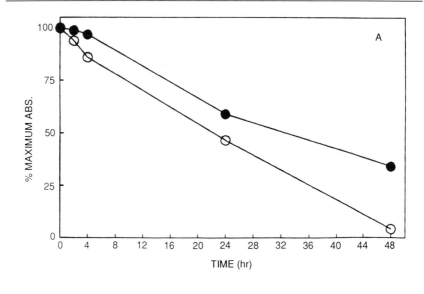

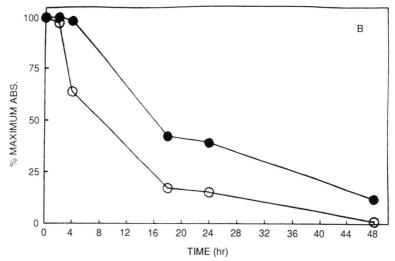

Fig. 2. Rate of degradation of a toluene solution of β-carotene exposed to different light sources in the presence (●) or in the absence (○) of butylated hydroxytoluene. Solutions of BC (4 μM) were put in a 200-ml beaker and exposed to ultraviolet light (A) or to fluorescent light (B) for different times. At each time point an aliquot was taken and the BC content was determined by spectrophotometer at 463 nm in toluene. Each data point is the mean of two determinations, which differed by no more than 5% of the mean.

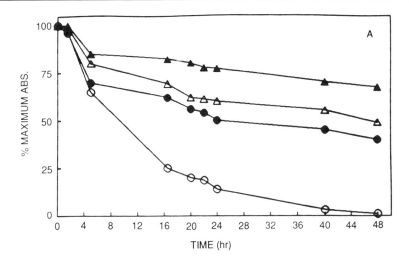
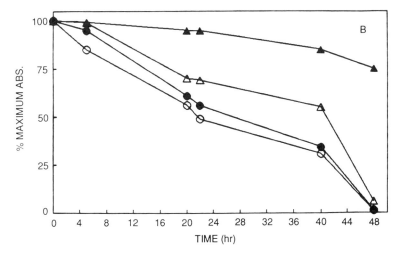

FIG. 3. Effect of different concentrations of α-tocopherol on BC degradation by air–light exposure. Solutions of BC (4 μM) in toluene were exposed to UV light (A) or to fluorescent light (B) in the presence of different concentrations of α-tocopherol: control, no α-tocopherol (○); 0.1 mM α-tocopherol (●); 1 mM α-tocopherol (△); 10 mM α-tocopherol (▲). The BC content was determined by measuring the absorbance at 463 nm in toluene. Each value is the mean of two determinations, which differed by no more than 4% of the mean.

occurs after 17 hr compared to 8 hr in the absence of antioxidant) to more than 140% (50% loss after 33 hr compared to 23 hr in the absence of antioxidant), respectively, on UV light and fluorescent light exposure. The effect of the most potent natural antioxidant, α-tocopherol, is shown in Fig. 3. There was a concentration-dependent effect in both conditions assayed. At the same concentration of BHT (1 mM) the 50% BC loss occurs by 48 and 40 hr, respectively, on UV and fluorescent light exposure, showing that α-tocopherol had a much stronger relative potency to prevent BC degradation than BHT under our exposure conditions. The protective effect of α-tocopherol and BHT was also shown by Peiser and Yang[9] in an experimental system in which BC destruction was induced by free radicals formed from the transformation of sulfite to sulfate. This might suggest that even in the present incubation, BC degradation could be a free radical-mediated process that could be promoted by light absorption either by BC or by the toluene used as a solvent. Nevertheless, because a cis-trans isomerization has been suggested to occur in the autooxidation process of carotenoids,[12,13] further investigation will be required to assess the relevance of this interconversion in reducing the maximum absorbance used as monitoring signal of BC degradation in the present assay.

Interestingly, there is a marked difference in the disappearance curves for BC under UV and fluorescent illumination. The BC concentration, in the absence of antioxidant, decreased in a hyperbolic fashion when exposed to UV light and linearly under fluorescent illumination. Thus UV light appears to operate by a first-order kinetic model, while the action of fluorescent light is best described as zero order. Speculation about this difference in terms of mechanism requires further investigation, although the direct absorption of fluorescent light by BC ($\lambda_{max} = 450$) may account for such a difference.

The stability of BC exposed to fluorescent light in relation to different rates of air-solution exchange was also studied to investigate the effect of oxygen. The rate of gas-liquid exchange is affected by the size of the surface of the solution exposed to air. An experiment was carried out comparing the degradation rate of BC solution in a test tube (exchange surface of 0.785 cm^2) and in a 200-ml beaker (exchange surface of 45.3 cm^2). As shown in Fig. 4 the 50% loss occurs after 72 hr for the solution in the test tube and after 23 hr for the beaker.

[12] R. T. Holman, *Arch. Biochem.* **21**, 51 (1949).
[13] P. Budowski and A. Bondi, *Arch. Biochem. Biophys.* **89**, 66 (1960).

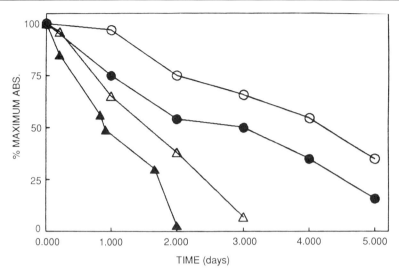

FIG. 4. Rate of degradation of toluene solution of β-carotene exposed to fluorescent light under conditions of varying air exchange. Solutions of BC (4 μM) in toluene in the presence (filled symbols) or in the absence (open symbols) of 0.025% BHT were put into test tubes (exchange surface, 0.785 cm^2: open and filled circles) or a 200-ml beaker (exchange surface, 45.3 cm^2: open and filled triangles) and exposed to fluorescent light. At each time point the BC content was analyzed as described in Fig. 3. The values are the mean of two determinations, which differed by no more than 6% of the mean.

Stability of BC under Storage Conditions

To test the stability under conditions of prolonged storage, BC in THF solution was kept at $-20°$, avoiding any illumination by using a double brown container (two brown bottles, one inside the other), in a nitrogen atmosphere. The concentration was analyzed at different times both by liquid chromatographic and spectrophotometric systems. The rate of BC loss ranged from 1.1%/month in the presence of BHT to 1.55%/month in the absence of antioxidant during the 4-month experiment as shown in Fig. 5. This clearly strengthens the importance of controlling air, light exposure, and temperature to reduce any degradative processes. An antioxidant like BHT can effectively improve the storage life of BC even when the degradative factors are minimized.

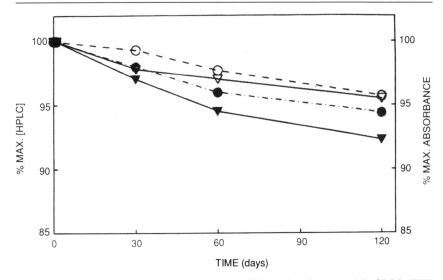

FIG. 5. Stability of β-carotene under storage conditions. Solutions (1 mM) of BC in THF in the presence (open symbols) or in the absence (filled symbols) of 0.025% BHT were kept at $-20°$ in a double brown bottle in the dark under N_2. At each time point an aliquot was taken and analyzed for BC content both by a liquid chromatographic system (circles) and a spectrophotometric monitoring system at 452 nm in benzene (triangles). Percentage of maximum peak area on HPLC separation was calculated on the basis of the nanomoles (5 nmol) of BC detected by the liquid chromatographic system. Each data point is the mean of two determinations, which differed by no more than 4% of the mean.

Conclusion

The present work is to my knowledge the first attempt at systematic investigation of the factors that could affect the stability of BC under the common laboratory conditions generally used to evaluate its biological role and to investigate its metabolic pathway.

BC remained fairly stable during 48 hr of what is defined as "incubation conditions" (37°, 5% CO_2, in the dark), generally used for cell-free or cell culture model systems. Besides, as great emphasis has been put on the anticarcinogenic properties of BC, independently of a vitamin A-mediated action, it was important to show that, under the condition used, the integrity of the molecule is preserved. BC is rapidly degraded by ultraviolet and visible light in the presence of atmospheric oxygen, the degradation rate being lower with a lower air–oxygen exchange. This points to the following laboratory working conditions to preserve BC solutions from degradation: use of a subdued light environment and the use of a dark container with a small air–surface exchange. A significant improvement in

BC stability is also given by the routine use of antioxidants such as the two well-known radical trapping agents, BHT and vitamin E. This, together with the absence of any sensitizer (to rule out the involvement of singlet oxygen), suggests that the light-induced degradation of this pigment could involve a chain-radical mechanism. Finally, the relative stability of BC was confirmed when the degradative factors were strictly controlled and antioxidants were present, suggesting conditions to be used to prolong its storage life.

Acknowledgments

I thank Dr. George Wolf for a careful review and stimulating discussion of the manuscript. I am grateful to Dr. H. N. Bhagavan (Hoffmann-La Roche, Inc.) for supplying the β-[^{14}C]carotene. Supported by USDA Grant No. 87-CRCR-1-2593.

[17] Carotenoid Reversed-Phase High-Performance Liquid Chromatography Methods: Reference Compendium

By NEAL E. CRAFT

Introduction

High-performance liquid chromatographic (HPLC) methods for the separation of carotenoids were first reported in 1970.[1] A positive pressure of nitrogen gas was applied to a traditional open column system to speed the separation and reduce oxygen exposure. By the standards of today this would constitute low-pressure liquid chromatography. The sophistication of carotenoid HPLC methods has evolved as HPLC columns and equipment have improved. The first report of carotenoids separated using high-pressure pumping systems was in 1971, when Stewart and Wheaton[2] employed gradient elution normal-phase HPLC to separate 23 different carotenoid peaks from citrus peel using a zinc carbonate column over a period of 5 hr. Their accomplishment constituted a phenomenal improvement over previous separations. By comparison, most current HPLC separations of carotenoids are performed using short (<20 min) isocratic conditions with commercially prepared reversed-phase columns.[3-6] This is

---

[1] J. P. Sweeney and A. C. Marsh, *J. Assoc. Off. Anal. Chem.* **53**, 937 (1970).
[2] I. Stewart and T. Wheaton, *J. Chromatogr.* **55**, 325 (1971).
[3] W. Driskell, M. Bashor, and J. Neese, *Clin. Chem.* **29**, 1042 (1983).

especially true for the analysis of serum carotenoids, which span a narrower polarity range than many plant carotenoids.

As in-depth discussion of HPLC theory and principles is beyond the scope of this chapter. For background or more detailed information, the reader is referred to books by Johnson and Stevenson[7] or Snyder and Kirkland.[8] Carotenoid separations have been performed by both normal-phase HPLC (NPLC) and reversed-phase HPLC (RPLC). NPLC encompasses adsorptive phases (such as silica and alumina) and polar bonded phases (e.g., alkylamine, alkylnitrile, and alkyl glycol) in conjunction with nonpolar mobile phases. Polar sites on the carotenoid molecules compete with solvent modifiers for adsorptive sites on the stationary phase. During this form of chromatography, the least polar carotenoids (hydrocarbons or carotenes) elute first while oxygenated carotenoids (xanthophylls) are retained longer. RPLC incorporates nonpolar bonded phases [such as octyl (C_8) and octyldecyl (C_{18})] and polymer phases (e.g., polystyrene–divinyl benzene and polymethacrylate) in conjunction with polar mobile phases. Carotenoids partition between the nonpolar stationary phase and the polar mobile phase. During RPLC, xanthopylls partition more effectively into the mobile phase and, therefore, elute first while the carotenes partition preferentially into the stationary phase and elute later. Both NPLC and RPLC can be run with the same solvent throughout (isocratic elution), or with solvent composition changing during the run (gradient elution). These distinctions are noted because the subsequent methods of analysis are limited to isocratic and gradient elution RPLC. Also, the focus will be on methods published after 1983 because previous volumes have summarized earlier HPLC methods.[9,10]

Reversed-Phase HPLC Separations

Reversed-phase packings have the advantage of being very "forgiving." They are not as sensitive to moisture content as adsorptive packings and reequilibrate rapidly following gradient elution. Table I summarizes the

---

[4] A. B. Barua, R. O. Batres, H. C. Furr, and J. A. Olson, *J. Micronutr. Anal.* **5**, 291 (1989).
[5] D. Milne and J. Botnen, *Clin. Chem.* **32**, 874 (1986).
[6] L. Kaplan, J. Miller, and E. Stein, *J. Clin. Lab. Anal.* **1**, 147 (1987).
[7] E. L. Johnson and R. Stevenson, "Basic Liquid Chromatography." Varian, Palo Alto, California, 1978.
[8] L. R. Snyder and J. J. Kirkland, "Introduction to Modern Liquid Chromatography." Wiley, New York, 1979.
[9] N. I. Krinsky and S. Welankiwar, this series, Vol. 105, p. 155.
[10] M. Ruddat and O. H. Will III, this series, Vol. 111, p. 189.

conditions employed in a variety of RPLC methods published in the literature. The most common column packing used for carotenoid separations is octyldecylsilane (ODS or C_{18}). Since 1985, the National Institute of Standards and Technology (NIST) has coordinated a quality assurance program for laboratories measuring fat-soluble vitamins in serum, including carotenoids. Of more than 50 laboratories involved internationally, only 1 laboratory uses a column other than C_{18} for carotenoid measurements.[11] Reasons for the popularity of C_{18} packings include compatibility with most solvents, usefulness for the entire polarity range of carotenoids, and wide commercial availability. Shorter alkyl-modified packings are not typically used to separate carotenoids due to their poorer selectivity and weak retention with organic solvents. Similarly, polymer-based columns have not been used primarily due to lower column efficiencies. Therefore, all reversed-phase HPLC separations reported here use ($\geq C_{18}$) modified silica.

Although the vast majority of RPLC methods use C_{18} columns, not all C_{18} packings are the same. Factors such as particle size and shape, pore diameter, surface coverage (carbon load), end capping, and monomeric vs polymeric synthesis influence the resultant separation.[12] The effects of many of these factors will be discussed below, in some cases, with reference to methods cited in Table I. Due to frequent changes in column manufacturing processes, which may alter column performance, all chromatograms illustrated in this chapter were performed in the author's laboratory using recent production lots of the columns cited in the original publications.

General Considerations

Many of the isocratic separations can be performed using a single high-pressure pump and premixed solvents. This approach results in stable baselines and reproducible retention times, assuming that (1) the mixing is performed accurately and consistently, and (2) the solvent does not contain small amounts of highly volatile modifiers that can preferentially evaporate during the course of a day. Solvents should be degassed in order to minimize baseline noise by reducing solvent outgassing at the detector flow cell. This can be accomplished using ultrasonic agitation, vacuum filtration, or helium sparging. Again, one must be aware of preferential evaporation of volatile components, which may change the desired mobile phase composition. The changes in solvent composition resulting from preferential

---

[11] J. M. Brown-Thomas, N. E. Craft, M. C. Kline, R. G. Christensen, W. A. MacCrehan, R. Schaffer, and W. E. May, *NIST J. Res.* submitted (1992).

[12] L. C. Sander and S. A. Wise, *LC-GC* **8**, 378 (1990).

TABLE I
REVERSE-PHASE HPLC SYSTEMS USED FOR CAROTENOID SEPARATIONS: ISOCRATIC ELUTION

Carotenoids separated[a]	Source	Column packing[b]	Mobile phase[a] (v/v)	Flow (ml/min)	Temperature[a] (°C)	Ref.
L, Z, Cx, βC, E, Ly, T, α, β	Human serum, algae, brine shrimp	Zorbax ODS, 5 μm	70 CH$_3$CN/20 CH$_2$Cl$_2$/10 CH$_3$OH	1.0	Ambient	c
α, β	Human serum	μBondapak C$_{18}$, 10 μm	89 CH$_3$CN/11 CHCl$_3$	2.0	Ambient	d
Ly, α, β, 1 unk	Human serum	Supelcosil LC-18, 3 μm	58.5 CH$_3$CN/35 CH$_3$OH/6.5 THF	2.0	Ambient	e
Z, Ly, α, β, 2 unk	Human plasma	Radial-Pak C$_{18}$, 10 μm	60 CH$_3$CH/25 CH$_3$OH/15 CHCl$_3$	1.5	Ambient	f
Cx, βC, α, β, γ, isomers	Vegetables	Vydac 201TP, 5 μm	90 CH$_3$OH/10 CHCl$_3$	1.0	Ambient	g
Cx, βC, α, β, γ, isomers			90 CH$_3$OH/10 THF			
Cx, βC, α, β, γ, isomers			52 CH$_3$OH/40 CH$_3$CN/8 THF			
Cx, βC, α, β, γ, isomers, Ly		Vydac 218TP, 5 μm	52 CH$_3$OH/40 CH$_3$CN/8 THF			
Cx, βC, Ly, γ, α, β		Nova-Pak C$_{18}$, 4 μm	58 CH$_3$CN/35 CH$_3$OH/7 THF			
Cx, βC, Ly, γ, α, β		Zorbax ODS, 5 μm	50 CH$_3$CN/35 CH$_3$OH/15 THF			
αC, βC, Zn, α, β, isomers	Orange juice	Vydac 201TP, 5 and 10 μm	90 CH$_3$OH/10 CHCl$_3$	1.0	Ambient	h
Ly, α, β	Human plasma	Hypersil ODS, 3 μm	100 CH$_3$OH	0.9	35	i
Z, Cx, βC, Ly, α, β	Human serum	Biophase ODS, 5 μm	78 CH$_3$CN/16 CHCl$_3$/3.5 IPA/2.5 H$_2$O	2.0	Ambient	j
N, V, L, Z, α, β	Leaf protein	Zorbax ODS, 5 μm	88 CH$_3$CN/12 EtOAc/0.1 decanol	1.6	Ambient	k
N, V, L, β	Turf grass	IBM C$_{18}$, 5 μm	83 CH$_3$CN/15 CHCl$_3$/2 H$_2$O	2.0	Ambient	l
L/Z, βC, Ly, α, β, 2 unk	Human serum	Nova-Pak C$_{18}$, 4 μm	50 CH$_3$OH/45 CH$_3$CN/5 THF	1.0	Ambient	m
L/Z, βC, Ly, α, β, Pf	Human serum	Microsorb C18, 5 μm	50 CH$_3$CN/50 ethanol	1.0	40	n
L, βC, Ly, α, β	Human serum	Spherisorb ODS2, 5 μm	85 CH$_3$OH/15 hexane	1.0	Ambient	o

				CH$_3$OH/3 H$_2$O/0.1 octanol	Ambient	
	Neat standards	Vydac 201TP, 5 µm		95 CH$_3$OH/5 THF	1.0	20
L, Z, βC, E, α, β, 13cβ, 9cβ, Ly						
β, 3 unk						

[a] A, Antheraxanthin; A5, antheraxanthin 5; A6, antheraxanthin 6; α, α-carotene; αC, α-cryptoxanthin; αT, α-tocopherol; Ax, auroxanthin; β, β-carotene; βC, β-cryptoxanthin; βd, β-carotene diepoxide; βm, β-carotene monoepoxide; β8′, β-apo-8′-carotenal; C, capsanthin; cβ, cis-β-carotene; Cc, cryptocapsin; cCx, cis-canthaxanthin; Cd, Cryptoxanthin diepoxide; Ce, cryptoxanthin 5,6-epoxide; Cfa, cryptoflavin a; Cfb, cryptoflavin b; CH$_3$CN, acetonitrile; CH$_3$OH, methanol; CH$_2$Cl$_2$, dichloromethane; CHCl$_3$, chloroform; cLy, cis-lycopene; cN, cis-neoxanthin; Cn, capsorubin; cR, cis-rhodoxanthin; Cr, cryptoxanthin; Cv, cis-violaxanthin; Cx, canthaxanthin; E, echinenone; esters, xanthophyll esters; EtOAc, ethyl acetate; Il/T, isolutein and taroxanthin; IPA, 2-propanol; isomers, geometric isomers; λ, γ-carotene; γ-T, γ-tocopherol; Le, lutein-5,6-epoxide; L, lutein; Lx, luteoxanthin; Ly, lycopene; L/Z, lutein and zeaxanthin; M, mutatoxanthin; N, neoxanthin; NA, neoxanthin A; NB, neoxanthin B; Nc, neochrome; NeA, neolutein epoxide A; nLA, neolutein A; nLB, neolutein B; Pf, phytofluene; Ret, retinol; R, rhodoxanthin; T, torulene; THF, tetrahydrofuran; tV, trans-violaxanthin; unk, unknown; V, violaxanthin; Z, zeaxanthin; Zn, zeinoxanthin; 9cC, 9-cis-capsanthin; 9cL, 9-cis-lutein; 13cC, 13-cis-capsanthin; 13cL, 13-cis-lutein; 13cZ, 13-cis-zeaxanthin; 15cβ, 15,15′-cis-β-carotene; 9cβ, 9-cis-β-carotene; 13cβ, 13-cis-β-carotene.

[b] Zorbax (Mac-Mod Analytical, Inc., Chadds Ford, PA); μBondapak, Nova-Pak, Radial-Pak, Resolve (Waters Chromatography Division, Milford, MA); Supelcosil (Supelco, Inc., Bellefonte, PA); Vydac (Vydac, Hesperia, CA); Hypersil (Shandon Scientific, Ltd., Runcorn, Cheshire, England); Biophase (Bioanalytical Systems, Inc., West Lafayette, IN); IBM (Jones Chromatography, Ltd., Mid-Glamorgan, England); Microsorb (Rainin Instrument Co., Inc., Woburn, MA); Spherisorb (Phase Separations, Ltd., Deeside, Clwyd, England); LiChrosorb (E. Merck, Darmstadt, Germany); Adsorbosphere (Alltech Associates, Inc., Deerfield, IL); and Bakerbond (J. T. Baker, Inc., Phillipsburg, NJ).

[c] H. J. C. F. Nelis and A. P. De Leenheer, Anal. Chem. 55, 270 (1983).
[d] W. Driskell, M. Bashor, and J. Neese, Clin. Chem. 29, 10442 (1983).
[e] C. C. Tangney, J. Liq. Chromatogr. 7, 2611 (1984).
[f] K. W. Miller and C. S. Yang, Anal. Biochem. 145, 21 (1985).
[g] R. J. Bushway, J. Liq. Chromatogr. 8, 1527 (1985).
[h] F. W. Quackenbush and R. L. Smallidge, J. Assoc. Off. Anal. Chem. 69, 767 (1986).
[i] D. Milne and J. Botnen, Clin. Chem. 32, 874 (1986).
[j] L. Kaplan, J. Miller, and E. Stein, J. Clin. Lab. Anal. 1, 147 (1987).
[k] D. R. Lauren, D. E. McNaughton, and M. P. Agnew, J. Assoc. Off. Anal. Chem. 70, 428 (1987).
[l] B. H. Chen and C. A. Bailey, J. Chromatogr. 393, 297 (1987).
[m] M. Stacewicz-Sapuntzakis, P. E. Bowen, J. W. Kikendall, and M. Burgess, J. Micronutr. Anal. 3, 27 (1987).
[n] A. L. Sowell, D. L. Huff, E. W. Gunter, and W. J. Driskell, J. Chromatogr. 431, 424 (1988).
[o] G. Cavina, B. Gallinella, R. Porra, P. Pecora, and C. Suraci, J. Pharm. Biomed. Anal. 6, 259 (1988).
[p] A. B. Barua, R. O. Batres, H. C. Furr, and J. A. Olson, J. Micronutr. Anal. 5, 291 (1989).
[q] N. E. Craft, S. A. Wise, and J. H. Soares, Jr., J. Chromatogr. 589, 171 (1992).

evaporation can be avoided by using an HPLC equipped with high- or low-pressure solvent mixing. However, these types of mixing are not always efficient or consistent and may lead to baseline noise due to solvent outgassing, refractive index effects, and poor mixing.

When analyzing extracts from complex matrices, a guard column of similar pore size and chemistry as the analytical column should be incorporated between the injection valve and the column. Small-bore tubing (≤ 0.010 in.) should be used in all connections from the injection valve through to the detector to minimize dead volume and band broadening. Also, a back pressure restrictor (spring loaded or narrow diameter tubing) should be incorporated after the detector to minimize solvent outgassing. Replacement of deuterium detector lamps with tungsten lamps can increase detector sensitivity by an order of magnitude.[13]

Injection volumes should be kept to a minimum to prolong column life and reduce band broadening. The injection solvent should be compatible with the HPLC mobile phase. If the injection solvent is much stronger than the mobile phase and nearly saturated with carotenoid, the carotenoids will precipitate on injection into the mobile phase, leading to peak tailing, or will form broad bands and doubled peaks by remaining with the injection solvent as it passes through the column. Stronger, miscible solvents can be injected if the volume is small ($\leq 10 \mu$l) and the concentrations of carotenoids and other extract components are not greatly in excess of their solubility in the mobile phase. If the injection solvent is too weak, the sample extract will not completely dissolve in the solvent. Agitation using an ultrasonic bath for approximately 30 to 60 sec is highly recommended to facilitate the dissolution of extracted residues into the injection solvent.

Last, appropriate standardizaton is essential for quantitative carotenoid measurements. Sample preparation usually involves multiple steps during which losses and errors can be introduced. The incorporation of a volume correction standard (frequently referred to as an internal standard) is highly recommended because it accounts for volume losses due to incomplete recovery, evaporation, and injection variability. Concentrations should be value assigned to calibration standards spectrophotometrically by applying Beer's law and then corrected for HPLC purity.[14] Published absorptivities are listed in works by Davies[15] and De Ritter and Purcell.[16] Where available, standard reference materials (SRMs) containing known

[13] F. W. Quackenbush and R. L. Smallidge, *J. Assoc. Off. Anal. Chem.* **69,** 767 (1986).
[14] N. E. Craft, L. C. Sander, and H. F. Pierson, *J. Micronutr. Anal.* **8,** 209 (1991).
[15] B. H. Davies, in "Chemistry and Biochemistry of Plant Pigments" (T. W. Goodwin, ed.), Vol. 2, pp. 38–165. Academic Press, New York, 1976.
[16] E. De Ritter and A. E. Purcell, in "Carotenoids as Colorants and Vitamin A Precursors: Technical and Nutritional Applications" (J. C. Bauernfeind, ed.), pp. 815–923. Academic Press, New York, 1981.

concentrations of carotenoids should be analyzed to validate methods and value assign in-house quality control materials.[11] An example of such a material is SRM 968 (fat-soluble vitamins in human serum), which has certified values for total β-carotene, retinol, and α-tocopherol.

Equipment

During the course of these experiments low-pressure mixing, high-pressure mixing, and premixing of solvents were used. Pumps consisted of single-piston, dual-piston, and triple-piston arrangements equipped with either low- or high-pressure solvent mixers. Samples were introduced via manual or automatic injection valves. Carotenoids were detected in most cases using high-sensitivity programmable wavelength, UV/VIS detectors equipped with tungsten lamps to improve sensitivity in the visible region.[13] In some cases in which sensitivity was not a factor, filter photometry or photodiode array detection was used. Data were recorded using a computer-controlled data system. Specific HPLC conditions are given in the figure captions.

Column Parameters

Particle Size and Shape

Historically, chromatographers have improved the separation efficiency by using smaller particle size. As the importance of small uniform particle size and packing techniques was established, columns were packed with 10-μm, then 5- and 3-μm particles using sophisticated high-pressure packing techniques. Particle shape was changed from irregular to spherical to assure uniform particle size and thereby yield more reproducible columns. Advantages of small particle size include high efficiency and lower solvent consumption because a comparable separation can be obtained using a shorter column. The disadvantages of small particles include decreased column life due to plugging and higher back pressures that increase pump wear. These disadvantages can be diminished by filtering all samples and using protective guard columns. The majority of current carotenoid separations are performed using 5-μm spherical particles packed in 25-cm columns.

Pore Diameter and Surface Coverage

The effects of pore diameter and surface coverage on carotenoid separations are not well documented. Usually carotenoid chromatographers attempt to use columns with high surface coverage because it increases retention and thereby allows the use of stronger solvents with a wider range of modifiers.[17,18] One way of increasing carbon load is to increase the surface area of the base silica by using silica with a small pore diameter. The drawback is that pore size may become so narrow that the carotenoid molecules do not penetrate the pores to interact with the C_{18} phase within the pores. Based on molecular modeling, *trans*-β-carotene is approximately 33 Å long. Columns packed with 60-μm pore diameter silica, which is occluded by approximately 17 μm on either side when the C_{18} chain is chemically bonded, will result in pore diameters that β-carotene can enter only lengthwise.[19] The effect of pore diameter on the separation of geometric isomers of β-carotene is illustrated in Fig. 1. This suggests that there exists an optimum pore diameter/surface area combination that will provide good retention, yet will permit interaction of the analyte with the interior of the pores.

End Capping

End capping is a means of covering unreacted silanol groups on the silica surface to minimize polar interactions with analytes and thereby improve column reproducibility. The majority of commercial RPLC columns are end capped for the above reasons. Unfortunately, the additional interaction of silanols was beneficial for the separation of xanthophylls.[20] The widely accepted method of Nelis and De Leenheer[17] is an example of the beneficial effect of silanol residues on carotenoid separations. However, the Zorbax ODS column (Mac-Mod Analytical) used in their method is no longer available in a non-end-capped form. The effect of end capping on the separation of a carotenoid mixture, especially lutein and zeaxanthin, is illustrated in Fig. 2. The mobile phase is the same as used by Nelis and De Leenheer and the column packing materials are identical except for the end capping.

Stationary Phase Synthesis

Another very important factor in reversed-phase separations of carotenoids is the form of synthesis used to bond the C_{18} phase to the base silica. Two principal techniques are used. The most common technique is referred to as monomeric synthesis, in which a monolayer of C_{18} is bonded to the silica surface using monochlorosilanes. The second technique is referred to as polymeric synthesis, and results in a polymeric layer of C_{18} that is bound to the silica surface using trichlorosilanes in the presence of

[17] H. J. C. F. Nelis and A. P. De Leenheer, *Anal. Chem.* **55**, 270 (1983).
[18] F. Khachik, G. R. Beecher, and N. F. Whittaker, *J. Agric. Food Chem.* **34**, 603 (1986).
[19] L. C. Sander, C. J. Glinka, and S. A. Wise, *Anal. Chem.* **62**, 1099 (1990).
[20] Z. Matus and R. Ohmacht, *Chromatographia* **30**, 318 (1990).

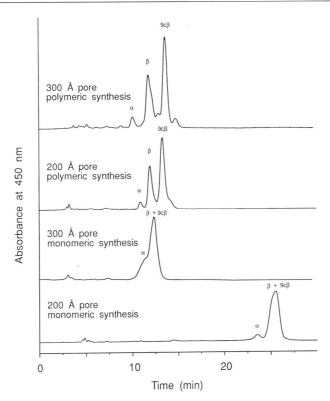

FIG. 1. Separation of a mixture of β-carotene isomers using monomeric and polymeric C_{18} phases, prepared on 200- and 300-Å pore diameter silicas. Mobile phase, 100% methanol; flow rate, 1.5 ml/min; detection, 450 nm. α, α-carotene; β, *trans*-β-carotene; 9cβ, 9-*cis*-β-carotene. (Column synthesis is from Ref. 21.)

restricted amounts of water. These two types of bonding are graphically illustrated in Fig. 3. The former technique is most common because it is simpler and has been reported to result in a more reproducible stationary phase.[21] On the other hand, polymerically synthesized phases (not to be confused with polymer-based packings) have been shown to offer unique selectivity for structurally similar compounds.[12,22,23] This is evidenced by the ability of polymeric C_{18} phases to separate geometric isomers of carotenoids.[14,22,23] The separation of β-carotene isomers using monomeric and polymeric columns synthesized from the same silica is illustrated in Fig. 1.

[21] L. C. Sander and S. A. Wise, *J. Chromatogr.* **316**, 163 (1984).
[22] F. W. Quackenbush, *J. Liq. Chromatogr.* **10**, 643 (1987).
[23] R. J. Bushway, *J. Liq. Chromatogr.* **8**, 1527 (1985).

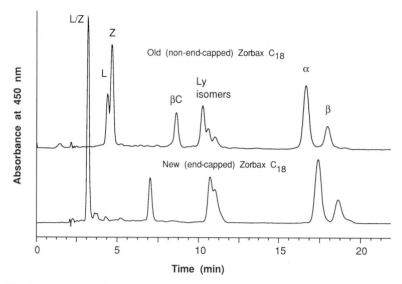

FIG. 2. Separation of a mixture of carotenoid standards using end-capped and non-end-capped Zorbax C_{18} HPLC columns. Mobile phase, 70:20:10 (v/v) acetonitrile–dichloromethane–methanol; flow rate, 1.5 ml/min; detection, 450 nm. L, Lutein; Z, zeaxanthin; αC, α-cryptoxanthin; βC, β-cryptoxanthin; E, echinenone; Ly, lycopene; α, α-carotene; β, β-carotene; cisβ, β-carotene cis-isomers; 9cβ, 9-cis-β-carotene; 13cβ, 13-cis-β-carotene.

The monomeric columns fail to resolve the structurally similar isomers. Khachik et al.[24] attributed the superior selectivity of polymeric C_{18} phases to the low carbon loading. Figure 1 demonstrates that the separation of geometric isomers of β-carotene is superior using the polymeric C_{18} phase compared to the monomeric C_{18} phase when all other variables are controlled.[21] The primary reason for poor acceptability of polymeric C_{18} HPLC columns among carotenoid chromatographers is weak carotenoid retention resulting from the low carbon load associated with the wide-pore, low surface area silica from which these columns are prepared. The mobile phase usually consists of >90% methanol, which limits the solubility of many carotenoids.[22,23,25] Sander and Wise[26] have been able to circumvent this limitation by bonding polyfunctional silanes to high surface area silica or increasing the alkyl chain length to between 20 and 30 carbons. Figure 4 illustrates the separation of serum carotenoids using a polymeric C_{30} phase.

[24] F. Khachik, G. R. Beecher, and W. R. Lusby, J. Agric. Food Chem. **37**, 1465 (1989).
[25] K. S. Epler, L. C. Sander, R. G. Ziegler, S. A. Wise, and N. E. Craft, J. Chromatogr. **595**, 89 (1992).
[26] L. C. Sander and S. A. Wise, Anal. Chem. **59**, 2309 (1987).

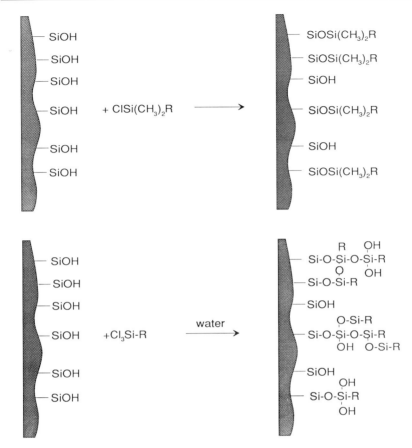

FIG. 3. Schematic representation of monomeric (top) and polymeric (bottom) surface modification schemes for silanization of silica used in C_{18} stationary phase synthesis. (From Ref. 12.)

Temperature

Temperature is a frequently neglected factor in the chromatographic separation of carotenoids. The influence of temperature on column selectivity appears to be primarily related to modification of the stationary phase to enable recognition of differences in the shape of molecules,[27,28] but it is also important for day-to-day reproducibility. Figure 5 illustrates

[27] L. C. Sander and S. A. Wise, *Anal. Chem.* **61**, 1749 (1989).
[28] E. Lesellier, C. Marty, C. Berset, and A. Tchapla, *HRC CC, J. High Resolut. Chromatogr. Chromatogr. Commun.* **12**, 447 (1989).

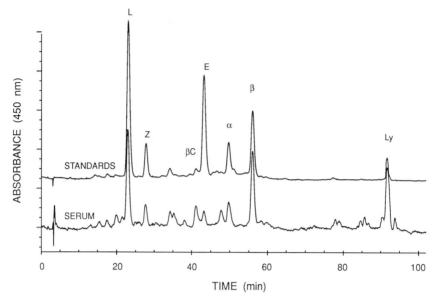

FIG. 4. Separation of carotenoid standards and serum carotenoids using a C_{30} polymeric HPLC column. Mobile phase, solvent A = methanol, solvent B = THF; flow rate, 1.5 ml/min; detection, 450 nm; gradient, 95:5 A/B for 10 min, then concave from 95:5 A/B to 50:50 A/B in 50 min, convex to 35:65 A/B in 15 min, hold 25 min. See Fig. 2 for peak identifications. (Column synthesis is from Ref. 26.)

the separation of lutein from zeaxanthin and β-carotene isomers using the same column and conditions at -13 and $30°$.[29] At subambient temperatures the bonded phase is more rigid and therefore more sensitive to differences in molecular shape.[27] Temperature drift can cause a substantial fluctuation in carotenoid retention times, leading to erroneous peak identification and quantitation.

Mobile Phase

Thus far no mention has been made of the influence of mobile phase on carotenoid separations. This is the variable that is most easily altered and therefore the one frequently responsible for "new" methods. However, most mobile phase manipulations do not dramatically alter carotenoid selectivity and the majority of methods are only subtle modifications of a few basic solvent systems. Many of these modifications must be performed to alter analysis time or adjust selectivity to a new or different packing

[29] L. C. Sander and N. E. Craft, *Anal. Chem.* **62**, 1545 (1990).

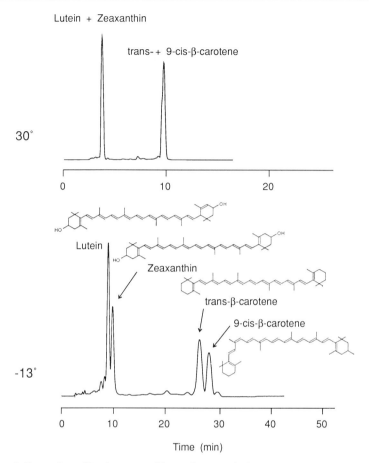

FIG. 5. Separation of lutein, zeaxanthin, and geometric isomers of β-carotene at 30 and −13° using a monomeric C_{18} HPLC column. Mobile phase, 70:20:10 acetonitrile–dichloromethane–methanol; flow rate, 1.5 ml/min; detection, 450 nm. (From Ref. 29.)

material. Carotenoids are practically insoluble in water so it should be avoided or only be incorporated sparingly as a solvent modifier. The primary solvent should be a weak organic solvent with low viscosity, thereby allowing use with a range of modifiers, yielding adequate carotenoid solubility, and low back pressure. These criteria limit the choices to acetonitrile and methanol. Generally acetonitrile has been used because of its lower viscosity and slightly improved selectivity in the xanthophyll region when using monomeric C_{18} columns.[4,18] In a report evaluating columns, methanol-based solvents yielded higher recoveries of carotenoids

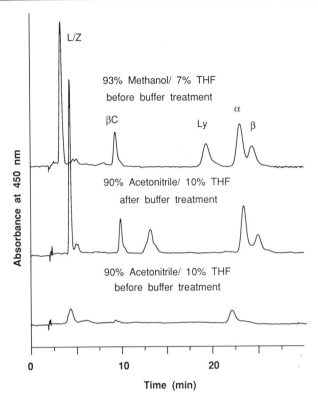

FIG. 6. Separation of a mixture of carotenoid standards using a methanol-based mobile phase and an acetonitrile-based mobile phase before and after treating the column with buffer. Mobile phase, as given in figure; flow rate, 2.0 ml/min; detection, 450 nm. See Fig. 2 for peak identifications.

than acetonitrile-based solvents for 60 out of 65 columns evaluated.[25] Additionally, both the cost and toxicity of methanol are less than that of acetonitrile. In the column evaluation study, it was found that the recovery of carotenoids from the columns in acetonitrile-based solvents could be improved by adding an ammonium acetate buffer to the mobile phase.[25] This observation has been reported by others.[30,31] Buffer selection is limited for nonaqueous mobile phases and very few published methods incorporate buffers.[32-34] The combination of these factors provides good incentive

[30] G. J. Handelman, B. Shen, and N. I. Krinsky, this volume [31].
[31] A. M. Gilmore and H. Y. Yamamoto, *J. Chromatogr.* **543**, 137 (1991).
[32] Y. M. Peng, J. Beaudry, D. S. Alberts, and T. P. Davis, *J. Chromatogr.* **273**, 410 (1983).
[33] W. A. MacCrehan and E. Schönberger, *Clin. Chem.* **33**, 1585 (1987).
[34] N. Krinsky, M. Russett, G. Handelman, and M. Snodderly, *J. Nutr.* **120**, 1654 (1990).

to use methanol-based solvents. Figure 6 illustrates the separation of carotenoid standards from the same column in acetonitrile, buffered acetonitrile, and in methanol.

Modifiers are added to the mobile phase to achieve the desired retention, increase solubility, and alter selectivity. Chlorinated solvents (chloroform and dichloromethane) have been used as modifiers for carotenoid separations most frequently[4,6,35,36] due to their good solvent characteristics and effects on selectivity. Prior to the advent of modern HPLC, chlorinated solvents were avoided in carotenoid analysis due to carotenoid degradation resulting from traces of HCl in the solvents. As indicated in Table I, other strong solvents such as tetrahydrofuran (THF), ethyl acetate, and acetone have been employed as modifiers. The mobile phase used by Zakaria *et al.*[35] is the predecessor of most RPLC mobile phases currently used to separate carotenoids. Many mobile phases now include three of four solvents to achieve the desired selectivity. This presents a problem by making the method more complicated, and the different volatilities of the solvents cause variations in retention times during the course of a day. If adequate separation can be obtained with a simple mobile phase, then the mobile phase should not be complicated by a "touch" of this and a "dash" of that. Figure 7 illustrates the effect of different modifiers on the separation of carotenoid standards.[37] The separation of carotenoid standards and serum carotenoids, using the simple isocratic method of Craft *et al.*,[37] is illustrated in Fig. 8 and indicates that the right combination of column, mobile phase, and temperature can circumvent complicated manipulations of solvent composition.

Gradient Elution

Altering the mobile phase composition during the course of an HPLC run introduces unlimited flexibility into the methodology. The advantages of gradient elution include a wider range of analytes, improved sensitivity, improved selectivity, and elution of all strongly retained compounds. The disadvantages of gradient elution include increased complexity, requirements for additional equipment, column reequilibration after each analysis, exclusion of some detectors, and greater variability. Gradient elution is not recommended if the analysis can be performed using isocratic elution.

Table II summarizes the conditions reported for a variety of gradient RPLC methods. The equipment and considerations discussed for isocratic

---

[35] M. Zakaria, K. Simpson, P. R. Brown, and A. Krstulovic, *J. Chromatogr.* **176**, 109 (1979).
[36] C. S. Yang and M.-J. Lee, *J. Nutr. Growth Cancer* **4**, 19 (1987).
[37] N. E. Craft, S. A. Wise, and J. H. Soares, Jr., *J. Chromatogr.* **589**, 171 (1992).

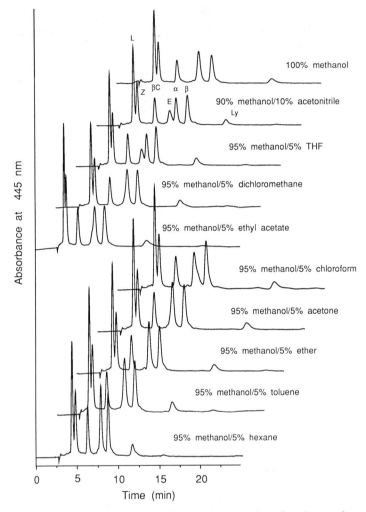

FIG. 7. Influence of mobile phase modifiers on the separation of a mixture of carotenoid standards. Column, Vydac 201TP C_{18}, 5 μm (4.6 × 250 mm); mobile phase, as given in figure; flow rate, 1.0 ml/min; detection, 445 nm. See Fig. 2 for peak identifications. (From Ref. 37.)

RPLC are also applicable to gradient elution. Additionally, (1) the column must be returned to the original solvent composition; (2) the column should be reequilibrated with approximately 10 column volumes of the original solvent; (3) good solvent miscibility is essential to prevent baseline disturbances due to outgassing and refractive index effects; and (4) con-

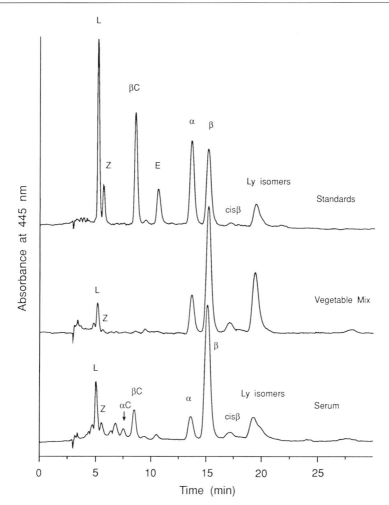

Fig. 8. Isocratic separation of carotenoids: serum, food, and standards. Column, Vydac 201TP C_{18}, 5 μm (4.6 × 250 mm); mobile phase, 95:5 methanol–THF; flow rate, 1.0 ml/min; temperature, 20°; detection, 445 nm. See Fig. 2 for peak identifications. (From Ref. 37.)

taminants present in a weaker solvent may appear as unwanted peaks in the chromatogram as the gradient proceeds (not usually a problem in the visible wavelengths).

Figure 9 illustrates the separation of serum carotenoids and carotenoid standards using a gradient system developed[38] employing the observations

[38] K. S. Epler, R. G. Ziegler, and N. E. Craft, *J. Chromatogr.* submitted (1992).

TABLE II
REVERSED-PHASE HPLC SYSTEMS USED FOR CAROTENOID SEPARATIONS: GRADIENT ELUTION[a]

Carotenoids separated	Source	Column packing	Mobile phase	Flow (ml/min)	Temperature (°C)	Ref.
NA, NB, tV, Lx, Cx, A, M, Le, L/Z, ll/T, Cd, Ce, Cfa, $\beta 8'$, Cfb, βC, ζ, α, β	Citrus	LiChrosorb RP-18, 10 μm, 2 in series	A = 90 CH$_3$OH/10 H$_2$O; B = acetone; 5% B to 75% B in 38 min, linear	1.0	Ambient	b
Ret, αT, γT, L, Z, Cr, α, β, cβ	Human serum	Vydac 201TP, 5 μm	A = 15 H$_2$O/75 CH$_3$OH/10 n-butanol; B = 2 H$_2$O/88 CH$_3$OH/10 n-butanol; 100% A for 3 min, to 100% B in 15 min, linear	1.5	Ambient	c
N, cN, V, Nc, Le, NeA, L, nLB, nLA, $\beta 8'$, β, 15cβ, tV, cV, Lx, CV, Ax, ~15 esters	Vegetables	Microsorb C$_{18}$, 5 μm	A = 15 CH$_3$OH/75 CH$_3$CN/5 CH$_2$Cl$_2$/5 hexane; B = 15 CH$_3$OH/40 CH$_3$CN/22.5 CH$_2$Cl$_2$/22.5 hexane; 100% A for 12 min, to 100% B in 15 min, linear	0.5	Ambient	d
Cn, V, C, 9cC, 13cC, A, Z, Cc, Cr, β, ~34 esters	Paprika	Zorbax ODS, 5 μm	A = 75 acetone/25 H$_2$O, B = 75 acetone/25 CH$_3$OH; 100% A to 65% B in 10 min, to 80% B by 30 min, to 100% B by 60 min	1.0	Ambient	e
L, Z, R, cR, cCx, 9cL, Cx, βC, Ly, γ, α, β, 2 unk, 9 esters	Marigold, primrose	Supelcosil LC-18-S, 5 μm	A = 33 dichloroethane/67 IPA; B = 100 CH$_3$CN; 95% B to 90% B in 10 min, to 70% B by 15 min, to 50% B by 30 min, to 35% B by 50 min	1.5	Ambient	f

Cr, β, 38 identified esters, 11 unk, Cn, C, Ce, β8'	Fruit	Resolve C_{18}, 5 μm	A = 100 CH_3OH; B = 100 EtOAc; 100% A to 50% B in 15 min concave, to 100% B by 20 min	1.0	Ambient	g, h
N, V, A5, A6, Le, A, L, Z, βd, αC, βC, βme, α, β	Neat standards	Custom-made C_{18}	A = 12 H_2O/88 CH_3OH; B = CH_3OH; C = 10 acetone/90 CH_3OH; 100% A for 2 min, to 50% A/50% B by 10 min, to 100% B by 18 min, hold 2 min, to 100% C by 27 min, hold 6 min		Ambient	i
L, Z, 13cL, 13cZ, αC, βC, Ly, cLy, α, β, 4 unk	Human plasma, food	Adsorbosphere HS C18, 3 μm	A = 85 CH_3CN/15 CH_3OH + 0.01% ammonium acetate; B = 2-propanol; 100% A for 10 min, step to 70 A/30 B	1.0	25	j
L, Z, αC, βC, Ly isomers, α, β, 13cβ, 9cβ	Human serum	Bakerbond C_{18}, 5 μm	A = 90 CH_3CN/10 EtOAc; B = 90 CH_3OH/10 EtOAc + 100 mM ammonium acetate; 100% A for 1 min, linear to 50% B by 6 min, linear to 100% B by 20 min	2.0	24	k

[a] See footnotes a and b in Table I for abbreviations.
[b] G. Noga and F. Lenz, *Chromatographia* **17**, 139 (1983).
[c] W. A. MacCrehan and E. Schönberger, *Clin. Chem.* **33**, 1585 (1987).
[d] F. Khachik, G. R. Beecher, and N. F. Whittaker, *J. Agric. Food Chem.* **34**, 603 (1986); F. Khachick and G. R. Beecher, *J. Agric. Food Chem.* **36**, 929 (1988).
[e] C. Fisher and J. A. Kocis, *J. Agric. Food Chem.* **35**, 55 (1987).
[f] F. Zonta, B. Stancher, and G. P. Marletta, *J. Chromatogr.* **403**, 207 (1987).
[g] T. Philip and T. Chen, *J. Food Sci.* **53**, 1720 (1988).
[h] G. K. Gregory, T. Chen, and T. Philip, *J. Food Sci.* **52**, 1071 (1987).
[i] Z. Matus and R. Ohmacht, *Chromatographia* **30**, 318 (1990).
[j] N. Krinsky, M. Russett, G. Handelman, and M. Snodderly, *J. Nutr.* **120**, 1654 (1990).
[k] K. S. Epler, R. G. Ziegler, and N. E. Craft, *J. Chromatogr.* submitted (1992).

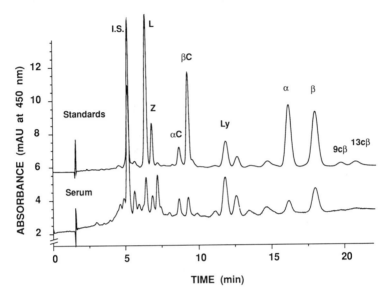

Fig. 9. Gradient separation of carotenoid standards and carotenoids in human serum. Column: Bakerbond C_{18}, 5 μm (4.6 × 250 mm); mobile phase, A = 90:10 acetonitrile-ethyl acetate, B = 90:10 methanol-ethyl acetate containing 100 mM ammonium acetate; flow rate, 2.0 ml/min; detection, 450 nm; gradient, 100% A for 2 min, linear to 50% A/50% B in 4 min, linear to 100% B in 14 min. See Fig. 2 for peak identifications. (From Ref. 38.)

made during a recent column evaluation study.[25] A narrow-pore polymeric C_{18} column with "biocompatible" frits is used. Because acetonitrile was necessary to achieve the desired selectivity, the methanolic portion of the gradient was buffered with ammonium acetate to minimize on-column losses of carotenoids. Total run time, including reequilibration, was 30 min. This method results in good retention of the polar carotenoids and excellent separation of lycopene and β-carotene geometric isomers, which should also make it applicable to food analysis.

Comments

An overview of many principles and factors involved in the RPLC separation of carotenoids has been presented. The separation characteristics of all C_{18} HPLC columns are not the same and variability exists among different lots of columns from a single manufacturer. As the separations have become more sophisticated, yielding resolution of more carotenoids, the influence of solvent purity and modifiers, column materials, temperature regulation, and stationary phase packings has become more critical.

Carotenoids appear to be much more sensitive to changes in HPLC conditions than many other classes of compounds, such as polycyclic aromatic hydrocarbons. The mechanisms responsible for many of the observed behaviors are not fully elucidated. Thus, even reproducing all aspects of a published method does not guarantee an identical separation. Similarly, no single HPLC separation is capable of separating all natural carotenoids. Therefore, more than one technique may be necessary to accomplish the degree of separation required. A representative selection of RPLC methods and essential information is given in Tables I and II for the convenience of the reader.

Acknowledgments

This work was supported, in part, by the National Cancer Institute, Division of Cancer Etiology, and Division of Cancer Prevention and Control, Bethesda, MD.

[18] Separation and Quantification of Carotenoids in Human Plasma

By FREDERICK KHACHIK, GARY R. BEECHER, MUDLAGIRI B. GOLI, WILLIAM R. LUSBY, and CHARLES E. DAITCH

Introduction

A number of studies have reported an inverse relationship between the high consumption of certain fruits and vegetables and a lower incidence of several types of cancer.[1-3] The experimental[4] and human epidemiological evidence[5] for the nutritional prevention of cancer has been reviewed. Carotenoids, which are one of the largest naturally occurring groups of pigments found in these foods, are absorbed by the body, circulate in plasma, and are found in several tissues.[3] There is keen interest in quanti-

---

[1] National Research Council, "Diet, Nutrition, and Cancer," pp. 358–370. National Academy, Washington, D.C., 1982.
[2] R. Peto, R. Doll, J. D. Buckly, and M. B. Sporn, *Nature (London)* **290,** 201 (1981).
[3] R. B. Shekelle, M. Lepper, S. Liu, P. Oglesby, A. M. Shryock, and J. Stamler, *Lancet* **2,** 1185 (1981).
[4] P. M. Newberne, T. F. Schrager, and M. W. Conner, *in* "Nutrition and Cancer Prevention: Investigating the Role of Micronutrients" (T. E. Moon and M. S. Micozzi, eds), pp. 33–82. Dekker, New York, 1989.
[5] D. J. Hunter and W. C. Willett, *in* "Nutrition and Cancer Prevention: Investigating the Role of Micronutrients" (T.E. Moon and M. S. Micozzi, eds), pp. 83–100. Dekker, New York, 1989.

fying each of the various carotenoids in plasma because of the potential function of carotenoids as anticarcinogenic agents.

The first successful separation of carotenoids from an extract of human serum was reported by Nelis and De Leenheer,[6] using nonaqueous reversed-phase chromatographic conditions on a Zorbax ODS column, with a mixture of acetonitrile, dichloromethane, and methanol (70:20:10, v/v) as eluent. Under these conditions six carotenoids were separated and identified in an extract from human serum: lutein, zeaxanthin, β-cryptoxanthin, lycopene, α-carotene, and β-carotene. Although many other researchers have developed nonaqueous reversed-phase high-performance liquid chromatography (HPLC) conditions employing a variety of organic solvents and various HPLC columns,[7-11] the HPLC conditions of Nelis and De Leenheer provide better separation of carotenoids. This has been demonstrated by Bieri et al.,[12] who successfully employed this chromatographic procedure to separate carotenoids from extracts of human plasma on a C_{18} reversed-phase HPLC column. The method of Nelis and De Leenheer for the analysis of carotenoids in human plasma is a rapid and isocratic HPLC procedure that provides information on the nature and the levels of a limited number of carotenoids.

The number of dietary carotenoids available for absorption, metabolism, and/or utilization by the human body is in excess of 40.[13] Therefore the existing HPLC procedures for detailed separation and detection of carotenoids in extracts of plasma may be inadequate.

In this chapter we describe methodologies for detailed separation and quantitation of 18 carotenoids as well as vitamin A and 2 forms of vitamin E in human plasma by HPLC on reversed-phase and silica-based nitrile-bonded columns.[13,14]

Nomenclature

For convenience only the trivial names of the carotenoids have been used throughout this chapter. The semisystematic names and chemical structures of carotenoids have been tabulated by Pfander.[15] In cases in

[6] H. J. C. F. Nelis and A. P. De Leenheer, *Anal. Chem.* **55**, 270 (1983).
[7] W. J. Driskell, M. M. Bashor, and J. W. Neese, *Clin. Chem.* **29**, 1042 (1983).
[8] N. Katrangi, L. A. Kaplan, and E. A. Stein, *J. Lipid Res.* **25**, 400 (1984).
[9] A. L. Sowell, D. L. Huff, E. W. Gunter, and W. J. Driskell, *J. Chromatogr.* **431**, 424 (1988).
[10] L. R. Cantilena and D. W. Nierenberg, *J. Micronutr. Anal.* **6**, 127 (1989).
[11] C. Fukasawa, *Tokyo Ika Daigaku Zasshi* **47**, 419 (1989).
[12] J. G. Bieri, E. D. Brown, and J. C. Smith, Jr., *J. Liq. Chromatogr.* **8**, 473 (1985).
[13] F. Khachik, G. R. Beecher, M. B. Goli, and W. R. Lusby, *Pure Appl. Chem.* **63**, 71 (1991).
[14] F. Khachik, G. R. Beecher, M. B. Goli, W. R. Lusby, and J. C. Smith, *Anal. Chem.* submitted (1992).
[15] H. Pfander, in "Key to Carotenoids," 2nd Ed. Birkhäuser, Basel, 1987.

which the definite geometrical configuration of the *cis*-carotenoids are not known, the term *cis*-carotene has been used to distinguish these carotenoids from their all-trans counterpart. According to the new nomenclature, the terms all-trans and -cis for in-chain geometric isomers of carotenoids should be referred to as all-E and -Z, respectively. However, for clarity of presentation, the terms all-trans and -cis, which are still being used by some scientists, have also been used throughout this text.

Experimental Procedures

Instrumentation. For gradient chromatography a ternary solvent delivery system, injector, and a rapid-scanning ultraviolet UV/visible photodiode array detector facilitates the separation and identification of carotenoids in the extracts from human plasma. For isocratic separations only a single pump is required.

Chromatographic Conditions. Chromatographic conditions for the separation of carotenoids in the extracts from human plasma are summarized in Table I. HPLC system A, which is a gradient system and employs a reversed-phase column, is identical to the HPLC system universally employed for the separation of carotenoids in foods (see Chapter [32] in this volume). This HPLC system separates most of the carotenoids found in human plasma, but it fails to resolve the more polar xanthophylls, such as

TABLE I
CHROMATOGRAPHIC CONDITIONS FOR SEPARATION OF PLASMA CAROTENOIDS

Parameter	Conditions			
HPLC system A, gradient				
Column	Rainin Microsorb 5-μm C_{18}, 25 cm × 4.6 mm, protected with Brownlee guard cartridge 5-μm C_{18}, 3 cm × 4.6 mm			
Mobile phase		Time 0–10 min		Time 40 min
	Acetonitril	85%	Linear gradient	45%
	Dichloromethane-hexane (1:1)	5%		45%
	Methanol	10%	Over 30 min.	10%
Flow rate	0.70 ml/min			
HPLC system B, isocratic				
Column	Silica-based nitrile-bonded (Regis Chemical Co., Morton Grove, IL:) 5-μm, 25 cm × 4.6 mm, protected with Brownlee nitrile-bonded guard cartridge 5-μm, 3 cm × 4.6 mm			
Mobile phase	Hexane, 74.65%			
	Dichloromethane, 25.00%			
	Methanol, 0.25%			
	N,N-Diisopropylethylamine, 0.10%			
Flow rate	1.0 ml/min			

lutein and zeaxanthin, as well as several newly identified oxygenated carotenoids. HPLC system B, which employs a silica-based, nitrile-bonded column and isocratic elution, effectively accomplishes the separation of these polar xanthophylls and several of their geometric isomers from human plasma. The HPLC separations with system A were monitored at five different wavelengths (470, 445, 400, 350, and 290 nm), whereas the HPLC separations with system B were monitored at 325 and 445 nm.

Reference Samples of Plasma Carotenoids. With the exception of lycopene, α-carotene, and β-carotene, the reference samples for most carotenoids are not commercially available. Therefore, carotenoid standards either should be isolated from natural sources[15] or prepared by total[16] and/or partial[17] synthesis. In Table II we have provided the most convenient source for individual carotenoids found in human plasma.[18-31] However, some of the carotenoids found in human plasma are either absent in common foods or they may be present at very low concentrations. Therefore, total or partial synthesis of these carotenoids is inevitable. Alternatively, several of these carotenoids may also be isolated from natural products other than foods.[15]

Internal Standards. In HPLC quantitation of plasma carotenoids, the use of an internal standard is critical because the residue from the extraction of human plasma may not be evaporated to complete dryness.[14] Therefore, it is essential to employ an internal standard that can be fully

---

[16] H. Mayer and O. Isler, in "Carotenoids" (O. Isler, ed), pp. 325–575. Birkhäuser, Basel, 1971.
[17] S. Liaaen-Jensen, in "Carotenoids" (O. Isler, ed), pp. 61–188. Birkhäuser, Basel, 1971.
[18] E. Widmer, R. Zell, H. Grass, and R. Marbet, *Helv. Chim. Acta* **65**, 958 (1982).
[19] F. Khachik, G. R. Beecher, M. B. Goli, W. R. Lusby, and J. C. Smith, *Anal. Chem.* submitted (1992).
[20] S. Liaaen-Jensen and S. Hertzberg, *Acta Chem. Scand.* **20**, 1703 (1966).
[21] R. P. Ritacco, D. B. Rodriguez, G. Britton, T. C. Lee, C. O. Chichester, and K. L. Simpson, *J. Agric. Food Chem.* **32**, 296 (1984).
[22] F. Khachik, G. R. Beecher, and W. R. Lusby, *J. Agric. Food Chem.* **36**, 938 (1988).
[23] R. Buchecker, C. H. Eugster, and A. Weber, *Helv. Chim. Acta* **61**, 1962 (1978).
[24] R. Buchecker and C. H. Eugster, *Helv. Chim. Acta* **62**, 2817 (1979).
[25] E. Widmer, M. Soukup, R. Zell, E. Broger, H. P. Wagner, and M. Imfeld, *Helv. Chim. Acta* **73**, 861 (1990).
[26] M. Soukup, E. Widmer, and T. Lukac, *Helv. Chim. Acta* **73**, 868 (1990).
[27] D. E. Loeber, S. W. Russel, T. B. Toube, B. C. L. Weedon, and J. Diment, *J. Chem. Soc. C* p. 404 (1971).
[28] F. Khachik, G. R. Beecher, and W. R. Lusby, *J. Agric. Food Chem.* **37**, 1465 (1989).
[29] F. Khachik, M. B. Goli, G. R. Beecher, J. Holden, W. R. Lusby, M. D. Tenorio, and M. R. Barrera, *J. Agric. Food Chem.* **40**, 390 (1992).
[30] F. Khachik and G. R. Beecher, *J. Agric. Food Chem.* **35**, 732 (1987).
[31] H. Pfander, A. Lachenmeier, and M. Hadorn, *Helv. Chim. Acta* **63**, 1377 (1980).

TABLE II
SOURCE FOR REFERENCE SAMPLES OF PLASMA CAROTENOIDS

Carotenoids	Source	Ref.
ε,ε-Caroten-3,3'-dione	Total synthysis	18
3'-Hydroxy-ε,ε-caroten-3-one	Total synthesis	19
3-Hydroxy-β,ε-caroten-3'-one	Total synthesis	19
	Partial synthesis	20
5,6-Dihydroxy-5,6-dihydrolycopene	Partial synthesis	21
Lutein	Isolation from green fruits and vegetables	13
	Isolation from squash	22
3'-Epilutein	Partial synthesis	23
	Isolation from flowers of *Caltha palustris*	24
Zeaxanthin	Total synthesis	25, 26
2',3'-Anhydrolutein	Partial synthesis	19
	Isolation from squash	22
α-Cryptoxanthin	Total synthesis	27
	Isolation from flowers of *Caltha palustris*	24
β-Cryptoxanthin	Total synthesis	27
	Isolation from mango and papaya	13
	Isolation from peaches	28
Lycopene	Isolation from tomato paste/products	29
Neurosporene, γ-carotene, ζ-carotene, phytofluene, and phytoene	Isolation from tomato paste/products	29
	Isolation from pumpkin, mango, and papaya	13
	Isolation from apricots and peaches	28
α-Carotene and β-carotene	Commercially available; isolation from carrots, red palm oil, sweet potato	30
Enthyl-β-apo-8'-carotenoate (internal standard)	Commercially available from Fluka Chemical Co. (Ronkonkoma, NY)	—
(3R)-8'-Apo-β-carotene-3,8'-diol (internal standard)	Total synthesis	31

recovered at the end of the extraction of the human plasma and can be used to calculate the final concentration of carotenoids in plasma. Such internal standards can also be employed to monitor the losses of carotenoids as a result of extraction and sample preparation of plasma. In such a case, the recovery of the internal standard is critical, because losses of the internal standard during extraction and sample preparation can also contribute to the analytical errors. Studies in our laboratory have shown that two internal standards, namely, ethyl-β-apo-8'-carotenoate and (3R)-8'-apo-β-carotene-3,8'-diol, can be effectively employed for the quantitation of plasma carotenoids by HPLC on reversed-phase and nitrile-bonded phase columns, respectively.[14] Both of these internal standards are synthetic and their HPLC peaks do not result in a major interference with those of the plasma carotenoids. However, the HPLC peak of (3R)-8'-apo-β-carotene-3,8'-diol interferes with that of a minor component in human plasma, namely, 9'-*cis*-lutein. Therefore in HPLC separations of plasma carotenoids in which the evaluation of 9'-*cis*-lutein is of particular interest, the addition of this second internal standard may be omitted. We have examined many synthetic carotenoids as potential internal standards for the HPLC separation of plasma carotenoids on the nitrile-bonded column and have found (3*R*)-8'-apo-β-carotene-3,8'-diol to be the most suitable. Most importantly, the solubility and chromatographic behavior of (3*R*)-8'-apo-β-carotene-3,8'-diol and ethyl-β-apo-8'-carotenoate in the extracting solvents and the HPLC eluents of systems A and B are similar to those of plasma carotenoids, resulting in recoveries of more than 90%. While ethyl-β-apo-8'-carotenoate is commercially available, the elaborate synthesis of the second internal standard, namely, (3*R*)-8'-apo-β-carotene-3,8'-diol from its precursors has contributed to lack of commercial availability for this compound.[31]

Extraction of Human Plasma. For a detailed HPLC analysis of plasma carotenoids, the volume of plasma needed for extraction is in the range of 1–3 ml. This is mainly due to the fact that the photodiode array detectors, which are required to monitor the HPLC separation of the various classes of carotenoids at several wavelengths simultaneously, are less sensitive than single-wavelength detectors. Modified photodiode array detectors have been shown to improve the sensitivity for the detection of carotenoids by HPLC. As a result, detailed carotenoid analysis of human plasma by HPLC may be obtained from extraction of 0.5 ml of plasma with such detectors. However, in most clinical laboratories, with conventional detectors, limited HPLC analysis of six to eight carotenoids can be accomplished from extraction of only 0.1–0.20 ml of plasma. The application of other sensitive detectors for the detailed HPLC analysis of plasma carotenoids is discussed below. These detectors can be employed with the HPLC conditions described in this chapter to provide complete analyses of carotenoids in a small volume of plasma (0.5 ml), reducing the volume of whole blood required for extraction to 1 ml. The procedure described below allows an efficient extraction of the various classes of carotenoids in human plasma. However, this extraction technique may be modified to accommodate a large number of samples within a reasonably short period of time, as is desirable in many clinical laboratory studies. In a typical extraction, blood (5 ml) for extraction is collected using all-plastic syringes in a 5-ml Vacutainer tube containing 4.5 units of sodium or lithium heparin per milliliter of whole blood. The tube is centrifuged at 2000 g at 4° for 15 min. The plasma (about 2.5 ml) is removed and transferred into a 20-ml glass culture tube. Ethanol (2.5 ml) and the two internal standards, ethyl-β-apo-8'-carotenoate (0.045 μg) and (3*R*)-8'-apo-β-carotene-3,8'-diol (0.24 μg), are added; the tube is shaken vigorously for 20 sec. The mixture is treated with diethyl ether (5 ml) and shaken vigorously, and the precipitated proteins are allowed to settle by gravity. The supernate is removed, and the solid is further extracted twice with 5 ml diethyl ether. At the end of the last diethyl ether wash, the mixture is centrifuged for 5 min to remove the proteins. The combined ethereal layers are evaporated to dryness under a stream of nitrogen. The colored residue is dissolved in a small volume (1–2 ml) of dichloromethane and filtered through a 0.45-μm disposable filter assembly (American Scientific Products, McGaw Park, IL). Dichloromethane is evaporated under a stream of nitrogen, and 100 μl of HPLC solvents (mobile phase, HPLC system B) is added to the remaining oil. Samples (20 μl) are injected in duplicate for analysis on HPLC systems A and B. The mobile phase of HPLC system B has been found to be an appropriate injection solvent for the HPLC analyses of carotenoids in human plasma[14] on both systems A and B and does not produce chromatographic artifacts.[32]

Quantification of Carotenoids, Vitamin A, and Vitamin E in Human Plasma. The carotenoids, vitamin A (retinol), and vitamin E (γ- and α-tocopherol) in the extracts from human plasma must be quantified from the HPLC response factors of the isolated or synthetic reference compounds at five or six different concentrations, employing HPLC systems A and B. Standard curves for plasma carotenoids may be obtained according to similar procedures described in Chapter [32] of this volume. The coefficient of variation (CV) of less than 5% for construction of the calibration curves (i.e., area response at various concentrations) of each of the reference samples is within experimental error and is normally acceptable. The HPLC peak area of *cis*-carotenoids that are not resolved, and appear as a trailing shoulder on their all-trans counterpart, may be combined with the HPLC peak area of all-*trans*-carotenoids. In the case of lutein and zeaxanthin, where the cis isomers are well separated from their all-trans counterparts (HPLC system B), the HPLC peak area of these cis carotenoids may be reported separately from their all-trans compounds. Plasma carotenoids are quantified by correlating their HPLC peak area to that of their corresponding standard curves. In cases in which the reference samples of *cis*-carotenoids are not available, the concentration of *cis*-carotenoids in plasma may be determined from the standard curve of their all-trans counterpart. The recoveries of the two internal standards, (3R)-8'-apo-β-carotene-3,8'-diol and ethyl-β-apo-8'-carotenoate, from extraction of plasma samples should be determined by the HPLC peak area of these internal standards before and after extraction and workup procedures.

---

[32] F. Khachik, G. R. Beecher, J. T. Vanderslice, and G. Furrow, *Anal. Chem.* **60**, 807 (1988).

Recoveries of more than 90% for the internal standards are desirable and they are an indication of reliability of sample preparation and the HPLC procedures.

The concentration of plasma carotenoids is traditionally expressed in micrograms per deciliter. International clinical standards (SI units) have been introduced that report data in molar terms, with the liter as the reference volume.[33] The corresponding values for plasma carotenoids are expressed in nanomoles per liter.[14] Similarly, values for vitamin A and the two forms of vitamin E (γ- and α-tocopherol) are expressed in micrograms per deciliter and micromoles per liter.

Detailed Separation of Carotenoids in Human Plasma by HPLC

Detailed analysis of plasma carotenoids may be accomplished by HPLC separation of extracts on reversed-phase C_{18} and a silica-based nitrile-bonded columns employing HPLC systems A and B, respectively. Although most of the plasma carotenoids can be best separated on C_{18} reversed-phase columns, several of the early eluting oxygenated carotenoids (i.e., ketocarotenoids, lutein, 3′-epilutein, zeaxanthin) because of their high polarity and low affinity for the C_{18} reversed-phase adsorbents, cannot be resolved on such columns. On the other hand, because the HPLC separation of plasma carotenoids on the nitrile-bonded columns results in excellent resolution for the oxygenated carotenoids, additional chromatographic separation employing HPLC system B becomes necessary. Under these HPLC conditions, most of the nonpolar carotenoids that are well separated on a C_{18} reversed-phase column coelute early on the nitrile-bonded column, resulting in no interference with the separation of the carotenoids of interest. We have been unable to develop a single HPLC system for the detailed analysis of plasma carotenoids employing these two columns and column switching techniques. This is because the HPLC solvents of systems A and B are not compatible and result in rapid elution of carotenoids of interest. Alternatively, a single HPLC system can be employed; following each injection of an extract from human plasma under HPLC system A conditions, the column is replaced and the system is reequilibrated under HPLC system B conditions. The major disadvantage of such an arrangement is the long time needed to achieve equilibrium with each column and eluent for consistent chromatographic results. Therefore it is most appropriate to designate two independent HPLC systems, such as A and B, that can be simultaneously operated to provide detailed analysis of plasma carotenoids on a continuous basis. In the

[33] D. S. Young, *Ann. Intern. Med.* **106,** 20 (1987).

following section the detailed separation of plasma carotenoids by HPLC employing C_{18} reversed-phase and nitrile-bonded columns is discussed.

Separation of Plasma Carotenoids by HPLC on C_{18} Reversed-Phase and Nitrile-Bonded Columns. The chromatographic profiles of an extract of human plasma injected on a C_{18} reversed-phase column (HPLC system A, mobile phase A) and a nitrile-bonded column (HPLC system B, mobile phase B) are shown in Figs. 1 and 2, respectively. Eighteen carotenoids, as well as vitamin A and 2 forms of vitamin E (γ- and α-tocopherol), are shown to be present in human plasma. These components (see Table III) have each been isolated by preparative thin-layer chromatography (TLC) and HPLC from an extract of a large volume of plasma and have been identified by comparison of their HPLC retention times and their UV/visible absorption and mass spectra with those of synthetic compounds.[19] The chemical structures of these carotenoids are shown in Fig. 3. The later peaks in the chromatogram shown in Fig. 1 (peaks 11–18) are the hydrocarbon carotenoids, which are lycopene (peak 11), neurosporene (peak 12), γ-carotene (peak 13), ζ-carotene (peak 14), α-carotene (peak 15), β-carotene (peak 16), phytofluene (peak 17), phytoene (peak 18), and noncarotenoid. The noncarotenoid is cholesteryl oleate, eluting immediately after phytoene. Cholesteryl oleate in human plasma has been identified by mass spectrometry and comparison of its HPLC retention time with that of a reference sample of this compound. Detection of cholesteryl

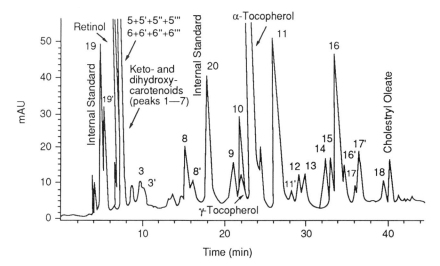

FIG. 1. Carotenoid HPLC profile of an extract from human plasma on a C_{18} reversed-phase column employing HPLC system A (eluent A). For peak identification see Table III.

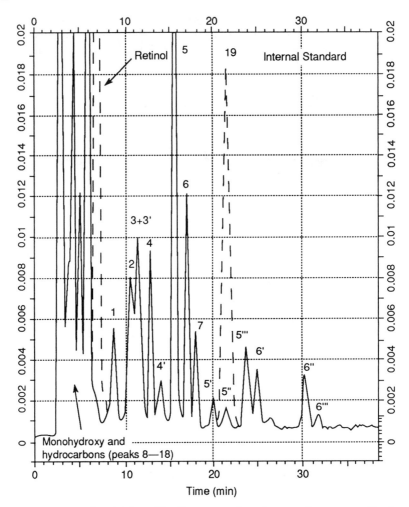

FIG. 2. Carotenoid HPLC profile of an extract from human plasma on a silica-based nitrile-bonded column employing HPLC system B (eluent B). For peak identification see Table III.

oleate by HPLC is mainly due to a weak absorption in the UV spectrum of this compound at the monitoring wavelength of 290 nm. Several geometric isomers of lycopene (peak 11'), β-carotene (peak 16'), and phytofluene (peak 17') are also found in plasma along with their all-trans compounds. Hydrocarbon carotenoids in human plasma are of dietary origin, namely common fruits and vegetables.[13] The more polar components with shorter retention times than the hydrocarbons in the chromatogram shown in Fig.

TABLE III
HPLC Peak Identification and Wavelength Maxima of Internal Standards and Carotenoids[a]

Peaks	Carotenoids[b]	Wavelength (nm)
		Eluent B
1	ε,ε-Caroten-3,3'-dione	442
2	3'-Hydroxy-ε,ε-caroten-3-one	442
3	all-*trans*-5,6-dihydroxy-5,6-dihydrolycopene	460
3'	*cis*-5,6-Dihydroxy-5,6-dihydrolycopene	458
4	3-Hydroxy-β,ε-caroten-3'-one	448
4'	*cis*-3-Hydroxy-β,ε-caroten-3'-one	442
5	all-*trans*-Lutein	448
6	all-*trans*-Zeaxanthin	454
7	3'-Epilutein	448
5'	9-*cis*-Lutein	442
5''	9'-*cis*-Lutein	444
5'''	13-*cis* and 13'-*cis*-Lutein	440
6'	9-*cis*-Zeaxanthin	450
6''	13-*cis*-Zeaxanthin	446
6'''	15-cis-Zeaxanthin	450
		Eluent A
8	all-*trans*-2',3'-Anhydrolutein	446
8'	*cis*-2',3'-Anhydrolutein	440
9	β,ε-Caroten-3-ol (α-cryptoxanthin)	446
10	3-Hydroxy-β-carotene (β-cryptoxanthin)	454
11	all-*trans*-Lycopene	472
11'	*cis*-Lycopene	468
12	all-*trans*-Neurosporene	440
13	γ-Carotene	462
14	ζ-Carotene	400
15	α-Carotene	446
16	all-*trans*-β-Carotene	454
16'	*cis*-β-Carotene	446
17	all-*trans*- or *cis*-Phytofluene	350
17'	*cis*- or all-*trans*-Phytofluene	350
18	Phytoene	286
19	(3R)-8'-Apo-β-carotene-3,8'-diol (internal standard)	426
19'	*cis*-(3R)-8'-Apo-β-carotene-3,8'-diol	422
20	Ethyl-β-apo-8'-carotenoate (internal standard)	458

[a] Separated from an extract of human plasma; wavelength maxima similar to chromatographic elution with eluent B (HPLC system B) and eluent A (HPLC system B).

[b] *cis*-Carotenoids have been designated the same number as their all-trans isomers, but distinguished from their all-trans compounds by prime symbols.

1) 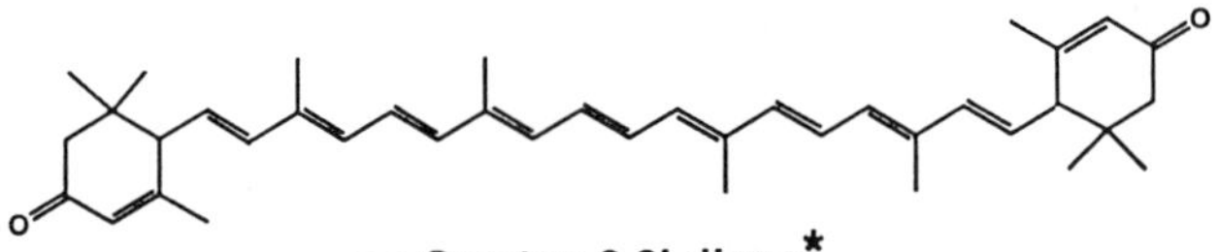 ε,ε-Caroten-3,3'-dione*

2) 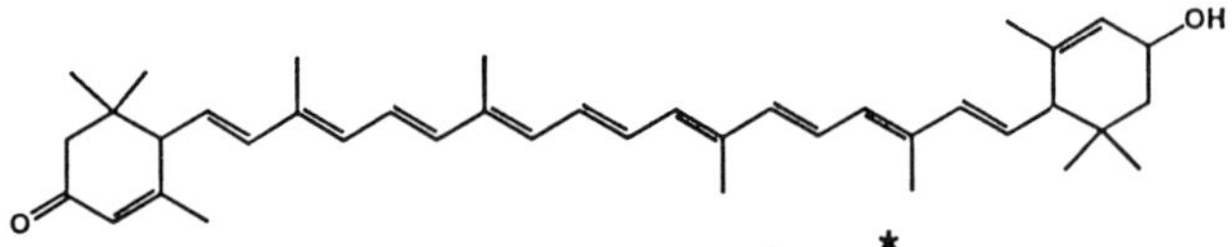 3'-Hydroxy-ε,ε-caroten-3-one*

3) 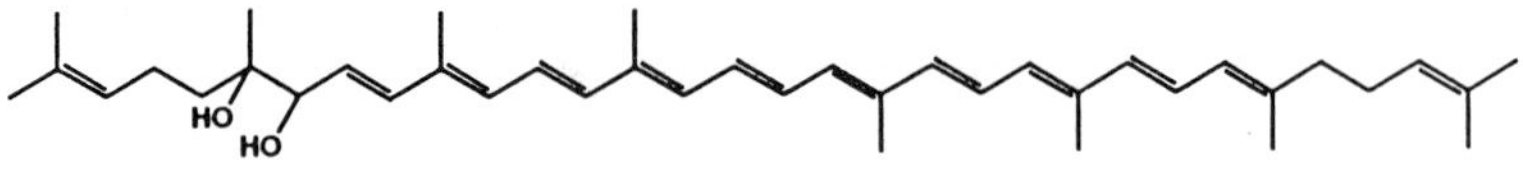 5,6-Dihydroxy-5,6-dihydrolycopene*

4) 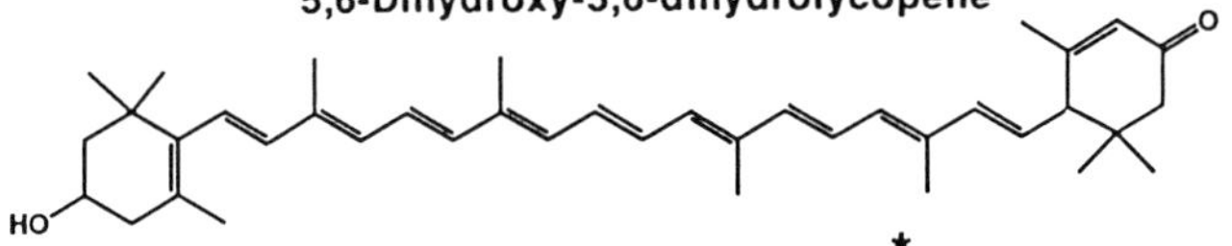 3-Hydroxy-β,ε-caroten-3'-one*

5) 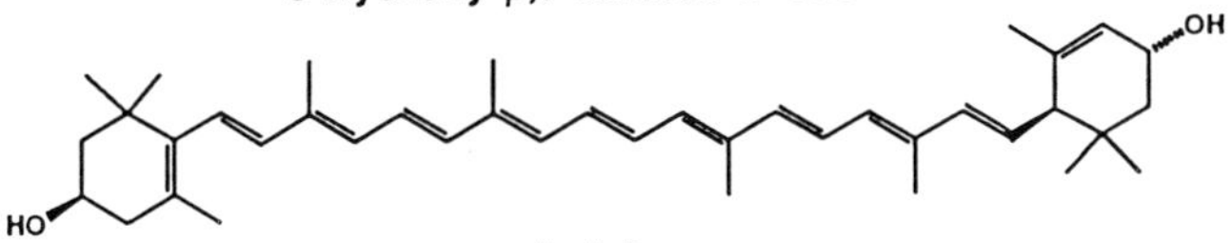 Lutein

6) 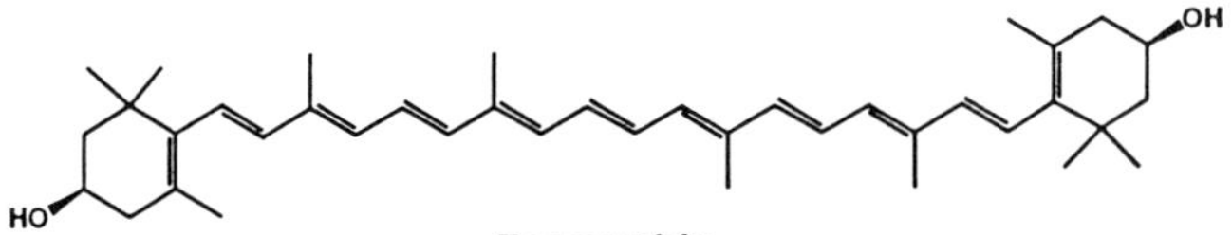 Zeaxanthin

7) 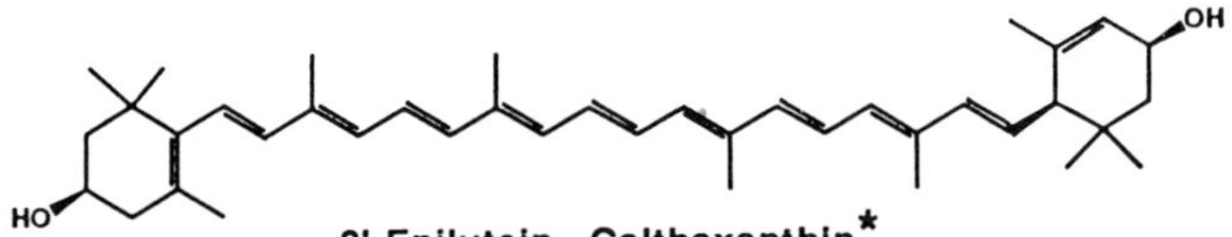 3'-Epilutein, Calthaxanthin*

8) 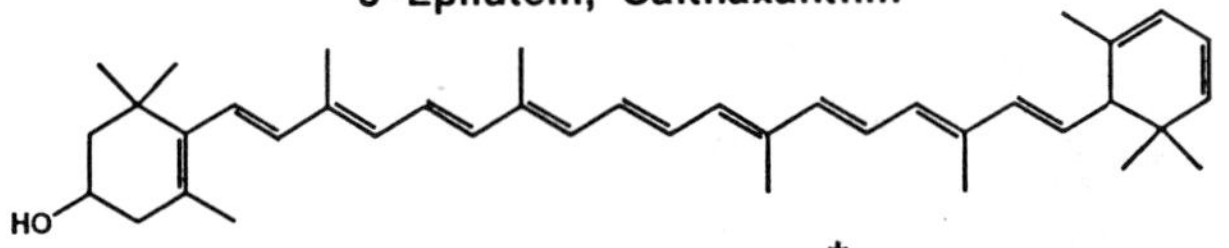 2',3'-Anhydrolutein*

9) 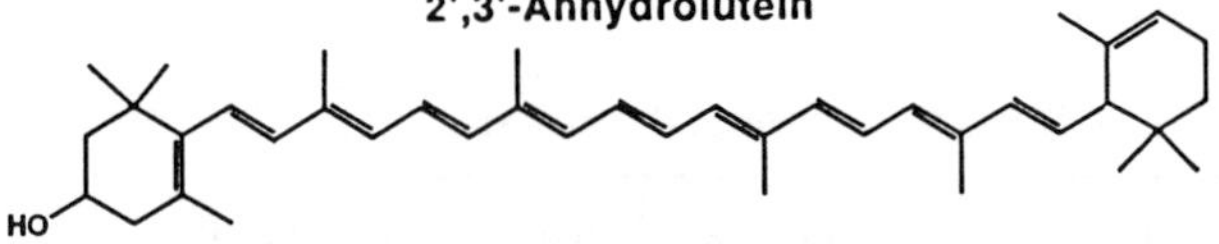 α-Cryptoxanthin, zeinoxanthin

10) 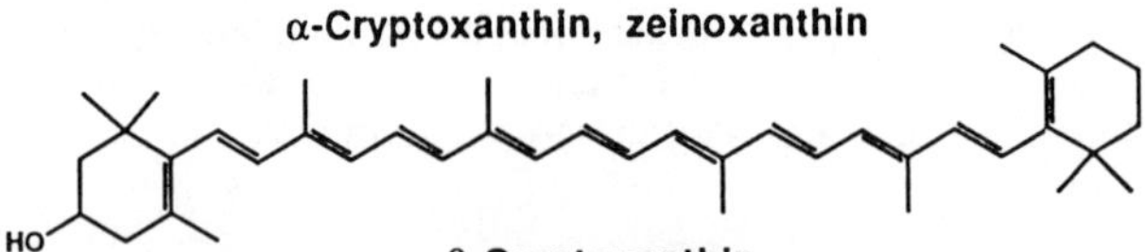 β-Cryptoxanthin

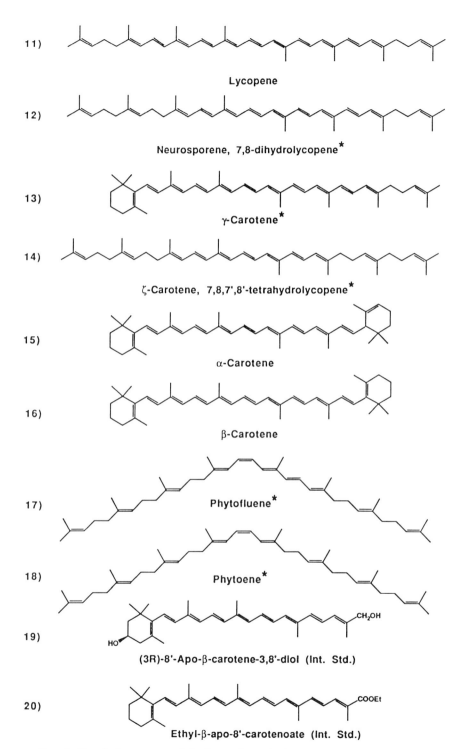

FIG. 3. Chemical structures of the internal standards (Int. Std.) and the carotenoids isolated and characterized from extracts of human plasma. *, Newly identified carotenoids.

1 are α-cryptoxanthin (peak 9), β-cryptoxanthin (peak 10), γ-tocopherol, and α-tocopherol. Although the HPLC peaks of β-cryptoxanthin and γ-tocopherol coelute (Fig. 1), the absorption spectra of these compounds in the HPLC solvents with maxima at 454 and 292 nm, respectively, allow accurate determination of these two components in the extracts from human plasma. There are two possible chemical structures for α-cryptoxanthin, depending on whether the 3-hydroxyl group is located on the ε or β end groups.[15] We have now established that the hydroxyl group in α-cryptoxanthin in human plasma is substituted in the β rather than the ε end group.[13,19] From the more polar carotenoids in the HPLC profile shown in Fig. 1, we have separated and identified 3-hydroxy-2',3'-didehydro-β,ε-carotene (2',3'-anhydrolutein, peak 8). Dietary sources of 2',3'-anhydrolutein are limited, as this compound has been detected only in one variety of squash.[22,34] Another newly identified carotenoid in human plasma is 5,6-dihydroxy-5,6-dihydrolycopene (peak 3, Fig. 1). We have shown that the dietary sources of this compound are tomato paste and concentrated tomato products.[29]

The most polar fraction of the extracts from human plasma consists of several components that have short retention times (6–12 min) on a C_{18} reversed-phase column (see Fig. 1). These components are vitamin A alcohol (retinol), lutein, and zeaxanthin, together with several newly identified carotenoids, which are only partially separated because of their high polarity and low affinity for the C_{18} reversed-phase column. Therefore, as pointed out earlier, in addition to chromatography on the C_{18} reversed-phase column, extracts from human plasma should also be chromatographed on a silica-based nitrile-bonded column to resolve these early eluting carotenoids.

The separation of the plasma carotenoids on the nitrile-bonded column (Fig. 2) results in coelution of nonpolar carotenoids that are well separated on the C_{18} reversed-phase column (peaks 8–18, Fig. 1) within the first 6 min. This is then followed by the elution of retinol and several newly identified carotenoids, namely, ε,ε-caroten-3,3'-dione (peak 1, Fig. 2), 3'-hydroxy-ε,ε-caroten-3-one (peak 2), 5,6-dihydroxy-5,6-dihydrolycopene (peak 3), 3-hydroxy-β,ε-caroten-3'-one (peak 4) and its cis isomer (peak 4'), (3R,3'R,6'R)-lutein (peak 5), zeaxanthin (peak 6), and (3R,3'S,6'R)-lutein [known as 3'-epilutein (peak 7)]. Several geometric isomers of (3R,3'R,6'R)-lutein and zeaxanthin, which have been identified from their 400-MHz ^{1}H-NMR, UV/visible absorption, and CD spectra (see Fig. 2), are also present in the extracts of human plasma.[19,35] Although most of the

[34] F. Khachik and G. R. Beecher, *J. Agric. Food Chem.* **36**, 929 (1988).
[35] F. Khachik, G. Englert, C. E. Daitch, G. R. Beecher, W. R. Lusby, and L. H. Tonucci, manuscript in preparation (1992).

carotenoids of interest are definitively separated using the nitrile-bonded column, the HPLC peaks of 3'-hydroxy-ε,ε-caroten-3-one (peak 2) and 5,6-dihydroxy-5,6-dihydrolycopene (peak 3) are only partially resolved (see Fig. 2).

In such complex and detailed HPLC separations of carotenoids in human plasma, it is desirable to monitor the chromatograms of carotenoids by a photodiode array detector; this allows confirmation of the structure of carotenoids by comparison of the HPLC retention times with those of reference samples. The structural assignments can be further ascertained from the UV/visible absorption spectrum of the individual carotenoids. As described earlier, one of the major disadvantages of photodiode array detectors in the HPLC separation of carotenoids is the lower sensitivity of these detectors in the visible region compared with single-wavelength detectors. As a result, appropriate concentration of plasma carotenoids that lie well above the detection limit of photodiode array detectors can be obtained only by increasing the volume of human plasma needed for extraction. Typically, HPLC analysis of plasma carotenoids by single-wavelength detectors can be performed with the extraction of only 0.1 ml of human plasma,[12] whereas in similar separations with photodiode array detectors as much as 1 to 3 ml of plasma may be necessary. A compromise between these two detector extremes is the use of a variable-wavelength detector that permits several wavelengths to be monitored sequentially, yet with respectable sensitivity. The capability of acquiring spectra, however, is lost in this situation. The monitoring of coeluting peaks with divergent absorption maxima is also not possible with a single- or variable-wavelength detector but can be accomplished by adding a second detector in series with the first detector. The routine analysis of carotenoids in human plasma on the nitrile-bonded column (HPLC system B), in which the chromatogram is monitored at 445 nm, can be accomplished with either a single-wavelength or variable-wavelength detector. Instrument manufacturers are constantly increasing the sensitivity of HPLC detectors, especially photodiode array detectors. Scientists planning to measure carotenoids in human plasma are well advised to ascertain the sensitivity and other characteristics of detectors with actual separation of carotenoids in plasma extracts prior to selecting a specific detector. This is particularly important when the sample size is extremely limited.

Acknowledgments

We thank F. Hoffmann-La Roche & Co. (Basel, Switzerland) and BASF (Ludwigshafen, Germany) for their gift of carotenoid reference samples and precursors. We thank Dr. J. C. Smith and Anne Dulin (U.S. Department of Agriculture, Vitamin and Mineral Laboratory) for their technical support throughout this project. Partial support by the National Cancer Institute through reimbursable agreement Y01-CN-30609 is acknowledged.

[19] Measurement of Carotenoids in Human and Monkey Retinas

By GARRY J. HANDELMAN, D. MAX SNODDERLY, ALICE J. ADLER, MARK D. RUSSETT, and EDWARD A. DRATZ

Introduction

A prominent, bright yellow spot has long been known to exist in the center of the typical primate retina. This spot is called the *macula lutea,* or simply the macula. This region of the primate retina is specialized for acute central vision, and contains the highest density of cone photoreceptors. In elderly humans, the macular region may deteriorate, with partial or complete loss of central vision, causing difficulty with such everyday tasks as reading or driving a car. This disorder (called age-related macular degeneration, or AMD) affects about 10% of humans over 70 years of age, and is more prevalent in those with light iris pigmentation.[1]

Wald inferred in 1949[2] that the yellow macular pigment was a polar carotenoid, from the absorption spectrum of extracts and from the solvent partitioning properties of the pigment. In 1985, Bone and co-workers[3] demonstrated by high-performance liquid chromatography (HPLC) that two carotenoids, lutein and zeaxanthin, were present in the human macula. In an analysis of distinct regions of the retina, Bone and Landrum[4] and Handelman and co-workers[5] showed that zeaxanthin is concentrated in the macula and that lutein is contained in the macula as well as in the peripheral retina. Human plasma contains a complex mixture of polar and nonpolar carotenoids, with relatively low levels of zeaxanthin and high levels of β-carotene.[6] The retina selectively accumulates lutein and zeaxanthin from this complex mixture of carotenoids found in the plasma by an unknown mechanism.

The function of the carotenoid pigments in the macula has not yet been determined. The pigments may serve to limit the effects of chromatic

---

[1] L. Hyman, A. M. Lilienfeld, F. L. Ferris III, and S. L. Fine, *Am. J. Epidemiol.* **118**, 213 (1983).
[2] G. Wald, *Documenta Ophthalmol.* **3**, 94 (1949).
[3] R. A. Bone, J. T. Landrum, and S. L. Tarsis, *Vision Res.* **25**, 1531 (1985).
[4] R. A. Bone, J. T. Landrum, L. Fernandez, and S. L. Tarsis, *Invest. Ophthalmol. Visual Sci.* **29**, 843 (1988).
[5] G. J. Handelman, E. A. Dratz, C. C. Reay, and F. J. G. M. van Kuijk, *Invest. Ophthalmol. Visual Sci.* **29**, 850 (1988).
[6] N. I. Krinsky, M. D. Russett, G. J. Handelman, and D. M. Snodderly, *J. Nutr.* **120**, 1654 (1990).

aberration by selectively absorbing blue light.[7] Other roles proposed include selective elimination of phototoxic blue light from the sensitive central cone photoreceptors,[8] and quenching of potentially damaging active oxygen species.[9] The consistent occurrence of these pigments in a wide variety of different primate species[10-12] suggests an important function for these carotenoids in the retina.

Dissection of Human Autopsy Retinas

The most readily available primate retinas are those of humans whose eyes have been donated to eye banks. It is preferable to obtain these specimens within 24 hr postmortem, but even shorter postmortem times may be necessary for good preservation of the macula.

The dissection of the primate retina is technically demanding, and the investigator should anticipate the need for practice to attain mastery of these procedures. Helpful suggestions can be found in the volume on retinal whole mounts by Stone,[13] including different ways of flattening the spherical surface of the retina for dissection.

Figure 1 shows the human eye in cross-section, labeled to emphasize anatomical structures relevant to the dissection of the retina. The macula lutea is defined as the yellow pigmented spot in the back of the retina. The spot is about 2 mm in diameter in humans. The primate retina is thin at the macula and hence requires gentle handling. We describe a procedure for dissecting the retina free of the rest of the eyeball, and then removing the macula from the retina with a circular punch.

There are minimal criteria that must be met for eyes included in a study of macular pigment. We found it impossible to properly dissect eyes that had been frozen and thawed. Our experience indicates that human retinas more than 24 hr postmortem do not dissect well, because of postmortem deterioration of the retinal tissues. Even eyes only 8 hr postmortem can have holes in the center of the macula. Whether this constitutes a loss of

---

[7] G. J. Walls and H. D. Judd, *Br. J. Ophthalmol.* **17,** 641 (1933).
[8] K. Kirschfeld, *Proc. R. Soc. London, B* **216,** 71 (1982).
[9] N. I. Krinsky, *Free Radicals Biol. Med.* **7,** 617 (1989).
[10] D. M. Snodderly, P. K. Brown, F. C. Delori, and J. D. Auran, *Invest. Ophthalmol. Visual Sci.* **25,** 660 (1984).
[11] G. J. Handelman, D. M. Snodderly, N. I. Krinsky, M. D. Russett, and A. J. Adler, *Invest. Ophthalmol. Visual Sci.* **32,** 257 (1991).
[12] D. M. Snodderly, G. J. Handelman, and A. J. Adler, *Invest. Ophthalmol. Visual Sci.* **32,** 268 (1991).
[13] J. Stone, "The Whole Mount Handbook: A Guide to the Preparation and Analysis of Retinal Whole Mounts." Maitland, Sydney, Australia, 1981.

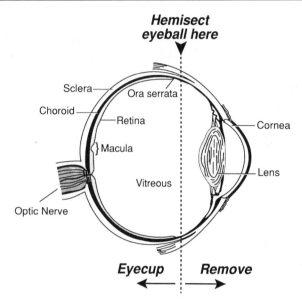

FIG. 1. Cross-sectional diagram of the human eye. After being detached from its connections to the eyecup, the retina can be pinned to a flat surface and the macula removed with a trephine.

tissue or simply a tear in the retina is not known. The effects of postmortem time need to be studied systematically.

Latex gloves should be worn during the dissection for personal protection. The first step in the dissection is to grasp the eye gently between the center of the cornea and the posterior pole (near the optic nerve) with the thumb and forefinger. The edge of the eye is nicked with a sharp scalpel, just in front of the equator, as shown in Fig. 1. The eye should be cut in front of the equator, ahead of the ora serrata, to retain all the retina in the eyecup. A pair of sharp scissors should be used to cut completely around the eyeball at the level of the initial scalpel cut (Roboz #555 scissors; Roboz Instrument, Washington, D.C.). The vitreous should then be sectioned with a pair of scissors, and the front half of the eye removed, as in Fig. 1.

For the next step, a plastic petri dish, 2.5 cm deep and 10 cm in diameter, is filled with buffer [isotonic phosphate-buffered saline, pH 7.4, containing 1 mM ethylenediaminetetraacetic acid (EDTA)] to a depth of 2 cm. The unfixed posterior eyecup is immersed in the buffer, and forceps (Roboz) are used to gently free the retina from its attachment at the ora serrata (see Fig. 1). The optic nerve connection is carefully snipped with a pair of small curved scissors (Miltex). The fully detached retina usually floats free of the eye cup. The free-floating retina is moved to one side in the dish and the eyecup is discarded.

When dissecting monkey eyes, which can often be obtained more quickly after death, the retina may still be attached to the eyecup. In this case, the retina can be detached from the eyecup by gentle manipulation with a fine artist's paintbrush.

A second petri dish is prepared with a 5-mm layer of paraffin wax in the bottom of the dish. About 10 layers of 25-mm-diameter, soft Whatman (Clifton, NJ) number 1 filter paper is secured to the wax with 3 dissection pins. The second petri dish is then filled with buffer to a depth of 1 cm. An alternative to paraffin wax is silicone rubber (RTV-12; General Electric, Schenectady, NY), which can be poured into the petri dish and allowed to harden as a backing material. If this is used to pin down the retina, only a single layer of filter paper need be used.

The retina is then transferred from the first petri dish to the second with a pouring motion. The retina is then oriented concave (vitreous) side up; it may be necessary to invert the retina gently with a blunt forceps. With a pair of dissecting scissors, four radial incisions are made in four quadrants, reaching half the distance from the edge of the retina to the center. These incisions allow the retina to lie flat on the filter paper. The retina is then pinned to the filter paper with four dissection pins. The buffer is removed with a Pasteur pipette, leaving the damp retinal specimen pinned to the filter paper.

At this stage, there may still be some fragments of vitreous attached to the specimen. With experience the investigator will develop appropriate methods for removing the bulk of the vitreous during the dissection procedure. After the retina is pinned down for dissection of the macula, the remaining fragments of vitreous should not be disturbed, because attempts to remove these vitreal fragments will often tear the retina. Adhering vitreous does not interfere with the analysis by HPLC, and does not contain any carotenoid.

Carefully inspect the retina through a ×10 dissecting microscope to ascertain the exact location of the macula. The macula can then be punched out with a trephine. A 4-mm-diameter trephine is recommended (Biomedical Research Instruments, Rockville, MD). The 4-mm-diameter trephine is large enough to harvest the main peak of the macular pigment, even if there is a slight error in the placement of the trephine.

Typically some of the filter paper is collected along with the macula, but the filter paper does not interfere with the analysis. However, it is necessary to avoid collecting any paraffin wax because it will carry through the procedure and impair recovery of pigments at the HPLC step. The specimen is removed from the trephine with a fine pointed forceps (Dumont, #5). The specimen should be picked up *with the adhering filter paper* and transferred to a 10-ml glass homogenizer (Wheaton model #358039; Fisher Scientific, Fairlawn, NJ).

Other regions of the retina can be removed with dissection scissors or punched out with the trephine and transferred to the homogenizer for analysis. Alternatively, the entire undissected retina can be placed in the homogenizer and analyzed. The dissection procedure described is fully satisfactory for the smaller eyes of monkeys, but requires a high level of proficiency.

Dissection of Retinas from Monkeys Perfused after Death with Aldehyde Fixatives

Monkeys are often sacrificed for non-vision-related studies, and then systemically perfused immediately after death with aldehyde fixatives to preserve the tissues. We have conducted experiments that demonstrate that the retinas of perfusion-fixed animals can be reliably used for macular pigment measurements.[11,12] The eyes of monkeys perfused in this manner are frequently available to other investigators through the Primate Supply Information Clearing House (University of Washington, Seattle, WA).

For the dissection of the perfusion-fixed monkey eye, the entire eyeball is hemisected, as with the human autopsy specimens. However, the retina may not separate easily from the back of the eyecup. To remove the macular region, several radial cuts are made in the entire posterior eyecup so that it can be pinned to a flat surface. Then the macular region is examined under bright light and punched out with the 4-mm trephine, along with the choroid and sclera beneath the macula. Because the nonretinal tissues contribute little carotenoid, this entire punch specimen can be homogenized and analyzed, as described below, to estimate the macular pigment content of the specimen.

Other, more refined procedures can be used to remove the adherent nonretinal tissue and to microdissect the retina and macula for determination of the carotenoid content of very small regions. Details of these procedures have been described previously.[11,12,14]

Extraction of Carotenoids from Tissue Samples

The dissected specimen is homogenized for 30 sec in 0.5 ml of 100% ethanol and 0.5 ml of buffer, using a tight-fitting Teflon piston and a motor drive operated at about 600 rpm. The ethanol used for these mea-

---

[14] D. M. Snodderly and R. S. Weinhaus, *J. Comp. Neurol.* **297**, 145 (1990).

surements contains 10 μg/ml of butylated hydroxytoluene (BHT), as antioxidant. The tissue homogenate is transferred to an 8-ml glass scintillation vial (Fisher Scientific) using a Pasteur pipette. The homogenizer is rinsed with 0.5 ml of 100% ethanol and 0.5 ml of buffer using 10 sec of homogenization and the rinse solution is transferred to the 8-ml sample vial. One milliliter of hexane (HPLC grade), containing a known amount of suitable carotenal oxime internal standard,[5] is added quantitatively, using a Pasteur pipette connected with Tygon tubing to a Gilson Pipetteman (Rainin Instruments, Woburn, MA). The accuracy and reproducibility of the method depend critically on the repeatable delivery of internal standard at this step. The accuracy of solvent volume delivery may be enhanced by prewetting the pipette tip with hexane. The weight of solvent delivered may be checked on a pan balance.

The carotenal oxime derivative used in this report is synthesized from β-apo-10′-carotenal (Hoffmann-La Roche, Nutley, NJ) and O-ethylhydroxylamine, as described elsewhere in this volume.[15] The carotenal oxime solution is diluted to a final OD_{450} of 0.006. Depending on the HPLC column employed, it may be necessary to evaluate different carotenal oximes to obtain an internal standard with a retention time about 1.5 times that of zeaxanthin. For this purpose, several of these carotenal oximes, each with a different retention time, can be easily prepared.[5,15]

Three milliliters of hexane, free of internal standard, is added to the sample, the vial is tightly capped, and the sample is mixed on a vortex mixer for 60 sec. The extract is centrifuged for 30 sec at 500 g, and the hexane layer is transferred to a clean 8-ml scintillation vial. The hexane extract is stable at room temperature for several hours, without loss of pigment.

For calibration, we prepare stock solutions of lutein and zeaxanthin (both from Hoffmann-La Roche) in methanol, each at about 4 μg/ml (8 nM). The exact concentration of the stock solutions is determined by measurement of optical density. For lutein, we use $\varepsilon_{446} = 137,000$ liter M^{-1} cm^{-1},[16] and for zeaxanthin we use $\varepsilon_{454} = 139,000$ liter M^{-1} cm^{-1}.[16] These stock solutions are diluted with methanol to 0.5 μg/ml. Using an SMI-positive displacement micropettor (American Scientific Products, McGaw Park, IL), 50 μl of each of these dilute stock solutions (for a total of 25 ng of each carotenoid) is added to a vial containing 1 ml of ethanol (containing 10 μg/ml BHT) and 1 ml buffer. This calibration mixture is then carried through the analysis in parallel with the tissue homogenates.

[15] G. J. Handelman, B. Shen, and N. I. Krinsky, this volume [31].
[16] J. Szabolcs, in "Carotenoids: Chemistry and Biology" (N. I. Krinsky, M. M. Mathews-Roth, and R. F. Taylor, eds.), Proc. 8th Int. Symp. Carotenoids, p. 39. Plenum, New York, 1989.

HPLC Analysis

The hexane phase from the tissue extraction is evaporated with a stream of dry nitrogen while the vial is held in a warm-water bath. Then 40 µl of 100% methanol is added to dissolve the residue, and 20 µl is injected onto the HPLC. It is very important that methanol is used for an injection solvent as a number of other injection solvents degrade reversed-phase resolution of these carotenoid compounds.[15]

Chromatography is carried out with an Alltech C_{18} Adsorbosphere-HS 3-µm column (15 × 0.46 cm; Alltech Associates, Deerfield, IL). The mobile phase is acetonitrile-methanol (85:15, v/v) containing 0.01% (w/v) ammonium acetate. The flow rate is 1 ml/min. This column is very effective at resolving lutein and zeaxanthin, using the mobile phase described. Other combinations of column and mobile phase may be suitable, but the ability of the system to separate lutein and zeaxanthin must be determined. To evaluate a separation, the carotenoid pigment extracted from corn meal (using the same procedure as described for the retina) is a useful reference material, as corn meal contains substantial amounts of both lutein and zeaxanthin.[6,17] For reversed-phase analyses, the inclusion of ammonium acetate is recommended because it prevents on-column degradation of the carotenoids.[11,15] The ammonium acetate (obtained as the crystalline material from several suppliers) can be dissolved in methanol, at 5%, and then diluted into the mobile phase to a final concentration of 0.01%.

The measurement of about 2 ng of each pigment can be carried out at 450 nm on a single-wavelength HPLC detector, with baseline noise of about 1×10^{-4} AU. The HPLC detector equipped with a mercury lamp and 436-nm filter gives excellent performance, with baseline noise of about 2×10^{-5} AU. Useful spectral information about constituents of the macula can be obtained with an on-line diode array HPLC detector, but 10 ng or more of each carotenoid pigment is usually required for reliable spectra.

Nonpolar carotenoids, such as β-carotene, if present, are eluted from the column by a step gradient to 30% 2-propanol, executed after the lutein and zeaxanthin elute. Analyses have often been done without the step gradient, because nonpolar hydrocarbon carotenoids are not observed in carefully dissected maculas, which are free of blood contamination.

The column temperature can be stabilized with a water jacket, and we routinely work at a column temperature of 25°. This is helpful to achieve maximum retention time stability.[15] The reliability of this method for tissue extraction and analysis, including the step of dissolving the tissue

---

[17] E. J. Weber, *J. Am. Oil Chem. Soc.* **64**, 1129 (1987).

extract in methanol for HPLC measurement, has been carefully validated in previous work.[5,11]

Findings in the Macula and Peripheral Retina

Macula

The human macula contains substantial amounts of lutein and zeaxanthin (each with two hydroxyl groups), in the range of 5–40 ng of each pigment. Figure 2 shows the HPLC analysis of the macula from a human retina, from a 67-year-old male donor. This macular region (contained within the 4-mm-diameter punch) has 17.4 ng of zeaxanthin and 8.8 ng of lutein. The macula of the monkey may contain only about one-half as much pigment, but the amounts are quite sufficient for precise HPLC analysis.

The biological features of the human macula that lead to the predominance of zeaxanthin at the macular center have not yet been elucidated. This same predominance of zeaxanthin is observed in macaques (an Old World monkey), but some squirrel monkeys (a New World monkey) show

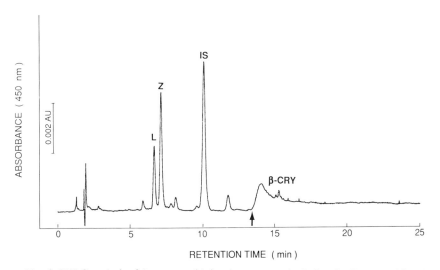

FIG. 2. HPLC analysis of the carotenoids in a human macula. L, Lutein; Z, zeaxanthin; IS, internal standard; β-CRY, β-cryptoxanthin. The arrow marks the refractive index change during the solvent gradient change to 30% 2-propanol. Column: C_{18} Adsorbosphere-HS, 3 μm, 15 × 0.46 cm (Alltech Associates). Mobile phase, acetonitrile–methanol (85:15, v/v), with 0.01% ammonium acetate; flow rate 1 ml/min. Addition of 30% 2-propanol to mobile phase at 10 min. Detection at 450 nm.

about equal amounts of lutein and zeaxanthin at the macular center.[12] Understanding of these differences may come from a better understanding of the distribution of different cone phenotypes in the macula of various primate species.

The major cis isomers of lutein and zeaxanthin have not been identified in these specimens. A trace of β-cryptoxanthin (with one hydroxyl group) is present in this analysis, eluting at 15.2 min. The hydrocarbon carotenoids lycopene, α-carotene, and β-carotene are not present at detectable levels in the human macula,[5] despite their abundance in human plasma.[6] With a detection limit of 0.25 ng, we estimate that β-carotene, if present, could not be more than 1% of the sum of lutein and zeaxanthin, which are the dominant macular pigments.

All specimens of retina contain retinoids, usually some mixture of vitamin A alcohol and vitamin A aldehyde. The aldehyde, if present, generates a peak at 3.8 min, if the analysis is carried out at 450 nm. In the specimens shown here, all the vitamin A is in the form of the alcohol, and no peak is seen at 450 nm. The aldehyde has been detected in other specimens of retinal tissue.[5,11]

The peak at 5.9 min has been seen in most human and monkey maculas and retinas, but has not been identified. The unidentified peak at 11.8 min has been seen in several human specimens, but in no monkey

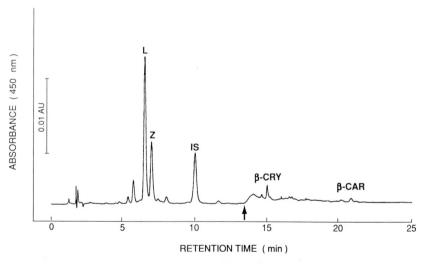

FIG. 3. HPLC analysis of carotenoids in a human peripheral retina. β-Carotene (β-CAR) has been detected at trace levels in some human peripheral retina specimens, but never in the human macula. Other peak labels and analytical conditions are as in Fig. 2.

specimens. The peak at 8.0 min in Figs. 2 and 3 has recently been identified as a contaminant derived from large amounts of BHT (> 100 µg), which was used as the antioxidant for these analyses. If the total amount of BHT used in the analysis is 10 µg, this contaminant is not observed.

Peripheral Retina

Figure 3 shows the HPLC analysis of the remainder of the same retina used for the macular analysis, shown in Fig. 2. The peripheral retina contains 60.3 ng of lutein (eight times the amount found within the 4-mm macular zone) and 27.4 ng of zeaxanthin, which is 1.5 times the amount in the macula. This predominance of lutein in the periphery is typical of all primate retinas reported in the literature.[4,5,12] The biochemical and anatomical basis for partitioning of zeaxanthin and lutein in the retina is still not understood. This needs to be addressed in future studies.

Several of the minor carotenoid constituents observed in the macula (Fig. 2) are also present in the peripheral retina (Fig. 3). Unlike the macula, there are also small traces of the hydrocarbon carotenoids lycopene and β-carotene, which may be contributed by blood present in the retinal vasculature of the autopsy specimen. The specimen of peripheral retina analyzed in Fig. 3 contains 2 ng of β-carotene. Because the peripheral retina contains 2% of its volume as whole blood, this amount of blood could account for the β-carotene present. The data in Figs. 2 and 3 were acquired from an eyeball that was removed at 3.5 hr postmortem, and dissected to obtain macula and peripheral retina 4 hr later. We have not been consistently able to obtain retinal tissues within this short postmortem period.

Discussion

Proficiency with the dissection method described here requires practice and familiarity with the structure of the eyeball. Once mastered, the method takes approximately 30 min/specimen, and eight specimens per day can reasonably be dissected. After removal from the eyecup and separation of the macula from the retina, the specimens can be safely frozen at $-80°$ in buffer for subsequent HPLC analysis. If the sample is stored frozen, the vial should be rinsed well with 1.0 ml of ethanol–buffer (50:50, v/v) to pick up fragments of tissue that might be left in the vial. This additional rinse liquid should be homogenized and pooled with the rest of the homogenate. The total volumes in this case will need to be adjusted to maintain the solvent proportions of one part ethanol to one part buffer to four parts hexane.

The routine procedure of fixation by immersion of the specimen in aldehyde solutions has not been explored for the purpose of macular pigment analysis. This method needs to be investigated, because it might minimize postmortem deterioration, thereby making more retinal samples available for reliable analysis.

The dissection method described here employs a 4-mm trephine to collect the macula from the retina. The zeaxanthin is very highly localized in the center of the macula,[4,5,12] and the 4-mm punch described here will probably collect almost all of the macular zeaxanthin, even if the placement of the punch is slightly eccentric to the true macular center. However, the lutein is much more dispersed. Therefore, slight eccentric placement of the punch might lead to variation in the amount of lutein collected. Further experiments are needed to determine the optimal punch size, which may be larger than 4 mm for human retinas.

The level of the macular pigment in normal humans varies over a large range (as much as 10-fold), as determined by psychophysical[18,19] and biochemical[4] techniques. Substantial variation has also been reported between individual monkeys of the same species.[11] This variation is not a random trait; the left and right maculas of a given monkey show close agreement both in total pigment and in anatomical pigment distribution,[11] indicating that the level of pigment is biologically regulated. However, no definite function for the pigment is yet established. The use of the procedures described here, in conjunction with psychophysical methods and other experimental approaches, may contribute to the future definition of the role of these macular carotenoids in human vision.

Acknowledgments

The work described was supported by a grant from Mr. Sam Williams, Walled Lake, MI (to A.J.A.), U.S. PHS Grant EY-00175 (to E.A.D.), and U.S. PHS Grant EY-04911 (to D.M.S.). Donations of human autopsy material from the New England Eye Bank are gratefully acknowledged.

[18] J. S. Werner, S. K. Donnelly, and R. Kliegl, *Vision Res.* **27**, 257 (1987).
[19] P. L. Pease, A. J. Adams, and E. Nuccio, *Vision Res.* **27**, 705 (1987).

[20] Isolation of Fucoxanthin and Peridinin

By Jarle André Haugan, Torunn Aakermann, and Synnøve Liaaen-Jensen

Introduction

The allenic carotenoids fucoxanthin (**I**) and peridinin (**II**) (Scheme I), are the two carotenoids biosynthesized in largest quantity on our planet.[1,2] Both carotenoids have the same end groups, five chiral centers, and one chiral axis. The structure of the central chain differs, however: fucoxanthin (**I**) is a C_{40}-skeletal carotenoid with an octaenone chromophore whereas peridinin (**II**) is a C_{37}-skeletal butenolide carotenoid.[3]

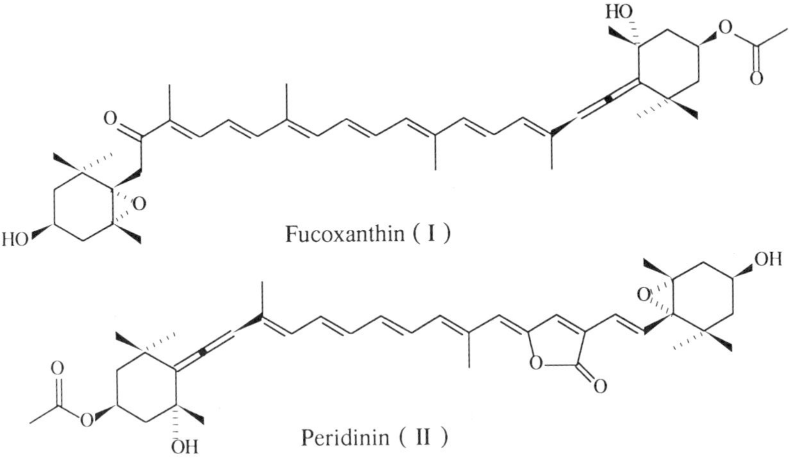

Scheme I. Structures of fucoxanthin (**I**) and peridinin (**II**).

Fucoxanthin (**I**) is produced *de novo* be several classes of algae within the Chromophyta, representing major carotenoids in the Chrysophyceae, Prymnesiophyceae, Bacillariophyceae, and Phaeophyceae.[1,4] The most convenient source is the wild macroalgae of the Phaeophyceae.

---

[1] S. Liaaen-Jensen, *in* "Marine Natural Products: Chemistry and Biological Perspectives" (P. Scheuer, ed.), Vol. 2, Ch. 1. Academic Press, New York, 1978.

[2] H. H. Strain, W. A. Svec, P. Wegfahrt, H. Rapoport, F. T. Haxo, S. Norgård, H. Kjøsen, and S. Liaaen-Jensen, *Acta Chem. Scand., Ser. B* **B30**, 109 (1976).

[3] O. Straub, *in* "Key to Carotenoids" (H. Pfander, ed.), 2nd Ed, p. 198. Birkhäuser, Basel, 1987.

[4] T. W. Goodwin, "The Biochemistry of the Carotenoids" 2nd Ed., Vol. 1. Chapman and Hall, London, 1980.

De novo synthesis of peridinin (**II**) is restricted to the Dinophyceae,[1,4] although peridinin (**II**) is encountered further along the marine food chain. Isolation of peridinin (**II**) is best effected from wild blooms of dinoflagellates or from dinoflagellates grown in pure culture.

Much interest has centered around these structurally complex carotenoids. Their chemistry continues to be actively investigated.[3,5-7] Interest has been focused on the chirality and isomerization of the allenic bond.[8] Both fucoxanthin (**I**) and peridinin (**II**) serve as light-harvesting pigments in photosynthesis.[4] In algae they occur as carotenoid–chlorophyll–protein complexes,[4] the structural details of which still remain to be revealed. Furthermore, they serve as useful chemosystematic markers for algae.[1,4]

Small-scale total synthesis of peridinin (**II**) has been reported,[9] whereas fucoxanthin (**I**) has not yet been prepared by chemical synthesis.

Because there are frequent demands for reference samples of these carotenoids we present in detail recommended procedures for convenient isolation from algae. Brief outlines have been given elsewhere.[10]

Strategy

The cooccurrence of carotenoids and chlorophylls in photosynthetic tissues complicates the isolation of the pure carotenoids. For alkali-stable carotenoids alkaline hydrolysis (saponification) is frequently employed, followed by separation of carotenoids and chlorophyll derivatives by partition between water and diethyl ether. This procedure is not applicable for the isolation of fucoxanthin (**I**) and peridinin (**II**), because both are alkali-labile carotenoids, yielding products with shorter chromophores.[6,11]

Previous isolation procedures for fucoxanthin (**I**) from brown algae suffered from various disadvantages. Use of fresh, brown algae[12,13] required large volumes of solvents. Grinding of fresh algae prior to drying[14-16] may

---

[5] J. A. Haugan, G. Englert, E. Glinz, and S. Liaaen-Jensen, *Abstr. Int. IUPAC Carotenoid Symp., 9th* p. 78 (1990).

[6] J. A. Haugan, G. Englert, and S. Liaaen-Jensen, *Abstr. Int. IUPAC Carotenoid Symp., 9th* p. 112 (1990).

[7] J. A. Haugan and S. Liaaen-Jensen, *Abstr. Int. IUPAC Carotenoid Symp., 9th* p. 113 (1990).

[8] T. Bjørnland, G. Englert, K. Bernhard, and S. Liaaen-Jensen, *Tetrahedron Lett.* **17**, 2577 (1989).

[9] M. Ito, Y. Hirata, Y. Shibata, and K. Tsukida, *Chem. Commun.* in press.

[10] J. A. Haugan and S. Liaaen-Jensen, *Phytochemistry* **28**, 2797 (1989).

[11] H. Kjøsen, S. Norgård, S. Liaaen-Jensen, W. A. Svec, H. H. Strain, P. Wegfahrt, H. Rapoport, and F. T. Haxo, *Acta Chem. Scand., Ser. B* **B30**, 157 (1976).

[12] R. Willstätter and H. Page, *Annalen* **404**, 237 (1914).

[13] I. M. Heilbron and R. F. Phipers, *Biochem. J.* **29**, 1369 (1935).

[14] P. Karrer, A. Helfenstein, H. Wehrli, B. Pieper, and R. Morf, *Helv. Chim. Acta* **14**, 614 (1931).

initiate enzymatic reactions. Time-consuming procedures may result in cis–trans isomerization. Calcium carbonate columns[17] have low capacity, and alumina used for chromatography[13,16] must be neutral.

Previously reported isolation procedures for peridinin (**II**) were based on chromatographic separation on sugar columns[2] of restricted capacity or by successive column chromatography on (1) polyethylene and (2) calcium carbonate, followed by thin-layer chromatography (TLC) on silica.[18]

The procedure recommended here for large-scale isolation of fucoxanthin (**I**) is based on acetone–methanol extraction of carefully dried, ground macroalgae, followed by partition between hexane and aqueous methanol to remove chlorophylls and less polar carotenoids, and subsequent column chromatography (CC) on silica (see flow diagram, Scheme II).

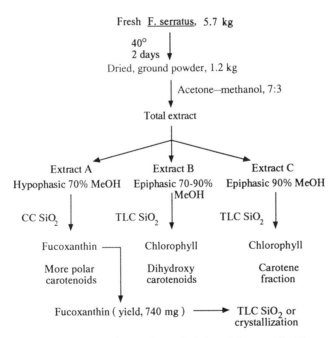

SCHEME II. Flow diagram for the isolation of fucoxanthin (**I**).

---

[15] P. Karrer and E. Jucker, "Carotinoide." Birkhäuser, Basel, 1948.
[16] R. Bonnett, A. K. Mallans, A. A. Spark, J. L. Tee, B. C. L. Weedon, and A. McCormick, *Chem. Soc. J.* **54**, 429 (1969).
[17] A. Jensen, "Carotenoids of Norwegian Brown Seaweeds and Seaweed Meals." Tapir, Trondheim, Norway, 1966.
[18] S. W. Jeffrey and F. T. Haxo, *Biol. Bull. (Woods Hole, Mass.)* **135**, 149 (1968).

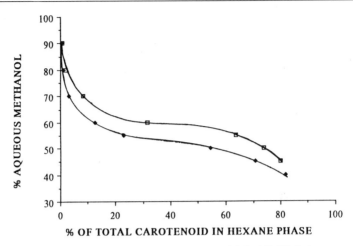

FIG. 1. Partition data for fucoxanthin (□) (I) and peridinin (◆) (II) in hexane–aqueous methanol. [From *Phytochemistry* **28**, 2797 (1989).[10]]

Identification criteria for all-*trans*-fucoxanthin (I), including TLC R_f values, high-performance liquid chromatography (HPLC) R_T values, visible absorption (VIS), Fourier transform infrared absorption (FT-IR), mass spectrum (MS), circular dichroism (CD), proton magnetic resonance (^{1}H NMR), and carbon-13 magnetic resonance (^{13}C NMR) data, are cited.

An HPLC procedure for the semipreparative isolation of all-*trans*-fucoxanthin (I) is also described, as well as an HPLC procedure for the qualitative detection of fucoxanthin (I) in brown algae.

The partition procedure may also be used for the separation of peridinin (II) from chlorophylls and less polar carotenoids. Fucoxanthin (I) and peridinin (II) have rather similar partition coefficients (see Fig. 1).[10]

A procedure (B, Scheme III) involving partition is outlined here. Alternatively a procedure (A, Scheme III) based on rough chromatographic separation of peridinin (II) from less strongly adsorbed chlorophylls and carotenoids and more strongly adsorbed glycosidic carotenoids by column chromatography on acetylated polyamide followed by column chromatography on silica is recommended (see flow diagram, Scheme III).

Identification criteria for all-*trans*-peridinin (I), including R_f, R_T, VIS, FT-IR, CD, MS, ^{1}H NMR, and ^{13}C NMR data, are cited.

An HPLC procedure for the qualitative detection of peridinin (II) in dinoflagellates is given. HPLC is also recommended as a purity test for peridinin preparations.

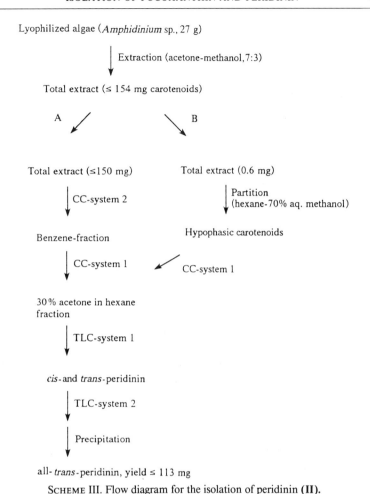

SCHEME III. Flow diagram for the isolation of peridinin (II).

Reagents and Instrumentation

Solvents. Technical-grade acetone and hexane are distilled prior to use. Diethyl ether is dried with metallic sodium and filtered through an alumina column in order to remove peroxides. Methanol, isopropyl acetate, carbon disulfide, toluene, and benzene are of pure, analytical grade quality.

Chromatographic Systems

Column chromatography:
 1. Silica [Kieselgel 60 G, Merck (Rahway, NJ) No. 9385]; particle size, 0.040–0.063 mm]
    Length: 15 cm; diameter: 4.5 cm
    Eluent: n-Hexane–acetone, concentration specified in each case

2. Acetylated polyamide (MN-polyamide SC 6-AC; Mackerey-Nagel, Düren, Germany)
   Length: 15 cm; diameter: 4.5 cm
   Eluent: Benzene or toluene

Thin-layer chromatography:
1. Silica plates, 0.5 mm [Kieselgel 60 G, Merck No. 7731 (50 g) plus distilled water (98 ml)]
   Eluent: n-Hexane–acetone (7:3)
2. Alkaline plates, 0.5 mm [Kieselgel 60 G, Merck No. 7731 (14 g); Kieselguhr, Merck, No. 8119 (16 g); Ca(OH)$_2$, Riedel-de Haen No. 12038 (Seelze, Germany) (9 g); and MgO, Baker (Phillipsburg, NJ) No. 1123 (9 g) plus distilled water (about 103 ml)]
   Eluent: n-Hexane–acetone (7:3)

High-performance liquid chromatography [Perkin-Elmer (Norwalk, CT) series 2 LC with a Pye 402 detector and a Merck Hitachi D 2000 Chromato-Integrator]:
1. CN column [Ultrasphere 5 CN (Wellington House, Cheshire, England): diameter, 4.6 mm; length, 25 cm; particle size, 5 μm]
2. Silica column [Silica Spheri 5 (Brownlee): diameter 4.6 mm; length, 25 cm; particle size, 5 μm]
3. CN column [Attex Ultrasphere TH Cyano (Bechman), semipreparative: diameter, 10 mm; length, 25 cm; particle size, 5 μm]
   General eluent: n-Hexane–isopropyl acetate–acetone–methanol (76:17:7:0.1)

Spectrometric Instruments

VIS: Perkin-Elmer 552 spectrophotometer (Norwalk, CT)
IR: Nicolet 20 SCX FT-IR spectrometer (Madison, WI)
MS: AE1 MS 902 spectrometer (Manchester, England) with direct inlet to the ion source
NMR: 500-MHz Bruker FT or 400-MHz Bruker FT instruments
CD: Jobin-Yvon dictograph, Mark IV (Langjumeau, France)

Procedures

General Precautions

All operations are carried out in subdued light, under an N$_2$ atmosphere (N$_2$ gas > 99.96 purity), and at the lowest possible temperature. Removal of solvents is effected under reduced pressure by a Rotavapor (Büchi, Switzerland). During manipulations extracts, columns, and TLC chambers are covered with black cloth. The water bath temperature of the Rotavapor must not exceed 35°. Extracts are flushed with N$_2$ gas and stored cold (−20°), preferably frozen in benzene.

Isolation of Fucoxanthin

Preparative Procedure. Fresh algae, *Fucus serratus,* are carefully dried at 40° for 2 days with good air circulation and ground on a Wiley mill model No. 3 (A. H. Thomas, Philadelphia, PA). The ground dry algae (1.2 kg) are extracted several times with acetone–methanol (7:3) in an Erlenmeyer flask (10 liter) and left standing at room temperature in darkness under an N_2 atmosphere, until the extract is colorless. The combined extracts are evaporated to dryness under reduced pressure at 30–35° on a Rotavapor and dissolved in methanol. The total amount of pigment is determined from the visible spectrum of the total extract.

The green extract is partitioned in a separatory funnel in portions containing about 50–100 mg fucoxanthin (**I**), between *n*-hexane (200 ml) and 90% aqueous methanol (v/v) (220 ml). This procedure is repeated three times. To the 90% aqueous methanol extract water is added until a 70% aqueous methanol concentration is reached, followed by extraction once with *n*-hexane (200 ml). This extract is discarded. Finally, fucoxanthin (**I**) is transferred from the aqueous hypophase to diethyl ether (150 ml). The extract is evaporated to dryness on a Rotavapor, and the pigments are dissolved in benzene or CS_2 (1–10 ml) and chromatographed on a silica column (CC system 1) two or three times.

Fucoxanthin (**I**) is isolated in a relatively pure state (95–100%) by elution with 30% acetone in *n*-hexane. Yield, 740 mg fucoxanthin (0.06% of the total dry weight).

The fucoxanthin fraction is further purified on silica plates (TLC system 1) and/or by crystallization from acetone–*n*-hexane. For crystallization an aliquot containing 70 mg fucoxanthin (**I**) is taken to dryness on a Rotavapor, dissolved in a minimum amount of acetone and *n*-hexane added until cloudiness occurs, followed by gradual cooling from 5 to −20°. The crystals are harvested after 2 days to 2 weeks using a Willstätt filter. The crystals are washed on a filter with cold *n*-hexane. Yield, 52 mg **I** (74%); crystallized **I**, mp 168–169°.[16]

Fucoxanthin (**I**) isolated by this procedure contains 5–20% mono-cis isomers together with the all-trans isomer. The all-trans isomer may be isolated 100% pure by semipreparative HPLC, using HPLC system 3, as described below.

Semipreparative HPLC Procedure. The HPLC system 3 was used. TLC-purified and/or crystallized fucoxanthin (**I**) is dissolved in a minimum amount of benzene. About 1 mg fucoxanthin (**I**) is injected per run; the flow rate is 4 ml/min. Both the detector, integrator, and the color of the eluate are employed as guidelines to decide when to collect the carotenoid.

By injecting preparative quantities, i.e., milligram scale, a mixture of compounds with about the same retention times seems to coelute if the detection wavelength is set to the λ_{max} values of the compounds (for carotenoids, about 450 nm) (see after third injection in Fig. 2). This may be avoided either by reducing the amount of injected sample or by changing the detection wavelength to a wavelength at which the absorption is much lower than the maximum value. Reducing the amount of injected sample increases the time consumed for the procedure. This is not the case for the second alternative. Consequently the detection wavelength used should be in the range 480–500 nm. The effect is clearly demonstrated in Fig. 2.

The time consumed for preparative HPLC is reduced by injecting samples with a time interval equal to the difference in retention time between the first and last compound in the mixture. This is particularly efficient when the retention time for the first compound in the mixture is high. The resulting chromatogram is shown in Fig. 2.

The collection flask is kept in darkness, under nitrogen at $<0°$. By this procedure 4 mg of all-*trans*-fucoxanthin (I) (100% pure) is obtained in less than 3 hr.

Characterization of Fucoxanthin

TLC: R_f (TLC system 1) = 0.26
HPLC: R_T (HPLC system 1; flow rate, 2.0 ml/min) = 6.5 min
R_T (HPLC system 2; flow rate, 2.0 ml/min) = 14 min

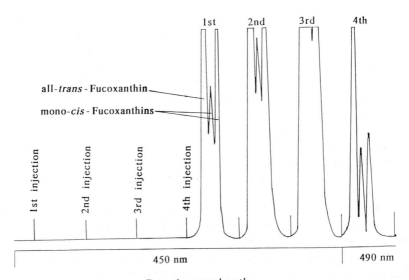

FIG. 2. Semipreparative HPLC (HPLC system 3) purification of fucoxanthin (I), illustrating the advantage of coordinated injection/collection and detection at 490 nm.

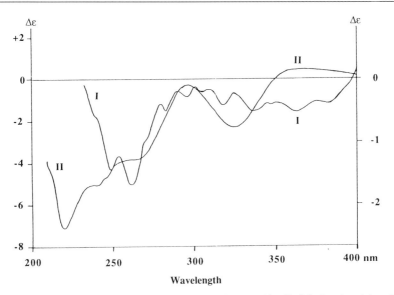

FIG. 3. CD spectra (EPA solution) of all-*trans*-fucoxanthin (I, right-hand scale) and of all-*trans*-peridinin (II, left-hand scale).

VIS: λ_{max} (*n*-hexane): (428), *446*, 475 nm; percent III/II[19] = 38
λ_{max} (acetone): (420), *444*, 467 nm; $E_{1\,cm}^{1\%} = 1660$
MS: (210°, 70 eV) m/z (ion, intensity relative to base peak in percent): 658 (M, 23), 640 (M − 18, 37), 622 (M − 18 − 18, 26), 580 (M − 18 − 60, 21), 578 (M − 80, 7) 562 (M − 18 − 18 − 60, 13), 560 (M − 18 − 80, 6), 488 (9), 484 (7), 482 (8), 237 (27), 221 (43), 212 (51), 197 (100)
CD: (EPA = diethyl ether−isopentane−ethanol, 5:5:2); nm ($\Delta\varepsilon$): 233 (−0.11), 250 (−1.36), 254 (−1.21), 261 (−1.65), 280 (−0.46), 281 (−0.47), 290 (−0.22), 293 (−0.24), 300 (−0.10), 305 (−0.20), 307 (−0.17), 316 (−0.28), 323 (−0.23), 337 (−0.48), 352 (−0.35), 362 (−0.49), 375 (−0.33), 378 (−0.34), 382 (−0.39), 396 (0), >396 (+) (see Fig. 3).
FT-IR (KBr, ν_{max} cm^{-1}): 3483 (OH stretch), 3030-2856 (CH stretch), 1930 (allene), 1732 (C=O, acetate), 1654 (conjugated C=O), 1607, 1576, 1530, 1471, 1456 (CH$_2$), 1385 and 1367 (*gem*-methyl), 1335, 1261, 1245 (C—O, acetate), 1201, 1175−1157, 1071, 1053, 1032, 958 (trans-disubstituted C=C), 917 (see Fig. 4).
^{1}H NMR (400 MHz, C^{2}HCl$_3$):
Coupling constants (J_{H-H}, Hz): $^2J_7$ 18.3, $^3J_{10-11}$ 11.7, $^3J_{11-12}$ 14.8, $^3J_{14-15}$ 11.8, $^3J_{15-15'}$ 15.0, $^3J_{14'-15'}$ 11.8, $^3J_{11'-12'}$ 15.1, $^3J_{10'-11'}$ 11.8

[19] B. Ke, F. Imsgard, H. Kjøsen, and S. Liaaen-Jensen, *Biochim. Biophys. Acta* **210**, 139 (1970).

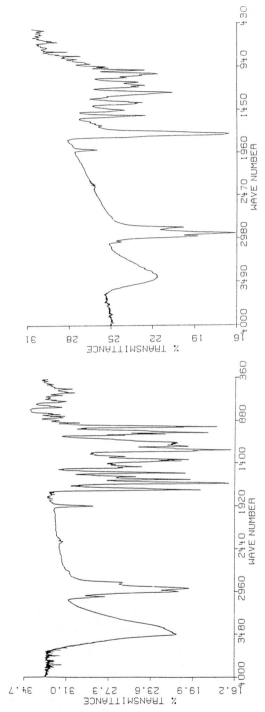

FIG. 4. FT-IR spectra in KBr disk of fucoxanthin (**I**, left) and peridinin (**II**, right).

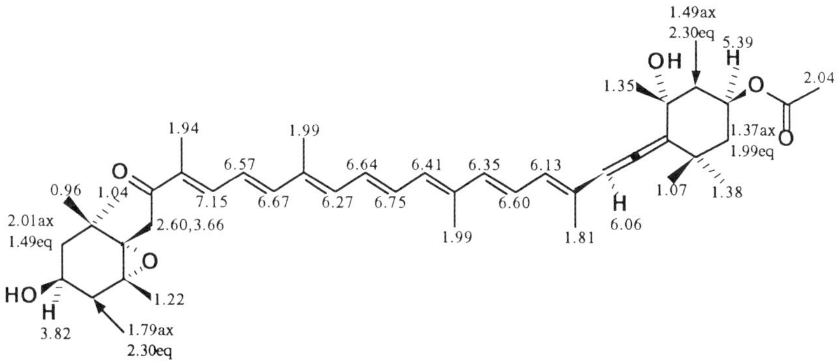

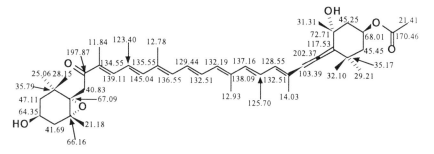

^{13}C NMR (100 MHz, C^{2}HCl$_3$)[20]:

HPLC of Pigments from Brown Algae

Fresh algae (22 g wet weight) are extracted in acetone–*n*-hexane (7:3) in darkness using a Waring blender. Solid CO$_2$ is added to keep the extract cold during extraction. After 5–10 min the extract is evaporated to dryness on a Rotavapor at less than 35° Azeotropic evaporation with acetone–benzene is required. The dry extract is dissolved in benzene, filtered, and analyzed by HPLC (HPLC systems 1 and 2). In Fig. 5 an HPLC chromatogram (HPLC system 1; flow rate, 2.0 ml/min) of the pigments in *Fucus serratus* is shown. Identification of the carotenoid peaks is based on visible spectra, cochromatography with authentic carotenoids, and isolation of the individual carotenoids, followed by determination of MS and ^{1}H NMR spectra.

Isolation of Peridinin

Ground lyophilized algae (*Amphidinium* spp., 27 g) are extracted with acetone–methanol (7:3) on a sintered glass filter until the extract is colorless. The extract is flushed with N$_2$ gas, evaporated to dryness on a Rotava-

[20] G. Englert, T. Bjørnland, and S. Liaaen-Jensen, *Magn. Reson. Chem.* **28**, 519 (1990).

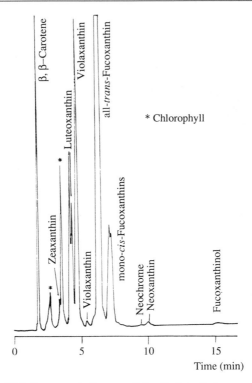

FIG. 5. HPLC (HPLC system 1, flow rate 2.0 ml/min) of pigments from brown algae *(Fucus serratus)*.

por, and dissolved in acetone. The total amount of pigment is roughly determined from the visible spectrum. Two different isolation procedures are used, as follows:

1. The total extract is dissolved in CS_2 (about 3 ml) and chromatographed on an acetylated polyamide column (CC system 2) to remove glycosidic carotenoids. The eluate is collected, evaporated on a Rotavapor, dissolved in benzene (about 3 ml), and chromatographed on a silica column (CC system 1). Chlorophylls and less polar carotenoids are eluted with 5–20% (v/v) acetone in *n*-hexane. Peridinin is isolated relatively pure with 30% acetone in *n*-hexane. The peridinin fraction is further purified on silica plates (TLC system 1), while the cis and all-trans isomers are separated on alkaline plates (TLC system 2). All-*trans*-Peridinin is precipitated from $CHCl_3$–hexane (1 : 15).[2] Crystallized II has a melting point of 128–132°.[2]

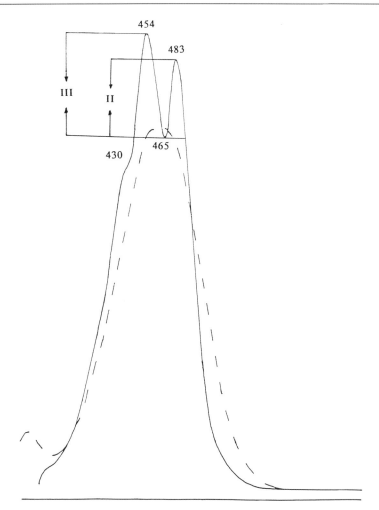

FIG. 6. Absorption spectra in visible light of peridinin (II) in hexane (—) and acetone (---).

2. An aliquot of the total extract (0.6 mg) is dissolved in n-hexane (5 ml), transferred to a separatory funnel, and partitioned between n-hexane and 70% aqueous methanol (7 ml). The hypophasic carotenoids are transferred to diethyl ether, the extract evaporated to dryness on a Rotavapor, dissolved in benzene (3 ml), and chromatographed on a silica column (CC system 1). Further purification of peridinin is obtained by TLC as described in the preceding procedure.

Yield: 113 mg (0.4% of the dry weight, 73% of total carotenoid content).

Characterization of Peridinin

TLC: R_f (TLC system 1) = 0.2
HPLC: R_T (HPLC system 1; flow rate, 1.5 ml/min) = 11.39 min (all-trans)
R_T (HPLC system 2; flow rate, 2.0 ml/min) = 24.90 min (all-trans)
VIS: λ_{max} (acetone): *465* nm (see Fig. 6); λ_{max} (*n*-hexane) (430) *454, 483* nm; percent III/II[19] = 74 (see Fig. 6); λ_{max} (ethanol) *475* nm; $E_{1\ cm}^{1\%} = 1325^2$
MS: (210°, 70 eV); m/z (ion, intensity relative to base peak): 630 (M, 36), 612 (M − 18, 30), 594 (M − 18 − 18, 2), 586 (M − 44, 1), 570 (M − 60, 9), 568 (M − 18 − 44, 2), 552 (612 − 60, 8), 538 (M − 92, 6), 221 (10), 181 (33)
CD: (EPA); nm ($\Delta\varepsilon$): 220 (−7.1), 241 (−4.9), 270 (−4.0), 296 (−0.2), 328 (−2.3), 350 (0), 365 (0.5), 404 (0.3) (see Fig. 3)
FT-IR (KBr, v_{max} cm^{-1}): 3427 (OH) 3023-2958, 2853 (CH) 1927 (allene), 1739 (C=O, acetate), 1640, 1617, 1596, 1521, 1450, (CH$_2$), 1377 and 1364 (*gem*-CH$_3$), 1247 (C—O, acetate), 1182, 1161, 1124, 1070, 1046, 1029, 963 (trans-disubstituted C=C), 956, 940, 904 (see Fig. 4)
^{1}H NMR (500 MHz, C^{2}H$_3$O^{2}H): coupling constants[21]:

[21] J. Krane, T. Aakermann, and S. Liaaen-Jensen, *Magn. Res. Chem.* in press (1992).

^{13}C NMR (500 MHz, C^{2}H$_3$O^{2}H):

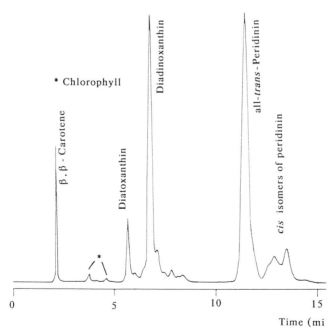

HPLC of Dinoflagellate Pigments

An aliquot of the crude, total extract (Scheme III) is taken to dryness on the Rotavapor and dissolved in CS$_2$ prior to injection. HPLC systems 1 and 2 cause similar resolution of the individual chlorophylls and carotenoids (see Fig. 7) (HPLC system 1; flow rate, 1.5 ml/min).

The carotenoid peaks are identified as described above for pigments from brown algae.

FIG. 7. HPLC (HPLC system 1, flow rate 1.5 ml/min) of pigments from dinoflagellates (*Amphidinium* spp.).

[21] Preparative High-Performance Liquid Chromatography of Carotenoids

By GEORGE W. FRANCIS

Introduction

The need for large amounts of pigment to carry out chemical reactions and indeed spectroscopic measurements required the development of open column and preparative thin-layer chromatographic methods for carotenoid work.[1-3] Although these methods are still in widespread use, advances in microchemical methods and spectroscopic techniques allow thorough investigations on smaller quantities of pigment and make other methods of separation viable alternatives. This is well demonstrated by the work of a Japanese group[4] that identified by nuclear magnetic resonance (NMR) spectroscopy individual carotenoid isomers after having separated cis–trans mixtures by high-performance liquid chromatography (HPLC) on laboratory-packed calcium hydroxide columns. In our laboratories, we are currently using commercial self-packing columns having readily available chromatographic adsorbents and standard solvent systems. Carotenoids (1–100 mg) are conveniently separated[5,6] using either normal- or reversed-phase systems, depending on the chromatographic behavior of the sample.

HPLC Chromatography

Equipment. The HPLC chromatography is carried out on a Miniprep LC (Jobin-Yvon, Longjumeau, France). This system (diagram, Fig. 1) is built about a vertical stainless steel tube (50 cm, i.d. 20 mm) that forms the chromatography column. The upper end of the column is fitted with a removable head containing a steel frit to allow passage of liquids. The lower end of the column is closed by means of a hydraulically driven piston, which again is fitted with a steel frit. During operation the piston holds the column compressed. Solvent is forced by air pressure from a

---

[1] S. Liaaen-Jensen, *in* "Carotenoids" (O. Isler, ed.), p. 61. Birkhäuser, Basel, 1971.
[2] G. Britton, *Nat. Prod. Rep.* **1**, 67 (1984).
[3] B. H. Davies, *in* "Chemistry and Biochemistry of Plant Pigments" (T. W. Goodwin, ed.), Vol. 2, p. 38. Academic Press, New York, 1976.
[4] N. Katayama, H. Hashimoto, Y. Koyama, and T. Shimamura, *J. Chromatogr.* **519**, 221 (1990), and references therein.
[5] M. Isaksen and G. W. Francis, *Chromatographia* **27**, 325 (1989).
[6] S. Scalia and G. W. Francis, *Chromatographia* **28**, 129 (1989).

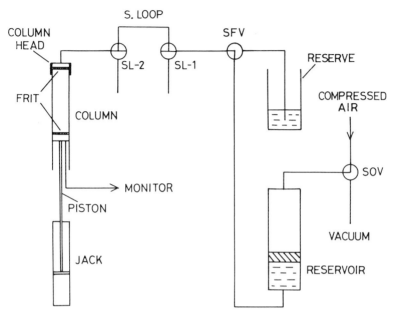

FIG. 1. Diagram of the equipment used for HPLC of carotenoids. Valves are indicated as follows: solvent operation valve (SOV), solvent feed valve (SFV), solvent loop valves (SL-1 and SL-2). The column head is shown in position, and valves are set as during elution.

solvent reservoir, through a sample loop onto the top of the column, through which it then passes. The eluate may then be analyzed by passing it through a suitable visible light absorption monitor and collected by means of a fraction collector. We use a type 6 optical unit (ISCO, Lincoln, NE) operating with a 435-nm filter as monitor and a 2070 Ultrarac II (LKB Instruments, Rockville, MD) fraction collector, although any number of suitable systems are available.

Choice of Chromatographic System. The only restrictions set by the Miniprep system are the availability of the stationary phase in a suitable form (particle size 5–50 μm) and a homogeneous solvent system during operation. Thus, the large amount of data available on thin-layer chromatography can readily be used as a basis for developing new systems. In practice we have found that most separations can be carried out on either silica gel (silica gel 60H 15 μm, No. 11695; Merck, Rahway, NJ) or in the case of reversed-phase separation RP-18 (Lichoprep RP-18 15–25 μm, No. 13901; Merck).

Suitable solvent systems for normal phase separation are provided by mixtures of petroleum ether (PE) with acetone or tert-butanol. The solvent composition found to give the best separation for TLC on commercial silica gel layers (No. 5721; Merck) may be directly applied to the HPLC system.[7]

Choice of solvent for reversed-phase separations has proved more difficult, but may be based on prior investigation of the TLC behavior of the carotenoid mixture, here on RP-18 layers (No. 15423; Merck).[8] Solvent systems consisting of mixtures of PE, acetonitrile, and methanol give good results on TLC. However, direct application of the optimal TLC-based solvent mixture generally leads to very large retention volumes for the least polar components, carotenes, and less polar esters. When only free carotenoids are present the choice of a slightly less polar solvent within the system normally suffices. However, particular difficulties occur with carotenoid esters and modification to replace some of the methanol in the solvent mixture with tetrahydrofuran has proved to be the method of choice. Starting with a solvent mixture of PE–acetonitrile–methanol in the proportions 10:15:75 (v/v), stepwise modification to a system consisting of PE–acetonitrile–methanol–tetrahydrofuran in the proportions 10:15:60:15 (v/v) provides useful eluants. Monoesters are separated with a tetrahydrofuran content from 5 to 10%, while diesters normally require from 10 to 15% tetrahydrofuran for separation. Where additional polar functional groups are present somewhat lower tetrahydrofuran contents are adequate.

Operation. The first stage of operation is to fill the solvent reservoir. This is done by attaching a vacuum line (water pump) at the solvent operation valve (SOV, Fig. 1) and switching the solvent feed valve (SFV) to the solvent reserve. After filling is complete (2 liters), the vacuum line is removed, the first sample loop valve is closed (SL-1), and the feed valve (SFV) is switched to the column position. The solvent is then pressurized (~5 atm) through the solvent operation valve (SOV).

The stationary phase is prepared by making a slurry of the required material (50 g) in the solvent to be used. A final volume of 120 ml provides a full-length column. Residual air may be removed from the slurry by placing it in an ultrasonic bath for up to 5 min.

The column head, if in place, is removed. Pressure (5 atm) is now applied to the hydraulic jack and the piston slowly withdrawn while the stationary phase slurry is fed into the upper end of the column, thus avoiding trapping air in the column. When the piston reaches the fully

---

[7] G. W. Francis and M. Isaksen, *J. Food Sci.* **53**, 979 (1988).
[8] M. Isaksen and G. W. Francis, *J. Chromatogr.* **355**, 358 (1986).

withdrawn position, the column head is replaced, and the sample loop valves (SL-1 and SL-2) set to allow flow from the column through the loop to the sample container. Pressure is then applied to the column jack to raise the piston and thus compress the column. Pressure is increased in stages to give a final valve of 10–12 atm. Excess solvent from the column escapes through the sample loop.

Changing the setting of the first sample loop valve (SL-1) now allows application of solvent to the column. Solvent pressure is adjusted to a flow rate of 5–10 ml/min and about two column volumes of solvent passed to ensure equilibrium within the system.

The sample is dissolved in a suitable volume of solvent: 1–10 ml depending on the nature of the sample and the sample loop size. The loop valves (SL-1 and SL-2) are set to isolate the loop from the remainder of the equipment and the sample applied using either injection into the loop or the siphon principle. Care should be taken to avoid trapping air in the sample loop.

Elution is now started by setting the sample loop valves (SL-1 and SL-2) to allow flow from the sample reservoir to the column head.

On completion of the separation the first sample loop valve (SL-1) is set to isolate the reservoir, and the hydraulic piston withdrawn to decompress the column. The column head is removed and the piston raised to extrude the column material. Excess solvent may be removed by setting the solvent feed valve (SFV) to the solvent reserve.

Examples. Two examples of separations obtained by this method are seen in Fig. 2. These show a normal-phase separation of a saponified parsley *(Petroselinum crispum)* extract and a reversed-phase separation of an unsaponified extract of French marigold *(Tagetes erecta)* petals. Conditions are given in Fig. 2.

Comments

1. The Miniprep LC has two parallel solvent reservoirs. In practice the second reservoir was not used in our work on carotenoids and therefore it was omitted from both Fig. 1 and the discussion.

2. Analytical-grade solvents are used throughout.

3. Several separations can be conducted on the same column. In this case two column volumes of solvent should be passed through the column after elution of the final component of the mixture and prior to application of the new sample.

4. The reversed-phase packing material may be regenerated after use by washing with a 1:1 mixture of acetone–petroleum ether.

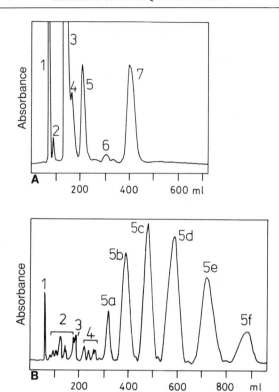

FIG. 2. HPLC separations of carotenoids. (A) Parsley *(Petroselinum crispum)*, 35 mg total saponified carotenoid; stationary phase, silica; mobile phase, acetone–petroleum ether (bp 40–60°) (40:60, v/v); flow rate, 6.0 ml/min. Compound identities, established by standard methods, are as follow: β-carotene (1), cryptoxanthin (2), lutein (3), unidentified (4), violaxanthin (5), *cis*-neoxanthin (6), and neoxanthin (7). (B) French marigold *(Tagetes erecta)* petals, 31 mg unsaponified carotenoid; stationary phase, RP-18; mobile phase, petroleum ether (40–60°)–acetonitrile–methanol–tetrahydrofuran (10:15:65:10, v/v); flow rate, 6.0 ml/min. Compound identities, established by standard methods, are as follow: lutein (1), monoesters of lutein, violaxanthin, and neoxanthin (2), β-carotene (3), diesters of violaxanthin and neoxanthin (4), diesters of lutein (5a–f). The esterifying acids were $C_{14:0'}$ $C_{14:0}$ for peak 5b, $C_{14:0'}$ $C_{16:0}$ for peak 5c, and $C_{16:0'}$, $C_{16:0}$ for peak 5d.

Acknowledgments

The contributions to the development of this method by my co-workers Morten Isaksen (Bergen) and Dr. Santo Scalia (Ferrara) are gratefully acknowledged. Thanks are due to Fanny Schnelle's Research Fund (Bergen), which has supported this work by a grant for consumables.

[22] Profiling and Quantitation of Carotenoids by High-Performance Liquid Chromatography and Photodiode Array Detection

By A. P. DE LEENHEER and H. J. NELIS

Introduction

Several chapters in this volume have highlighted high-performance liquid chromatography (HPLC) as the current method of choice for the separation and quantitation of carotenoids. Gas chromatography[1] of perhydrogenated carotenoids, although showing considerable potential for mass spectrometric identification studies, has never found practical application. Thin-layer chromatography is also superseded by HPLC in that it is less conveniently coupled on-line with spectrometric techniques[2] (absorption spectrometry and mass spectrometry) used for compound identification.

Proper peak identification is indeed of paramount importance in connection with the profiling of carotenoids in biological materials. In daily HPLC practice this is too often done by simply comparing retention times of unknown peaks with those of authentic reference substances.[3] However, only a limited number of synthetic carotenoids is commonly available. Alternatively, fractions can be collected from the HPLC column and compounds characterized off-line by spectrometric techniques.[4] This approach may be seriously hampered by the incomplete resolution between peaks in the complex carotenoid patterns that are often encountered in biological extracts. In case a multichannel absorption detector with scanning capability[5] is not available, absorbance ratios at selected wavelengths can be determined before and after specific chemical reactions.[6] However, this requires repeated injections of the sample to provide detection at each individual wavelength. The most sophisticated scanning absorption detector presently available is the photodiode array detector. This versatile instrument permits the on-line recording of the complete absorption spectrum of a chromatographic peak in less than 1 sec. It owes this unique

---

[1] R. F. Taylor and M. Ikawa, this series, Vol. 67, p. 233.
[2] K. L. Busch, *in* "Handbook of Thin-Layer Chromatography" (J. Sherma and B. Fried, eds.), p. 183. Dekker, New York, 1991.
[3] H. S. Nam, S. Y. Cho, and J. S. Rhee, *J. Chromatogr.* **448**, 445 (1988).
[4] H. J. Korthals and C. L. M. Steenbergen, *FEMS Microbiol. Ecol.* **31**, 177 (1985).
[5] A. C. Palmisano, S. E. Cronin, and D. J. Des Marais, *J. Microbiol. Methods* **8**, 209 (1988).
[6] Z. Matus, M. Baranyai, G. Tóth, and J. Szabolcs, *Chromatographia* **14**, 337 (1981).

performance to a reverse-optics configuration, the presence of an array of photodiodes each dedicated at a particular band of wavelengths, and its connection to a powerful computer for data handling.[7]

Because carotenoids possess highly characteristic absorption spectra and because small structural modifications already induce detectable changes in their spectra, they can be considered as model components for (tentative) identification by the combined technique of HPLC-photodiode array detection. Typical applications include the determination of carotenoids in fruits and vegetables,[8-12] flowers,[13] orange juice,[14] algae,[15] palm oil,[16] and bacteria.[17] The aim of this chapter is to illustrate, on the basis of selected examples from the authors' own experience, the great potential of HPLC-photodiode array detection for carotenoid profiling in a variety of biological materials.

Biological Samples

Brevibacterium KY-4313 was a gift from T. Oka (Kyowa Hakko Kogyo Co., Tokyo, Japan). It was grown in brain heart infusion broth. *Rhodomicrobium vannielii* (ATCC 1700, American Type Culture Collection, Rockville, MD) and *Micrococcus roseus* (ATCC 516) were grown as described previously.[18,19] Pollen of the ice plant, *Dorotheanthus bellidiformis,* was provided by A. Veerman (University of Amsterdam, The Netherlands). *Artemia* cysts and live *Dunaliella (Du.) tertiolecta* algae came from the Laboratory of Aquaculture (P. Sorgeloos, University of Gent, Belgium).

Standard Compounds

Lycopene and β-carotene were purchased from Sigma (St. Louis, MO). Lutein came from Extrasynthèse (Genay, France). Canthaxanthin, echinenone, isozeaxanthin, and isocryptoxanthin were donated by Hoffmann-

[7] J. C. Miller, S. A. George, and B. G. Willis, *Science* **218**, 241 (1982).
[8] F. Khachik, G. R. Beecher, and N. F. Whittaker, *J. Agric. Food Chem.* **34**, 603 (1986).
[9] F. Khachik and G. R. Beecher, *J. Agric. Food Chem.* **35**, 732 (1987).
[10] F. Khachik and G. R. Beecher, *J. Agric. Food Chem.* **36**, 929 (1988).
[11] F. Khachik, G. R. Beecher, and W. R. Lusby, *J. Agric. Food Chem.* **36**, 938 (1988).
[12] F. Khachik, G. R. Beecher, and W. R. Lusby, *J. Agric. Food Chem.* **37**, 1465 (1989).
[13] P. Rüedi, *Pure Appl. Chem.* **57**, 793 (1985).
[14] J. F. Fisher and R. L. Rouseff, *J. Agric. Food Chem.* **34**, 985 (1986).
[15] R. F. C. Mantoura and C. A. Llewellyn, *HRC CC, J. High Resol. Chromatogr. Chromatogr. Communic.* **7**, 632 (1984).
[16] J. H. Ng and B. Tan, *J. Chromatogr. Sci.* **26**, 463 (1988).
[17] H. J. Nelis and A. P. De Leenheer, *Appl. Environ. Microbiol.* **55**, 3065 (1989).
[18] L. Segers and W. Verstraete, *Experientia* **41**, 99 (1985).
[19] J. J. Cooney and R. A. Berry, *Can. J. Microbiol.* **27**, 421 (1981).

La Roche (Basel, Switzerland). Neoxanthin, violaxanthin, and antheraxanthin were gifts from J. Szabolcs (Pecs, Hungary). cis-Canthaxanthin was synthesized as described previously.[20]

Chromatography

Liquid Chromatographic Instrumentation

The instrument used in the authors' laboratory consisted of a Varian 5020 pump (Varian Associates, Palo Alto, CA), a Valco N60 valve injector (Valco, Houston, TX) fitted with a 100-μl loop, and an HP 1040A photodiode array detector (Hewlett Packard, Palo Alto, CA). The latter was further equipped with an HP 85 computer, an HP 9121 dual disk drive, and an HP 7470A plotter.

Liquid Chromatographic Conditions

Carotenoids can be separated by normal- and reversed-phase chromatography.[21] Both techniques have their own applications and are complementary.[21] Nonaqueous reversed-phase chromatography (NARP)[22] is used for the initial carotenoid separation. When required, certain fractions can be further differentiated by normal-phase chromatography. This is notably required when the presence of coeluting geometric (cis–trans) isomers is suspected. Silica and alumina are the preferred sorbents for the separation of isomers,[21] although calcium hydroxide may be a superior choice to achieve this end.[23]

Nonaqueous Reversed-Phase Chromatography.[22] A commercial 5-μm Zorbax ODS column (15 × 0.46 cm) (Du Pont, Wilmington, DE) is used. The eluents consist of mixtures of methanol–acetonitrile–dichloromethane in variable ratios, depending on the polarity of the compounds to be separated (Table I). The flow rate is 1 ml/min.

Normal-Phase Chromatography. cis/trans isomers of β-carotene can be separated on a 5-μm Spherisorb AY5 Alumina column (25 × 0.46 cm) (Chrompack, Middelburg, The Netherlands), eluted with dichloromethane–hexane (23:77, v/v), at a flow rate of 2 ml/min. For the differentiation of geometric isomers of canthaxanthin a 10-μm CP-Spher

---

[20] H. J. C. F. Nelis, P. Lavens, L. Moens, P. Sorgeloos, J. A. Jonckheere, G. R. Criel, and A. P. De Leenheer, *J. Biol. Chem.* **259**, 6063 (1984).
[21] W. E. Lambert, H. J. Nelis, M. G. M. De Ruyter, and A. P. De Leenheer, in "Modern Chromatographic Analysis of the Vitamins" (A. P. De Leenheer, W. E. Lambert, and M. G. M. De Ruyter, eds.), p. 1. Dekker, New York, 1985.
[22] H. J. C. F. Nelis and A. P. De Leenheer, *Anal. Chem.* **55**, 270 (1983).
[23] H. Hashimoto, Y. Koyama, and T. Shimamura, *J. Chromatogr.* **448**, 182 (1988).

TABLE I
COMPOSITION OF MOBILE PHASES USED IN NONAQUEOUS REVERSED-PHASE CHROMATOGRAPHY

Sample	Compounds to be determined	Mobile phase composition (%)		
		Acetonitrile	Methanol	Dichloromethane
Brevibacterium KY-4313	Ketocarotenoids	40	50	10
Rhodomicrobium vannielii	Acyclic xanthophylls	70	15	15
Micrococcus roseus	Unknown	70	15	15
Dorotheanthus bellidiformis pollen	β-Carotene + lutein esters	60	—	40
Dunaliella tertiolecta algae	Various xanthophylls and carotenes	41	50	9
Artemia	Canthaxanthin, echinenone, β-carotene, lutein	{ 40 { 41	50 50[a]	10 9

[a] Plus 0.15% triethylamine (final concentration in the eluent).

silica column (25 × 0.46 cm) (Chrompack) is used. The eluent is dichloromethane–2-propanol (99.4:0.6, v/v) and the flow rate is 1 ml/min.

Preparation of Extracts

Each sample of biological material requires different extraction conditions for maximum recovery of the pigments. We found that some matrices, e.g., bacteria, are rather refractory toward a complete release of carotenoids. Sometimes unusual procedures (treatment with liquefied phenol) are necessary to ensure exhaustive extraction. Other authors used trichloroacetic acid,[24] potassium hydroxide,[25] or phenol–glycerol[26] to treat the bacteria prior to extraction.

Bacteria

To about 10 mg of freeze-dried cells of *Brevibacterium* KY-4313 is added, under vigorous vortex mixing, 0.3 ml of potassium hydroxide (60%, w/v) followed by 3 ml of methanol. After centrifugation (3000 rpm for 5 min at ambient temperature) and isolation of the supernatant, the same

---

[24] J. Downs and D. E. F. Harrison, *J. Appl. Bacteriol.* **37**, 65 (1974).
[25] R. F. Taylor, *Adv. Chromatogr.* **22**, 157 (1980).
[26] K. Merritt and N. J. Jacobs, *J. Clin. Microbiol.* **8**, 105 (1978).

step is repeated once more. The residue is treated with 0.4 ml of liquefied phenol (vortex mixing for no longer than 30 sec) and extracted with 4 ml of diethyl ether. The ethereal supernatant is combined with the methanolic extracts. After addition of 5 ml of sodium chloride (5%, w/v) and 4 ml of diethyl ether the carotenoids are extracted in the upper organic phase. The latter is isolated, successively washed with potassium hydroxide (2%, w/v) and water, dried over anhydrous sodium sulfate, and evaporated to dryness under nitrogen. The residue is redissolved in the chromatographic solvent (0.2–2 ml, depending on the concentration of the carotenoids) and a 100-μl sample is injected on the chromatographic column. For the isolation of the carotenoids from phototrophic bacteria (e.g., *R. vannielii*) and *M. roseus*, 10 mg of cells can be directly treated with 0.4 ml of liquefied phenol and extracted with 3 ml of methanol. Two to three steps are usually required to ensure complete bleaching of the residue. Three milliliters of 60% (w/v) potassium hydroxide is added to the combined extracts and the mixture is left at room temperature for 2 hr to destroy chlorophyll and hydrolyze esters (phototrophic bacteria). Extracts of *M. roseus* can be worked up immediately. After the addition of 6 ml of sodium chloride (5%, w/v) the pigments are reextracted in 10 ml of diethyl ether. The ethereal extract is further treated as described for *Brevibacterium* KY-4313.

Pollen

Ten to 100 mg of pollen of the ice plant, *D. bellidiformis*, is homogenized in 6 ml of a mixture of acetone–hexane (1:1, v/v), using a Potter–Elvehjem (Egilabo, Edegen, Belgium) device. After the addition of 10 ml of an ammonium sulfate solution (38%, w/v) the organic layer is isolated and washed with 10 ml of the same solution. The lower phase is discarded and the organic phase (hexane) is partitioned against 3 ml of 95% methanol. After drying over anhydrous sodium sulfate the hexane is evaporated to dryness and the residue is redissolved in 250 μl of the chromatographic solvent (acetonitrile–dichloromethane, 6:4, v/v). The solution is filtered over a 0.45-μm Millipore (Bedford, MA) filter and 100 μl is injected on the liquid chromatographic column.

Dunaliella tertiolecta Algae

About 5 mg of algal matter is homogenized in 2–3 ml of the chromatographic solvent (acetonitrile–methanol–dichloromethane, 41:50:9, v/v) using a Potter–Elvehjem device. After centrifugation (3000 rpm for 5 min at ambient temperature) a 100-μl aliquot of the supernatant is injected.

Artemia Cysts

Ten milligrams of *Artemia* cysts is thoroughly homogenized in a Potter Elvehjem device with 0.3 ml of sodium deoxycholate (5%, w/v). Three milliliters of methanol is added and the homogenization is continued. After centrifugation (3000 rpm for 5 min at ambient temperature) and isolation of the supernatant the same procedure is once more applied to the residue. The supernatants are combined, filtered over a 0.45-μm Gelman (Ann Arbor, MI) Acro LC 13 filter, and a 100-μl aliquot is injected.

Reduction of Ketocarotenoids

When ketocarotenoids in an extract are suspected, the presence of keto groups in conjugation with the polyene chain can be confirmed by reducing them with $LiAlH_4$. To this end, the chromatographic peak of interest is trapped from the column, the solvent is removed under nitrogen, and 0.1 ml of a 1 M solution of $LiAlH_4$ in diethyl ether is added. After 10 min, 3 ml of water is added and sufficient phosphoric acid (85%, w/v) to dissolve the precipitate. The reduced carotenoid is extracted in 5 ml of diethyl ether. The organic layer is dried over anhydrous sodium sulfate, evaporated to dryness, and the residue is reconstituted with 150 μl of the chromatographic solvent (acetonitrile–methanol–dichloromethane, 41:50:9, v/v). A 100-μl aliquot is injected.

Qualitative Aspects: Pigment Profiling

Definitive Peak Identification

Figure 1 represents a pigment profile of a nonphotosynthetic bacterium (*Brevibacterium* KY-4313) in which the major peaks can be definitively characterized. Two minimal requirements are to be met to conclude on peak identity, i.e., agreement between the retention time of the unknown and that of the corresponding authentic compound and coincidence of their respective absorption spectra, preferably in more than one chromatographic system (i.e., reversed phase and normal phase). On the basis of these criteria peaks 1, 4, and 5 (Fig. 1) can be identified as all-*trans*-canthaxanthin (λ_{max} 476 nm), echinenone (λ_{max} 465 nm), and β-carotene (λ_{max} [430], 453, and 481 nm), respectively. In addition, the identity of the two ketocarotenoids, canthaxanthin and echinenone, can be confirmed by reduction with $LiAlH_4$. The reduced carotenoids display the same retention times and absorption spectra as the authentic equivalents of the expected reaction products, isozeaxanthin (λ_{max} 453 and 478 nm) and isocryptoxanthin (λ_{max} 453 and 478 nm), respectively. The shifts occurring

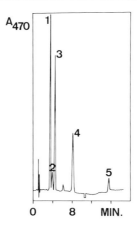

FIG. 1. Chromatogram of an extract of *Brevibacterium* KY-4313. Column, 5-μm Zorbax ODS, 15 × 0.46 cm; mobile phase, acetonitrile–methanol–dichloromethane, 40:50:10 (v/v); flow rate, 1 ml/min; detection wavelength, 470 nm. Peak 1, all-*trans*-canthaxanthin; peak 2, *cis*-canthaxanthins; peak 3, β-apo-8'-carotenal (internal standard); peak 4, echinenone; peak 5, β-carotene.

in retention and spectral properties after reduction of canthaxanthin to isozeaxanthin are illustrated in Figs. 2 and 3.

Similarly, most carotenoids present in an extract of *Du. tertiolecta* algae (Fig. 4) can be identified by comparison with authentic reference components. This is the case for neoxanthin, violaxanthin, antheraxanthin, lutein, and β-carotene. Retention and spectral data for these compounds are listed in Table II.

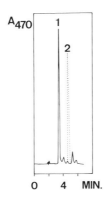

FIG. 2. Chromatogram of the reduction product of all-*trans*-canthaxanthin (peak 1, full line) and of unreduced all-*trans*-canthaxanthin (peak 2, dotted line). Conditions as in Fig. 1.

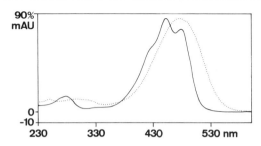

FIG. 3. Absorption spectrum of peak 1 (Fig. 2), corresponding to isozeaxanthin (reduction product of canthaxanthin)(—), and of peak 2 (Fig. 2) (all-*trans*-canthaxanthin)(· · ·).

Tentative Peak Identification

In the absence of authentic substances peaks can often be tentatively identified based on their absorption spectra. Thus, in chromatograms of *R. vannielii* (Fig. 5), six major peaks can be assigned (Table III). Only one reference component (lycopene), corresponding to peak 8, is commercially available. However, with literature data as a guideline[27-29] the absorption spectra of the five other major peaks can be matched with those of their probable authentic equivalents (Table III). In addition, the elution sequence of these presumed compounds is rationalized by referring their retention times and partition coefficients (hexane–95% methanol) to those of lycopene (Table III).

In a given carotenoid profile the systematic monitoring of the absorption spectra with the photodiode array detector may also dismiss some peaks as noncarotenoids. For example, the absorption spectrum of peak 6 in Fig. 4 (presented in Fig. 6) most probably suggests a pigment of a different chemical class. In some cases, a tentative peak identification based exclusively on a retention time (comparison with a reference component) may even be refuted. This is illustrated in Fig. 7, showing an extract of *M. roseus*. According to a series of literature reports,[19,30-34]

[27] T. W. Goodwin, "The Biochemistry of the Carotenoids," 2nd Ed., Vol. 1, p. 320. Chapman and Hall, London, 1980.
[28] G. Britton, R. K. Singh, T. W. Goodwin, and A. Ben-Aziz, *Phytochemistry* **14**, 2427 (1975).
[29] K. Schmidt, *in* "The Photosynthetic Bacteria" (R. K. Clayton and W. R. Sistrom, eds.), p. 729. Plenum, New York, 1978.
[30] J. J. Cooney, H. W. Marks, and A. M. Smith, *J. Bacteriol.* **92**, 342 (1966).
[31] G. E. Ungers and J. J. Cooney, *J. Bacteriol.* **96**, 234 (1968).
[32] E. H. Schwartzel and J. J. Cooney, *Can. J. Microbiol.* **20**, 1007 (1974).
[33] E. H. Schwartzel and J. J. Cooney, *Can. J. Microbiol.* **20**, 1015 (1974).
[34] S. M. Dieringer, J. T. Singer, and J. J. Cooney, *Photochem. Photobiol.* **26**, 393 (1977).

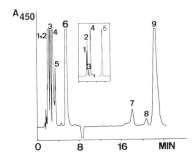

FIG. 4. Chromatogram of an extract of *Dunaliella tertiolecta*. Column, 5-μm Zorbax ODS, 15 × 0.46 cm; mobile phase, acetonitrile–methanol–dichloromethane, 41:50:9 (v/v); flow rate, 1 ml/min; detection wavelength, 450 nm. Peak 1, neoxanthin; peak 2, violaxanthin; peak 3, antheraxanthin; peak 4, all-*trans*-lutein; peak 5, *cis*-lutein(s) (tentative); peak 6, unknown; peak 7, unidentified (possibly γ-carotene); peak 8, α-carotene (tentative); peak 9, β-carotene. (Inset) Detail of the separation between early eluting peaks.

canthaxanthin is the predominant pigment in this species. At first sight, the chromatogram in Fig. 7 seems to confirm this, as it contains a peak with a retention time close to that of all-*trans*-canthaxanthin. However, the corresponding absorption spectrum (Fig. 8) of the presumed canthaxanthin peak is totally inconsistent with this structural assignment.

Determination of Peak Purity

The photodiode array detector is also capable of recognizing peak heterogeneity, i.e., the coelution of different compounds. Such peak impurity may become evident from even slight deviations between the absorp-

TABLE II
IDENTIFICATION OF PRINCIPAL CAROTENOIDS IN EXTRACT OF *Dunaliella tertiolecta*

Peak number[a]	Unknown		Reference component		Identification
	k'	λ_{max} (nm)	k'	λ_{max} (nm)	
1	0.47	414, 437, 468	0.45	414, 437, 468	Neoxanthin
2	0.57	414, 441, 471	0.55	417, 441, 471	Violaxanthin
3	0.83	(425), 447, 476	0.84	(425), 447, 476	Antheraxanthin
4	1.14	425, 447, 475	1.14	425, 447, 475	Lutein
5	1.55	331, 418, 439, 468	—	—	*cis*-Lutein[b]
9	16.2	457, 483	16.2	457, 483	β-Carotene

[a] See Fig. 5. Only six major peaks assigned.
[b] Tentative identification (no reference component).

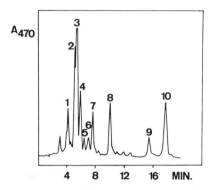

FIG. 5. Chromatogram of an extract of *Rhodomicrobium vannielii*. Column, 5-μm Zorbax ODS, 15 × 0.46 cm; mobile phase, acetonitrile–methanol–dichloromethane, 70:15:15 (v/v); flow rate, 1 ml/min; detection wavelength, 470 nm; tentative peak identification, see Table III.

TABLE III
TENTATIVE IDENTIFICATION OF MAJOR CAROTENOIDS IN EXTRACTS OF *Rhodomicrobium vannielii*[a]

Peak No.	λ_{max}[b]	λ_{max}[c]	p[d]	Identity	Number of double bonds	Functional group(s)
2	460, 487, 519	460, 488, 522	66:34	Rhodovibrin	12	OH, OCH$_3$
3	448, 475, 508	448, 474, 506	76:24	Rhodopin	11	OH
4	470, 497, 532	468, 499, 533	84:16	Spirilloxanthin	13	OCH$_3$, OCH$_3$
7	461, 489, 521	460, 485, 520	94:6	Anhydrorhodovibrin	12	OCH$_3$
8	448, 475, 508	448, 474, 505	100:0	Lycopene[e]	11	—
9	418, 443, 473	416, 440, 470	100:0	Neurosporene	9	—

[a] See Fig. 5. Peaks 1, 5, and 6 are unidentified; peak 10 is β-carotene (internal standard).
[b] Experimentally determined in the chromatographic solvent CH_2Cl_2–CH_3OH–CH_3CN (15:15:70, v/v).
[c] Absorption maxima in acetone, except for neurosporene (ethanol). [From E. De Ritter and A. E. Purcell, in "Carotenoids as Colorants and Vitamin A Precursors: Technical and Nutritional Applications" (J. C. Bauernfeind, ed.), p. 815. Academic Press, New York, 1981.]
[d] Partition coefficient hexane–95% methanol. [From F. H. Foppen, *Chromatogr. Rev.* **14**, 133 (1971).]
[e] Definitive identification versus reference substance.

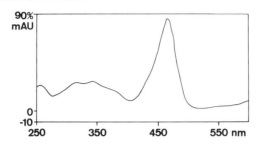

FIG. 6. Absorption spectrum of peak 6 in the chromatogram of *Dunaliella tertiolecta* (Fig. 4).

tion spectrum of the unknown and that of the corresponding reference component. In an extract of pollen (Fig. 9) peak 1 coelutes with synthetic β-carotene. However, the respective absorption spectra do not completely coincide (Fig. 10). The main absorption maximum in the unknown spectrum is slightly hypsochromically shifted and some increase in absorption in the 350- to 370-nm region is noted. Both phenomena are indicative of the presence of cis isomers. This is confirmed by rechromatographing the collected peak on alumina (Fig. 9, inset). Two cis isomers can be separated from all-*trans*-β-carotene, and the latter peak now yields an absorption spectrum that is totally identical with that of reference all-*trans*-β-carotene.

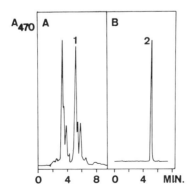

FIG. 7. (A) Chromatogram of an extract of *Micrococcus roseus* (ATCC 516). (B) Chromatogram of synthetic all-*trans*-canthaxanthin. Conditions as in Fig. 5. Peak 1, unknown (absorption spectrum, see Fig. 8); peak 2, all-*trans*-canthaxanthin (synthetic).

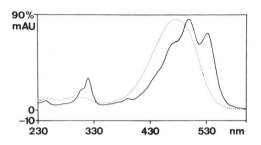

FIG. 8. Absorption spectra of peak 1 from Fig. 7A (—) and of authentic all-*trans*-canthaxanthin (· · ·) (peak 2, Fig. 7B).

A convenient way to assess peak heterogeneity is to record spectra at three points of the peak, i.e., the upslope, apex, and downslope.[7] When, after normalization by the computer the three spectra are overlaid they should coincide if the peak consists of a single compound. The pigment fraction of *Artemia* cysts is composed of all-*trans*- and *cis*-canthaxanthins.[20] Reversed-phase chromatography is only capable of separating the all-trans from a "total" cis peak (Fig. 11). When the latter peak was first detected in *Artemia*[20] it became rapidly clear that in fact it was composite. The absorption spectra recorded at three different points of the peak agreed remarkably well in the main band (468 nm), but showed slight discrepancy in the 360-nm band (Fig. 12). It is well known that profound

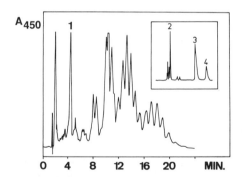

FIG. 9. Chromatogram of pollen of the ice flower, *Dorotheanthus bellidiformis*. Column, 5-μm Zorbax ODS, 15 × 0.46 cm; mobile phase, acetonitrile–dichloromethane, 6:4 (v/v); flow rate, 1 ml/min; detection, 450 nm. Peak 1, β-carotene; all other peaks, esters of lutein (λ_{max} 443–449 nm). (Inset) Rechromatography of the β-carotene peak on 5-μm Spherisorb alumina (25 × 0.46 cm). Mobile phase, hexane–dichloromethane, 77:23 (v/v); flow rate, 1 ml/min; detection, 450 nm. Peaks 2 and 4, mono-*cis*-β-carotenes; peak 3, all-*trans*-β-carotene.

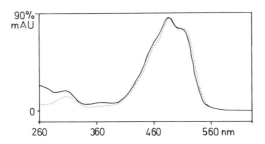

FIG. 10. Absorption spectra of peak 1 (Fig. 9) (—) and of synthetic all-*trans*-β-carotene (· · ·).

differences occur in the intensity of the "cis band" of these geometric isomers, depending on the position of isomerism (9-cis, 13-cis, or 15-cis). Again, normal-phase chromatography proves the correctness of the observation in that indeed the presumed three mono-*cis*-canthaxanthins, each characterized by their own typical spectrum, can be resolved (Fig. 13).

Quantitative Aspects

In its quantitative performance the photodiode array detector does not differ essentially from any variable- or fixed-wavelength detector. The absolute sensitivity of the former is often even lower. Quantitation of

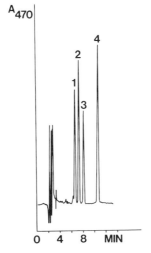

FIG. 11. Chromatogram of an extract of *Artemia* cysts. Conditions as in Fig. 1. Peak 1, all-*trans*-canthaxanthin; peak 2, *cis*-canthaxanthins; peak 3, β-apo-8′-carotenal (internal standard); peak 4, β-apo-8′-carotenoic acid ethyl ester (alternative internal standard).

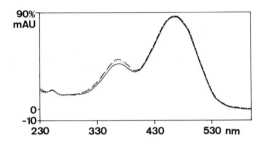

FIG. 12. Absorption spectra of peak 2 (Fig. 11) taken at three positions on the peak, i.e., upslope (—), apex (· · ·), and downslope (- - -).

carotenoids in biological extracts relies on calibration with authentic substances. For example, all-*trans*-canthaxanthin and *cis*-canthaxanthin can thus be determined in *Brevibacterium* KY-4313 and *Artemia*. The former compound is commercially available whereas the latter must be synthesized. Precision is much improved by the addition of an internal standard at an early stage of the analysis. The requirements of such internal standard are physicochemical similarity to the compound of interest, elution in an "open window," and sufficient stability. Internal standards used in this work are β-apo-8′-carotenal or β-apo-8′-carotenoic acid ethyl ester (quantitation of canthaxanthins, Figs. 1 and 11), C_{37}, C_{45}, and C_{50} analogs of β-carotene (quantitation of β-carotene),[7,8] or β-carotene (quantitation of

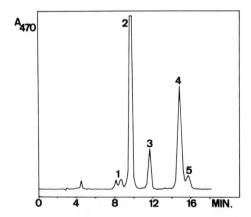

FIG. 13. Chromatogram of an extract of *Artemia* cysts (Macau). Column, 10-μm CP-Spher Si (25 × 0.46 cm); mobile phase, dichloromethane–2-propanol (99.4:0.6, v/v); flow rate, 1 ml/min. Peak 1, di-*cis*-canthaxanthins; peak 2, all-*trans*-canthaxanthin; peak 3, 9-*cis*-canthaxanthin; peak 4, 13-*cis*-canthaxanthin; peak 5, 15-*cis*-canthaxanthin. All identities except for all-*trans*-canthaxanthin are tentative.

carotenoids in *R. vannielii*, Fig. 5). Either peak height or peak area ratios (compound of interest versus internal standard) can be used, but we prefer peak height measurements. In the absence of authentic substances only a semiquantitative determination can be carried out. This can be done using a structural analog for calibration. For example, the major xanthophylls in *R. vannielii* were determined using lycopene as a standard for calibration[17] (Fig. 5). Corrections must be made on the basis of the respective molar absorption coefficients of the analyte and the calibration standard. Hence, at least a tentative identification of the former remains necessary. Again the absorption spectrum will be a key element in this respect.

Acknowledgment

H.J.N. acknowledges his position of Research Associate from the Belgian Foundation for Scientific Research (NFWO). This work was supported by FKFO Contract 2.0012.82 and by FGWO Contracts 3.0011.81 and 3.0048.86.

[23] Mammalian Metabolism of Carotenoids Other Than β-Carotene

By MICHELINE M. MATHEWS-ROTH AND NORMAN I. KRINSKY

Interest in carotenoids that lack provitamin A activity has been increasing as reports have appeared indicating that these pigments can protect against photosensitization, can serve as antioxidants, and may function as antimutagenic and anticarcinogenic compounds. However, with the exception of some studies in fish and birds, virtually nothing is known about the metabolism of these compounds. We have focused on the study of two such carotenoids, lycopene and canthaxanthin.[1] Lycopene is a common component of the human diet, coming primarily from tomato products,[2] and in many people occurring at plasma levels equal to or greater than those of β-carotene.[3-5] Canthaxanthin, which is present in

[1] M. M. Mathews-Roth, S. Welankiwar, P. K. Sehgal, N. C. G. Lausen, M. Russett, and N. I. Krinsky, *J. Nutr.* **120**, 1205 (1990).
[2] T. W. Goodwin, "The Biochemistry of the Carotenoids," 2nd Ed., Vol. 1. Chapman and Hall, London, 1980.
[3] D. B. Milne and J. Botnen, *Clin. Chem.* **32**, 874 (1986).
[4] A. B. Barua, R. O. Batres, H. C. Furr, and J. A. Olson, *J. Micronutr. Anal.* **5**, 291 (1989).
[5] N. I. Krinsky, M. D. Russett, G. J. Handelman, and D. M. Snodderly, *J. Nutr.* **120**, 1654 (1990).

some strains of mushrooms[2] but is present in human plasma only when used as an oral supplement,[6,7] has been found to have antitumor[8-10] and immunoenhancement activity[11,12] and to inhibit mutagenesis,[13] malignant transformation,[14] and chromosome instability.[15] Both pigments have been shown to have antioxidant effects,[16,17] to protect against photosensitization,[18] and to quench singlet oxygen.[19,20]

The metabolism of canthaxanthin and lycopene has been studied in fish and birds[21,22] and in rats and monkeys.[1] We chose to use ^{14}C-labeled carotenoids, rather than ^{3}H-labeled carotenoids, because of the stability of the carbon label in the molecule.

Materials

To make the diets used in the study on canthaxanthin metabolism in rats, beadlets containing 10% (w/w) canthaxanthin in a mixture of peanut oil, starch, ascorbyl palmitate, α-tocopherol, gelatin, and sucrose (Hoffmann-La Roche, Nutley, NJ) are mixed with a nonpurified diet (powdered Purina 5001 rodent chow; Ralston Purina C., St. Louis, MO) just before use to obtain the desired concentration of carotenoid. The diets

---

[6] H. H. Gunson, A. H. Merry, G. Britton, and F. Stratton, *Clin. Lab. Haematol.* **6,** 287 (1984).
[7] J. C. Meyer, H. P. Grundmann, B. Seeger, and U. W. Schnyder, *Dermatologica* **171,** 76 1985).
[8] L. Santamaria, A. Bianchi, A. Arnaboldi, and L. Andreoni, *Med. Biol. Environ.* **9,** 113 (1981).
[9] M. M. Mathews-Roth, *Oncology* **39,** 33 (1982).
[10] J. Schwartz and G. Shklar, *Nutr. Cancer* **11,** 35 (1988).
[11] A. Bendich and S. S. Shapiro, *J. Nutr.* **116,** 2254 (1986).
[12] J. L. Schwartz, D. Sloane, and G. Shklar, *Tumor Biol.* **10,** 297 (1989).
[13] Y. He and T. C. Campbell, *Nutr. Cancer* **13,** 243 (1990).
[14] A. O. Pung, J. E. Rundhaug, C. N. Yoshizawa, and J. S. Bertram, *Carcinogenesis* **9,** 1533 (1988).
[15] H. F. Stich, S. S. Tsang, and B. Palcic, *Mutat. Res.* **241,** 387 (1990).
[16] N. I. Krinsky and S. M. Deneke, *JNCI, J. Natl. Cancer Inst.* **69,** 205 (1982).
[17] J. Terao, *Lipids* **24,** 659 (1989).
[18] M. M. Mathews-Roth, *Photochem. Photobiol.* **40,** 63 (1984).
[19] P. Di Mascio, S. Kaiser, and H. Sies, *Arch. Biochem. Biophys.* **274,** 532 (1989).
[20] M. R. Rich and S. S. Brody, *Photochem. Photobiol.* **49,** 83 (1989).
[21] T. Matsuno, in "Carotenoids: Chemistry and Biology" (N. I. Krinsky, M. M. Mathews-Roth, and R. F. Taylor, eds.), Proc. 8th Int. Symp. Carotenoids, p. 59. Plenum, New York, 1989.
[22] K. Scheidt, in "Carotenoids: Chemistry and Biology" (N. I. Krinsky, M. M. Mathews-Roth, and R. F. Taylor, eds.), Proc. 8th Int. Symp. Carotenoids, p. 247. Plenum, New York, 1989.

consist of either 0.1% (w/w) canthaxanthin, 1% (w/w) canthaxanthin, or 10% (w/w) placebo beadlets (beadlets lacking canthaxanthin).

The radioactive pigments (also supplied by Hoffmann-La Roche) are prepared by dissolving a weighed amount of either 15,15'-[^{14}C]canthaxanthin (specific activity of 32.4 μCi/mg) or 6,7,6',7'-[^{14}C]lycopene (specific activity of 101 μCi/mg) in hexane.[1] Olive oil containing 1 mg/ml of α-tocopherol is added to the hexane solution, to give a final concentration of 20 μCi of [^{14}C]canthaxanthin or [^{14}C]lycopene/ml olive oil. The hexane is removed by a gentle stream of nitrogen at 35°, and the resulting solution of pigment in olive oil is divided into small aliquots and stored at $-70°$ until needed.

Animals

For the study of canthaxanthin metabolism in rats, adult (4-month-old) Sprague-Dawley rats (Charles River Breeding Co., Wilmington, MA) are divided into three groups, each group consisting of three male and three female rats. Group I receives the placebo diet, group II receives the 0.1% canthaxanthin diet, and group III receives the 1% canthaxanthin diet. Each rat is housed in a separate cage. The room temperature is maintained between 23 and 25°, the humidity between 45 and 50%, and the room lights are on 12 hr a day. The rats receive their assigned diets for 8 weeks, and are offered food and water *ad libitum*. After this period, each rat is then given by gavage a single dose of 20 μCi of [^{14}C]canthaxanthin in 1 ml olive oil containing α-tocopherol. The rats are not fasted prior to dosage. After gavage, the three groups are given the nonpurified diet and water *ad libitum:* no additional pigment is administered. Blood is drawn into tubes containing ethylenediaminetetraaetic acid (EDTA) before gavaging, and at 4, 8, 24, 48, and 72 hr after gavage. Plasma is separated by centrifugation at 2700 g at room temperature for 5 min, and then stored at $-20°$ until analysis, which for all studies is done within 4 months, as it is known that carotenoids are stable for this period at $-20°$.[23] The animals are killed at 72 hr and the organs are removed and stored at $-20°$ until analysis.

An identical protocol is used for the studies of lycopene metabolism in rats. However, pigment prefeeding is not done because large amounts of lycopene are not available. Blood is drawn, the animals killed, and the blood specimens and organs treated as in the study with canthaxanthin described above.

Ten female rhesus monkeys are divided into 2 groups of 5 each. The first group is given by gavage 50 μCi of [^{14}C]canthaxanthin in 2.5 ml of olive oil containing α-tocopherol, and the second group receives by gavage a similar solution containing 50 μCi of [^{14}C]lycopene. The monkeys do not receive a pretreatment with carotenoids, but receive the usual primate center diet,[24] which does include some carotenoid-containing foods. They are not fasted before receiving the [^{14}C]carotenoids. In these experiments, blood is drawn into EDTA before the gavage, and at 2, 4, 8, 24, 48, and 72 hr after dosing. Plasma is separated and stored as described above. The monkeys are killed at 48 hr (lycopene experiment), or 72 hr (canthaxanthin experiment) and their organs stored at $-20°$ until analyzed.

Carotenoid Pigment Analysis

The organs are thawed, washed free of blood, and weighed. With respect to the rat experiments, preliminary studies showed that the extracts of many individual rat organs (except for the livers) do not contain sufficient ^{14}C-labeled pigment to be analyzed individually. Thus, the individual organs are pooled according to diet group and organ type, irrespective of sex, and the extracts separated by high-performance liquid chromatography (HPLC).

Monkey organs on the other hand, are analyzed individually. All values reported for monkey tissue represent individual determinations for plasma and organ samples.

HPLC Analysis

The HPLC system of Krinsky and Welankiwar[25] is used for separating the extracts obtained in the study of the metabolism of canthaxanthin in the rat. This system (system A) uses a 5-μm ODS-Hypersil (C_{18}) column (4.6 × 100 mm) (Hewlett-Packard Co., Avondale, PA) with an initial flow rate of 0.5 ml/min. The starting solvent is acetonitrile–methanol––H_2O (72.25:12.75:15, v/v), which is changed over a 5-min linear gradient to acetonitrile–methanol–H_2O (74:13:13, v/v). From 5 to 8 min a linear gradient is formed to give acetonitrile–methanol–H_2O (76.5:13.5:10, v/v), and from 8 to 10 min dichloromethane is introduced to form an acetonitrile–methanol–H_2O–dichloromethane solvent (63.75:11.25:15:10, v/v). From 10 to 15 min the H_2O is replaced with dichloromethane to give the final solvent of acetonitrile–methanol–dichloromethane (63.75:11.25:25, v/v). A modification of the above pro-

---

[24] N. I. Krinsky, M. M. Mathews--Roth, S. Welankiwar, P. K. Sehgal, N. C. G. Lausen, and M. Russett, *J. Nutr.* **120**, 81 (1990).

[25] N. I. Krinsky and S. Welankiwar, this series, Vol. 105, p. 155.

cedure[26] is used for the studies of lycopene in the rats, and of both pigments in the monkeys. This modified system (system B) uses a 5-μm Econosphere (C_{18}) column (4.6 × 250 mm) (Alltech Associates, Inc., Deerfield, IL) with an initial flow rate of 0.7 ml/min. The starting solvent consists of acetonitrile–methanol (85:15, v/v) containing 1.3 mM ammonium acetate. At 15 min the flow rate is increased to 1.0 ml/min and a step gradient to 2-propanol–acetonitrile–methanol (40:51:9, v/v) is introduced to separate the hydrocarbon carotenoids. In all cases, 0.5-min fractions are collected for determination of radioactivity. In the case of rat livers, we had to modify the separation because of the large amounts of lipid that had to be chromatographed; we used system C with the Econosphere column. The initial solvent was acetonitrile–H_2O (75:25, v/v) containing 50 mM ammonium acetate. At 25 min, the solvent was changed to acetonitrile–dichloromethane–methanol (70:20:10, v/v).

Results

Neither canthaxanthin nor lycopene were present in the plasma samples of either rats or monkeys before gavage: these pigments are not components of animal feeds. Maximal absorption of radioactive canthaxanthin and lycopene in rats was similar, occurring between 4 and 8 hr after administration. The monkeys absorbed canthaxanthin and lycopene with considerable individual variation, the maximal absorption occurring between 8 and 48 hr.[1]

The rats accumulated canthaxanthin and lycopene into most of the organs examined, although the degree of canthaxanthin accumulated in most cases varied inversely with the amount of carotenoid in the diet prior to receiving the [^{14}C]canthaxanthin (Figs. 1 and 2). Our findings that accumulation of [^{14}C]canthaxanthin was lower in animals previously fed pigment confirmed our findings with prefeeding of β-[^{14}C]carotene,[24] and also support the observations of Shapiro et al.[27] that the absorption of β-carotene in rats decreased as the dose of administered pigment increased. These observations also parallel the finding that the absorption of some fat-soluble micronutrients decreases as the administered dose is increased.[28] At what time in the 8-week prefeeding period the decrease in absorption occurred was not determined by our experiment.

[26] G. J. Handelman, E. A. Dratz, C. C. Reay, and F. J. G. M. van Kuijk, *Invest. Ophthalmol. Visual Sci.* **29**, 850 (1988).
[27] S. S. Shapiro, D. J. Mott, and L. J. Machlin, *J. Nutr.* **114**, 1924 (1984).
[28] G. J. Handelman, L. J. Machlin, K. Fitch, J. J. Weiter, and E. A. Dratz, *J. Nutr.* **115**, 807 (1985).

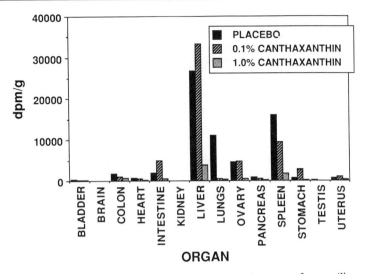

FIG. 1. The distribution of [^{14}C]canthaxanthin in the pooled extracts of organs (livers were extracted individually) of six rats receiving diets containing 1 or 0.1% canthaxanthin or placebo, and killed 72 hr after the oral administration of 20 μCi of [^{14}C]canthaxanthin. (From Ref. 1. Reproduced with permission of the American Institute of Nutrition.)

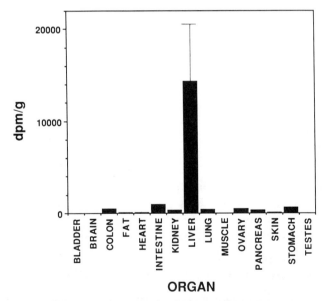

FIG. 2. The distribution of [^{14}C]lycopene in the pooled extracts of organs of rats killed 72 hr after the oral administration of 20 μCi of [^{14}C]lycopene. Livers were extracted individually. Error bar indicates 1 SD. (From Ref. 1. Reproduced with permission of the American Institute of Nutrition.)

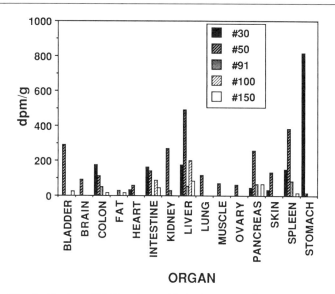

FIG. 3. The distribution of [^{14}C]canthaxanthin in the organs of monkeys killed 72 hr after the oral administration of 50 μCi of [^{14}C]canthaxanthin. (From Ref. 1. Reproduced with permission of the American Institute of Nutrition.)

There was wide variation among the individual monkeys as to the absorption of canthaxanthin and lycopene into the different organs (Figs. 3 and 4). In both monkeys and rats, the liver contained the largest amount of [^{14}C]canthaxanthin and [^{14}C]lycopene. We also found small amounts of lutein, zeaxanthin, α- and β-cryptoxanthin, and β-carotene in the rat livers and in the livers and most other organs of the monkey. These pigments are present in the diet of the animals, but we were unable to detect measurable amounts of carotenoids in the other organs of the rats.

We found no evidence of any significant metabolites of canthaxanthin in the plasma or any of the organs in either rats or monkeys. We did find the presence of some *cis*-canthaxanthin in the spleen extract from a rat on the 1% canthaxanthin diet. This observation indicated that some isomerization of the carotenoid took place, because the canthaxanthin fed contained no cis isomer. This phenomenon has been seen by other workers.[29] No *cis*-canthaxanthin was found in the livers of either rats or monkeys. Additionally, we found that lycopene, in spite of being stored at $-70°$ in the presence of α-tocopherol, deteriorated and formed a radioactive compound that eluted immediately before retinol. There had been a 7-month

[29] S. T. Mayne and R. S. Parker, *J. Agric. Food Chem.* **36,** 478 (1988).

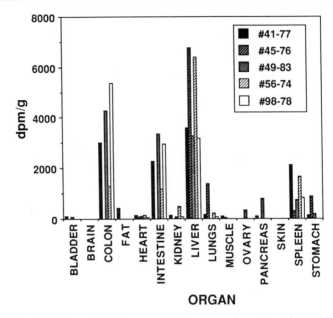

FIG. 4. The distribution of [^{14}C]lycopene in the organs of monkeys killed 48 hr after the oral administration of 50 μCi of [^{14}C]lycopene. (From Ref. 1. Reproduced with permission of the American Institute of Nutrition.)

lag between the experiments with lycopene in monkeys and the use of the same [^{14}C]lycopene solution in the rat studies; it was not until the completion of these latter studies that we detected the deterioration. No such deterioration occurred in the canthaxanthin/oil solutions. As a result of this experiment, we recommend that lycopene solutions be made up just before their use.

In summary, our studies[1] demonstrate that monkeys and rats can indeed absorb single doses of canthaxanthin and lycopene into their plasma and organs, with similar time courses for the appearance in plasma of both pigments in each species (4–8 hr in the rat; 8–48 hr in the monkey). We were unable to detect any retinoid metabolic products formed from these pigments, although we found the formation of cis-canthaxanthin in the spleen of one rat. We offer an important reminder: lycopene decomposes readily, and solutions should be made immediately before use, and not stored (even at −70° in the presence of antioxidants) for prolonged periods of time.

[24] Extraction and Analysis by High-Performance Liquid Chromatography of Carotenoids in Human Serum

By ARUN B. BARUA and HAROLD C. FURR

Introduction

The carotenoid pattern of human plasma or serum is not as complex as in vegetables and fruits. The major carotenoids occurring in human plasma are lutein, zeaxanthin, α-cryptoxanthin,[1,2] β-cryptoxanthin, lycopene, α-carotene, and β-carotene. Although the trans forms predominate, cis forms of these carotenoids often occur in blood.[1,2] A number of procedures for the extraction of the carotenoids from plasma or serum and their analysis by high-performance liquid chromatography (HPLC) can be found in the literature; for a review, see Ref. 3.

The following criteria are important and critical in selecting the method for analysis.

1. Large quantities of human blood are seldom available. Most of the published procedures describe the use of 200 μl or more of plasma or serum for extraction, and 100 μl for analysis. In large field studies, however, only a few capillaries filled with whole blood are available. Hence, the analysis should be possible with as little blood as possible.

2. It should be possible to analyze the whole range of individual carotenoids in one assay to get a clear picture of the whole carotenoid profile.

3. It would be advantageous if other micronutrients, such as vitamin A and vitamin E, could be simultaneously determined in the same extract and same assay.

4. If possible, there should be baseline separation of the individual carotenoids to ensure analytical accuracy.

5. It should be possible to use internal standards that have close similarity in structure with the nutrients under assay, yet not interfere with the separation of the individual peaks during HPLC.

6. An isocratic HPLC procedure will be preferred over gradient analysis

---

[1] N. I. Krinsky, M. D. Russet, G. J. Handelman, and D. M. Snodderly, *J. Nutr.* **120**, 1654 (1990).
[2] F. Khachik, G. R. Beecher, M. B. Goli, and W. R. Lusby, *Pure Appl. Chem.* **63**, 71 (1992).
[3] H. C. Furr, A. B. Barua, and J. A. Olson, *in* "Modern Chromatographic Analysis of the Vitamins" (A. P. De Leenheer, W. E. Lambert, and H. S. Nelis, eds.), 2nd Ed, p. 1. Dekker, New York, 1992.

in order to obtain a stable baseline, especially at low attenuation, and to avoid solvent equilibration between runs.

7. A nonaqueous solvent system will be preferred over an aqueous system, as the presence of water in the solvent mixture tends to result in loss of carotenoids on the HPLC column.

As a result of our continued work on the HPLC separation of carotenoids and retinoids during the past several years, we have been able to develop an extraction procedure and an isocratic HPLC procedure that enables simultaneous determination of all the major carotenoids as well as of retinol, retinyl esters, and the tocopherols in very small volumes of human serum. The details of the procedure are described below.

Working Standards

Commercial preparations of lutein, lycopene, β-carotene, retinal, retinyl palmitate, α-tocopherol, and γ-tocopherol can be purchased from Sigma Chemical Company (St. Louis, MO). β-Apo-8'-carotenal and β-carotene can be purchased from Fluka Chemical Corporation (Ronkonkoma, NY). Retinol and β-apo-8'-carotenol are prepared by reduction of retinal and β-apo-8'-carotenal, respectively, with sodium borohydride.[4] The internal standards, retinyl hexanoate and β-apo-8'-carotenyl decanoate, are prepared by reaction of retinol and β-apo-8'-carotenol in triethylamine with hexanoic anhydride and n-decanoic anhydride (Sigma), respectively.[6]

If the carotenoids used as standards are not available commercially, a plant source rich in the carotenoid of interest can be selected for extraction. For example, carrots are a rich source of α- and β-carotenes, ripe tomatoes for lycopene, red bell pepper for β-cryptoxanthin, and mustard green leaves (or corn meal) for lutein, zeaxanthin, and β-carotene as well.[5] The carotenoids are extracted from 5 to 10 g of the fresh material by grinding first with acetone and then with hexane several times. The extracts are pooled and diluted with water until two phases separate. The upper hexane phase is separated, dried over anhydrous sodium sulfate, and then evaporated to dryness in a rotary evaporator. The residue is dissolved in a small volume of either acetone (for TLC) or a mixture of $CH_3OH-CH_2Cl_2$ (4:1, v/v) (for HPLC).

If the carotenoid composition of the crude extract is complex (as in red bell pepper extract), a preliminary separation of the major carotenoids can

[4] R. Hubbard, P. K. Brown, and D. Bownds, this series, Vol. 18, p. 615.
[5] A. B. Barua, R. O. Batres, H. C. Furr, and J. A. Olson, *J. Micronutr. Anal.* **5**, 291 (1989).
[6] A. C. Ross, this series, Vol. 123, p. 68.

be done by subjecting the extract to thin-layer chromatography (TLC).[5] For this, the concentrated extract is applied as a strip on a TLC plate (silica G, 250 μm thick, 20 × 20 cm, supplied by Universal Scientific, Inc., Atlanta, GA), and developed with hexane–acetone (4:1, v/v). The colored band of interest is immediately scraped off the plate, and the carotenoid is eluted with diethyl ether. The solvent is evaporated under argon or nitrogen, and the residue is dissolved in a mixture of methanol–dichloromethane (4:1, v/v) for HPLC. If the carotenoid composition of the crude extract is simple (as in carrot extract), the carotenoids can be purified directly by HPLC as described previously.[5] Because the amounts of carotenoids needed for preparation of standard curves are small, purification by HPLC can be carried out on an analytical column. It is a good idea to use an older column that is still good, but no longer used for analytical work. A reversed-phase Resolve 5-μm spherical C_{18} column (3.9 mm × 15 cm; Waters Associates, Milford, MA) is used. The composition of solvent mixtures, the flow rate, and the retention times during purification of the carotenoids by HPLC are shown in Table I.

If a 15-cm Resolve column is not available, the procedure and conditions described for the HPLC analysis of serum carotenoids using a 30-cm column can be used for purification of the individual carotenoids. Because standard carotenoids degrade even after careful storage and handling, they are checked for purity by HPLC as described above and purified, whenever necessary, by HPLC prior to use. For purification both of commercial

TABLE I
COMPOSITION OF SOLVENT MIXTURES, FLOW RATE, AND RETENTION TIME OF CAROTENOIDS DURING HPLC[a]

Carotenoid	Solvent mixture (v/v)	Flow rate (ml/min)	Retention time (min)
Lycopene	$CH_3OH-CH_2Cl_2$ (4:1)	0.5	10.7
α-Carotene	$CH_3OH-CH_2Cl_2$ (4:1)	0.5	12.2
β-Carotene	$CH_3OH-CH_2Cl_2$ (4:1)	0.5	13.0
β-Cryptoxanthin	$CH_3CN-CH_2Cl_2$ (85:15)	1.0	8.8
Neoxanthin	CH_3OH-H_2O (9:1)	1.0	8.0
Violaxanthin	CH_3OH-H_2O (9:1)	1.0	10.8
Lutein	CH_3OH-H_2O (9:1)	1.0	31
Zeaxanthin	CH_3OH-H_2O (9:1)	1.0	33

[a] Waters C_{18}, 5 μm, 15-cm-long Resolve column.

TABLE II
Spectral Characteristics of Carotenoids, Retinoids, and Tocopherols Frequently Found in Human Blood

Compound	Absorption maxima	$E_{1\,cm}^{1\%\,a}$	Solvent	Ref.
α-Carotene	422, 444, 474	2800	Petroleum ether	b
β-Carotene	425, 453, 479	2592	Petroleum ether	b
Lycopene	444, 472, 502	3450	Petroleum ether	b
Zeaxanthin	426, 452, 479	2348	Petroleum ether	b
Lutein	421, 445, 475	2550	Ethanol	b
β-Cryptoxanthin	425, 452, 479	2386	Petroleum ether	b
Retinol	325	1810	Hexane	c
Retinyl palmitate	325	940	Ethanol	d
α-Tocopherol	292	76	Ethanol	e
γ-Tocopherol	294	91	Ethanol	e

[a] $E_{1\,cm}^{1\%}$ values are given for the major underlined wavelengths.

[b] E. De Ritter and A. E. Purcell, in "Carotenoids as Colorants and Vitamin A Precursors: Technical and Nutritional Applications" (J. C. Bauernfeind, ed.), p. 883. Academic Press, New York, 1981.

[c] R. Hubbard, P. K. Brown, and D. Bownds, this series, Vol. 18, p. 615.

[d] U. Schwieter and O. Isler, in "The Vitamins" (W. H. Sabrell, Jr., and R. S. Harris, eds.), 2nd Ed., Vol. 1, p. 5. Academic Press, New York, 1967.

[e] P. Schudel, H. Mayer, and O. Isler, in "The Vitamins" (W. H. Sebrell, Jr., and R. S. Harris, eds.), 2nd Ed., Vol. 5, p. 168. Academic Press, New York, 1967.

samples and of crude plant extracts, a much higher attenuation (9–10, AUFS 0.5–1.0) and concentrated solutions of the standards are used. The relevant peak is collected from several HPLC injections of concentrates, and the pooled solution is evaporated, reconstituted with appropriate solvent in which the extinction coefficient ($E_{1\,cm}^{1\%}$) is known (Table II), and the amount of the standard present is determined by spectrophotometric methods.[3] For example, if a solution of β-carotene in petroleum ether gives an absorbance reading (OD) of 0.53 at 453 nm, the concentration of β-carotene in the solution is 0.53/0.2592 = 2.04 μg/ml. The standard solution is kept at −20° under argon or nitrogen, and is checked for purity by HPLC prior to subsequent use.

Standard curves based on integrator peak areas are prepared for α-carotene, β-carotene, lycopene, β-cryptoxanthin, and lutein/zeaxanthin. The amount of each of these carotenoids present in human serum is calculated from these curves. The concentrations of α-cryptoxanthin[1,2] and the unidentified carotenoids (X and Y)[2,5] are calculated by reference to the β-cryptoxanthin and lutein/zeaxantin standard curves, respectively.[5]

Capacity factor k_1, which is defined as a corrected retention time [$k_1 = (t_R - t_M)/t_M$, where t_R is the observed retention time and t_M is the chromatographic dead time], and relative retention indices (retentions relative to alkylphenone standards, with acetophenone as 800) are calculated as described previously.[7]

Sample Extraction

All extractions are carried out in laboratories lighted with gold fluorescent lamps (F40 Gold).

For routine analysis, serum (100 μl) is transferred by means of a 200-μl Gilson pipette (Rainin Instruments, Woburn, MA) to a 13 × 100 mm culture tube. To the aliquot of serum is added 10 μl each of methanolic solutions of the internal standards, retinyl hexanoate (13 μM = 5 ng/μl), and β-apo-8'-carotenyl decanoate (5 μM = 3 ng/μl), followed by ethanol (200 μl) and ethyl acetate (500 μl). We found that ethyl acetate can extract the hydrocarbon carotenoids in serum more efficiently than hexane alone.[5] The mixture is vortexed (30 sec) and centrifuged at room temperature (30 sec at 2000 rpm). The supernatant solution is saved, and the pellet is broken and extracted with ethyl acetate (500 μl) twice and with hexane (500 μl) once by vortexing and centrifuging as before. All the organic phases are pooled, water (500 μl) is added, and the sample is vortexed and centrifuged as before. The upper organic phase is transferred carefully by means of a Pasteur pipette to a 13 × 100 mm culture tube. (Often it is easier not to pipette out the organic phase completely, but to leave a trace, and then to add about 100 μl of hexane, and to pipette out as much of the organic phase as feasible and mix it with the rest in the culture tube. This eliminates the possibility of pipetting traces of water with the organic phase.) The solvent is then evaporated under a stream of argon, and the residue is dissolved in ice-cold dichloromethane (40 μl) (to minimize loss due to evaporation) followed by methanol (60 μl). The solution is transferred to a 500-μl glass conical tube, placed inside the autosampler (WISP; Waters Associates) vial for injection, and capped tightly. Aliquots of 2–50 μl are analyzed immediately.

When large volumes of serum are not available, 20 μl of serum is used for extraction. To 20 μl of serum is added 10 μl of each of the internal standards, 40 μl of ethanol, and 100 μl of ethyl acetate. The solution is vortexed and then centrifuged as described above. The pellet is broken up and extracted with 100 μl of ethyl acetate twice, and finally with 100 μl of hexane as described above. The pooled organic phase is vortexed with 100 μl of water and then centrifuged. The organic phase is evaporated

[7] H. C. Furr, *J. Chromatogr. Sci.* **27**, 216 (1989).

under argon. The residue is dissolved in a mixture of dichloroethane (40 μl) and methanol (20 μl), and 30 μl (equivalent to 10 μl of serum) is injected. Dichloroethane, which has a much higher boiling point than dichloromethane, is used to minimize solvent loss.

High-Performance Liquid Chromatography

The HPLC system includes a Waters Associates model 510 pump and an automated injection system WISP (Waters Associates), connected in series to two ISCO (Lincoln, NE) model V^4 detectors, the one nearest to the pump set at 450 nm (for carotenoids) and the other set at 300 nm (for retinol and tocopherols) or 325 nm (for retinol and retinyl esters). Retinyl esters normally occur only in small quantities in the blood, and tend to escape detection if the detector is set at 300 nm. A Waters C_{18} 5-μm Resolve column (30 cm × 3.9 mm i.d.) is preceded by a C-130 guard column (Upchurch Scientific, Oak Harbor, WA). The detectors are connected to a dual-channel Shimadzu CR-4A integrator (Shimadzu Corp., Columbia, MD). Detector sensitivity is routinely set at 0.004 AUFS to detect carotenoids at 450 nm, and at 0.016 AUFS to detect retinol and tocopherols at 300 nm, or at 325 nm to detect retinol and retinyl esters. The mobile phase for chromatography is a mixture of acetonitrile–dichloromethane–methanol–1-octanol (90:15:10:0.1, v/v/v/v). HPLC-grade acetonitrile and methanol and reagent-grade dichloromethane and 1-octanol (Fisher Scientific Co., Fairlawn, NJ) are used. The solvents are filtered through a 0.45-μm nylon filter (Millipore Corp., Bedford, MA) before mixing. The flow rate is 1.0 ml/min. Identification of the various carotenoids, retinol, and tocopherols is confirmed by comparing relative retention times to known standards. Furthermore, the identity of individual peaks obtained during HPLC of larger volumes of serum (0.5–1 ml) is confirmed from their characteristic ultravioilet/visible (UV/VIS) absorbance spectra.

Results

Figure 1B shows the isocratic reversed-phase HPLC chromatogram obtained after injection of an extract of human serum equivalent to 10 μl of serum. It can be seen that baseline separation of all the major carotenoids, namely, β-carotene, α-carotene, lycopene, β-cryptoxanthin, α-cryptoxanthin, and another unidentified carotenoid, is achieved. Lutein and zeaxanthin, two other major carotenoids present in human serum, although well separated from other carotenoids, do not resolve well from each other. Furthermore, fair to good separation of cis and trans isomers of

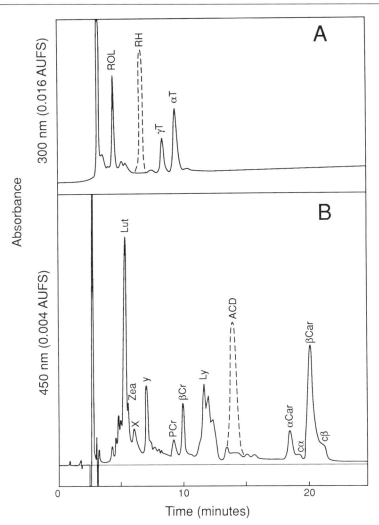

FIG. 1. Chromatograms obtained with isocratic nonaqueous reversed-phase HPLC of 10 μl of human serum extract. Column, 5-μm Waters Resolve C_{18} (30 cm × 3.9 mm i.d.); eluent, acetonitrile–dichloromethane–methanol–1-octanol (90:15:10:0.1, v/v/v/v); flow rate, 1 ml/min. Two detectors (ISCO V^4) in series were used; integrator, Shimadzu CR4A. (A) Detection at 300 nm (retinol and tocopherols); (B) detection at 450 nm (carotenoids). Peaks: (A) ROL, retinol; RH, retinyl hexanoate (internal standard); γT, γ-tocopherol; αT, α-tocopherol; (B) Lut, lutein; Zea, zeaxanthin; X, unidentified; Y, unidentified; PCr, α-cryptoxanthin; βCr, β-cryptoxanthin; Ly, lycopene; ACD, apo-8'-β-carotenyl decanoate (internal standard); αCar, α-carotene; cα, cis-α-carotene; βCar, β-carotene; cβ; cis-β-carotene.

α-carotene and β-carotene is achieved. Even partial resolution of the cis–trans isomers of lycopene is observed. The internal standard β-apo-8'-carotenyl decanoate (Fig. 1B, broken line) elutes after lycopene, and does not interfere in the quantitation of any of the major carotenoids.

With simultaneous detection at 300 nm, the detection and quantitation of retinol and the tocopherols (Fig. 1A) is possible in the same human serum extract equivalent to 10 μl of serum. When butylated hydroxytoluene (BHT) is added in ethanol during the extraction of lipids from the serum, retinol often elutes at the tail end of the BHT peak, and may result in some error. However, it has been shown by others[8,9] that addition of BHT as an antioxidant during extraction of carotenoids and retinoids in serum has no effect, possibly due to the presence of large quantities of tocopherols present in serum that themselves serve as antioxidants. If retinol is to be analyzed, BHT may not be used during the extraction step. The internal standard retinyl hexanoate (Fig. 1A, broken line) elutes after retinol, and does not interfere with the tocopherols. It is yet to be determined if the small peak next to the internal standard is due to β-tocopherol.

By this new procedure, it is possible to quantitate picomole quantities of the micronutrients. Limits of detection, compared with typical ranges in serum, are given in Table III for major analytes. Capacity factors and relative retention indices (RRIs) are presented in Table III. RRIs are found to be constant for a given type of column.

For a sample analyzed in quadruplicate (100 μl each) on three successive days, recovery of retinyl hexanoate averages 98.7% [relative standard deviation (rsd) 9.6%], recovery of β-apo-8'-carotenyl decanoate averages 95% (rsd 8.4%). For other analytes relative standard deviations are given as: retinol (1.63 μM), 1.1%; γ-tocopherol (6.77 μM), 7.3%; α-tocopherol (32.6 μM), 2.5%; lycopene (0.73 μM), 15.7%; α-carotene (0.33 μM), 7.3%; β-carotene (0.42 μM), 12.2%. Retinyl esters were not detected in this particular fasting blood sample. However, retinyl esters in serum can be resolved and analyzed. The following t_R values are observed for the different retinyl esters present: retinyl linoleate, 11.7 min; retinyl oleate, 16.7 min; retinyl palmitate, 18.4 min; and retinyl stearate, 26.8 min.

---

[8] D. W. Nierenberg, Y. Peng, and D. S. Alberts, in "Nutrition and Cancer Prevention: Investigating the Role of Micronutrients" (T. E. Moon and M. S. Micozzi, eds.), p. 181. Dekker, New York, 1989.

[9] L. R. Cantilena and D. W. Nierenberg, J. Micronutr. Anal. **6**, 127 (1989).

TABLE III
CHROMATOGRAPHIC CHARACTERISTICS OF CAROTENOIDS, RETINOIDS, AND TOCOPHEROLS

Compound	Capacity factor	Relative retention index[a]	Limit of detection		Reported ranges	
			pmol	ng	μM	ng/ml
α-Carotene	6.61	2,835	0.8	0.45		
β-Carotene	7.28	2,885	0.8	0.45	0.3–1.1	151–616
Lycopene	3.99	2,563	ND[b]			
β-Cryptoxanthin	3.43	2,468	ND			
Lutein	0.96	1,858	ND			
Zeaxanthin	1.43	2,006	ND			
Retinol	0.70	1,664	0.21	0.06	1.6–3.4	456–973
Retinyl palmitate	8.59	2,966	1.8	0.5		
α-Tocopherol	3.02	2,410	6.4	2.7	16–36	6,800–15,600
γ-Tocopherol	2.58	2,325	3.6	1.5		

[a] Retention relative to alkylphenone standards; with acetophenone as 800.
[b] ND, Not determined.

Acknowledgments

We are grateful to Dr. James A. Olson for financial support to carry out the work, and for his help and constructive criticism in preparing the manuscript. This work was supported by the Competitive Research Grants Program, Science and Education Administration, U.S. Department of Agriculture (87-CRCR-1-2320), and by the National Institutes of Health (DK 32793, DK 39733, and CA 46406).

[25] Analysis of Apocarotenoids and Retinoids by Capillary Gas Chromatography–Mass Spectrometry

By HAROLD C. FURR, ANDREW J. CLIFFORD, and A. DANIEL JONES

Retinoids and carotenoids are particularly labile under conventional gas chromatographic (GC) conditions, being sensitive to hot injector surfaces and partly inactivated column packings. Ninomiya *et al.*[1] found that retinol, retinyl acetate, and retinyl palmitate are dehydrated to anhydroretinol under conditions of conventional packed-column gas chromatography. Dunagin and Olson[2] used column pretreatment with injections of

[1] T. Ninomiya, K. Kidokoro, M. Horiguchi, and N. Higosaki, *Bitamin* **27**, 349 (1963).
[2] P. E. Dunagin and J. A. Olson, this series, Vol. 15, p. 289.

β-carotene to achieve packed-column GC of retinol and retinyl acetate; pretreatment was not required for anhydroretinol, methyl retinyl ether, retinaldehyde, and methyl retinoate. Cullum et al.[3] took advantage of the on-column dehydration of retinol to anhydroretinol for gas chromatographic–mass spectrometric (GC–MS) analysis of deuterated retinol. Vecchi et al.[4] found that the trimethylsilyl derivative of retinol was stable on QF-1 (fluorosilicone)-packed columns. The syn and anti conformers of retinal methoxime were separated on a short QF-1-packed column.[5]

In another approach, retinol and retinyl acetate were hydrogenated over PtO_2 before GC analysis.[6] Gas chromatography of hydrogenated carotenoids has been reviewed by Taylor and Ikawa.[7] Those authors tabulated retention times of perhydrogenated carotenoids relative to squalene, on silicone stationary phases in packed columns.

Methyl retinoate is much more stable to GC conditions than other retinoids; it can be readily prepared from retinoic acid by treatment with diazomethane. These applications have been reviewed.[8] The pentafluorobenzyl ester of an aromatic retinoic acid analog was analyzed by capillary GC–MS with column switching (peak transferred from an SE-54 column to an OV-240 column); the tetradeuterated analog was used as internal standard, and negative-ion chemical-ionization mass spectrometry with ammonia as reagent gas was used for detection.[9]

Bonded-phase capillary columns are inherently much more inert than packed columns, and provide much sharper peaks (hence better resolution and improved lower limits of detection). Smidt et al.[10] demonstrated that on-column injection at room temperature onto bonded-phase capillary columns allowed GC and GC-MS of underivatized retinol. A subsequent study used this approach to chromatograph and determine Kovats indices of a series of aporetinoids, retinoids, and an apocarotenoid.[11] Application of this technique to determine vitamin A status by stable-isotope dilution has been reviewed.[12]

[3] M. E. Cullum, J. A. Olson, and S. W. Veysey, Int. J. Vitam. Nutr. Res. 53, 3 (1983).
[4] M. Vecchi, W. Vetter, W. Walther, S. F. Jermstad, and G. W. Schutt, Helv. Chim. Acta 50, 1243 (1967).
[5] A. P. De Leenheer and M. G. M. De Ruyter, Anal. Biochem. 63, 169 (1975).
[6] T. W. Fenton, H. Vogtmann, and D. R. Clandinin, J. Chromatogr. 77, 410 (1973).
[7] R. F. Taylor and M. Ikawa, this series, Vol. 67, p. 233.
[8] A. P. De Leenheer and W. E. Lambert, this series, Vol. 189, p. 104.
[9] H.-J. Egger, U. B. Ranalder, E. U. Koelle, and M. Klaus, Biomed. Environ. Mass Spectrom. 18, 453 (1989).
[10] C. R. Smidt, A. D. Jones, and A. J. Clifford, J. Chromatogr. 434, 21 (1988).
[11] H. C. Furr, S. Zeng, A. J. Clifford, and J. A. Olson, J. Chromatogr. 527, 406 (1990).
[12] A. J. Clifford, A. D. Jones, and H. C. Furr, this series, Vol. 189, p. 94.

Equipment

In these studies, either Hewlett-Packard (Palo Alto, CA) (model 5830A) or Carlo Erba (Valencia, CA) (model 4130) gas chromatographs were used, with minor modifications to permit use of capillary columns.[10,11] On-column injectors (J & W Scientific, Folsom, CA, or Scientific Glass Engineering, Austin, TX) were used with fused-silica, wall-coated, open-tubular capillary columns; typically, 15 m × 0.25-mm i.d. columns were used, with 0.25 µm methylsilicone bonded-phase (Durabond-1; J & W Scientific). For some studies with less volatile compounds, thin-film short methylsilicone columns (5 m × 0.25-mm i.d.; 0.1-µm film thickness) were used. The more stable compounds, methyl retinoate and anhydroretinol, were also chromatographed on a more polar cyanopropylphenylmethylsilicone column (DB-225, 30 m × 0.25-mm i.d.; 0.15-µm film thickness). Hydrogen was used as a carrier gas, at 50 cm/sec (35 kPa, 5 psi), or helium was used at 25 cm/sec (35 kPa, 5-psi). For some studies, flame-ionization detection was used; in others, the chromatographs were connected to quadrupole mass spectrometers (ZAB-HS-2F; VG Analytical, Wythenshawe, England; or SpectrEL, EXTREL, Pittsburgh, PA). Direct interfacing (extending the capillary GC column up to, but not into, the mass spectrometer ion source) was used. The mass spectrometers were routinely used in electron-impact ionizing mode at 70 eV, source temperature 180°.

Kovats retention indices were determined under appropriate isothermal conditions for each compound (capacity factor between 1 and 10) by comparison of retention times with those of n-alkane standards.[13]

Results

On the Hewlett-Packard gas chromatograph, flame-ionization detector responses to retinol were linear over the range 1 to 1000 ng injected, demonstrating that retinol was not adsorbed significantly to active sites in the injector or in the column. Although cis isomers of retinol apparently can be resolved (small peaks eluting shortly before and after all-*trans*-retinol have very similar mass spectra), chromatography and mass spectrometry of cis isomers have not yet been studied carefully. Geometric isomers of retinal oxime can be chromatographed and resolved by this technique, but individual retention indices or mass spectra have not been studied (H. C. Furr, unpublished observations, 1991).

Direct interfacing of the gas chromatograph with the mass spectrometer

---

[13] H. C. Furr, *J. Chromatogr. Sci.* **27**, 216 (1989).

avoids potentially reactive sites within a jet separator, and provides all of the sample to the mass spectrometer for improved limits of detection. The column segment (or transfer line) should be maintained at a somewhat higher temperature than the GC oven to avoid loss of sample by condensation.

Figure 1 shows the GC–MS ion current chromatograms for retinol (m/z 286) and α-tocoperol (m/z 430). Figure 2 shows total ion current chromatograms for retinol, retinal, and retinyl acetate, with mass spectra for each. With the mass spectrometer used in that study, the parent ion is not a highly abundant peak but is clearly recognizable. A characteristic peak in the mass spectra of retinol and its esters is at m/z 268 ($M^+ - H_2O$). It has been found that this ion does not arise from fragmentation of the molecular ion, but rather from dehydration of retinol occurring within the ion source.[14] The relative abundances of this peak and of the parent peak are dependent on the geometry and the cleanliness of the ionizer, not only on the ionizing voltage.

Decreasing ionizing energy from the conventional 70 eV gives different effects on different mass spectrometers. In general, reducing ionizing energy from 70 eV will increase the intensity of the parent peak relative to the total ion intensity; however, the total ion current decreases so rapidly with decreasing ionization voltage that the limit of detection may be severely compromised. On the SpectrEL mass spectrometer used in these studies, an ionization voltage of 50 eV gave an acceptable compromise between improved parent ion relative abundance and decreased ion yield (H. C. Furr, unpublished observations, 1991); the optimum ionizing voltage should always be established for the individual instrument.

Although chemical ionization usually gives much improved relative abundances of parent ions, it is not recommended for isotope dilution studies using deuterium-labeled retinoids, because dehydration often occurs as a result of ion–molecule collisions within the ion source. Attempts to use thermospray for high-performance liquid chromatography (HPLC)–MS analysis of retinol have given similar results (A. D. Jones and A. J. Clifford, unpublished observations, 1991). Chemical ionization conditions also produce high concentrations of hydrogen-containing radicals that may cause hydrogen–deuterium exchange and inaccurate estimates of isotope enrichment (A. J. Clifford and A. D. Jones, unpublished observations, 1991).

Table I gives the Kovats retention indices (retention relative to n-alkanes) of a series of aporetinoids and retinoids. As illustrated in Fig. 3,

[14] A. J. Clifford, A. D. Jones, Y. Tondeur, H. C. Furr, H. R. Bergen III, and J. A. Olson, *Proc. Annu. Conf. Mass Spectrom. Allied Top., 34th* p. 327 (1986).

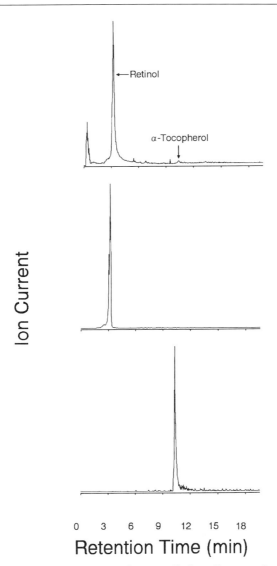

FIG. 1. GC–MS ion chromatograms of a saponified rat liver sample. *Top:* total ion current (TIC). (Middle and bottom) Reconstructed ion chromatograms for m/z 286 (retinol) and m/z 430 (α-tocopherol), respectively. Column: wall-coated open tubular methylsilicone (DB-1, 15 m × 0.25 mm i.d., 0.25-μm film thickness); temperature gradient from 220 to 270° at 3°/min, held 20 min. Carrier gas: hydrogen at 50 cm/sec (5 psi). Detection by electron-impact MS (ZAB-HS-2F quadrupole mass spectrometer), 70-eV ionizing voltage, monitored at m/z 286 and 268 (for retinol) and m/z 430 (for α-tocopherol). [From C. R. Smidt, A. D. Jones, and A. J. Clifford, *J. Chromatogr.* **434,** 21 (1988), with permission of the publisher.]

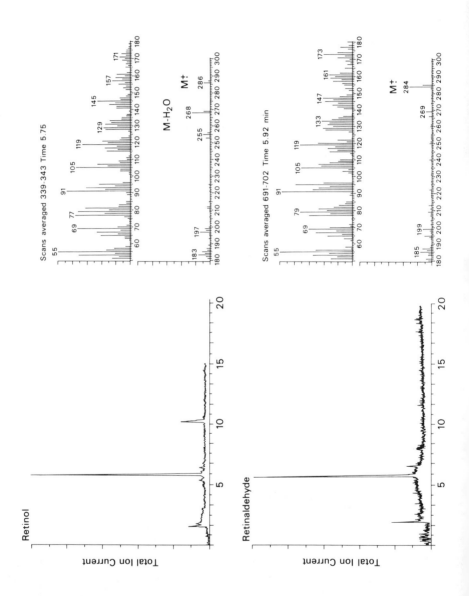

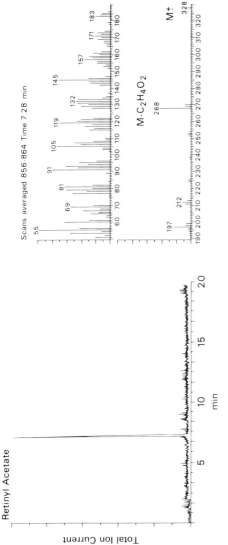

FIG. 2. Capillary GC–MS of retinoids, with on-column injection. (Top) retinol; (middle) retinaldehyde; (bottom) retinyl acetate. Column: wall-coated open tubular methylsilicone (DB-1, 15 m × 0.25 mm i.d., 0.25-μm film thickness); linear temperature gradient from 200 to 300° at 5°/min. Carrier gas: helium at 25 cm/sec (2.5 ml/min, 5 psi). Detection by electron-impact MS (SpectrEL quadrupole mass spectrometer), 70–eV ionizing voltage; scan range 50–300 or 50–350 amu. Left-hand side: total ion current chromatogram; right-hand side: mass spectra of retinoid peaks. (The peak at 10.3 min in the chromatogram of retinol has been identified as a phthalate ester.) [From H. C. Furr, S. Zeng, A. J. Clifford, and J. A. Olson, *J. Chromatogr.* **527**, 406 (1990), with permission of the publisher.]

TABLE I
KOVATS RETENTION INDICES OF APORETINOIDS AND RETINOIDS[a]

Column	Compound	Column temperature	Kovats index
DB-1 (15 m × 0.25 mm; 0.25-μm film thickness)	Neral	100	1216
	Geranial	100	1244
	α-Ionone	145	1416
	β-Ionol	145	1406
	β-Ionone	145	1469
	β-Ionyl acetate	150	1525
	Anhydro C_{15}-alcohol	150	1569
	C_{15}-Alcohol	165	1732
	C_{15}-Aldehyde	165	1754
	C_{15}-Acetate	180	1857
	C_{15}-Acid ethyl ester	180	1885
	C_{18}-Alcohol	180	2029
	C_{18}-Ketone	180	2089
	C_{18}-Acetate	180	2161
	Axerophthene	220	2148
	Anhydroretinol	240	2233
	Retinol	240	2453
	Retinaldehyde	240	2466
	Methyl retinoate	240	2528
	Retinyl acetate	240	2578
DB-1 (5 m × 0.25 mm; 0.10-μm film thickness)	Retinyl acetate	190	2531
	Retinyl butanoate	190	2738
	Retinyl hexanoate	220	2970
	Retinyl octanoate	220	3157
	Retinyl decanoate	220	3359
	Retinyl dodecanoate	240	3577
	β-Apo-12′-Carotenal	240	3040
DB-225 (30 m × 0.25 mm; 0.15-μm film thickness)	Anhydroretinol	220	2468
	Methyl retinoate	220	3200

[a] From H. C. Furr, S. Zeng, A. J. Clifford, and J. A. Olson, *J. Chromatogr.* **527**, 406 (1990), with permission of the publisher.

retention indices increase linearly with chain length for these compounds. Similarly, retention indices of retinyl esters increase linearly with increasing fatty acyl chain length. (The difference in retention index, 1.8%, of retinyl acetate determined on two different methylsilicone columns may be due to the differences in film thickness between the two columns; this effect has not been studied further.) For a given alkyl chain length (excluding carbon atoms in ester moieties), the order of elution is anhydro < alcohol < aldehyde/ketone < acetate. The retention indices of anhy-

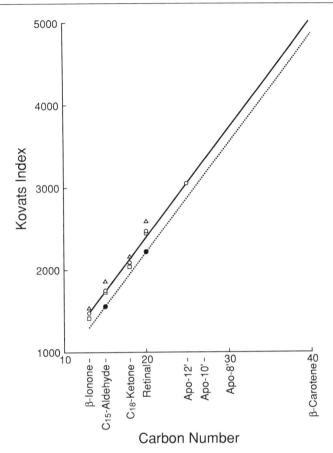

FIG. 3. Relationship of Kovats retention index to carbon number for aporetinoids and retinoids: (□), alcohol; (○), aldehyde or ketone; (△), acetate; (●), anhydro forms. [From H. C. Furr, S. Zeng, A. J. Clifford, and J. A. Olson, *J. Chromatogr.* **527,** 406 (1990), with permission of the publisher.]

droretinol and methyl retinoate are increased on the more polar cyanopropylphenylmethylsilicone (DB-225) column as compared with the methylsilicone column (DB-1), presumably because of increased interactions of the polyene chain with the stationary phase.

The limitation on analysis of longer-chain apocarotenoids and carotenoids has been the lack of volatility of these compounds coupled with the temperature limitations of GC columns, ovens, and detectors. Attempts to chromatograph ethyl-β-apo-8′-carotenoate (retention index estimated as 3200 from Fig. 3) gave a broad peak with approximate retention index

3100 (H. C. Furr, unpublished observations, 1991); similarly, attempts to chromatograph long-chain esters of retinol (retinyl tetradecanoate, hexadecanoate, and octadecanoate) at high temperatures gave broad peaks, presumably caused not by degradation of the compounds but by their condensation at the detector and/or by thermal isomerization during chromatography.[11] Extrapolation of retention indices of aporetinoids and apocarotenoids suggests that β-carotene should have a Kovats retention index of approximately 5000 (i.e., retention of a C_{50} alkane).

It is not yet known at what operating temperature the thermal isomerization of these polyene compounds will limit usefulness of this technique. In preliminary studies on a high-temperature carborane column, capillary gas chromatography of β-carotene produced a broad peak, which we attribute to cis–trans isomerizations occurring during the separation (A. J. Clifford and A. D. Jones, unpublished observations, 1991). Differences in partition coefficients of the various isomers would lead to peak broadening if rapid interconversion of isomers occurred during the separation. Several lines of evidence support this conclusion: (1) we have observed (A. D. Jones, unpublished results, 1991) similar peak broadening during gas chromatographic separations of several isomers of derivatives of dihydroxyeicosatetraenoic acids at even lower column temperatures; (2) the mass spectra of β-carotene did not exhibit significant variation across the width of the peak, and a distinct molecular ion for β-carotene was observed throughout; (3) condensation at cold spots is not believed to have occurred in that study because similar peak widths were obtained on two different chromatographs. Improvements in the chromatography of β-carotene and other carotenoids may yet be obtained, particularly if conditions can be found that minimize the rates of isomerization.

Acknowledgments

These studies were encouraged and advised by Dr. James A. Olson. This work was supported by the U.S. Department of Agriculture (87-CRCR-1-2320) and the National Institutes of Health (DK-32793, DK-39733, DK-43098, and CA-46406).

[26] Separation of Isomers of cis-β-Carotenes

By KIYOSHI TSUKIDA

Introduction

Although 272 geometric isomers of β-carotene can exist theoretically, 12 cis forms in total have been recorded and analyzed.[1] Two main cis isomers were designated neo-β-carotene U and B, the former being adsorbed above all-*trans*-β-carotene (I) on a lime column, and the latter below it; they were interpreted as having a peripheral 9-mono-cis or 9,13'-di-cis configuration, respectively, mainly based on careful inspection of ultraviolet and infrared spectral data.[2] Growing interest in biomolecular functions of *cis*-carotenoids[3] prompted us to report our results obtained by ^{1}H and ^{13}C nuclear magnetic resonance (NMR) analysis, including a series of selective INEPT experiments showing that the former is 9-mono-*cis*-β-carotene (II) and the latter is 13-mono-*cis*-β-carotene (III).[4] Multicomponent mixtures of *cis*-β-carotenes were further extended to high-performance liquid chromatography (HPLC) separation using a lime column as a new tool,[5] and to an NMR structural determination study.[6]

Stereoisomerization

Chemicals. all-*trans*-β-Carotene is a gift from Dr. T. Takahashi (Sumitomo Chemical Co., Ltd., Osaka, Japan). Lime (calcium hydroxide, guaranteed reagent) is purchased from Kishida Chemical Co., Ltd. (Osaka, Japan) or Nakarai Chemical Industries (Kyoto, Japan). A batch of lime is selected at random and particles in the range of 200–300 mesh are used. All solvents and reagents are of analytical grade and used without further purification.

Thermal Isomerization. Synthetic all-*trans*-β-carotene is isomerized by immersing a thin-walled evacuated tube containing the crystals in a bath at 190–200° for 15 min. The melt is rapidly solidified in ice–water and dissolved in *n*-hexane.

---

[1] L. Zechmeister, "*cis–trans* Isomeric Carotenoids, Vitamins A, and Arylpolyenes." Springer-Verlag, Vienna, 1962.
[2] K. Lunde and L. Zechmeister, *J. Am. Chem. Soc.* **77**, 1647 (1955).
[3] K. Tsukida, *Bitamin* **58**, 185 (1984).
[4] K. Tsukida, K. Saiki, and M. Sugiura, *J. Nutr. Sci. Vitaminol.* **27**, 551 (1981).
[5] K. Tsukida, K. Saiki, T. Takii, and Y. Koyama, *J. Chromatogr.* **245**, 359 (1982).
[6] K. Tsukida and K. Saiki, *J. Nutr. Sci. Vitaminol.* **29**, 111 (1983).

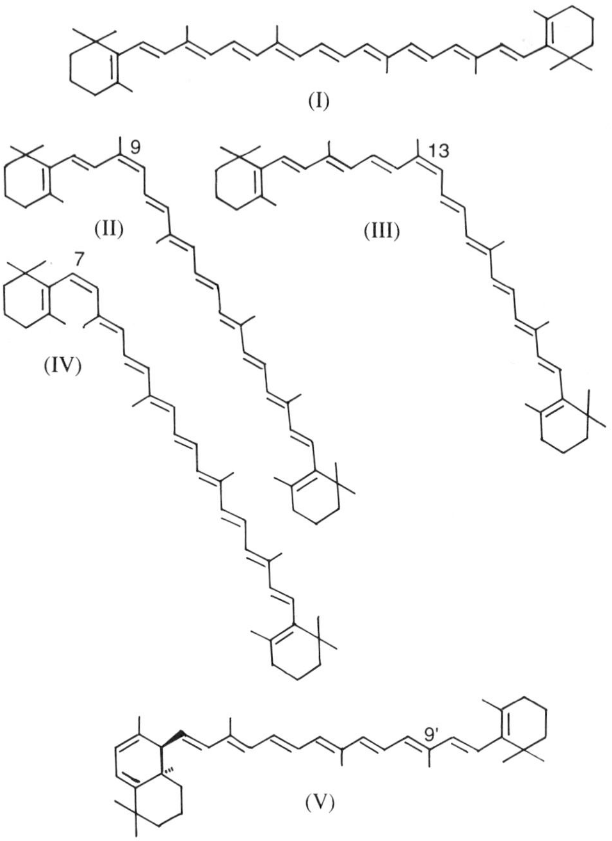

(I)

(II) 9

(III) 13

(IV) 7

(V) 9'

Iodine-Catalyzed Photoisomerization. A solution containing all-*trans*-β-carotene (about 1.2 mg in 10 ml) and iodine (2% of the weight of the pigment, in *n*-hexane) is irradiated with a fluorescent lamp (20 W, 40 cm long) from a distance of 20 cm for 10 min at room temperature to produce a quasi-equilibrium mixture. The solution is evaporated to dryness and the residue is dissolved in *n*-hexane.

Other Isomerization Methods. Other methods of artificial stereoisomerization of the all-*trans* form are described elsewhere.[1,4]

Resolution of cis Isomers

Classical Adsorption Chromatography. An *n*-hexane solution of a multicomponent mixture of *cis*-β-carotene is poured on a lime column (45 × 4.5-cm i.d. for 15 mg, 75 × 7.5-cm i.d. for 100 mg of β-carotene), then developed with 1–2% acetone in *n*-hexane. Then a column is extruded and each chromatogram is cut out separately and eluted with acetone. After water and n-hexane are added, the isomer is transferred into the organic layer and is characterized by NMR or estimated photometrically.

Preparative TLC. A mixture can be developed and eluted with n-hexane on an aluminum oxide $60F_{254}$ (type E) plate (0.25 mm thick).

HPLC Separation. A Shimadzu LC2 liquid chromatograph is equipped with an ultraviolet/visible light (UV/VIS) detector (450 or 335 nm). In stainless steel tubing, 250 × 4-mm i.d., a lime column is prepared with a high-pressure packing apparatus. First, ultrasonic degassing (5 min) is used to obtain a stable suspension of lime (5 g) and tetrachloroethylene (37 ml); the slurry is then packed and rinsed with n-hexane (80 ml) from the solvent reservoir under a pressure not exceeding 200 kg/cm^2. Injected samples are eluted at ambient temperature with 0.1–2% acetone in n-hexane at a flow rate of 1 ml/min. The acetone content in an eluant depends on the carotenoids and the time required for analysis. Each HPLC peak is fractionated, accumulated, and then identified by spectroscopic methods, mainly by UV/VIS or ^{1}H or ^{13}C NMR.

Structural Assignment. Structural assignment of a particular cis isomer is straightforward, if highly resolved ^{1}H and ^{13}C NMR data are considered in addition to UV/VIS spectral data.[6] Isomerization shift values ($\delta_{cis} - \delta_{trans}$) in NMR are especially important from diagnostic points of view. Details are described in Chapter [27] in this volume.

Scope and Limitation of Method

trans → cis Isomerization. Typical results for the photochemical and thermal isomerization of all-*trans*-β-carotene are given in Table I. The percentage composition of the cis–trans isomeric mixture is determined by estimating the corrected peak area of each isomer on an HPLC chart. Because sterically hindered (7-mono-cis and 7,13'-di-cis) isomers are recognized only in the thermally isomerized mixture, the thermal isomerization technique may be useful for obtaining sterically hindered cis isomer, although the 11-cis isomer has not yet been detected in our experiments.

Separation of cis-β-Carotenes. Among several procedures used for separation, classical adsorption column chromatography can still be useful and is especially valuable for large-scale separation of trans–cis isomeric carotenoids.[1] Although this operation is quite a laborious task even for a one-run operation, this traditional method is the most effective and practical way of dealing with isomeric mixtures of carotenoid in quantity. Preparative thin-layer chromatography (TLC) may be used successfully for preliminary fractionation of a mixture or purification of a crude carotenoid ingredient.

TABLE I
PERCENTAGE COMPOSITION OF CIS/TRANS-ISOMERIZED MIXTURES
FROM all-*trans*-β-CAROTENE[a]

β-Carotene isomer	Content (%)	
	Thermal isomerization	Photoisomerization
all *trans* (unchanged) (I)	26.7	47
Mono-*cis*	46.4	40
7-*cis* (IV)	1.7	—
9-*cis* (II)	22.4	20
13-*cis* (III)	19.5	18.5
15-*cis*	2.8	1.5
Di-*cis*	19.3	9.8
13,15-*cis*	2.3	1.3
9,13-*cis*	6.2	4.0
9,13'-*cis*	3.3	—
9,15-*cis*	7.5	4.5
7,13'-*cis*	Traces	—
Unidentified	7.6	3.2

[a] Isomerization procedures are described in text.

HPLC methods for carotenoids are now available.[7,8] It is the analytical method of choice for separation, quantitation, and structural characterization of naturally occurring carotenoids. In general, delicate separation may be obtained using a chart recorder, but running the experiment should still be repeated several times in a dark room to accumulate sufficient labile specimen for spectral measurements. For this reason classical adsorption column chromatography alone is a more preferable and practical technique in carotenoid chemistry.

A complicated mixture of *cis*-β-carotenes has been resolved on a recording chart of HPLC using a lime column with acetone in *n*-hexane as eluant at ambient temperature (Fig. 1).[5] Vecchi *et al.*[9] independently succeeded in separating 12 *cis*-β-carotenes (3 mono-cis, 4 di-cis, 1 tri-cis, and 4 unidentified cis) by HPLC. They employed an alumina column (Spherisorb A5Y) with *n*-hexane as eluant under a "moisture control device." Compared with our procedure, their method takes a longer time for one run (about 80 min at 20°) and each retention time seems to depend

[7] M. Ruddat and O. H. Will III, this series, Vol. 111, p. 189.
[8] G. R. Beecher and F. Khachik, *in* "Nutrition and Cancer Prevention: Investigating the Role of Micronutrients" (T. E. Moon and M. S. Micozzi, eds.), p. 105. Dekker, New York, 1989.
[9] M. Vecchi, G. Englert, R. Maurer, and V. Meduna, *Helv. Chim. Acta* **64**, 2746 (1981).

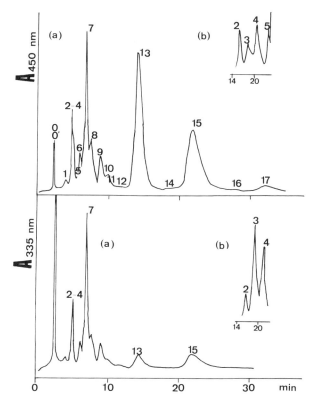

FIG. 1. HPLC separation of thermally isomerized products derived from all-*trans*-β-carotene on a lime column. Eluent: (a) 0.5% acetone in *n*-hexane, (b) 0.1% acetone in *n*-hexane. Flow rate: 1 ml/min. Peak 0, electrocyclized β-carotene (V); peak 0′, 9′-*cis*-electrocyclized β-carotene; peak 2, 13,15-di-*cis*-β-carotene; peak 3, 15-mono-*cis*-β-carotene; peak 6, 9,13′-di-*cis*-β-carotene; peak 7, 13-mono-*cis*-β-carotene (III); peak 8, 9,15-di-*cis*-β-carotene; peak 9, 9,13-di-*cis*-β-carotene; peak 12, 7,13′-di-*cis*-β-carotene; peak 13, unchanged all-*trans*-β-carotene (I); peak 15, 9-mono-*cis*-β-carotene (II); peak 17, 7-mono-*cis*-β-carotene (IV).

markedly on the temperature and the moisture content of the mobile phase. The lime-packed column we used is inexpensive, is durable enough for the present purposes, and we have had no serious trouble with this column throughout our experiments. Although this lime column is not commercially available, unmoistened lime is generally acceptable as in the case of classical adsorption chromatography with a lime column.

From a thermally isomerized mixture of all-*trans*-β-carotene, 17 cis isomers including 4 mono-cis, 5 di-cis, and 1 tri-cis isomer have been well resolved and identified by this method. Although a few supplementary and revised assignments of the peaks were proposed (the 7,9-di-cis isomer was

TABLE II
ULTRAVIOLET/VISIBLE SPECTRUM CHARACTERISTICS OF MONO-cis-β-CAROTENES

β-Carotene isomer	HPLC peak number	UV/VIS in n-hexane				Ratio of ε at λ_{max} to ε at cis peak
		λ_{max} (nm)	ε ($\times 10^{-4}$)^a	cis peak λ_{max} (nm)	ε ($\times 10^{-4}$)^a	
all-trans (I)	13	450	13.90	337	0.80	17.4
Mono-cis						
7-cis (IV)	17	446	13.87	337	0.86	16.1
9-cis (II)	15	445	12.97^b	338	1.13^b	11.5
13-cis (III)	7	442	10.73^b	336	3.88^b	2.8
15-cis	3	448	10.32	335	5.46	1.9
Di-cis						
7,13'-cis	12	440	11.17	334	2.81	4.0

^{a} Calculated from the values for (I) and the quasiequilibrium mixture therefrom at a definite temperature.
^{b} Reported values[4] were revised.

newly added[10]) our method has proved useful for the separation and analysis of nonpolar as well as weak polar carotenoid isomers in nature and in foodstuffs,[11] i.e., in addition to β-carotene isomers[10,12] and its monoepoxide,[13] isomeric canthaxanthins,[14] and neurosporenes[15,16] as well as spirilloxanthins[17] were well analyzed on an HPLC lime column.

An advantage of this method is that rapid and selective resolution can be obtained even in a higher retention time (Rt) region. Sterically hindered isomers possessing the 7-cis geometry (7-mono-cis and 7,13'-di-cis) were thus characterized. They were recognized in a stable state and had UV/VIS characteristics similar to those of the all-trans and 13-cis pigment, respectively (Table II and Fig. 2). Existence of this novel geometry was estab-

[10] Y. Koyama, M. Hosomi, A. Miyata, H. Hashimoto, S. A. Reames, K. Nagayama, T. Kato-Jippo, and T. Shimamura, *J. Chromatogr.* **439**, 417 (1988).
[11] L. A. Chandler and S. J. Schwartz, *J. Food Sci.* **52**, 669 (1987).
[12] Y. Koyama, T. Takii, K. Saiki, and K. Tsukida, *Photobiochem. Photobiophys.* **5**, 139 (1983).
[13] I. Ashikawa, M. Kito, K. Satoh, H. Koike, Y. Inoue, K. Saiki, K. Tsukida, and Y. Koyama, *Photochem. Photobiol.* **46**, 269 (1987).
[14] H. Hashimoto, Y. Koyama, and T. Shimamura, *J. Chromatogr.* **448**, 182 (1988).
[15] Y. Koyama, M. Kanaji, and T. Shimamura, *Photochem. Photobiol.* **48**, 107 (1988).
[16] N. Katayama, H. Hashimoto, Y. Koyama, and T. Shimamura, *J. Chromatogr.* **519**, 221 (1990).
[17] Y. Koyama, I. Takatsuka, M. Kanaji, K. Tomimoto, M. Koto, T. Shimamura, J. Yamashita, K. Saiki, and K. Tsukida, *Photochem. Photobiol.* **51**, 119 (1990).

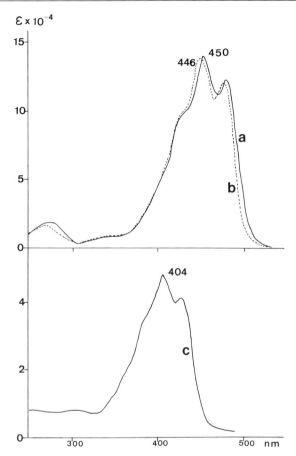

FIG. 2. UV/VIS spectra of all-trans and peculiar cis isomers of β-carotene in n-hexane. a, all-*trans*-β-carotene (**I**); b, 7-mono-*cis*-β-carotene; c, electrocyclized β-carotene (**V**).

lished by formation and separation of electrocyclized β-carotene and by spectral data.[6,18] It was also noted that not only the ratio of ε at λ_{max} to ε at the cis peak, but also the relative adsorption affinity of each mono-cis isomer on HPLC seems to be a function of the molecular shape and hence of the spatial configuration. Thus, the 15-mono cis isomer has the most bent shape and the least adsorptivity, whereas the 7-mono-cis isomer has the most adsorptivity (Table II).

Formation of 6e-Electrocyclized β-Carotene. It is noteworthy to observe the formation of novel thermocyclized isomer (all-trans and 9′-cis) by

[18] K. Tsukida and K. Saiki, *J. Nutr. Sci. Vitaminol.* **28**, 311 (1982).

thermal isomerization of all-*trans*-β-carotene crystals. These isomers were obtained and separated by employing column chromatography, followed by pTLC purification. The 6e (5 → 10)-electrocyclized 5,10-*trans*-all-*trans*-β-carotene (V) and its 9′-mono-cis structures as well as their stereochemistry were determined by spectral data.[6] It is evident that the electrocyclized isomer can be produced from all-*trans*-β-carotene via the 7-cis isomer intermediate involving the 5-, 7-, and 9-C=C double bonds. This is the first report of the formation of the 7-cis isomers as well as the electrocyclized products in the field of highly conjugated alicyclic carotenoids.

It is hoped that our method can offer an important means for studying the general thermochemistry and other geometric isomers of conjugated polyenes.

[27] Proton Nuclear Magnetic Resonance and Raman Spectroscopies of *cis–trans*-Carotenoids from Pigment–Protein Complexes

By YASUSHI KOYAMA

Introduction

The most reliable method, at present, to determine the cis–trans configuration of a carotenoid bound to a pigment–protein complex consists of the following three steps: (1) isolation of the isomeric carotenoid, (2) determination of its configuration by proton nuclear magnetic resonance (^{1}H NMR) spectroscopy, and (3) comparison of its Raman spectra between the extracted and bound states in order to confirm that the configuration is not altered on extraction and to detect minor structural change, if any. X-Ray crystallography does not provide a definitive structure of a low molecular weight pigment ($M_r \sim 10^2$) such as carotenoid, which is bound to the high molecular weight apo complex (10,000–100,000). It is most important to avoid or to minimize artificial isomerization of the isomeric carotenoid during each step of isolation and of NMR and Raman measurements. Therefore it is necessary to develop a high-performance liquid chromatography (HPLC) technique to analyze the cis–trans isomer of the particular carotenoid in order to monitor possible isomerization of the isomer in question.

Isolation of Carotenoids

Extraction Procedure

The extraction procedure is as follows: (1) Acetone is added to the suspension of the pigment–protein complex and shaken; (2) n-hexane is added to the mixture and shaken; (3) water is added to the mixture, shaken, and centrifuged; and (4) the n-hexane layer is collected. The volume ratio of the suspension–acetone–n-hexane–water is typically 1:5:2:2, and it must be modified when perfect separation of the n-hexane layer, the lipid paste, and the water layer is not obtained after centrifugation.[1-3] Saponification is more appropriate for complete extraction of the carotenoid, but it causes isomerization.

Pigment–protein complexes from photosynthetic systems contain comparative amounts of chlorophylls (chlorophylls or bacteriochlorophylls), and chlorophylls function as strong triplet sensitizers. As a result, carotenoids in solution are extremely sensitive to photoisomerization when even a trace amount of chlorophyll is present. On the other hand, purified carotenoids do not isomerize so efficiently, and carotenoids bound to pigment–protein complexes are also resistant to isomerization. Therefore, the crucial point to avoid isomerization resides in the primary extraction procedure, where both carotenoid and chlorophyll are dissolved in solution. The procedure should be conducted below 4° in complete darkness. It is desirable to keep all solutions on ice and to practice every procedure in the dark. (Be cautious about near-infrared light, which we cannot see but is absorbed by chlorophylls.) In the case of a mixture of carotenoid and bacteriochlorophyll, scattered dim light from a He–Ne laser (632.8 nm) can be used.

Complete removal of contaminated chlorophyll and traces of water from the n-hexane layer is essential for subsequent analytical and preparative HPLC procedures and spectroscopic measurements. The n-hexane layer should be passed through a short column packed with sodium sulfate and calcium hydroxide as quickly as possible.

The above extraction procedure should first be attempted in the light to confirm that each step proceeds properly.

---

[1] Y. Koyama, M. Kanaji, and T. Shimamura, *Photochem. Photobiol.* **48,** 107 (1988).
[2] Y. Koyama, I. Takatsuka, M. Kanaji, K. Tomimoto, M. Kito, T. Shimamura, J. Yamashita, K. Saiki, and K. Tsukida, *Photochem. Photobiol.* **51,** 119 (1990).
[3] I. Ashikawa, A. Miyata, H. Koike, Y. Inoue, and Y. Koyama, *Biochemistry* **25,** 6154 (1986).

Analytical and Preparative High-Performance Liquid Chromatography

Adsorption chromatography using calcium hydroxide has been most successful to analyze isomers of nonpolar and less polar carotenoids. The packing material can be purchased from Kishida Chemicals (Osaka, Japan) or from Nakalai Tesque (Kyoto, Japan) as chemicals, but a batch that results in good HPLC resolution must be selected by trial and error. Selection of the particles in the range of 200–300 mesh enables better resolution. The packing material is suspended by sonication in tetrachloroethylene (6 g in 25–27 ml) and packed at the pressure 300 kg/cm^2, and then washed with n-hexane. The eluent depends on the kind of carotenoids: 0.1–2% (v/v) acetone in n-hexane for β-carotene,[4,5] 1% acetone and 17% benzene in n-hexane[1] or 12–15% benzene in n-hexane[6] for neurosporene, 5% n-hexane in benzene for canthaxanthin,[7] and 0.1–1% acetone in benzene for spirilloxanthin.[2]

A LiChrosorb CN-5 column (Merck) and an eluent of 15% acetone in n-hexane or 1% methanol and 10% acetone in n-hexane can be used for the polar carotenoid, peridinin.

The isomer in question can be collected by preparative HPLC. (See Chapter [26] in this volume for the details of the above HPLC techniques.)

^{1}H NMR Spectroscopy

The olefinic ^{1}H signals of carotenoids are used to determine the cis–trans configurations. A set of values of their chemical shifts read from a 400-MHz ^{1}H NMR spectrum usually suffice for configurational determination, but a 500- to 600-MHz ^{1}H NMR spectrum is needed for a carotenoid having a longer conjugated chain consisting of many olefinic protons in a similar environment.

Methods to Avoid Isomerization

Sampling and NMR measurements should be done under red light or in the dark to prevent photoisomerization. The carotenoid sample should be completely free from contaminated chlorophyll as described above. Chloroform, a polar solvent, causes thermal isomerization of some isomeric carotenoids during NMR measurements even at subzero temperatures. Best results have been obtained for the solvent, benzene, at 7° (just

---

[4] K. Tsukida, K. Saiki, T. Takii, and Y. Koyama, *J. Chromatogr.* **245**, 359 (1982).
[5] Y. Koyama, M. Hosomi, A. Miyata, H. Hashimoto, S. A. Reames, K. Nagayama, T. Kato-Jippo, and T. Shimamura, *J. Chromatogr.* **439**, 417 (1988).
[6] N. Katayama, H. Hashimoto, Y. Koyama, and T. Shimamura, *J. Chromatogr.* **519**, 221 (1990).
[7] H. Hashimoto, Y. Koyama, and T. Shimamura, *J. Chromatogr.* **448**, 182 (1988).

above its freezing point). To prevent thermal isomerization, the sample solution should be kept frozen whenever possible.

Assignment of Olefinic ^1H Signals and Configurational Determination

A set of assignments of the olefinic ^1H signals can be obtained, on a purely experimental basis, by the use of (1) the splitting patterns, the values of coupling constants, and the area under each peak of a one-dimensional (1D) spectrum and (2) the correlation peaks of two-dimensional (2D) spectra, i.e., $^1H,-^1H$ and $^1H-^{13}C$ correlation spectroscopy (COSY), distortionless enhancement by polarization transfer (DEPT), and $^1H-^1H$ rotating frame nuclear Overhauser effect spectroscopy (ROESY).

The absolute values of the chemical shifts of the olefinic ^1H signals vary depending on the kind of carotenoid and the solvent, but the direction (sign) and relative values of major "isomerization shifts," i.e., changes in the values of chemical shifts in reference to those of the all-trans isomer when a cis bend is introduced, are not so strongly dependent on those factors and can be used to identify each cis configuration.[8] Table I lists the values of isomerization shifts of β-carotene,[9] canthaxanthin,[7] neurosporene,[1,6] and spirilloxanthin[2] in benzene solution (the values of chemical shifts are given for the all-trans isomers). Table I also shows the values of isomeric β-carotene in chloroform solution for comparison.[9] Configurational determination is straightforward once two sets of chemical shifts (and as a result, the isomerization shifts) are obtained for a cis isomer in question and for the all-trans isomer, which is most stable and can be purified in quantity by recrystallization.

Raman Spectroscopy

Resonance Raman Process

When a Raman process is induced by radiation, the photon energy ($h\nu$) of which is close to that of the $E_v \leftarrow E_i$ electronic transition of the molecule and, as a result, the Raman process is in resonance with the particular transition, it is called "a resonance Raman process." (E_i and E_v stand for the initial and virtual electronic states in a Raman process.) The Raman scattering is strongly enhanced under resonance conditions, by a factor of 10^4-10^6 when compared to an "off-resonance" condition. The vibrational modes that appear in a resonance Raman spectrum are those taking place within the chromatophore, which is responsible for the $E_v \leftarrow E_i$ transition.

---

[8] G. Englert, in "Carotenoid Chemistry and Biochemistry" (G. Britton and T. W. Goodwin, eds.), Proc. 6th Int. Symp. Carotenoids, p. 107. Pergamon, Oxford, 1982.

[9] Y. Koyama, M. Hosomi, H. Hashimoto, and T. Shimamura, *J. Mol. Struct.* **193**, 185 (1989).

TABLE I
CHEMICAL SHIFTS FOR ALL-TRANS ISOMER AND ISOMERIZATION SHIFTS[a] FOR CIS ISOMERS

	Benzene solution															Chloroform solution				
	β-Carotene					Canthaxanthin				Neurosporene				Spirilloxanthin			β-Carotene				
	all-trans	7-cis	9-cis	13-cis	15-cis	all-trans	9-cis	13-cis	15-cis	all-trans	9-cis	13'-cis	15-cis	all-trans	13-cis	15-cis	all-trans	7-cis	9-cis	13-cis	15-cis
7 H / 7' H	6.33	−0.39	+0.04			6.15	+0.03			6.66				6.79			6.17	−0.34			
8 H / 8' H	6.40	−0.14	+0.69	+0.03		6.38	+0.75	+0.03	−0.03	6.42	+0.66			6.53	−0.05		6.12		−0.03	+0.54	
10 H / 10' H	6.35	+0.07	−0.14			6.29	−0.07			6.34 / 6.14	−0.18		+0.03	6.39	−0.06		6.14		+0.07	−0.10	
11 H / 11' H	6.79		+0.26			6.73	+0.20			6.75 / 6.66	+0.26		+0.04	6.78	−0.04[b] / −0.04[b]		6.64	−0.08	+0.10		
12 H / 12' H	6.48		−0.04	+0.60		6.50	−0.08	+0.55		6.48 / 6.42	−0.06	+0.56	+0.05	6.54	+0.52 / −0.03		6.34			+0.53	+0.07
14 H / 14' H	6.32		−0.04	−0.19	+0.54	6.36	−0.03	−0.20 / +0.04	+0.54	6.33[b]		−0.23	+0.57[b]	6.38	−0.25	+0.54	6.24			−0.15	+0.41
15 H / 15' H	6.68		−0.06 / −0.03	+0.25 / −0.07	−0.22	6.70	−0.04	+0.24 / −0.07	−0.23	6.69[b]		−0.10 / +0.26	−0.22[b]	6.73	+0.21 / −0.12	−0.26	6.62		+0.17 / −0.07		−0.24

[a] Isomerization shifts for which absolute values are larger than 0.03 are shown; major isomerization shifts are underlined.
[b] Approximate values.

In the case of carotenoids, vibrational modes taking place in the conjugated polyene chain, which is responsible for the $\pi-\pi^*$ transition, appear. Furthermore, those vibrational modes, in which the displacements of the atoms are in accord with the geometric changes of the molecule on the $E_v \leftarrow E_i$ transition, are strongly enhanced. Therefore, a resonance Raman spectrum is simpler than its counterpart, the infrared absorption spectrum. The above characteristics are regarded as the most important advantages of resonance Raman spectroscopy of a carotenoid bound to a pigment–protein complex. When a probing wavelength in the 400- to 550-nm region is used, only a limited number of vibrational modes of the carotenoid (in-plane vibrations sensitive to the configuration) is seen. Here, the constituents of the apo complex, such as peptides, chlorophylls, and pheophytins, are under off-resonance conditions. Thus, information concerning the configuration of the polyene backbone of the particular carotenoid can be obtained.

Resonance Raman spectroscopy has its intrinsic problems: First, self-absorption of the Raman scattering by the sample solution, due to the $E_v \leftarrow E_i$ transition, can distort the Raman spectrum. The effect can be revealed by concentration dependence of the relative intensities of the Raman lines. Second, the absorption of the photon energy by the $E_v \leftarrow E_i$ transition can cause isomerization and even degradation of the carotenoid. Therefore, resonance Raman spectroscopy of carotenoids is a contradictory technique, because it necessitates recording of a photosensitive pigment by the use of a laser beam. Thus, methods to overcome this problem need to be developed.

Methods to Avoid Effect of Isomerization

First of all, carotenoids should be pure and free from potential sensitizers such as chlorophylls and the degradation products of carotenoids.

Freezing the sample solution in a glass tube by dipping it into liquid nitrogen and recording the Raman spectrum with low power (<30 mW) without tight focusing usually prevents isomerization. Most of the isomers except for extremely unstable ones can last for hours. A drop (0.3 ml) of a 10^{-5}–10^{-3} M solution is enough for this measurement.

For measurements at room temperature, the flow method, i.e., the use of a flow cell or a jet stream squirted from a nozzle, is strongly recommended; isomerization does take place, but the isomerized sample is continuously replaced by fresh sample. The sample solution can be circulated; when the total volume of the solution is large enough, the isomerized sample is diluted and is regarded as negligible for a certain period of time. (Isomerization can be estimated by HPLC.) Typically, 20 ml of a 10^{-5}–10^{-3} M solution is necessary for this type of measurement.

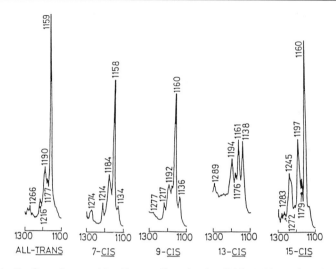

FIG. 1. Configuration-sensitive Raman lines in the 1300–1100 cm^{-1} region: Raman spectra of isomeric β-carotene in n-hexane recorded at liquid nitrogen temperature by the use of the 488.0 nm CW line from an Ar$^+$ laser. [Reproduced with permission from Y. Koyama, in "Carotenoids: Chemistry and Biology" (N. I. Krinsky, M. M. Mathews-Roth, and R. F. Taylor, eds.), Proc. 8th Int. Symp. Carotenoids, p. 207. Plenum, New York, 1989.]

A rotating Raman cell or its alternative, i.e., a rotating ampoule or a rotating glass tube, can be used for room temperature measurements of a stable isomer or an isomer bound to a pigment–protein complex. However, these rotating cells are not recommended for free carotenoids, because the laser beam irradiates the same portion of the sample solution repeatedly when the cell turns around. Typically, 1 ml of a 10^{-5}–10^{-3} M solution is necessary for this measurement.

Raman Measurements and Determination of Configuration

The freezing or flow method mentioned above should be used for carotenoids free in solution, while the freezing or rotating method is usually used for a carotenoid bound to a pigment–protein complex. A conventional scanning-type Raman spectrometer, for which ~30 min is necessary for one recording, can be used only for the freezing method. A Raman spectrometer equipped with a multichannel detector, i.e., an image-intensified diode array or a charge coupled device (CCD) detector, is necessary for the flow or rotating method in order to reduce the time for recording; a spectrum with a very high signal-to-noise (S/N) ratio is obtained in less than 1 min of exposure. (The combination of the flow method and a Raman spectrometer equipped with a multichannel detector is absolutely necessary for excited-state Raman measurements,[10] which are not covered in this chapter.)

Spectral comparison of an isomer isolated from a pigment–protein complex, the configuration of which has been determined by ^{1}H NMR spectroscopy, with the original isomer bound to the pigment–protein complex ensures that no artificial isomerization has taken place during the isolation procedures and spectroscopic measurements. It leads to a straightforward and definitive configurational assignment. Some distortion of the polyene backbone, due to the binding of the carotenoid, can be detected in this step as a difference in relative Raman intensities.[1,11]

Further, Raman spectroscopy can predict at least the all-trans and some mono-cis configurations. Figure 1 shows the spectral patterns in the 1300–1000 cm^{-1} region for isomeric β-carotene. A strong single peak around 1160 cm^{-1} is the key Raman line of an all-trans isomer. The relative intensity of the peak around 1140 cm^{-1} vs the one around 1160 cm^{-1} increases in the order 7-cis < 9-cis < 13-cis, and they are comparable for the 13-cis isomer.[12] A medium peak around 1240 cm^{-1} is the key Raman line of a 15-cis isomer.[1,2,13]

[10] H. Hashimoto and Y. Koyama, *J. Phys. Chem.* **92,** 2101 (1988).
[11] K. Iwata, H. Hayashi, and M. Tasumi, *Biochim. Biophys. Acta* **810,** 269 (1985).
[12] Y. Koyama, I. Takatsuka, M. Nakata, and M. Tasumi, *J. Raman Spectrosc.* **19,** 37 (1988).
[13] Y. Koyama, T. Takii, K. Saiki, and K. Tsukida, *Photobiochem. Photobiophys.* **5,** 139 (1983).

[28] Electron Paramagnetic Resonance Studies of Carotenoids

By HARRY A. FRANK

Introduction

It is well known that carotenoids act as protective devices against the irreversible photodestruction of the cells of bacteria, plants, and animals.[1] It is generally accepted that the triplet states of carotenoids are involved in the photoprotective mechanism.[2] The triplet states of carotenoid molecules

[1] N. I. Krinsky, *in* "Photophysiology" (A. C. Giese, ed.), Vol. III, pp. 123–195. Academic Press, New York, 1968.
[2] R. J. Cogdell and H. A. Frank, *Biochim. Biophys. Acta* **895,** 63 (1987).

are poorly understood, however, owing to the fact that the direct population of triplet states of isolated carotenoids via singlet-triplet intersystem crossing is not a very efficient process. The low yields of intersystem crossing make investigations of carotenoid triplets feasible only through the use of energy donors that sensitize the formation of the carotenoid triplet state.[3] The triplets are then detected using pulse radiolysis,[3] flash photolysis,[3,4] or electron paramagnetic resonance[5] (EPR) spectroscopic techniques. Pulse radiolysis and flash photolysis optical spectroscopic experiments have been used to determine triplet-triplet absorption spectra of carotenoids, quantum yields of carotenoid triplet formation, and relative triplet state energies (e.g., by a judicious choice of donors of known triplet energies).[3,4] EPR detection of carotenoid triplet states provides information about three other important factors.

1. The mechanism of carotenoid triplet state formation: This information is derived from the polarization pattern of the triplet state EPR signals. Because essentially all carotenoid triplets are formed via triplet energy transfer from a suitable triplet energy donor, the polarization pattern of the carotenoid triplet state signals reveals whether the donor triplet originated via the intersystem crossing mechanism or via the radical-pair mechanism of triplet energy formation.[5]

2. The structure of the carotenoid: The zero-field splitting parameters, $|D|$ and $|E|$, provide a sensitive probe of the molecular structure of the carotenoid by assessing the extent of dipolar interaction between the two electrons that comprise the triplet state.[6]

3. The geometry of the carotenoid molecule relative to other chromophores: Advantage can be taken of the inherent anisotropy of the dipolar interaction that exists in molecular triplet states to map the orientation of the carotenoid relative to other chromophores whose orientations have also been determined (e.g., bacteriochlorophyll).[7,8] This requires the use of an oriented sample (e.g., membranes dried onto Mylar strips or single crystals[7,9]) or polarized light (i.e., magnetophotoselection[8]).

This chapter illustrates how each of these three factors can be addressed

[3] R. Bensasson, E. J. Land, and B. Maudinas, *Photochem. Photobiol.* **23,** 189 (1976).
[4] P. Mathis and J. Kleo, *Photochem. Photobiol.* **18,** 343 (1973).
[5] H. A. Frank, J. D. Bolt, S. M. De B. Costa, and K. Sauer, *J. Am. Chem. Soc.* **102,** 4893 (1980).
[6] H. A. Frank, B. W. Chadwick, J. J. Oh, D. Gust, T. A. Moore, P. A. Liddell, A. L. Moore, L. R. Makings, and R. J. Cogdell, *Biochim. Biophys. Acta* **892,** 253 (1987).
[7] H. A. Frank, J. Machnicki, and P. Toppo, *Photochem. Photobiol.* **39,** 429 (1984).
[8] W. J. McGann and H. A. Frank, *Biochim. Biophys. Acta* **807,** 101 (1985).
[9] D. E. Budil, S. S. Taremi, P. Gast, J. R. Norris, and H. A. Frank, *Isr. J. Chem.* **28,** 59 (1988).

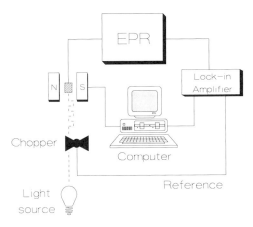

FIG. 1. Experimental diagram of the double-modulation EPR spectrometer used for detecting carotenoid triplet states. The output of the EPR is routed to a lock-in amplifier referenced to the frequency of the light modulation. The output of the lock-in is fed to a computer, which serves as the recorder and also sweeps the magnetic field. The light source is a 1000-W xenon arc lamp filtered by ~5 cm of water in a Pyrex bottle. Typically the magnetic field modulation frequency is 100 kHz, and the light modulation frequency is 10–100 Hz.

and uses examples from the photosynthetic literature, which is now replete with carotenoid triplet state EPR spectroscopic observations.[2]

Materials and Methods

The optimal method for the EPR detection of carotenoid triplet states is schematically represented in Fig. 1. The unfiltered DC output of the EPR magnetic field modulation amplifier is fed directly to an external lock-in amplifier that is referenced to the frequency of a modulated excitation light source. This experimental arrangement offers a substantial improvement in the sensitivity of triplet state detection over the traditional "light-minus-dark" spectral approach to observing light-induced signals and is necessary for the detection of carotenoid triplets, probably because of two factors: (1) The steady state triplet population of carotenoids is low owing to their typically very rapid (~10 μsec) decay back to the ground state;[10] and (2) carotenoid triplets may exhibit substantial spin-lattice relaxation between triplet spin sublevels, which may diminish the EPR signal amplitudes.[11] Thus, a "double-modulation" (field and light modulation) technique, although not obligatory, facilitates detection of carotenoid triplet states.

[10] S. S. Taremi, C. A. Violette, and H. A. Frank, *Biochim. Biophys. Acta* **973**, 86 (1989).
[11] W. J. McGann and H. A. Frank, *Chem. Phys. Lett.* **121**, 253 (1985).

Results and Discussion

The first EPR observation of a carotenoid triplet state was made on the B800–850 light-harvesting complex purified from the photosynthetic bacterium, *Rhodobacter* (previously *Rhodopseudomonas*) *sphaeroides,* wild-type strain 2.4.1.[5] Subsequently, carotenoid triplet state signals were detected in whole cells, chromatophores (photoactive membrane vesicles), several different light-harvesting and reaction center pigment–protein complexes prepared from a wide variety of photosynthetic organisms, and in synthetic, covalently linked carotenoporphyrin molecules.[6,12] The initial unambiguous assignment of the triplet state EPR signals to carotenoids was made in the photosynthetic bacterial reaction center and deduced from the following observations: (1) The triplet state EPR signals displayed the so-called "radical-pair" triplet state spin polarization pattern (*eaa eea,* where *e* denotes a signal in emission and *a* denotes a signal in absorption). This pattern (or its inverse, *aee aae*) is observed only when triplets are formed via a charge separation-recombination (i.e., radical-pair) mechanism or when they coherently quench triplets formed via this mechanism.[5] The reaction center-bound carotenoid had been shown from previous optical triplet-triplet absorption experiments to be quenching the primary donor triplet.[13] It is well known that the primary donor triplet state is born via the radical pair mechanism;[14] (2) no triplet state signals displaying zero-field splitting parameters similar to those observed in the *R. sphaeroides* wild-type strain 2.4.1 preparations were observed in samples from the carotenoidless mutant *R. sphaeroides* R-26;[5] (3) the temperature dependence of the triplet state EPR signals from the carotenoid-containing *R. sphaeroides* wild-type strain 2.4.1 was precisely the same as that observed for the optically detected, triplet-triplet absorption signals belonging to the reaction center-bound carotenoid and reported by Cogdell *et al.*[13] It was observed that the carotenoid triplet state signal amplitudes increased relative to the primary donor triplet signals as the temperature was raised from 10 to 100 K;[13] (4) the final piece of evidence that confirmed that the observed triplet state EPR signals were arising from carotenoids came from experiments carried out on reaction centers from the carotenoidless mutant *R. sphaeroides* R-26 that had been reconstituted with the carotenoid, spheroidene.[15] This sample displayed an EPR spectrum (see Fig. 2) identical to that of *R. sphaeroides* wild-type strain 2.4.1, which naturally con-

---

[12] H. A. Frank, J. Machnicki, and M. Felber, *Photochem. Photobiol.* **35,** 713 (1982).
[13] R. J. Cogdell, T. G. Monger, and W. W. Parson, *Biochim. Biophys. Acta* **408,** 189 (1975).
[14] A. J. Hoff, *Phys. Rep.* **54,** 75 (1979).
[15] B. W. Chadwick and H. A. Frank, *Biochim. Biophys. Acta* **851,** 257 (1986).

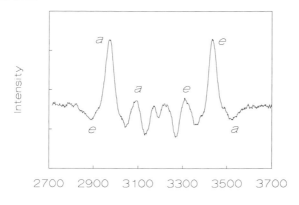

FIG. 2. Triplet state EPR signals from the carotenoid, spheroidene, reconstituted into reaction centers isolated from *R. sphaeroides* R-26. Note this spectrum displays the radical-pair polarization pattern, *eaa eea*. The experimental conditions were as follows: Magnetic field modulation frequency, 100 kHz; modulation amplitude, 25 G; microwave frequency, 9.054 GHz; microwave power, 20 mW; field modulation lock-in amplifier sensitivity, 2.5 mV; light modulation lock-in amplifier sensitivity, 10 mV; time constant, 3 sec light modulation frequency, 33 Hz; temperature, 90 K.

tains spheroidene, thus confirming that the EPR signal was due to the bound carotenoid.

Mechanism of Carotenoid Triplet State Formation

With few exceptions[2] all carotenoid triplet states are formed via energy transfer from a triplet energy donor. The reaction is

$$^3Donor^* + {}^1Car \rightarrow {}^1Donor + {}^3Car^*$$

where ^1Car and ^1Donor are the ground state singlets, and $^3Car^*$ and $^3Donor^*$ are the excited triplet states of the carotenoid and the energy donor, respectively. Because, the reaction involves a change in the spin multiplicity of each molecular state, the exchange (or Dexter) mechanism[16] is usually invoked to account for the process. Moreover, the polarization pattern observed for the carotenoid triplet state signals reveals whether the donor triplet was born via the intersystem crossing mechanism or via the radical-pair mechanism of triplet energy formation. Figure 2 shows the triplet state signals from spheroidene in the reaction center of *R. sphaeroides* wild-type strain 2.4.1. Note the polarization pattern is *eaa eea*,

---

[16] D. L. Dexter, *J. Chem. Phys.* **21**, 836 (1953).

indicating that it was transferred from the primary donor, whose triplet was formed via the radical-pair mechanism. Figure 3 shows the triplet state spectrum of the same molecule, spheroidene, in the B800–850 light-harvesting complex from *R. sphaeroides* wild-type strain 2.4.1. Note the polarization pattern in this triplet is *eae aea*, indicating that it was transferred from a triplet energy donor, presumably bacteriochlorophyll, whose triplet was formed by intersystem crossing. The polarization patterns offer an unambiguous assignment of the mechanism of formation of the triplet state that ultimately resides on the carotenoid. In whole cells of the photosynthetic bacteria, this provides a convenient method for deciding whether the observed carotenoid triplet signal originates in the reaction center or in the light-harvesting pigment–protein complexes.

Structure of Carotenoid

The zero-field splitting parameters of carotenoid triplet states correlate with the extent of π electron conjugation in their structures. For example, Table I gives the $|D|$ and $|E|$ values for the B800–850 light-harvesting complexes obtained from a series of photosynthetic bacterial preparations.[6] The data can be understood in terms of variations in the extent of dipole–

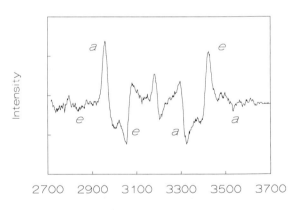

FIG. 3. Triplet state EPR signals from the carotenoid, spheroidene, in the B800–850 light-harvesting complex from *R. sphaeroides* wild-type strain 2.4.1. Note this spectrum displays an intersystem crossing mechanism polarization pattern, *eae aea*. The experimental conditions were as follows: Magnetic field modulation frequency, 100 kHz; modulation amplitude, 29 G; microwave frequency, 9.050 GHz; microwave power, 16 mW; field modulation lock-in amplifier sensitivity, 2.5 mV; light modulation lock-in amplifier sensitivity, 2.5 mV; time constant, 30 sec; light modulation frequency, 100 Hz; temperature, 100 K.

TABLE I
ZERO-FIELD SPLITTING PARAMETERS OF B800–850 COMPLEXES FROM DIFFERENT PREPARATIONS OF PHOTOSYNTHETIC BACTERIA[a]

Sample	$|D|$ (cm^{-2})	$|E|$ (cm^{-2})	Triplet assignment	Number of conjugated double bonds
Rhodobacter sphaeroides GA	0.0365 ± 0.0002	0.0035 ± 0.0002	Neurosporene	9
Rhodobacter sphaeroides wild type (anaerobically grown)	0.0324 ± 0.0002	0.0036 ± 0.0002	Spheroidene	10
Rhodobacter sphaeroides wild type (aerobically grown)	0.0318 ± 0.0002	0.0032 ± 0.0002	Spheroidenone	11
Rhodopseudomonas acidophila 7750	0.0279 ± 0.0003	0.0029 ± 0.0003	Rhodopin	13

[a] From Ref. 6.

dipole interaction arising from different amounts of π electron conjugation in the various carotenoids. The least conjugated carotenoid, neurosporene (nine conjugated double bonds), was found to have the largest $|D|$ value (largest dipole interaction). The $|D|$ values decrease in order of increasing π electron conjugation, with rhodopin (13 conjugated double bonds) having the smallest value. This demonstrates the utility of the zero-field splitting parameters in deducing the structure of the EPR-detected carotenoid.

Geometry of Carotenoid Molecule Relative to Other Chromophores

Owing to the inherent anisotropy of the dipolar interaction that exists in the triplet state, EPR spectra of carotenoids can be used to determine the orientations of carotenoids relative to other chromophores (e.g., bacteriochlorophylls). These experiments require the use of an oriented sample or polarized light. In the former case, it was shown that large orientation effects on the EPR spectra of spheroidene were observed from chromatophores of *R. sphaeroides* wild-type strain 2.4.1 dried onto Mylar strips.[7] The data yielded the projections of the carotenoid triplet state magnetic axes with respect to the normal to the membrane plane.[7] In the latter case, magnetophotoselection was used to determine the orientations of the primary donor and carotenoid magnetic axes with respect to each other.[8] These data are particularly important because the triplet state EPR spectrum of crystalline β-carotene has been reported, and the assignment of the triplet state magnetic axes with respect to the molecular symmetry of the carotenoid has been made.[17] Thus, it is now possible to deduce carotenoid molecular geometries from the orientation effects observed in their triplet state EPR spectra.

Conclusions

EPR detection of carotenoid triplet states continues to be an important spectroscopic tool for the elucidation of the mechanism of triplet state formation, the determination of carotenoid structures *in vivo,* and the specification of carotenoid molecular geometries. All of these factors are central to the issue of how carotenoids protect biological systems from photodestruction.[2] This brief chapter has not dealt with such questions as concern (1) the specific molecular features of carotenoids contributing to triplet energy transfer; (2) how the zero-field splitting parameters of carotenoids are modulated by the various stereochemical isomeric forms they can adopt; and (3) how different environments (i.e., different biological systems such as the photosynthetic pigment–protein complexes) affect the spectroscopic and functional properties of carotenoids. These questions remain for future investigations.

Acknowledgments

I thank Dr. John Bolt for sharing in the initial discovery of the triplet state EPR spectra of carotenoids, which was made in the laboratory of Professor Kenneth Sauer at the University of California, Berkeley. This work is currently supported by grants from the National Institutes of Health (GM-30353), the Competitive Research Grants Office of the U.S. Department of Agriculture (88-37130-3938), the NATO Scientific Affairs Division (880107), and the University of Connecticut Research Foundation.

[17] J. Frick, J. U. Von Schütz, H. C. Wolf, and G. Kothe, *Mol. Cryst. Liq. Cryst.* **183**, 269 (1990).

[29] Identification of Carotenoid Pigments in Birds

By JOCELYN HUDON and ALAN H. BRUSH

One extraordinary feature of avian carotenoids is their packaging. Feathers are unique to birds and form the interface between a remarkable metabolic machine and an unforgiving environment. The colors and patterns of their display are exceedingly important in individual and species recognition and in myriad aspects of communication. Historically, there has been an interest in both the molecules themselves and the nature of their metabolism, deposition, and regulation. Early workers attempted to resolve the cellular basis of deposition of the pigments then known simply as lipochromes. From lessons learned by the manipulation of the plumage color of display birds, interest moved to more biologically relevant studies of the nature of important plumage changes, and potential hormonal and enzymatic control. Progress with these issues has depended to various degrees on developments in peripheral areas (e.g., chromatography and spectrophotometry) and specific techniques to extract and recover pigments from the feathers. It is now possible to recover most of the carotenoids in feathers in a native state, and expeditiously characterize them spectrophotometrically and chemically. The convenience of thin-layer chromatography (TLC) and the resolution and reproducibility of high-performance liquid chromatography (HPLC) have greatly expedited the process of separation and identification of pigment mixtures, common in birds.

Carotenoids in birds are not limited to the feathers. The skin of various exposed parts, such as legs, feet, bill, caruncles, and comb, may contain carotenoids, and the skin color of chickens is a marketing consideration. The type and age of the feed influences the chicken skin color and can influence public acceptance of the product. Commercial poultry houses pay considerable attention to the relationship between feed and skin or yolk color.

Analysis of carotenoids falls under the broad theme of natural product chemistry. Traditionally, this activity has been important to the food industry, pharmaceuticals, cosmetics and animal feeds. For various reasons it has been necessary to identify or manipulate carotenoids in these products. Color can be changed directly or indirectly and many clever techniques have been developed to add and control carotenoid levels. Ornithologists have had a long-term interest in the nature of carotenoids in feathers. Workers in the 1930s were "natural product chemists"; among the most productive O. Völker and students. Their goal often was to identify the source of colors in feathers and relate this to microscopic features of anatomy. For many years this was a tedious process as they lacked chromatography and had only simple spectrophotometers. The application of information on carotenoid chemistry to biological problems and natural history events did not begin until the early 1960s with the work of D. L. Fox in the United States. Recently investigators have been able to structure questions that reflect biological processes and test specific hypotheses concerning both the evolution of carotenoid usage and the metabolic processes and regulation involved (reviewed by Brush[1]).

[1] A. H. Brush, *FASEB J.* **4**, 2969 (1990).

Carotenoids have been identified in integumentary structures such as feathers, bill, and feet of birds from about 19 families, and 10 of the extant orders,[2] and are presumed to be widespread in the group. Their study lags behind that of other groups of vertebrates, notably fish, because of difficulties with feather material. The conventional means of extraction, such as acetone, do not work, and rather harsh means of extraction of carotenoids have had to be devised. This is problematical as some feather carotenoids are sensitive to oxidation or extremes of pH. It is likely that investigators have misidentified pigments that were modified in the extraction process.

Investigators must balance the need to extract all the pigment with the potential for damage or chemical insult to the carotenoid products. This sets the study of feather carotenoids apart from work on other natural products. In the past it was necessary to destroy the feather structure totally to extricate the pigment. The classic guideline to the manipulation of biological material is that by Booth.[3] It was full of practical tips for extraction and identification of carotenoids, but did not deal with animal material such as feathers. Booth has been supplanted gradually by publications such as that by Manz,[4] produced for Hoffmann-La Roche (Nutley, NJ). This, and similar in-house publications have served various commercial and industrial research laboratories. The common methodologies used for carotenoids were evaluated by De Ritter and Purcell[5] and Britton.[6] These contributions have been extraordinarily helpful as the primary literature is widely scattered. The separation and purification of carotenoids, especially as applied to natural products, have been standardized by Hostettman *et al.*[7] and Ruddat and Will.[8] Together, these authors present a comprehensive assessment of useful techniques, methods of separation, and criteria for identification. With over 500 known carotenoids, the identification of metabolites, intermediates and products has become highly sophisticated.

Our approach has not involved special technical innovations. We have

[2] A. H. Brush, in "Carotenoids as Colorants and Vitamin A Precursors: Technical and Nutritional Applications" (J. C. Bauernfeind, ed.), p. 539. Academic Press, New York, 1981.

[3] V. H. Booth, "Carotene: Its Determination in Biological Materials." Heffer, Cambridge, England, 1957.

[4] U. Manz, "Methods for the Determination of Carotenoids in Carotenoid Preparations, Premixes, Feedstuffs and Egg Yolks." Hoffmann-La Roche, Basel, 1983.

[5] E. De Ritter and A. E. Purcell, in "Carotenoids as Colorants and Vitamin A Precursors: Technical and Nutritional Applications" (J. C. Bauernfeind, ed.), p. 815. Academic Press, New York, 1981.

[6] G. Britton, this series, Vol. 111, p. 113.

[7] K. Hostettman, M. Hostettman, and A. Marston, "Preparative Chromatographic Techniques: Applications in Natural Product Isolation." Springer-Verlag, New York, 1986.

[8] M. Ruddat and O. H. Will III, this series, Vol. 111, p. 189.

adapted standard techniques for the isolation and characterization of carotenoids to study a series of natural situations. For example, we have been able to document the chemical changes responsible for seasonal plumage change, sexual dimorphism, and subspecific differences in species selected because of their particular natural history. The chemical information has also been useful in the reconstruction of the underlying metabolic steps involved. This success has been subsequently extended to speculation regarding both metabolic regulation and genetic control. These are clearly not the only applications for these techniques, and they are now finding expression in the fields of endocrinology, behavior, and ecology.

Extraction of Feather Carotenoids

Prior to pigment extraction, the feathers should be washed with a commercial detergent in water (ca. 0.1%, v/v), followed by extensive rinsing with tap water, and, after drying, petroleum ether. Except in unusual cases, such as plumage blushes,[9] feather color should remain unaltered by the washes. It is also advised, especially for feather parts of low carotenoid content (e.g., pigmented feather shafts, as in *Colaptes auratus*), to wash the feathers further with acetone and methanol. The wash with methanol removes material that otherwise would transfer to the epiphase and may interfere with the characterization of the carotenoids. Minute amounts of carotenoids may be extracted during the latter washes. These should be saved and processed separately.

Carotenoids in feathers can be extracted in a number of ways. The preferred method seems to vary with the laboratory considered. Carotenoids have been extracted with warm acetone after feather treatment (incubation) in a saturated solution of lithium bromide (LiBr).[10] In our hands, extraction by this method was inconvenient and incomplete. An alternative procedure, long the standard for the extraction of feather carotenoids, employs alkaline ethanol (ca. 10% NaOH, w/v) over a steam bath.[10] This procedure is expeditious. However, some carotenoid structures may be chemically altered. For example, the acidogenic carotenoids are often partially transformed to their respective acid derivatives. We have found that alkaline ethanol also chemically alters the canary xanthophylls, although the precise molecular transformation was not elucidated. A more consequential problem with alkaline ethanol is that many extracted carotenoids transfer only with great difficulty to hexane in a separatory funnel (see below). This is especially problematical for the acidogenic carotenoids.

[9] J. Hudon and A. H. Brush, *Condor* **92**, 798 (1989).
[10] H. Brockmann and O. Völker, *Hoppe-Seyler's Z. Physiol. Chem.* **224**, 193 (1934).

In contrast to the above two methods, extraction with acidified pyridine provides high recovery and effects no apparent chemical change (unpublished observation, 1989). Compared to ethanol, pyridine mixes well with hexane and allows a smooth transfer of carotenoids, even acidogenic carotenoids, to hexane. Also, the procedure allows the recovery of feathers that are structurally intact. In contrast, alkaline ethanol completely destroys the feather structure. After extraction with acidified pyridine the decolorized feathers can be examined for remaining pigments (e.g., melanins), structural colors and other morphological features.

Acidified pyridine is prepared by adding about 4 drops of concentrated HCl per 100 ml of pyridine (final pH 2.5–3). The extraction is done on a hot plate at temperatures close to the boiling point (ca. 110°) of the solvent for about 4 hr, or until the feathers are decolorized of carotenoids. We routinely use beakers covered with watch glasses placed in a flow of air in a fume hood. Under these conditions, the solvent vapors are successively cooled, condensed, and distilled back to the solvent. The conditions do not appear to alter the chemical structure of the extracted pigments. However, the procedure is relatively untried, and some of its effects on carotenoids may be unknown.

Once extracted, the pigments are transferred rapidly to hexane (the epiphase) in a separatory funnel by addition of distilled water and vigorous shaking. The aqueous hypophase can be washed with further hexane or discarded, depending on the efficiency of pigment transfer. The carotenoid-containing epiphase is washed repeatedly with distilled water. Sometimes, fluff forms at the interface between the epiphase and the hypophase in the separatory funnel. This fluff can generally be eliminated, or at least greatly reduced, by adding salt (NaCl) to the funnel, or washing the feathers in methanol prior to extraction. The epiphase is finally concentrated under a flow of nitrogen, and the pigments stored over anhydrous sodium sulfate under nitrogen in the dark at 4°. With care, these preparations may last many months. The remaining feather material is washed with water until the pyridine is gone, and air dried. These feathers are then available for futher examination.

The extraction of carotenoids from soft parts can be accomplished by more conventional means such as acetone extraction. But acidified pyridine is also convenient. Carotenoids in soft parts differ from feather carotenoids in a few respects. Skin carotenoids are often esterified, and may include carotenes.[11] Saponification may then be necessary to free skin carotenoids (esterified hydroxycarotenoids, including acidogenic carotenoids) of chemically bonded fatty acids, and permit subsequent identifi-

---

[11] B. Czeczuga, *Comp. Biochem. Physiol. B* **62B**, 107 (1979).

cation. Saponification is accomplished in alkaline (ca. 5% KOH, w/v) methanol.[10] Feather carotenoids may also be esterified (unpublished observation). Sometimes, substances that interfere considerably with chromatography are extracted with the carotenoids in soft parts (e.g., the legs of *Fratercula arctica*) (unpublished observation). We have not been able to eliminate these contaminants from pigment extracts.

Quantitation

The total carotenoid content of feathers (mg/g of feather) can be calculated with the formula:

$$A_{450\,nm} \times \text{volume of extract (ml)} \times 10 / [E_{1\,cm}^{1\%} \times \text{feather mass } (g)]$$

where $A_{450\,nm}$ is the absorption at 450 nm of the pigment extract, and $E_{1\,cm}^{1\%}$, is the absorption coefficient of the carotenoids in the same solvent (an $E_{1\,cm}^{1\%}$ of 2500 is generally used for pigment mixtures in hexane[4]). The feathers should be weighed prior to extraction. Care should be taken to avoid sodium sulfate suspensions during spectrophotometry, because they adversely affect the absorption readings.

Separation of Mixtures

Carotenoid pigments can readily be isolated by preparative TLC. TLC, in general, provides both efficacious pigment separation and convenience. Preparative TLC, in addition, permits the isolation of small amounts (microgram range) of individual carotenoids. A solvent mixture of hexane–acetone (2:1, v/v) generally resolves most feather pigments on silica gel. The TLC plates should be prerun in the development solvent prior to pigment application. Following development the pigment spots are scraped off the plates, eluted off the sorbent with acetone, and transferred to hexane. For the isolation of large quantities (milligrams) of carotenoids, chromatographic columns should be used (see Britton[6]). Usually, these columns are less convenient and have lower resolving powers than TLC. For a discussion of the principles of carotenoid migration, the researcher is referred to the work of Britton.[6]

Identification

Carotenoid identifications are based on a combination of features, including relative mobility on TLC and HPLC, comparison with known standards, color, chemical tests, and spectral characteristics.

We routinely subject pigment extracts to TLC on both silica and aluminum oxide gels. Various proportions of hexane and acetone are generally sufficient to separate the pigments. The conditions can be varied to establish comigration of bands. The aluminum oxide plates are useful in the identification of 3-hydroxy-4-ketocarotenoids because the pigments bind tightly to the origin on the sheets, while the other pigments migrate as on silica. High atmospheric humidity can nullify this effect, however. Reversed-phase TLC plates are also available and often helpful. In our hands these special plates did not contribute significantly to carotenoid separations. We routinely use precoated plates mounted on flexible films. The support is not affected by any of the solvents used, and the plates are inexpensive and can be stored. The pigments can be visualized directly on TLC and a subjective color assigned.

HPLC offers greater resolving power and reproducibility than TLC. It is useful to separate unambiguously pigments that cannot be resolved on TLC. This is true, for example, of zeaxanthin and lutein. Unfortunately, HPLC is of little value to characterize unknown pigments, unless the equipment features a diode array detector. A diode array detector allows the visualization of the spectral characteristics of the pigments as they exit the HPLC column, and can provide important clues to the nature of the pigment chromophore (e.g., number of double bonds, and end-ring structure).

Successful separation of most feather carotenoids can be achieved with a Zorbax ODS (Du Pont Co., Wilmington, DE) column (4.6-mm i.d. × 25.0 cm), and a mixture of acetonitrile-dichloromethane-methanol (70:20:10, v/v)[12] (Fig. 1). Unfortunately, the system does not resolve well the 3-hydroxy-4-ketocarotenoids, which migrate as broad, unresolved peaks (astaxanthin in Fig. 1). This could be related to the incomplete inertness of the matrix, and the interaction of the pigments with that matrix.

Chemical tests used for identification include the reduction of carbonyl groups with sodium borohydride in methanol,[13] the acetylation of hydroxyl groups with acetic anhydride in dry pyridine,[14] and the oxidation of acidogenic carotenoids in alkaline (KOH) methanol.[10] Allylic hydroxyl groups of carotenoids with ε end groups can be tested through methylation in acidified (HCl) methanol.[15] For these reactions, the carotenoids in hexane are dried to a greasy residue and redissolved in the appropriate solvent. The chemical reactants are then added to the pigment solution. To establish the number of functional groups in a molecule, partial reactions

---

[12] H. J. C. F. Nelis and A. P. De Leenheer, *Anal. Chem.* **55**, 270 (1983).
[13] N. I. Krinsky and T. H. Goldsmith, *Arch. Biochem. Biophys.* **91**, 271 (1960).
[14] B. P. Schimmer and N. I. Krinsky, *Biochemistry* **5**, 1814 (1966).
[15] A. L. Curl, *Food Res.* **21**, 689 (1956).

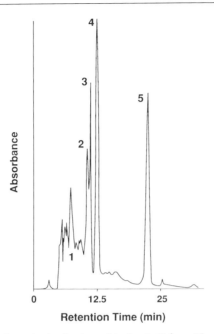

FIG. 1. Separation of standards of astaxanthin (peak 1) from *Homarus americanus,* and lutein (peak 2), zeaxanthin (peak 3), canthaxanthin (peak 4), and echinenone (peak 5) obtained from Hoffmann-La Roche on a Zorbax ODS (Du Pont) column (4.6 mm × 25.0 cm). Eluent: acetonitrile–dichloromethane–methanol (70:20:10). Flow rate, 0.5 ml/min; detection at 450 nm.

may be carried out. Depending on the identity of the two end rings, from one to three different products can be obtained. An extensive discussion of the use of chemical reactions for identification is presented by Liaaen-Jensen.[16]

Information on the structure of a pigment is provided directly by the absorption spectrum of the pigment. In particular, the number of conjugated double bonds can be determined from the position of the absorption peaks of the pigment. For spectrophotometric information on a variety of carotenoids, the researcher is referred to Britton.[6]

Carotenoids in Birds

Carotenoids produce the majority of bright yellow, orange, and red hues found in birds, and are often present as mixtures. Generally, plumage color is a good indicator of the kinds of carotenoids deposited. Typically,

---

[16] S. Liaaen-Jensen, *in* "Carotenoids" (O. Isler, ed.), p. 61. Birkhäuser, Basel, 1971.

4-ketocarotenoids produce the red hues, while their metabolic precursors, in particular lutein, produce the yellow hues. Other common yellow pigments are the canary xanthophylls, which we have identified as 3-keto-ε-carotenoids. Orange hues usually result from the combination of red and yellow pigments.

Astaxanthin, canthaxanthin, and phoenicoxanthin are common 4-ketocarotenoids of birds, with doradexanthin and echinenone as regular additions. Astaxanthin, phoenicoxanthin, and doradexanthin are acidogenic carotenoids, and will bind tightly to aluminum oxide TLC plates, On TLC (silica), a complex mixture of carotenoids would develop as follows (from fastest to slowest, with approximate R_f values): echinenone (0.75), canthaxanthin (0.50), ε,ε-carotene-3,3'-dione (0.43), phoenicoxanthin (0.41), 3'-hydroxy-ε,ε-caroten-3-one (0.37), astaxanthin (0.35), doradexanthin, (0.33), lutein or zeaxanthin (0.30).

Another red pigment seen in birds is rhodoxanthin. It is a 3-ketoretrodehydrocarotenoid, and is probably not metabolically related to the other red carotenoids. Rhodoxanthin is characterized on TLC by three bands (R_f values of 0.37, 0.39, and 0.41) of slightly different hues. These forms are interconvertible in solution.

The carotenoids in some birds do not match investigated pigments. We have encountered several unknown pigments. Red pigments in some carduelines *(Carpodacus roseus, Pyrrhula pyrrhula)* display mobilities on TLC unlike those of known carotenoids. An unidentified red pigment, and other pigments, are found in the legs of the pigeon *Columba livia.* Yellow pigments in passerines *(Coccothraustes verpertina)* and woodpeckers *(Colaptes auratus)* do not match known xanthophylls, including canary xanthophylls. These examples illustrate an incomplete knowledge of avian carotenoids.

Parrots present the most unusual condition among birds as they do not deposit carotenoids in their red and yellow feathers. This finding is both unexpected and remarkable. In all parrots so far investigated the pigments in the plumage differ chromatographically and spectrally from all known feather carotenoids. When compared to carotenoids, for a given feather color, the absorption peaks of the parrot pigments are shifted to shorter wavelengths. A typical pigment extract of red feathers *(Ara macao)* has peaks at 407, *430,* 458 nm in hexane, whereas canthaxanthin peaks at 462 nm (Fig. 2). The pigments of yellow feathers peak at about 385, *404,* 428 nm, compared to 421, *445,* 474 nm for lutein, and 416, *438,* 469 nm for the canary xanthophylls.

A scanty pigmented substance can be extracted with methanol from red feathers (such as those of *Ara macao*) that has an unusual spectrophotometric signature (356, 376, *398,* 420; Fig. 2). The pigment spectrum is

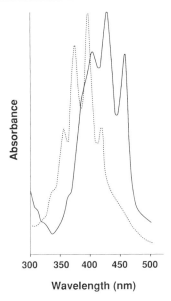

FIG. 2. Light absorption spectrum in hexane of a methanol extract (wash) of the red tail feathers of *Ara macao* (·····), and of an acidified pyridine extract (—) of the same feathers. The spectra were determined with a Perkin-Elmer (Norwalk, CT) 552 UV/VIS spectrophotometer.

unaffected by changes in the polarity of the solvent, which contrasts to the loss of fine structure for the copious pigments extracted with acidified pyridine. When the former substance was applied to silica it was chemically transformed to a substance spectrally similar to the copious pigments (although shifted hypsochromically at around 386, *407,* 432 nm). The parrot pigments are not structurally characterized. The pigments may be separated on HPLC with the Zorbax column, and a solvent mixture of methanol–0.01 M ammonium acetate (44:6, v/v).

[30] Fast-Atom Bombardment and Continuous-Flow Fast-Atom Bombardment Mass Spectrometry in Carotenoid Analysis

By HAROLD H. SCHMITZ, RICHARD B. VAN BREEMEN, and STEVEN J. SCHWARTZ

Certain carotenoids, including both provitamin A and nonprovitamin A carotenoids, have received considerable attention as having a possible role in the prevention and treatment of cancer and performing or enhancing other beneficial functions in the maintenance of human health.[1] Approximately 600 different carotenoids have been characterized,[2] many quite similar in structure, and for each of these compounds there may exist a number of geometric and stereospecific isomers. The similarity of physical properties within this group of compounds poses an analytical challenge in resolving and isolating individual components for structural elucidation. Currently, high-performance liquid chromatography (HPLC) is regarded as the method of choice for the separation, identification, and quantitation of carotenoids found in biological tissues.

Although HPLC is well suited for carotenoid analysis, the complex mixtures of these compounds often found in biological tissues can be difficult to resolve and identify with a reasonable degree of certainty. A minimum of identical retention time and spectrophotometric characteristics with known standards is necessary for tentative identification of carotenoids, although these two parameters alone are not considered sufficient for positive identification.[3] Factors such as cis-trans isomerization, varying solvent effects, the relative similarity of many absorbance spectra, and the difficulty of purifying carotenoids from complex mixtures render positive identification using absorption spectroscopy and retention characteristics alone questionable. Obtaining a mass spectrum of an individual carotenoid in conjunction with accurate spectrophotometric and chromatographic retention data does, however, provide enough information to identify positively the carotenoid.[3] It should be noted that more analytical data are required in addition to the above parameters to confirm the

---

[1] A. Bendich and J. A. Olson, *FASEB J.* **3**, 1927 (1989).
[2] O. Straub, *in* "Key to Carotenoids" (H. Pfander, ed.). Birkhäuser, Basel, 1987.
[3] R. F. Taylor, P. W. Farrow, L. M. Yelle, J. C. Harris, and I. G. Marenchic, *in* "Carotenoids: Chemistry and Biology" (N. I. Krinsky, M. M. Mathews-Roth, and R. F. Taylor, eds.), Proc. 8th Int. Symp. Carotenoids, p. 105. Plenum, New York, 1989.

identities of geometric and optical isomers of individual carotenoids.[4,5] Although conventional mass spectrometry (MS) and MS coupled with HPLC (LC-MS) have been reported for the analysis of many carotenoids,[3,4,6] the literature concerning the identification of carotenoids in biological samples does not often contain MS data.

Three MS techniques, electron impact (EI),[4] chemical ionization (CI),[6] and fast-atom bombardment (FAB),[7] have been used for the analysis of carotenoids. The main disadvantage of EI and CI techniques is their requirement for sample vaporization prior to ionization, increasing the likelihood of degradative product formation as a result of the heat-labile and nonvolatile nature of carotenoids.[4] FAB is an excellent technique for the analysis of thermally labile and nonvolatile compounds,[8] making this an appropriate method for carotenoid analysis. The use of MS-MS techniques, which employ collisional activation to promote fragmentation, followed by two or more MS analyzers in tandem, can provide characteristic fragmentation patterns that are quite useful for the identification of closely related carotenoids, such as the structural isomers α- and β-carotene.[3,7]

The importance of positively identifying certain carotenoids in various biological tissues cannot be overemphasized in light of the proposed beneficial effect(s) that carotenoids may have on human health. Furthermore, research has demonstrated that certain properties of individual carotenoids may differ significantly, such as the ability to quench singlet oxygen and the prevention and/or suppression of certain cancers.[1,9] Accordingly, the purpose of this chapter is to describe two methods, conventional FAB-MS of carotenoids separated using HPLC with diode array detection and on-line LC-MS, which can be used to identify carotenoids present in biological tissues.

Materials and Methods

Standards and Samples. An α-carotene standard is obtained from Sigma Chemical Company (St. Louis, MO), β-carotene from Hoffmann-La Roche (Nutley, NJ), and lycopene is isolated from tomatoes by open-

---

[4] W. Vetter, G. Englert, N. Rigassi, and U. Schwieter, *in* "Carotenoids" (O. Isler, ed.), p. 189. Birkhäuser, Basel, 1971.
[5] B. Renstrom, G. Borch, and S. Liaaen-Jensen, *Comp. Biochem. Physiol. B* **69B**, 621 (1981).
[6] F. Khachik, G. R. Beecher, and W. R. Lusby, *J. Agric. Food Chem.* **36**, 938 (1988).
[7] S. Caccamese and D. Garozzo, *Org. Mass Spectrom.* **25**, 137 (1990).
[8] C. Fenselau and R. J. Cotter, *Chem. Rev.* **87**, 501 (1987).
[9] P. Di Mascio, S. Kaiser, and H. Sies, *Arch. Biochem. Biophys.* **274**, 532 (1989).

column chromatography according to Simpson et al.[10] Purity of the standards is verified by using HPLC. Quantitation of carotenoid standards and carotenoids isolated from samples is carried out by using a UV-240 Shimadzu (Kyoto, Japan) recording spectrophotometer in conjunction with appropriate extinction coefficients from the literature.[10]

Extraction Procedures. Approximately 100 g of blueberries, 40 g of carrots, and 40 g of tomatoes are pureed separately in a Waring blender with 150 ml of methanol. The resulting puree is filtered through a Büchner funnel using Whatman (Clifton, NJ) number 1 filter paper. The puree is washed twice with 100 ml methanol, and the methanol discarded. The remaining plant tissue is placed under nitrogen in a beaker equipped with a magnetic stirring bar and extracted for 30 min by stirring in 100 ml of hexane. The hexane extract is filtered through a Büchner funnel using Whatman number 1 filter paper, and concentrated to approximately 10 ml using a rotary evaporator. Carrot and tomato extracts are then evaporated to dryness under a stream of nitrogen and stored at $-20°$ prior to analysis.

The blueberry extract requires further manipulation due to the presence of significant amounts of chlorophylls and other nonpolar components. The concentrated hexane extract is transferred to a 50-ml test tube and saponified by adding 10 ml of 10% KOH (w/v) in methanol and heating in a water bath at $65°$ under a stream of nitrogen until only one phase remains. This mixture is cooled and extracted twice with 10 ml of hexane. Addition of 5-ml aliquots of acetone and water is necessary to partition the carotenoids satisfactorily into the hexane phase. The hexane extract is evaporated to dryness under a stream of nitrogen and stored at $-20°$ until analysis. All extraction procedures are carried out either under yellow lighting or in the dark.

HPLC Equipment Used for Carotenoid Isolation. The HPLC system used to obtain diode array chromatograms and to purify carotenoid fractions for FAB-MS analysis consists of a Waters Associates (Milford, MA) model 501 or 590 pump, a model U6K sample injector, and a 990 photodiode array detector with a 990 plotter. The photodiode array detector is controlled by an NEC APCIV personal computer (Boxborough, MA). A reversed-phase Zorbax HPLC column (25 cm × 4.6 mm i.d.) packed with 5-μm ODS C_{18} protected by a Zorbax precolumn packed with ODS C_{18} (Du Pont, Wilmington, DE) is used in conjunction with the above HPLC system.

HPLC Procedures Used for Carotenoid Isolation. A mobile phase con-

---

[10] K. L. Simpson, S. C. S. Tsou, and C. O. Chichester, *in* "Methods of Vitamin Assay" (J. Augustin, B. P. Klein, D. A. Becker, and P. B. Venugopal, eds.), P. 185. Wiley, New York, 1985.

sisting of acetonitrile–methanol–ethyl acetate (61:14:25, v/v) is used in conjunction with the Zorbax column for the chromatography of blueberry, tomato, and carrot tissue extracts. A flow rate of 1.3 ml/min is used for the separation of carotenoids in the blueberry extract, while a flow rate of 1.1 ml/min is used for the separation of carotenoids in carrot and tomato extracts. HPLC solvents are reagent grade and are filtered and degassed prior to use.

Dried extracts are reconstituted in 500 μl of diethyl ether–methanol (50:50, v/v). Injection volumes range from 15 to 25 μl for carrot and tomato extracts and from 50 to 80 μl for blueberry extracts. Carotenoids absorbing in the visual spectrum are monitored at 451 nm, and colorless carotenoids are monitored at 314 nm. Individual carotenoids are collected from the ascending and descending slope of the absorption peak of interest.

FAB-MS and MS-MS Equipment. Positive-ion FAB mass spectra are obtained using a JEOL (Tokyo, Japan) JMS-HX110HF double-focusing mass spectrometer equipped with a JMA-DA5000 data system, collision cell in the first field-free region, and *B/E*-linked scanning capability. Xenon fast atoms at 6 kV are used for FAB ionization. The accelerating voltage is 10 keV, and the resolving power is 1000 for all measurements. A range of m/z 1–700 is scanned in approximately 8 sec for all mass spectometric analyses. MS-MS analyses are carried out using *B/E*-linked scanning, in which the magnetic field *(B)* is scanned in a constant ratio to the electrostatic analyzer voltage *(E)*. To increase the signal-to-noise ratio, three to five scans are averaged. During MS-MS, precursor ions formed during FAB ionization are fragmented by using helium collisional activation in the collision cell. The helium gas pressure is adjusted so that the abundance of the precursor ions is attenuated to 30%.

FAB-MS and MS-MS Procedures. Sample fractions collected during HPLC analysis are extracted from the mobile phase using hexane. The addition of 5-ml aliquots of acetone and water is necessary to facilitate the partition of lutein into the hexane phase for the collection of this carotenoid. These extracts are then concentrated for FAB-MS analysis under a stream of nitrogen.

A mixture of 3-nitrobenzyl alcohol (NBA) and glycerol (80:20, v/v) is used as the static FAB matrix for analysis of the more nonpolar carotenoids, while a matrix consisting of NBA–glycerol (50:50, v/v) is used for the analysis of lutein, a more polar carotenoid, from blueberry extract. In addition, 1 μl of trifluoroacetic acid (TFA) is added to the sample–matrix solution to facilitate ionization of lutein during FAB–MS analysis; Static FAB-MS and MS-MS analysis is carried out by applying 1 μl of extract containing at least 0.1 μg of individual carotenoid to the probe tip.

HPLC-MS Equipment and Procedures. The HPLC system used during

LC-MS consists of an Applied Biosystems (Foster City, CA) model 140A dual-syringe pump, a model 757 variable-wavelength ultraviolet/visible (UV/VIS) detector equipped with a capillary flow cell, a Rheodyne (Cotati, CA) model 8125 injector, and a Hewlett-Packard (Palo Alto, CA) model 3396 integrator. A Vydac (Hesparia, CA) TP218 narrow-bone HPLC column (15 cm × 2.1 mm) packed with C_{18} bonded to 5-μm diameter silica particles is used in conjunction with this system to obtain carotenoid separations during LC-MS analysis.

LC-MS analysis is carried out using a frit-FAB version of a continuous-flow FAB LC-MS interface. Compatability of the LC-MS interface with the vacuum system of the mass spectrometer is achieved by splitting the column eluent so that approximately 5-7 μl/min of mobile phase enters the absorbance detector and mass spectrometer. The temperature of the LC-MS ion source is maintained at 35° to prevent solvent from freezing in the frit.

HPLC-MS Procedures. An isocratic solvent system consisting of acetonitrile-methanol-ethyl acetate (50:30:20, v/v) is used at a flow rate of 100 μl/min to separate α-carotene, β-carotene, and lutein standards and carotenoids present in carrot extracts. The FAB matrix is introduced into the mobile phase by postcolumn addition of 1.4 μl/min of a 1% solution of NBA in water by using a T junction for the capillary tubing.

Standards and sample extracts are dissolved in diethyl ether-methanol (50:50, v/v). The injection volume for standards is 5 μl, containing approximately 4-5 μg of carotenoid. The injection volume of carrot extract is 2 μl. Column eluent, after the split and postcolumn addition of matrix, is monitored at 450 nm.

Identification of Selected Carotenoids. Carotenoids are identified by HPLC retention times, light absorption, and molecular weights determined by using mass spectrometry. The structural isomers α-carotene, β-carotene, and lycopene are further differentiated using MS-MS. Fragmentation patterns of these isomers are compared to standards and literature data.[3,7]

Results and Discussion

Identification of Carotenoids in Blueberries. Figure 1 demonstrates the reversed-phase HPLC separation of components present in a typical blueberry extract, with the eluent being monitored at 451 and 314 nm. Carotenoids identified include lutein, β-carotene, phytofluene, and phytoene. These carotenoids were tentatively identified based on retention and light absorption characteristics (Fig. 2). Mass spectrometric analysis of both the xanthophyll and carotene peak fractions isolated by HPLC yielded the molecular ions corresponding to the molecular weights of the carotenoids present, as well as to other eluting compounds (Fig. 3).

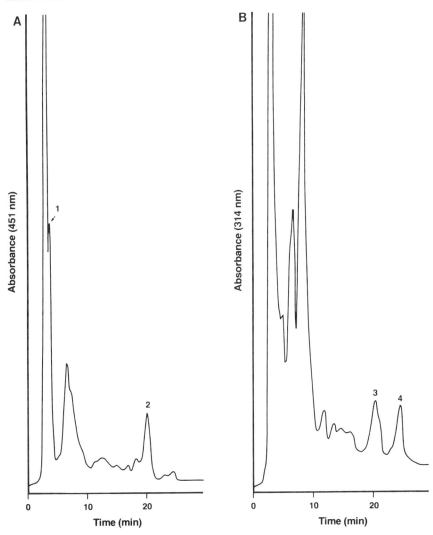

FIG. 1. HPLC separation of carotenoids present in blueberry extract monitored at 451 nm (A) and 314 nm (B). Peak 1, lutein; peak 2, β-carotene; peak 3, phytofluene; peak 4, phytoene.

It is of interest to note that FAB ionization of carotenoids, especially the hydrocarbon carotenoids, predominantly yielded odd-electron molecular ions, rather than even-electron protonated molecules, which are characteristic of positive-ion FAB mass spectra.[7] This is presumably because of the low ionization potential of carotenoids due to their conjugated diene structure, which can readily participate in an electron transfer from the

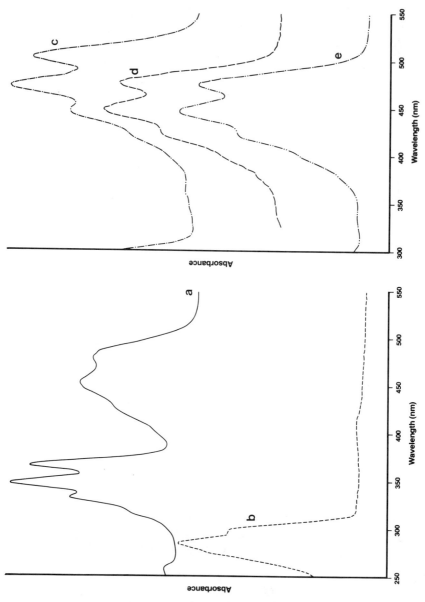

Fig. 2. UV/VIS light absorption spectra of carotenoids in blueberry, carrot, and tomato extracts. (a) β-Carotene and phytofluene; (b) phytoene; (c) lycopene; (d) α-carotene; (e) lutein.

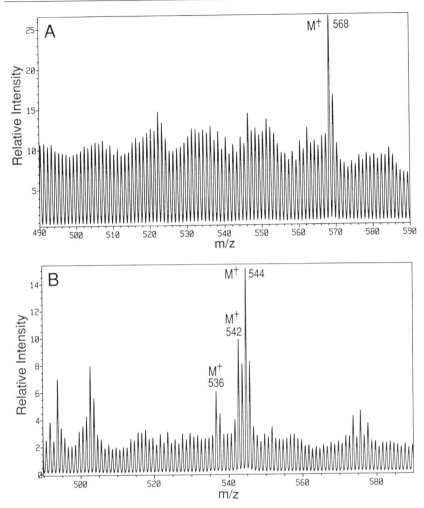

FIG. 3. FAB mass spectra of blueberry extract. (A) Xanthophyll HPLC fraction: m/z 568, molecular ion of lutein. (B) Carotene HPLC fraction: m/z 536, 542, and 544, molecular ions of β-carotene, phytofluene, and phytoene, respectively. The base peak in each mass spectrum (not shown) was a matrix ion at m/z 307.

molecule of interest to the matrix.[7] Hydroxylated carotenoids, such as lutein, exhibit more abundant even-electron protonated molecules because the more polar hydroxy groups are readily protonated, particularly when glycerol is used as the matrix. The hydrocarbon carotenoids lack sites that are good proton acceptors, thus inhibiting the formation of protonated molecules.

Special note should be given to Fig. 2, which contains an absorption spectrum for both β-carotene (absorption in the VIS region) and phytofluene (absorption in the UV region). These two carotenoids were not resolved by the reversed-phase HPLC system used in this study. Coelution is fairly commonplace when analyzing complex mixtures of carotenoids, and emphasizes the difficulty of positive identification via retention and light absorption characteristics alone.

Identification of Carotenoids in Carrots and Tomatoes. The reversed-phase HPLC separations of carrot and tomato extracts, monitored at 451 nm, are shown in Fig. 4. Analysis of retention and UV/VIS characteristics (Fig. 2) in comparison to standards resulted in the tentative identification of the more nonpolar carotenoids, α- and β-carotene in carrot extract and β-carotene and lycopene in tomato extract. Each of these HPLC fractions was analyzed by using FAB-MS and formed molecular ions at m/z 536. Typically, α- and β-carotene elute in close proximity on most reversed-phase HPLC systems, as observed in this study, and β-carotene and lycopene may elute close to one another on reversed-phase HPLC systems employing columns with polymeric packing in conjunction with certain mobile phases.[11] Therefore, it was desirable to further differentiate these structural isomers beyond that of molecular weight information.

FAB-MS-MS of Carotene Structural Isomers. A specific ion of interest formed during FAB ionization can be selected by using a double-focusing mass spectrometer, fragmented by collisional activation, and then analyzed via B/E-linked scanning to yield fragmentation data characteristic of the selected ion. Standards of the three carotene structural isomers, α-carotene, β-carotene, and lycopene, have been differentiated previously using B/E-linked scan FAB-MS,[7] and this procedure was modified to analyze the three carotene isomers present in carrot and tomato extracts.

The characteristic B/E-linked scans of the molecular ions at m/z 536 formed during FAB-MS for the three carotene isomers are shown in Fig. 5. The spectrum of β-carotene (Fig. 5A) shows an abundant fragmentation at m/z 444. In addition to m/z 444, the spectrum of α-carotene (Fig. 5B) displays characteristic fragments at m/z 480, 460, and 388, while the lycopene (Fig. 5C) B/E scan shows characteristic fragment ions at m/z 444 and 467. The mass spectral patterns obtained by MS-MS made it possible to further differentiate these structural isomers, and therefore can be used to confirm the identity of these carotenoids.

---

[11] C. S. Epler, N. Craft, L. C. Sander, S. A. Wise, and W. E. May, "Report of Analysis #552-90-058." U.S. Department of Commerce, National Institute of Standards, Gaithersburg, Maryland, 1990.

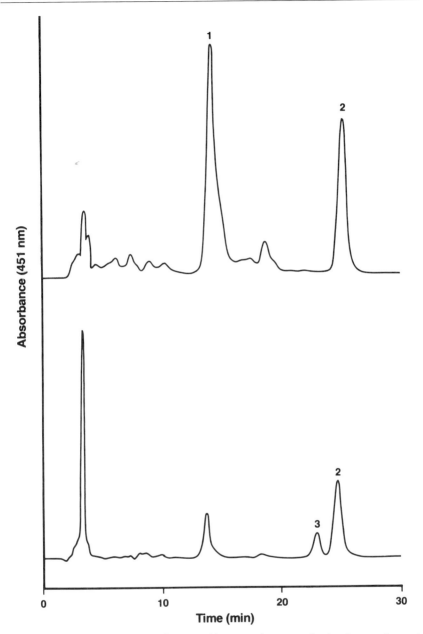

FIG. 4. HPLC chromatograms of carotenoids present in tomato *(top)* and carrot *(bottom)* extract. Peak 1, lycopene; peak 2, β-carotene; peak 3, α-carotene.

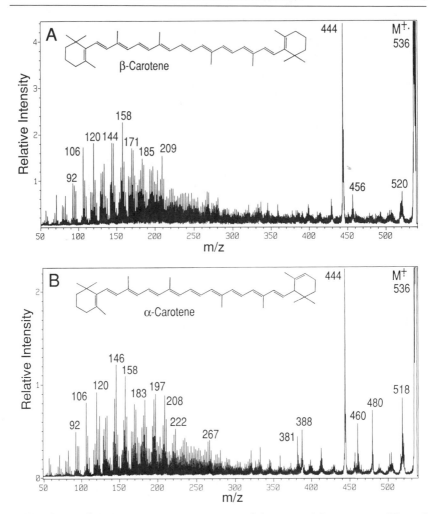

FIG. 5. B/E linked-scan FAB mass spectrum of β-carotene (A), α-carotene (B), and lycopene (C).

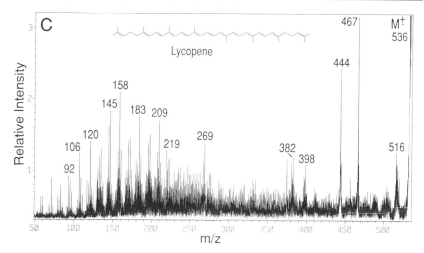

FIG. 5. *(continued)*

Selection of Matrix. Initially, glycerol was used as the FAB matrix during FAB-MS analysis. Because of the polarity of glycerol, the more hydrophobic carotenoids, such as β-carotene, were insoluble. This lack of solubility with the matrix resulted in all of the applied carotenoid sample remaining on top of the matrix, and FAB ionization could not be sustained. Caccamese and Garozzo[7] noted that glycerol by itself was not appropriate for FAB-MS analysis of carotenoid standards.

NBA proved to be a useful matrix for FAB-MS. Being less polar than glycerol, NBA facilitated greater interaction with the carotenoid sample, resulting in better dispersion of the carotenoid throughout the matrix. However, for some samples, a mixture of NBA–glycerol was found to provide greater signal to noise than either matrix alone during static FAB-MS while still facilitating the formation of a persistent signal. The addition of TFA increased the signal-to-noise ratio for lutein analysis.

Identification of Selected Carotenoids in Carrot Extract by Using LC-MS. Figure 6 shows the separation of carotenoids present in carrot extract, monitored by using continuous-flow FAB LC-MS with postcolumn addition of the aqueous matrix solution. The first three peaks are lutein, α-carotene, and β-carotene. Both the reconstructed total ion chromatogram (TIC) of the carrot extract, a summation of all ions detected by the mass spectrometer, as well as selected ion chromatograms showing the molecular ions of each carotenoid present are shown in Fig. 6. In addition to lutein, α-carotene, and β-carotene (m/z 568, 536, and 536, respectively), molecular ions were detected at m/z 542 and 544, which corresponded to

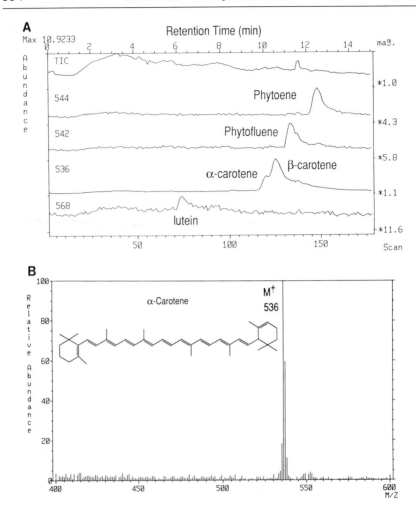

FIG. 6. (A) Reconstructed total ion chromatogram and selected ion chromatograms showing the LC-MS analysis of carrot extract. (B) Continuous-flow FAB mass spectrum of α-carotene obtained during the LC-MS analysis shown in (A). This mass spectrum has been background subtracted.

the molecular weights of phytofluene and phytoene. Retention and light absorption data confirmed the presence of these two compounds in the carrot extract. It should be noted that the ability to carry out MS-MS during LC-MS is possible, but was not done during these measurements.

Selection and Addition of Matrix to the Mobile Phase. Initially, glycerol was used as the matrix material and mixed into the mobile phase during

LC-MS. In addition, water, at a concentration of 15% by volume, was added to the mobile phase to obtain a stable FAB LC-MS signal. Next a mixture of lutein, α-carotene, and β-carotene standards were analyzed by using LC-MS, but only protonated lutein was detected at m/z 569.

The use of NBA as matrix mixed into the mobile phase reservoir at 0.5% by volume along with water at a minimum of 15% provided detectable m/z signals for all three of the carotenoids. However, the addition of NBA and water to the mobile phase resulted in coelution of α-carotene and β-carotene, yielding only one peak on the TIC corresponding to 536. Therefore, postcolumn addition of NBA and water to the mobile phase was necessary to obtain detectable molecular ions for both hydroxylated and hydrocarbon carotenoids, facilitate FAB mass spectrometric detection, and sufficiently resolve α-carotene and β-carotene. Postcolumn addition of a 50:50 mixture of NBA and glycerol in water provided no advantage over NBA in water for LC-MS.

Limits of Detection for FAB-MS and LC-MS. Although limits of detection were not determined, the smallest quantity of any carotenoid detected during this investigation was 30 ng of phytofluene using FAB-MS. During routine FAB-MS, at least 200 ng of the lutein sample, more than any other isolated carotenoid, was applied to the FAB probe, yet this carotenoid gave the weakest signal observed for any of the carotenoids during the analysis. Suppression of the lutein signal may have occurred due to the presence of more readily ionizable compounds present in the HPLC xanthophyll fraction. Further development of a matrix for carotenoid analysis might be needed to optimize the sensitivity for the detection of these compounds.

Although the sensitivity of LC-MS for carotenoid detection was decreased as a result of the 1:20 split of the column eluent necessary for compatibility with the vacuum system of the mass spectrometer, abundant molecular ions were detected for all carotenoid samples. The signal-to-noise ratio in the LC-MS spectra was routinely much greater than in the static FAB-MS analyses. Typically 400 ng of each carotenoid (after the split) produced abundant molecular ions.

Conclusions

A method for the identification of carotenoids in biological tissues using FAB-MS in conjunction with chromatographic retention and light absorption characteristics has been described here. In addition, the development of an HPLC-MS method for carotenoid analysis, which required postcolumn addition of matrix and water to the mobile phase, is shown. The application of FAB-MS-MS for the differentiation and confirmation of structural isomers of carotenoids found in biological tissues was also demonstrated. For the analysis of nonvolatile and thermally labile compounds, FAB-MS[8] is perhaps a better method of ionization than other techniques previously used for carotenoid analysis. Furthermore, FAB-MS can be readily interfaced with HPLC to provide a powerful analytical tool for carotenoid analysis.

[31] High Resolution Analysis of Carotenoids in Human Plasma by High-Performance Liquid Chromatography

By GARRY J. HANDELMAN, BINGHUA SHEN, and NORMAN I. KRINSKY

Introduction

The complex mixture of carotenoids in human plasma is a challenging analytical problem. For example, Khachik et al.[1] have shown that human plasma contains at least 14 carotenoids that elute in the highly polar xanthophyll region of the chromatogram. However, the high-performance liquid chromatography (HPLC) procedure of Khachik et al.[1] is not yet adapted to routine analyses of plasma samples.

Is it useful to increase the analytical resolution obtained in population surveys of human plasma carotenoids? Improved HPLC methods could allow the testing of several hypotheses that are listed below.

1. High plasma levels of lutein and zeaxanthin, which are selectively accumulated in the macular region of the human retina,[2,3] might be associated with a decreased risk of age-related macular degeneration.

2. Lutein may be protective against cancer in humans,[4,5] in addition to anticancer effects already established for β-carotene.[6]

3. The very polar ketocarotenoids that elute in the very early region of the chromatogram on reversed-phase HPLC[1,7] may derive either from

[1] F. Khachik, G. R. Beecher, M. B. Goli, and W. R. Lusby, *Pure Appl. Chem.* **63**, 71 (1991).
[2] R. A. Bone, J. T. Landrum, L. Fernandez, and S. L. Tarsis, *Invest. Ophthalmol. Visual Sci.* **29**, 843 (1988).
[3] G. J. Handelman, E. A. Dratz, C. C. Reay, and F. J. G. M. van Kuijk, *Invest. Ophthalmol. Visual Sci.* **29**, 850 (1988).
[4] D. I. Thurnham, *Lancet* **2**, 441 (1989).
[5] M. S. Micozzi, *in* "Nutrition and Cancer Prevention: Investigating the Role of Micronutrients" (T. E. Moon and M. S. Micozzi, eds.), p. 213. Dekker, New York, 1989.
[6] N. I. Krinsky, *in* "Carotenoids: Chemistry and Biology" (N. I. Krinsky, M. M. Mathews-Roth, and R. F. Taylor, eds.), Proc. 8th Int. Symp. Carotenoids, p. 279. Plenum, New York, 1989.
[7] N. I. Krinsky, M. D. Russett, G. J. Handelman, and D. M. Snodderly, *J. Nutr.* **120**, 1654 (1990).

dietary egg products,[1] or from metabolic conversion of lutein and/or zeaxanthin.

4. Humans and monkeys may accumulate dietary β,ε-carotenoids (such as lutein, α-cryptoxanthin, and α-carotene) with greater efficiency than the corresponding β,β-carotenoids (zeaxanthin, β-cryptoxanthin, and β-carotene).[7,8]

5. Studies of all-*trans*-β-carotene and natural β-carotene, which is reported to be rich in various cis isomers of β-carotene,[9] suggest that the latter preparation may have enhanced anticancer potency.[10,11]

To test these hypotheses, a high-resolution HPLC technique is needed to allow better quantitation of individual plasma carotenoids. This chapter describes such a method.

High-Resolution Chromatogram of Human Plasma Carotenoids

We have achieved the plasma carotenoid analysis shown in Fig. 1 with two Alltech (Deerfield, IL) C_{18} Adsorbosphere-HS 3-μm cartridge columns, connected in series. Full details of the method are given in the next section, Sample Preparation and Analysis.

Several observations can be made about the carotenoids eluting in the first portion of the chromatogram (10–20 min). Examination of the early-eluting peaks with the diode-array HPLC indicates that peaks 1 and 2 have the ε,ε-carotene chromophore, and peak 3 has the β,ε-carotene chromophore. We noted that these compounds are degraded by saponification with ethanolic potassium hydroxide.[12] The ketocarotenoid structures assigned by Khachik *et al.*[1] (see Fig. 1) are consistent with out observations. Some of the small peaks between peaks 1 and 3 (Fig. 1) may be cis isomers of these ketocarotenoids.

In order to achieve high resolution of carotenoids in the polar xanthophyll region (peaks 1–8, Fig. 1), it was necessary to dissolve the plasma extract in methanol before injection on the HPLC, as described in the next section.

Both all-*trans*-lutein (L) and all-*trans*-zeaxanthin (Z) are baseline re-

---

[8] D. M. Snodderly, M. D. Russett, R. I. Land, and N. I. Krinsky, *J. Nutr.* **120**, 1663 (1990).

[9] A. Ben-Amotz, A. Lers, and M. Avron, *Plant Physiol.* **86**, 1286 (1988).

[10] H. Nishino, J. Takayasu, T. Hasegawa, O. Kimura, E. Kohmura, M. Murakoshi, J. Okuzumi, N. Sugawa, N. Hosokawa, T. Sakai, T. Sugimoto, and J. Imanishi, *Kyoto-furitsu Ika Daigaku Zasshi* **97**, 1097 (1988).

[11] H. Nagasawa, R. Konishi, N. Sensui, K. Yamamoto, and A. Ben-Amotz, *Anticancer Res.* **9**, 71 (1989).

[12] G. J. Handelman, M. D. Russett, A. J. Adler, and N. I. Krinsky, unpublished results.

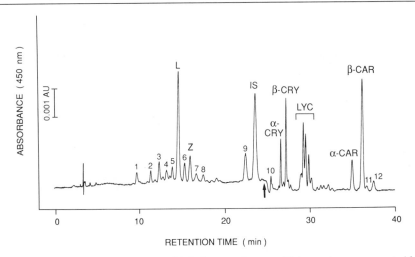

FIG. 1. HPLC analysis of carotenoids in human plasma, with two columns connected in series (other details are given in text). Peak identification: L, Lutein; Z, zeaxanthin; IS, internal standard; α-CRY, α-cryptoxanthin; β-CRY, β-cryptoxanthin; LYC, lycopene isomers; α-CAR, α-carotene; β-CAR, β-carotene; 1, ε,ε-carotene-3,3′-dione; 2, 3′-hydroxy-ε,ε-caroten-3-one; 3, 3-hydroxy-β,ε-caroten-3-one; 4 and 5, unknown; 6, 13-cis-lutein; 7, 13-cis-zeaxanthin; 8, unknown; 9, 2′,3′-anhydrolutein; 10, unknown; and 11 and 12, β-carotene cis isomers. The arrow marks the refractive index change following addition of 2-propanol to the mobile phase.

solved. Directly after lutein is 13-cis-lutein (peak 6), and after zeaxanthin is 13-cis-zeaxanthin (peak 7).

As observed with the diode array HPLC, the peak at 22.5 min (peak 9) has a β,ε-carotene chromophore, consistent with its identification by Khachik et al.[1] as 2′,3′-anhydrolutein.

At 23.6 min, the internal standard (IS) elutes. For this analysis, we synthesized β-apo-12′-carotenal-O-tert-butyl oxime, abbreviated carotenal oxime, as described in the next section.

Directly after the refractive index change from the addition of 2-propanol to the gradient (at the arrow), an additional saponification-sensitive carotenoid (peak 10) elutes[12] that has not yet been identified.

The cryptoxanthin group is resolved into α-cryptoxanthin and β-cryptoxanthin (α-CRY and β-CRY), and several possible cis isomers. The lycopene (LYC) group appears to be resolved into five separate constituents.

After α-carotene and β-carotene (α-CAR and β-CAR), there are two additional components (peaks 11 and 12), which we provisionally identify as cis isomers of β-carotene.

Sample Preparation and Analysis

Sample Preparation

The plasma sample is prepared from ethylenediaminetetraacetic acid (EDTA)-treated whole blood. The isolated plasma is digested with an enzyme reagent to hydrolyze nonpolar lipids. For hydrolysis of plasma triglycerides, the reagent contains 160 IU/ml of triglyceride hydrolase (lipase), from a microbial source (*Chromobacterium viscosum*; Calbiochem, San Diego, CA). For hydrolysis of plasma cholesterol esters, the reagent also contains 1 IU/ml of cholesterol esterase, from a microbial source (*Pseudomonas* species; Calbiochem). This reagent is prepared in 0.1 M sodium phosphate buffer, pH 7.0, containing 1 mM EDTA and 0.25% (w/v) Triton X-100, and is stable for at least 30 days at 4°.

For the analysis, 100 μl of plasma is mixed with 1 ml of the enzyme reagent in an 8-ml borosilicate glass vial, and held at ambient temperature in the dark for 1 hr. Then 100 μl of 5% sodium dodecyl sulfate (SDS) is added, and the sample is mixed vigorously for 5 sec with a vortex mixer. Then 1 ml of ethanol is added and the sample is vortexed for 60 sec.

To the sample is added 3 ml of 2:1 ether/hexane (stabilized with 1% ethanol). This is followed by 1 ml of hexane containing the internal standard, as prepared below. The sample is vortexed for 60 sec, centrifuged, and the upper layer transferred to a second 8-ml glass vial. The hexane–ether extract is evaporated under N_2 at 40°. The residue is dissolved in 40 μl of methanol, and 25 μl is injected onto the HPLC.

Suitable techniques for accurate pipetting of organic solvent volumes are described elsewhere in this volume.[13]

Synthesis of Internal Standard, β-apo-12'-carotenal-O-t-butyl-oxime

The synthesis of the internal standard used here is straightforward.[3] The compound is prepared from β-apo-12'-carotenal (courtesy of Dr. Peter Sorter, Hoffmann-La Roche, Nutley, NJ) and O-t-butyl-hydroxylamine (Sigma, St. Louis, MO). The β-apo-12'-carotenal is dissolved in methanol, at 1 mg/ml. The O-t-butyl-oxime, 0.1 M, is prepared in 0.1 M sodium acetate; the pH of this complete reagent is adjusted to 4.7. Then 1 ml of the carotenal solution is mixed with 100 μl of the oxime solution, in an 8 ml glass vial, and incubated for 12 hr at ambient temperature in the dark. About 80–90% conversion of carotenal to carotenal oxime is reliably obtained with a 12-hr incubation. For extraction, 1 ml of H_2O is added,

---

[13] G. J. Handelman, D. M. Snodderly, A. J. Adler, M. D. Russett, and E. A. Dratz, this volume [19].

and the carotenal oxime product is extracted with 4 ml of hexane. The material is purified by reversed-phase HPLC on an Alltech 5-μm C_{18} column, 25 × 0.46 cm, using isocratic elution in acetonitrile–methanol (85:15, v/v), as described earlier.[14] The internal standard is stored at $-20°$ in hexane, in which it remains stable for at least 1 year without degradation or isomerization.

For this experiment, the OD_{450} of the internal standard stock is measured, and the stock solution is diluted to 0.008 AU in hexane. One milliliter of this dilution is added to the sample after the enzyme treatment, as described above.

HPLC Equipment and Solvent Program

The plasma carotenoid analysis shown in Fig. 1 is carried out on an HP-1090 diode array HPLC (Hewlett-Packard, Avondale, PA). The detector wavelength is 450 nm. The two analytical columns, 15 cm × 0.46 cm diameter, are connected in series and packed with C_{18} Adsorbosphere-HS, 3-μm particle size material (Alltech). The analytical columns are protected with a C_{18} Adsorbosphere-HS, 5-μm particle size, 1 × 0.46 cm guard column (Alltech). To connect all three columns together with minimal losses in peak resolution, the analytical and guard columns are in the Alltech cartridge format, as described in the Alltech catalog.

The initial HPLC mobile phase, from 0 to 19 min, is acetonitrile–methanol (85:15, v/v), with 0.01% (w/v) ammonium acetate. At 19 min, a 1-min ramp is executed to the initial solvent plus 30% 2-propanol. The total run time is 40 min. At the end of the analysis, the column is returned to the initial solvent for 20 min to reequilibrate the column to the initial conditions. The flow rate throughout the run is 1 ml/min.

The column temperature is maintained at a fixed level with a 40-cm HPLC column water jacket (Alltech), and a constant-temperature circulator (Fisher model 910; Fisher Scientific, McGaw Park, IL). The analysis in Fig. 1 is carried out at a column temperature of 24°.

Technical Discussion

Use of Two Columns, Connected in Series, Packed with C_{18} Adsorbosphere-HS 3-μm Material

We have explored several different columns packed with 5-μm particle size, C_{18} material, from several firms, including Alltech, Beckman (Palo Alto, CA), and Rainin (Woburn, MA). The use of the 3-μm particle size,

---

[14] N. I. Krinsky and S. Welankiwar, this series, Vol. 105, p. 155.

C_{18} packing (Alltech) greatly improved the analysis when compared with the 5-μm materials. The analysis with a single 3-μm Adsorbosphere-HS column gives reliable separation of several major carotenoid pairs, including lutein and zeaxanthin.[7] The use of two columns in series allows additional constituents, including more cis isomers, to be resolved.

We cannot ascertain how many carotenoid components might be resolved at the theoretical limit of resolution. In a famous anecdote,[15] Keulemans described the analysis of coffee flavor constituents by gas chromatography (GC). Initially 20 constituents were resolved, with a few shoulders; with better packed columns, 100 peaks were resolved, with some shoulders. Finally, by capillary GC, about 1000 peaks were resolved—again, with a few shoulders.

The Adsorbosphere-HS (high surface) stationary phase, with 20 g of carbon/100 g of packing, separates carotenoid cis isomers from all-trans components much more effectively than the regular Adsorbosphere material, with 12 g of carbon/100 g of packing. The HS material is designed for virtually 100% coverage of the particle surface with the C_{18} functional group. This very hydrophobic surface differentiates the all-trans and cis isomers by some mechanism that requires further investigation.

Selection of Internal Standard

The methodology for synthesis of the internal standard is applicable to a variety of different carotenals and hydroxylamine reagents. As aldehyde reagent, β-apo-8'-, β-apo-10'-, and β-apo-12'-carotenal, and retinal can be utilized, as they all show substantial absorbance at 450 nm. Among hydroxylamine reagents, O-methyl-, O-ethyl-, O-*tert*-butyl-, and O-phenylhydroxylamine can be used. Each carotenal oxime addition product gives a different retention time,[3] and we have used several other derivatives in previous studies.[3,7,16]

The carotenal oxime used for this analysis, β-apo-12'-carotenal-*t*-butyloxime, was selected to elute in the vacant region of the chromatogram after peak 9 (Fig. 1).

Injection of Sample in Methanol

The choice of injection solvent is important for resolution of the polar carotenoids. The separation of the xanthophyll region of the chromatogram (consisting of carotenoids with several oxygen functions) is impaired

---

[15] D. H. Desty, *LC-GC* **9**, 414 (1991).
[16] G. J. Handelman, D. M. Snodderly, N. I. Krinsky, M. D. Russett, and A. J. Adler, *Invest. Ophthalmol. Visual Sci.* **32**, 257 (1991).

by dissolving the sample extract in ethanol, just prior to injecting the sample on the HPLC. Dissolving the extract in dichloromethane is even more injurious to the separation, and leads to complete loss of the xanthophyll resolution.[12]

The initial mobile phase for the analysis in Fig. 1 is acetonitrile–methanol (85:15, v/v). This polar HPLC solvent is a prerequisite for the separation of the polar carotenoids (peaks 1–8, Fig. 1). The injection of a relatively large amount (25 μl) of ethanol into the polar mobile phase may temporarily coat the packing material with ethanol and change the nature of the stationary phase, leading to a transient alteration in separation mechanisms.

Hydrolysis with Microbial Enzymes to Eliminate Triglycerides and Cholesteryl Esters

Full analytical recovery of β-carotene and other hydrocarbon carotenoids can be achieved when the plasma extract is solubilized in methanol, but only if the triglycerides and cholesteryl esters are first eliminated from the plasma sample. These nonpolar lipids are not soluble in methanol[17] and form droplets on the bottom of the sample vial. The β-carotene is trapped in the lipid droplets, and does not dissolve in the methanol. As a result, the β-carotene is not injected quantitatively onto the HPLC column with the rest of the sample.

We have selected microbial enzymes that hydrolyze these nonpolar lipids under very mild conditions: pH 7.0 and ambient temperature. The triglyceride lipase from *C. viscosum* was chosen because it accomplishes the complete hydrolysis of triglyceride to free fatty acids and glycerol.[18] By contrast, pancreatic lipase converts triglycerides only to monoglycerides, and might not be as effective for this procedure.

Ethanolic KOH (often at elevated temperatures) is frequently used to hydrolyze these lipids, but certain of the carotenoids in the sample (peaks labeled 1, 2, 3, and 10) are destroyed by ethanolic KOH. The use of KOH at elevated temperatures might also change the distribution of carotenoid cis isomers.[19]

---

[17] W. R. Morrison and L. M. Smith, *J. Lipid Res.* **5**, 600 (1964).
[18] T. Yamaguchi, *Agric. Biol. Chem.* **37**, 999 (1973).
[19] S. Liaaen-Jensen, in "Carotenoids: Chemistry and Biology" (N. I. Krinsky, M. M. Mathews-Roth, and R. F. Taylor, eds.), Proc. 8th Int. Symp. Carotenoids, p. 149. Plenum, New York, 1989.

Precise Control of Column Temperature

Changes in temperature change the HPLC carotenoid profile. Of special importance, the retention time of cis isomers may change in a different manner than the all-*trans*-carotenoids. Figure 2 shows the analysis of a lutein/zeaxanthin mixture at 20 and 24°. The peaks (peaks 2 and 4, Fig. 2) eluting directly after each major carotenoid (peaks 1 and 3, Fig. 2) are their corresponding 13-cis isomers. At 24°, the cis isomers elute in the region between the all-*trans* compounds. At 20°, the cis isomers are only partially resolved as shoulders of the main peaks. The analysis in Fig. 1, carried out at 24°, was optimized for the resolution of the lutein and zeaxanthin cis isomers in the plasma extract. The investigator will have to evaluate different temperatures experimentally to find the best resolution for a given application.

Variation in temperature also makes it difficult to obtain reproducible peak area and height measurements. Without temperature regulation, a 4° drift in laboratory temperature, observed over the course of a single day,

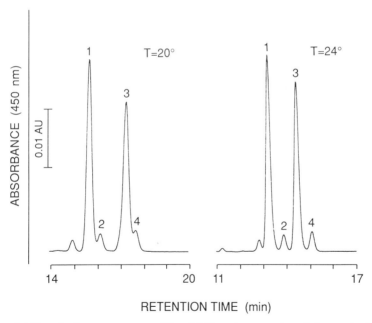

FIG. 2. Effect of column temperature (20 and 24°) on the separation of all-*trans*-lutein and all-*trans*-zeaxanthin, and their 13-cis isomers, which elute directly after the all-trans peaks. Other analytical conditions are as in Fig. 1.

can lead to a variation in peak retention time of 30 sec. With stable column temperature control, between-run retention time variation is typically 3 sec.

Further developments in chromatographic theory are needed to explain why the retention times of all-*trans*- and 13-*cis*-carotenoids respond differently to temperature. We hypothesize that the effect is partly due to conformational restraint imposed by the cis bond. The *cis*-carotenoids are more restricted in their range of molecular motions, and therefore the degree of molecular agitation of the cis isomers may not be affected as strongly by increases in temperature, as in the case of the all-*trans* isomers. If the overall molecular conformation is less temperature sensitive, that may be reflected in the temperature effects on retention time.

Ammonium Acetate in HPLC Mobile Phase

When carotenoids are analyzed on reversed-phase columns, the carotenoids may undergo degradation on the column. We have observed that the effect is most severe when small samples (< 5 ng) are chromatographed. By contrast, when several hundred nanograms is injected, the degradation still occurs, but is less prominent. This effect is seen on the right-hand side of Fig. 3.

The addition of 0.01% (w/v) ammonium acetate to the mobile phase allows reliable measurements of 1- to 100-ng amounts of carotenoids, without degradation of pigment. The protective effect is seen on the left-hand side of Fig. 3. The ammonium acetate is conveniently prepared from crystalline material (readily available from many suppliers) as a 5% stock solution in methanol, and then diluted into the final mobile phase to the 0.01% level. This amount of ammonium acetate has been effective with a variety of C_{18} columns, using an acetonitrile–methanol (85:15, v/v) mobile phase.

We added ammonium acetate to the mobile phase because of the hypothesis that it would buffer acidic uncapped silanols on the packing, and thereby prevent carotenoid destruction. However, we still lack a definite mechanism for the protective effect of ammonium acetate. Others have reported similar protective effects of Tris buffer in a water-containing

FIG. 3. Ammonium acetate in the mobile phase suppresses degradation of trace levels of zeaxanthin. Column: Alltech C_{18} Econosphere 5 μm, 25 × 0.46 cm. Mobile phase, acetonitrile–methanol (85:15, v/v); flow rate, 1 ml/min; detection, 450 nm. The left-hand side shows the series of injections (160, 16, and 4 ng of zeaxanthin) with 0.01% ammonium acetate in the mobile phase, and the right-hand side is the same series, but with no mobile-phase modifier added.

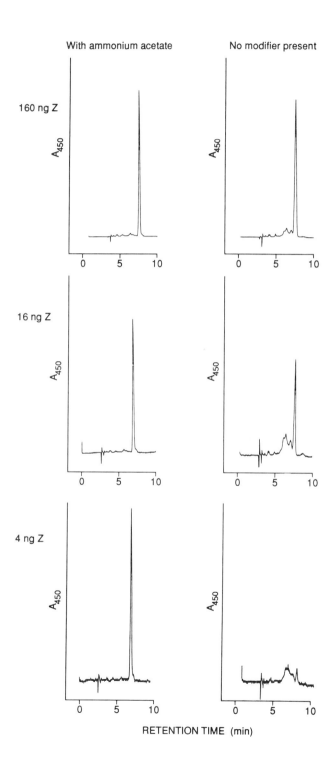

mobile phase,[20] which supports the hypothesis that the ammonium acetate is buffering a potentially destructive component.

But what is buffered? Is it the acidity of silanols on the packing material, or acidity in the mobile phase? Why is the "destructive" effect saturable, so that virtually all of a 4-ng injection of zeaxanthin is degraded (Fig. 3, right), but only a small fraction of a 160-ng injection? Commercial acetonitrile is reported to contain some acidity,[21] but it seems unlikely that 160 ng (320 pmol) of injected zeaxanthin could overcome the total acidity in several milliliters of the mobile phase. Interaction with occasional free silanols in the packing would appear to be a plausible mechanism.

Conclusions

The chromatographic separation described here may be applicable to the analyses of other tissue samples. Many human tissues contain negligible lipid other than cholesterol and phospholipid, and these lipids cause minimal interference with the procedure as described. The enzymatic hydrolysis method is suitable for elimination of neutral lipids in plasma samples, but in the future enzymatic hydrolysis may be adapted to eliminate the bulk triglyceride found in adipose tissue and milk.

Several of the procedures described (two columns in series, the use of ammonium acetate in the mobile phase, and control of column temperature) would be expected to have wide applicability to a variety of carotenoid analysis methods.

The HPLC solvent program is somewhat lengthy, requiring 40 min of analysis time per sample, and an additional 20 min for column reequilibration. Future developments in HPLC column technology may allow this type of high-resolution separation to be carried out in a shorter time period. However, if this HPLC procedure were adapted to an autosampler, it might be practical to use the method to analyze the large number of samples evaluated in typical epidemiological investigations.

Acknowledgments

The authors thank Alice J. Adler and Mark D. Russett for helpful discussions. Financial support was provided by U.S. PHS Grants EY-04911 and CA-51506.

---

[20] A. M. Gilmore and H. Y. Yamamoto, *J. Chromatogr.* **543**, 137 (1991).
[21] D. Siefermann-Harms, *J. Chromatogr.* **448**, 411 (1988).

[32] Separation and Quantitation of Carotenoids in Foods

By FREDERICK KHACHIK, GARY R. BEECHER, MUDLAGIRI B. GOLI, and WILLIAM R. LUSBY

Introduction

Consumption of certain fruits and vegetables has been associated with a lower incidence of several types of cancer.[1-3] As a result, several micronutrients common in the diets of humans have been suggested to be responsible for the reduction in the incidence of cancer or inhibition of carcinogen-induced neoplasia. Carotenoids are one of the most common dietary agents that have been studied as cancer-preventive agents. Although the number of naturally occurring carotenoids isolated from various sources is in excess of 600,[4] the number of carotenoids that are abundant in common foods is less than 50.[5] High-performance liquid chromatography (HPLC) has been shown to be the most efficient technique for the analysis of sensitive compounds in complex food extracts.[6] Numerous HPLC conditions have been developed to separate the carotenoids in extracts of natural products. However, HPLC procedures that can be universally used for the separation and quantification of carotenoids in extracts of foods are lacking. In this chapter we describe HPLC procedures that allow the effective separation and accurate quantitation of individual carotenoids in commonly consumed foods.

Nomenclature

For convenience only the common names of the carotenoids have been used throughout this text. The semisystematic names and chemical structures of these carotenoids have been tabulated by Pfander.[4] In cases in which definite geometric configuration of the *cis*-carotenoids is not known, the term *cis*-carotene has been used to distinguish these carotenoids from

---

[1] National Research Council, "Diet, Nutrition, and Cancer," pp. 358–370. National Academy, Washington, D.C., 1982.
[2] R. Peto, R. Doll, J. D. Buckley, and M. B. Sporn, *Nature (London)* **290**, 201 (1981).
[3] R. B. Shekelle, M. Lepper, S. Liu, P. Oglesby, A. M. Shryock, and J. Stamler, *Lancet* **2**, 1185 (1981).
[4] H. Pfander (ed.), "Key to Carotenoids," 2nd Ed. Birkhäuser, Basel, 1987.
[5] F. Khachik, G. R. Beecher, M. B. Goli, and W. R. Lusby, *Pure Appl. Chem.* **63**, 71 (1991).
[6] G. R. Beecher and F. Khachik, *in* "Nutrition and Cancer Prevention: Investigating the Role of Micronutrients" (T. E. Moon and M. S. Micozzi, eds), pp. 103–158. Dekker, New York, 1989.

their all-trans counterparts. According to the new nomenclature, the terms all-trans and -cis for in-chain geometric isomers of carotenoids should be referred to as all-E and -Z, respectively. However, for clarity of presentation, the terms all-trans and -cis, which are still being used by some scientists, have also been used throughout this text.

Experimental Procedures

Instrumentation. For gradient chromatography a ternary solvent delivery system, injector, and a rapid-scanning ultraviolet UV/visible photodiode array detector facilitate the separation and identification of carotenoids in food extracts. For isocratic separations only a single pump is required.

Chromatographic Conditions. Chromatographic conditions for the separation of carotenoids in various fruits and vegetables are summarized in Table I. The HPLC system A (gradient) can be universally applied to the separation of carotenoids and carotenol acyl esters isolated from various natural sources.[5,7,8] However, for chromatographic separation of hydrocarbon carotenoids found in certain yellow/orange fruits and vegetables, HPLC system B (isocratic) provides a simpler but effective HPLC procedure.[9] The HPLC system C (isocratic) is most applicable for the separation of geometric isomers of the hydrocarbon carotenoids.[8]

Carotenoid Standards. The reference samples of carotenoids, with the exception of lycopene, α-carotene, and β-carotene, are not commercially available. Therefore, carotenoid standards should be prepared either by total/partial synthesis or by isolation from natural sources. The total[10] and partial syntheses[11] of carotenoids have been fully investigated. A comprehensive list of the naturally occurring carotenoids, with appropriate references for the synthesis and isolation of more than 500 carotenoids from various classes, is available.[4] In cases in which the preparation of carotenoid standards requires elaborate organic synthesis, the isolation and purification of carotenoids from natural sources is often preferred as a more practical approach. In Table II we have summarized the source of standards for carotenoids and have provided appropriate references for the isolation of these compounds from common fruits and vegetables.[5,8,9,12–14]

---

[7] F. Khachik and G. R. Beecher, *J. Agric. Food Chem.* **36,** 929 (1988).
[8] F. Khachik, G. R. Beecher, and W. R. Lusby, *J. Agric. Food Chem.* **37,** 1465 (1989).
[9] F. Khachik and G. R. Beecher, *J. Agric. Food Chem.* **35,** 732 (1987).
[10] H. Mayer and O. Isler, *in* "Carotenoids" (O. Isler, ed), pp. 325–575. Birkhäuser, Basel, 1971.
[11] S. Liaaen-Jensen, *in* "Carotenoids" (O. Isler, ed), pp. 61–188. Birkhäuser, Basel, 1971.
[12] F. Khachik, G. R. Beecher, and N. F. Whittaker, *J. Agric. Food Chem.* **34,** 603 (1986).
[13] F. Khachik, G. R. Beecher, and W. R. Lusby, *J. Agric. Food Chem.* **36,** 938 (1988).

TABLE I
CHROMATOGRAPHIC CONDITIONS FOR SEPARATION OF CAROTENOIDS IN FOOD

Parameter	Conditions
HPLC system A, gradient	
Column	Rainin Microsorb 5-μm C_{18}, 25 cm × 4.6 mm, protected with Brownlee guard cartridge 5-μm C_{18}, 3 cm × 4.6 mm
Mobile phase	Acetonitrile 85%, Dichloromethane-hexane (1/1) 5% (Time 0–10 min) → Linear gradient → 45%, 45% (Time 40 min); Methanol 10% Over 30 min. 10%
Flow rate	0.70 ml/min
HPLC system B, isocratic	
Column	Rainin Microsorb 5-μm C_{18}, 25 cm × 4.6 mm, protected with Brownlee guard cartridge 5-μm C_{18}, 3 cm × 4.6 mm
Mobile phase	Acetonitrile 55%; Dichloromethane 23%; Methanol 22%
Flow rate	1 ml/min
HPLC system C, isocratic	
Column	Vydac (201 TP54) C_{18} with a low carbon loading (5-μm particles with a large pore size, 300 Å), 25 cm × 4.6 mm, protected with a Vydac guard cartridge, 3 cm × 4.6 mm, packed with the same adsorbent
Mobile phase	Acetonitrile 85%; Methanol 10%; Dichloromethane/hexane (1/1) 5%
Flow rate	0.70 ml/min

Even though the carotenoids listed in Table II can be isolated in high concentrations from many natural products,[4] the fruits and vegetables listed in Table II provide a readily available source of specific carotenoids.

Internal Standards. Because the preparation of carotenoid samples for HPLC analysis requires extensive extraction and work-up procedures that can be accompanied by various analytical errors, the use of an internal standard is essential. Several synthetic carotenoids can be used as internal standards for quantitation of carotenoids; these include β-apo-8'-carotenal,[12] ethyl β-apo-8'-carotenoate,[8] and nonapreno-β-carotene.[9] β-Apo-8'-carotenal and ethyl β-apo-8'-carotenoate are both commercially available (Fluka Chemical Corp., Ronkonkoma, NY); nonapreno-β-carotene, which is a synthetic C_{45} β-carotene, can be readily prepared from

---

[14] F. Khachik, M. B. Goli, G. R. Beecher, J. Holden, W. R. Lusby, M. D. Tenorio, and M. R. Barrera, *J. Agric. Food Chem.* **40**, 390 (1992).

TABLE II
SOURCE OF CAROTENOID STANDARDS IN COMMON FRUITS AND VEGETABLES

Carotenoids	Fruits and vegetables
Neoxanthin, violaxanthin, and lutein-5,6-epoxide	Green fruits and vegetables[a] (i.e., broccoli, spinach, kale, and Brussels sprouts)[b]
Neochrome, luteochrome, aurochrome, and flavoxanthin	Partial synthesis from neoxanthin, violaxanthin, and lutein epoxide isolated from green vegetables[b]
Lutein	Green fruits and vegetables,[a] and squash[c]
Zeaxanthin	Peaches[d]
2′,3′-Anhydrolutein	Squash[c]
β-Cryptoxanthin-5,6-epoxide	Papaya[a]
α-Cryptoxanthin	Oranges[a] (from saponified extracts)
β-Cryptoxanthin	Mango,[a] papaya,[a] and peaches[d]
Lycopene 1,2-epoxide, lycopene 5,6-epoxide, and lycopene	Tomato paste/products[e]
γ-Carotene	Apricots[d]
ζ-Carotene, phytofluene, and phytoene	Tomato paste/products,[e] apricots,[d] pumpkin,[a] mango,[a] papaya,[a] and peaches[d]
α-Carotene and β-carotene	(Carrots, red palm oil, and sweet potato)[f]

[a] Ref. 5.
[b] Ref. 12.
[c] Ref. 13.
[d] Ref. 8.
[e] Ref. 14.
[f] Ref. 9.

commercially available starting materials.[15] Other carotenoids for selected applications have also been used as internal standards.[7]

Preparation of Food Extracts. To prepare food extracts for the determination of carotenoids requires the use of appropriate organic solvents such as acetone, ether, petroleum ether, methanol, or tetrahydrofuran. In foods that contain significant amounts of water, it is desirable to use an organic solvent that is miscible with water in an attempt to facilitate the denaturing of carotenoid–protein complexes and to prevent the formation of emulsions. Acetone and tetrahydrofuran (THF) can be employed as extracting solvents; however, it is advisable to stabilize the extracting solvent with an antioxidant such as butylated hydroxytoluene (BHT) to prevent the promotion of epoxides. In cases in which dried fruits and vegetables are to be extracted, the extraction and removal of carotenoids can be facilitated if food samples are rehydrated for several hours prior to extraction. For a general homogenization procedure, foods are mixed with the extracting

[15] F. Khachik and G. R. Beecher, *J. Ind. Eng. Chem. Prod. Res. Dev.* **25,** 671 (1986).

solvent in the presence of sodium or magnesium carbonate using a blender, a known amount of the internal standard is added, and the resulting mixture is homogenized at a moderate speed. Extraction is normally carried out at 0° by immersing the homogenization vessel in an ice bath to prevent the degradation and isomerization of carotenoids. The function of sodium or magnesium carbonate is to neutralize trace levels of organic acids that are commonly present in some foods and can cause destruction and/or structural transformation of carotenoids. The resulting carotenoid extract is then filtered and the homogenization of the residue is repeated until the filtrate is devoid of yellow color. The combined filtrate is concentrated, and the carotenoids are partitioned into an appropriate organic solvent [i.e., petroleum ether, petroleum ether–85% (v/v) aqueous methanol, or dichloromethane] and water. The organic layer is removed, dried, and concentrated into the appropriate volume of selected solvent for HPLC analysis. During solvent evaporation, attempts must be made to avoid heating the carotenoid extracts above 40°. The exposure of the solution to air and direct light should be minimized. All carotenoid extracts of foods must be centrifuged or filtered through microfilters to ensure the removal of small particles, which will impair HPLC analysis. The injection solvent for the analysis of carotenoids by HPLC should be carefully selected so that chromatographic artifacts are not produced.[16] With HPLC systems A and B, acetonitrile–dichloromethane–hexane–methanol (40:20:20:20, v/v) is an effective solvent mixture for solubilizing carotenoids and can be used as the HPLC injection solvent without production of artifacts. For separations involving HPLC system B, the eluent used with this system is sufficiently strong to dissolve carotenoids isolated from various sources and as a result the same eluent may be used as the HPLC injection solvent. For detailed instructions on isolation and extraction procedures for carotenoids from natural sources see the publication by Liaaen-Jensen.[11]

Saponification. The saponification of carotenoid extracts can provide valuable information as to the nature of the carotenoids present in a specific food. The majority of carotenoids are stable toward alkaline treatment used to remove the unwanted chlorophylls and lipids from food extracts. For most fruits and vegetables, moderate saponification conditions can be employed. This involves treatment of the extract with 10% (w/v) methanolic potassium hydroxide for 2 to 3 hr at room temperature. However, for high-fat foods that contain a large concentration of lipids and sterols, more severe treatments may be required. Where carotenoids sensitive to saponification are present (i.e., astaxanthin and fucoxanthin) this step should be omitted. However, in some high-fat foods and some fruits

[16] F. Khachik, G. R. Beecher, J. T. Vanderslice, and G. Furrow, *Anal. Chem.* **60,** 807 (1988).

(citrus) with complicated carotenoid HPLC profiles (carotenol fatty acid esters) saponification may be a more practical approach. Saponification also converts carotenol acyl esters into their corresponding hydroxycarotenoids. HPLC conditions for separation of several of the naturally occurring straight-chain fatty acid mono- and bisesters of carotenoids (i.e., lutein, zeaxanthin, α- and β-cryptoxanthin, violaxanthin, and violeoxanthin) have been developed that allow the detection of these compounds in foods as they normally occur.[7,8,17] A comparison between the chromatographic profiles of the saponified and unsaponified food extracts should be a common practice in any carotenoid analysis of foods. A typical saponification procedure involves stirring a solution of the food extract in tetrahydrofuran and/or methanol with an equal volume of a 10% solution of methanolic KOH for 2 hr at room temperature under an inert atmosphere (i.e., nitrogen or helium) and subdued light. The mixture is then partitioned into dichloromethane and water, the organic phase is removed, washed with water until free of alkaline, and dried over sodium sulfate. The solution is evaporated to dryness and dissolved in the appropriate HPLC injection solvent and filtered through microfilters for chromatographic analysis.

Preparation of Standard Curves for Quantitation of Carotenoids. The carotenoid constituents in the extracts of various fruits and vegetables are quantified from the HPLC response factors (peak area or peak height) of individual carotenoids at five or six different concentrations, which are obtained under the appropriate HPLC conditions. For each carotenoid a standard curve is obtained by plotting HPLC peak area at various concentrations. The purity and accurate concentration of the carotenoid solutions are crucial for the construction of the standard curves. This can be determined from the UV/visible absorption spectrum of the carotenoid in question according to the analytical procedures described by De Ritter and Purcell.[18] Regression analysis of the data for the standard curve for each carotenoid provides several parameters with which to evaluate the validity of the standards and the instrumentation. First, the correlation coefficient should be greater than 0.9. Second, the intercept should be very close to zero, and finally, the relative standard deviation of the regression (standard error of the estimate divided by average concentration of standards multiplied by 100) should be less than 5%. If any of these statistical parameters are out of range, the standards as well as the HPLC instrumentation must

---

[17] F. Khachik and G. R. Beecher, *J. Chromatogr.* **449**, 119 (1988).
[18] E. De Ritter and A. E. Purcell, *in* "Carotenoids as Colorants and Vitamin A Precursors: Technical and Nutritional Applications" (J. C. Bauernfeind, ed), pp. 815–882. Academic Press, New York, 1981.

be carefully examined and the standard curve rerun. In cases in which the reference samples of *cis*-carotenoids are not available, the concentration of *cis*-carotenoids in food extracts may be determined from the standard curve of their all-trans counterpart. In separation of carotenoids from various food extracts, often the HPLC peak areas of *cis*-carotenoids are not resolved and appear as a trailing shoulder on their all-trans compounds. Therefore, in such separations, the concentration of certain carotenoids may be determined by combining the HPLC peak area of all-*trans*-carotenoids with that of their cis counterparts. Carotenoid constituents in food extracts are quantified by converting HPLC response to concentration employing the regression equation for the respective carotenoids. The recovery of the internal standard in the extracts from foods should be monitored, in an attempt to determine the loss of carotenoids due to extraction and workup procedures. The application of internal standards in quantitative determination of carotenoids has been discussed previously.[19]

Separation of Carotenoids of Fruits and Vegetables by HPLC

Chromatographic profiles of the common fruits and vegetables with respect to their carotenoid composition may be divided into three major categories[5]: (1) green fruits and vegetables with similar HPLC profiles; (2) yellow/red fruits and vegetables, containing predominantly hydrocarbon carotenoids; and (3) yellow/orange fruits and vegetables with complex HPLC profiles, mostly containing carotenol acyl esters. The separation of carotenoids from the extracts of fruits and vegetables from these three categories can be accomplished effectively by gradient chromatography employing HPLC system A. However, for selected separations of carotenoids from food extracts, isocratic chromatographic conditions employing HPLC systems B and C may provide more effective and less time-consuming alternatives.

Separation of Carotenoids of Green Fruits and Vegetables. Green fruits and vegetables all have similar chromatographic profiles showing three major groups of pigments. In the order of chromatographic elution from a C_{18} reversed-phase column these are (1) xanthophylls (oxygenated carotenoids), (2) chlorophylls and their derivatives, and (3) hydrocarbon carotenoids. A typical chromatogram of an extract from raw green beans is shown in Fig. 1. The HPLC peak identification of carotenoids is found in Table III. The separation of carotenoids by gradient chromatography allows the separation of all the major carotenoid and chlorophyll constitu-

---

[19] F. Khachik and G. R. Beecher, *J. Chromatogr.* **346**, 237 (1985).

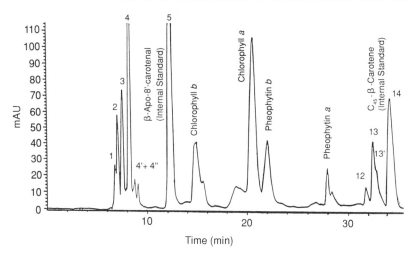

FIG. 1. Carotenoid HPLC profile of an extract from raw green beans employing HPLC system A (see Table I); for peak identification see Table III.

ents of the green fruits and vegetables without the need for saponification and removal of chlorophylls and their derivatives.[12] Two synthetic carotenoids, namely β-apo-8'-carotenal[12] and C_{45}-β-carotene (nonapreno-β-carotene),[9,14] have been used as internal standards for quantitative analysis of the xanthophylls and the carotenes, respectively. The major chlorophyll constituents of green fruits and vegetables are chlorophylls *b* and *a*. The general chromatographic profiles of the other green fruits and vegetables are very similar to that of raw green beans illustrated in Fig. 1. The major differences among the green fruits and vegetables appear to be the relative concentrations of the individual carotenoids.

Separation of Carotenoids in Yellow/Red Fruits and Vegetables, Containing Predominantly Hydrocarbon Carotenoids. The second category consists of a number of yellow/red fruits and vegetables that have similar chromatographic profiles and contain mostly hydrocarbon carotenoids. Some of the most common fruits and vegetables in this group are apricots, cantaloupe, carrots, grapefruit (pink), pumpkin, sweet potato, and tomatoes. A typical chromatogram (HPLC system A) of an extract from pumpkin is shown in Fig. 2. The major carotenoids in this vegetable include lutein, ζ-carotene, α-carotene, β-carotene, phytofluene, and phytoene. To ensure the detection of all the carotenoids, this chromatogram was monitored at four different wavelengths (450, 400, 350, and 286 nm) simultaneously using a photodiode array detector. According to the chromatogram shown in Fig. 2, there appears to be significant peak overlap between

TABLE III
IDENTIFICATION OF CAROTENOIDS IN FRUITS AND VEGETABLES SEPARATED BY HPLC IN THE ORDER OF ELUTION ON A C_{18} REVERSED-PHASE COLUMN

Peak[a]	Carotenoids[b]	λ_{max} (nm) in HPLC eluents of systems A, B, C
1	Neoxanthin	438
2	all-trans-Violaxanthin	442
2'	9-cis-Violaxanthin	438
3	Lutein 5,6-epoxide	442
4	all-trans-Lutein	446
4' + 4"	cis-Luteins	442
5	β-Apo-8'-carotenal (internal standard)	456
6	Ethyl β-apo-8'-carotenoate (internal standard)	456
7	β-Cryptoxanthin 5,6-epoxide	450
8	β-Cryptoxanthin	454
9	Lycopene	472
10	all-trans-γ-Carotene	464
10'	cis-γ-Carotene	460
11	ζ-Carotene	400
12	α-Carotene	446
13	all-trans-β-Carotene	454
13' + 13"	cis-β-Carotenes	450
14	C_{45}-β-carotene (internal standard)	478
15	all-trans- or cis-Phytofluene	350
15'	cis- or all-trans-Phytofluene	350
16	all-trans- or cis-Phytoene	286
16'	cis- or all-trans-Phytene	286

[a] See Figs. 1–5 for peaks.
[b] cis-Carotenoids have been designated with the same number as their all-trans isomers but distinguished from them by prime symbols.

ζ-carotene (peak 11) and α-carotene (peak 12) if these peaks were monitored at 400 and 450 nm, respectively. This interference can be minimized by monitoring ζ-carotene at 380 nm and α- and β-carotene at 475 nm. Although the separation of carotenoids in yellow/red fruits and vegetables can be accomplished employing gradient HPLC system A, much simpler isocratic HPLC conditions (i.e., systems B and C) can also be applied effectively. For example, Fig. 3 shows that the separation of α- and β-carotene from an extract of carrots can be readily achieved employing HPLC system B (isocratic).[9]

A major weakness of the C_{18} reversed-phase HPLC columns is the inadequate separation of the geometric (cis–trans) isomers of carotenoids.

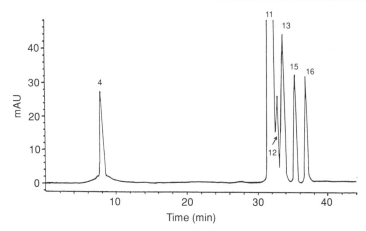

FIG. 2. Carotenoid HPLC profile of an extract from pumpkin employing HPLC system A (see Table I); for peak identification see Table III.

While C_{18} columns of different brands remain most popular for the separation of the geometric isomers of the oxygenated carotenoids from various foods, other HPLC columns have been found to be more effective for the separation of the geometric isomers of the hydrocarbon carotenoids. Separation of cis isomers of β-carotene are best achieved on an HPLC lime

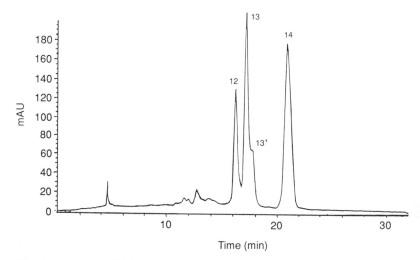

FIG. 3. Carotenoid HPLC profile of an extract from raw carrots employing HPLC system B (see Table I); for peak identification see Table III.

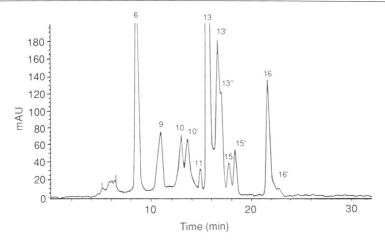

FIG. 4. Carotenoid HPLC profile of an extract from dried apricots employing HPLC system C (see Table I); for peak identification see Table III.

column as reported by Tsukida et al.[20] Although this column separates various cis isomers of β-carotene, it is not established if separation of other classes of carotenoids that are encountered in food extracts can also be simultaneously accomplished using this column. This is primarily due to the lack of commercial availability of the lime HPLC column. Meanwhile, modified C_{18} reversed-phase HPLC columns with a low carbon loading and large pore size (300 Å) show promise in the separation of the geometric isomers of the hydrocarbon carotenoids. For example, a Vydac HPLC column (201 TP54) has been shown to be effective in chromatographic separation of carotenoids and their stereoisomers in extracts from several fruits.[8] This is illustrated in the HPLC profile (HPLC system C) of an extract from dried apricots in Fig. 4, which demonstrates the presence of lycopene, γ-carotene, ζ-carotene, β-carotene, phytofluene, and phytoene. With the exception of all-*trans*-lycopene and ζ-carotene, the stereoisomers of γ-carotene, β-carotene, phytofluene, and phytoene in dried apricots are separated isocratically on a Vydac column employing HPLC system C. In the separation of carotenoids from various food extracts, the choice of the chromatographic conditions is entirely dependent on the carotenoid composition of foods. While HPLC system A can be universally used for the separation of various classes of carotenoids in complex food extracts, HPLC systems B and C can be used for simple and selected applications.

Separation of Carotenoids in Yellow/Orange Fruits and Vegetables, Containing Mainly Carotenol Acyl Esters. The third category of fruits and

[20] K. Tsukida, K. Saiki, T. Takii, and Y. Koyama, *J. Chromatogr.* **245**, 359 (1982).

vegetables consists of the yellow/orange foods, which usually contain carotenol acyl esters. Some common fruits and vegetables in this group are mango, papaya, peaches, prunes, squash, and oranges. The chromatographic profiles of the extracts from these fruits and vegetables are quite complex, because the naturally occurring hydroxycarotenoids (i.e., lutein, zeaxanthin, and β-cryptoxanthin) and violaxanthin (an epoxycarotenoid) are esterified with straight-chain fatty acids such as lauric, myristic, and palmitic acids. In certain extracts from fruits and vegetables that contain carotenoids with two hydroxyl groups (i.e., lutein, zeaxanthin, and viola-

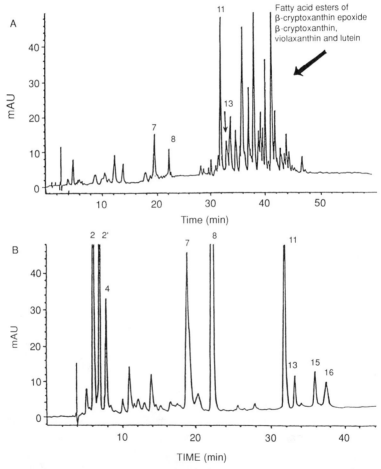

FIG. 5. Carotenoid HPLC profile of an extract from papaya before (A) and after (B) saponification employing HPLC system A (see Table I); for peak identification see Table III.

xanthin), carotenol esters with a mixture of fatty acid side chains are also present. Because of this complexity, carotenoid extracts from natural sources that are suspected to contain carotenol acyl esters are customarily saponified to remove the fatty acids and liberate the parent carotenoids. Typical chromatograms (HPLC system A) of an extract from papaya before and after saponification are shown in Figs. 5A and 5B, respectively. The major carotenoid constituents in the unsaponified extract (Fig. 5A) are ζ-carotene, β-carotene, phytofluene, phytoene, β-cryptoxanthin acyl esters, β-cryptoxanthin 5,6-epoxide acyl esters, violaxanthin acyl esters, and lutein acyl esters. Low levels of free β-cryptoxanthin 5,6-epoxide and β-cryptoxanthin are also present in the unsaponified extract. Unfortunately, under chromatographic conditions (system A) the HPLC peaks of phytofluene and phytoene are masked by the HPLC peaks of the acyl esters of violaxanthin and β-cryptoxanthin. Therefore, for quantitative analysis of carotenoids in papaya, saponification may provide a more satisfactory approach. The HPLC profile of the saponified extract of papaya (Fig. 5B) shows the presence of ζ-carotene, phytofluene, and phytoene. This chromatogram also confirms the liberation of free violaxanthin, lutein, β-cryptoxanthin 5,6-epoxide, and β-cryptoxanthin from their corresponding acyl esters. The HPLC profiles of extracts from other fruits and vegetables that are listed in Table III show similar separations for carotenol acyl esters. Although in certain cases the saponification of carotenoid extracts from foods may be inevitable, HPLC conditions such as system A can successfully be implemented to separate carotenol acyl esters in the extracts from most fruits and vegetables. Such HPLC procedures can be highly advantageous and provide valuable information on the identity and the levels of these compounds in their natural state in foods.

Acknowledgments

We thank Hoffmann-La Roche & Co. (Basel, Switzerland) and BASF (Ludwigshafen, Germany) for their gift of carotenoid reference samples and their precursors. Supported in part by the National Cancer Institute through reimbursable Agreement #Y01-CN-30609.

[33] Distribution of Macular Pigment Components, Zeaxanthin and Lutein, in Human Retina

By RICHARD A. BONE and JOHN T. LANDRUM

Introduction

The isomeric dihydroxycarotenoids, zeaxanthin and lutein, have been identified as the major constituents of the macular pigment of the human retina.[1] Although these pigments appear to be localized in a small area (~ 1- to 3-mm diameter) centered on the fovea, high-performance liquid chromatography (HPLC) has revealed their presence throughout the neural retina.[2,3] This technique is also sufficiently sensitive to quantify the pigments in neonatal and prenatal eyes, where their presence was previously in dispute.[4]

An alternative approach to determining the distribution of macular pigment carotenoids in the retina is through microspectrophotometry of fixed retinal tissue.[5-8] Beyond the immediate yellow spot, however, the pigment density appears to be too low for accurate quantitation by this technique. Neither does it allow the relative proportions of zeaxanthin and lutein to be determined, only the total amount. Both of these objections apply also to psychophysical methods, where pigment density is inferred from its effects on spectral sensitivity.[9-11]

---

[1] R. A. Bone, J. T. Landrum, and S. L. Tarsis, *Vision Res.* **25**, 1531 (1985).
[2] R. A. Bone, J. T. Landrum, L. Fernandez, and S. L. Tarsis, *Invest, Ophthalmol. Visual Sci.* **29**, 843 (1988).
[3] G. J. Handelman, E. A. Dratz, C. C. Reay, and F. J. G. M. van Kuijk, *Invest. Ophthalmol. Visual Sci.* **29**, 850 (1988).
[4] J. J. Nussbaum, R. C. Pruett, and F. C. Delori, *Retina* **1**, 296 (1981).
[5] D. M. Snodderly, P. K. Brown, F. C. Delori, and J. D. Auran, *Invest. Ophthalmol. Visual Sci.* **25**, 660 (1984).
[6] D. M. Snodderly, J. D. Auran, and F. C. Delori, *Invest. Ophthalmol. Visual Sci.* **25**, 674 (1984).
[7] G. L. Handelman, D. M. Snodderly, N. I. Krinsky, M. D. Russett, and A. J. Adler, *Invest. Ophthalmol. Visual Sci.* **32**, 257 (1991).
[8] D. M. Snodderly, G. J. Handelman, and A. J. Adler, *Invest. Ophthalmol. Visual Sci.* **32**, 268 (1991).
[9] P. L. Pease, A. J. Adams, and E. Nuccio, *Vision Res.* **27**, 705 (1987).
[10] J. S. Werner, S. K. Donnelly, and R. Kliegl, *Vision Res.* **27**, 257 (1987).
[11] R. A. Bone and J. M. B. Sparrock, *Vision Res.* **11**, 1057 (1971).

Tissue Handling

Donor eyes are often available from eye banks or may be obtained through the National Disease Research Interchange (Philadelphia, PA) either fresh or frozen. Dissection of the thawed eye is performed under subdued, incandescent light in 0.9% (w/v) saline solution in order to isolate the neural retina. After trimming the retina around the equator, it should be inspected for damage, including separation of retinal layers. A Lucite sphere (obtainable from hobby shops), with a diameter approximating that of the retina (1 in. for adults), is placed in the saline-filled dissecting dish. After maneuvering the inverted retina into position above it, the sphere is lifted from the solution and seated in a rubber-lined ring in the apparatus shown in Fig. 1. The retina will drape smoothly over the sphere.

The sphere is rotated in the ring until the fovea coincides with the cross-wires of a low-power microscope set concentrically in the upper aluminum block. The microscope is then replaced by a set of spaced, concentric trephines designed to slide telescopically relative to each other and to the upper block. By bearing down with these onto the sphere and applying a gentle back-and-forth rotation, the retina may be cut into a number of annuli concentric with the fovea. In our own study,[2] we ob-

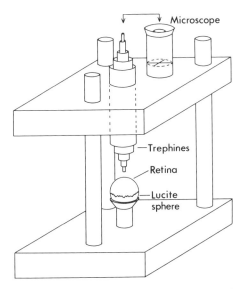

FIG. 1. Apparatus constructed for the purpose of dissecting retinas into annuli concentric with the fovea. (Adapted from Bone et al.,[2] with permission of *Investigative Ophthalmology & Visual Science.*)

tained a range of linear surface distances from the foveal center of 0–0.75, 0.75–1.6, 1.6–2.5, 2.5–5.8, 5.8–8.7, and 8.7–12.2 mm. Individual trephines of selected diameter may also be constructed for use with this apparatus.

Internal Standard

In order to accurately quantify the carotenoids in the tissue samples by HPLC, lutein monomethyl ether may be prepared as an internal standard.[12] This ether differs from lutein by the methylation of the allylic hydroxyl group of the ε ring. It is readily prepared by reaction of pure lutein (Sigma Chemical Company, St. Louis, MO) in methanolic HCl. Appropriate reaction conditions are 2 mg of lutein stirred for 7 min in 20 ml of the HCl–methanol solution at room temperature. The methanolic HCl is prepared by addition of aqueous concentrated HCl to methanol in the ratio of 1 to 9 on a volume-to-volume basis. The reaction is quenched by addition of aqueous saturated sodium bicarbonate. Workup is achieved in a separatory funnel by the addition of dichloromethane, in which the lutein monomethyl ether is preferentially soluble. The dichloromethane layer is washed several times with distilled water and dried under a stream of pure nitrogen.

The resulting products, believed to be two diastereomeric isomers of the monomethyl ether, appear as slightly overlapping peaks when analyzed by reversed-phase HPLC under conditions identical to those described below (see "Separation and Quantitation by HPLC"). The first eluting, and more abundant, isomer, with a yield of about 60%, is collected. Its retention time is significantly longer than that of lutein (24 vs 17 min) making it an ideal, noninterfering internal standard. Its similarity to the carotenoids being analyzed would be expected to render it equally susceptible to chemical degradation. Alternative internal standards, ethyloxime and methyloxime derivatives of β-apo-8'-carotenal, have also been used in this type of study.[3,7]

The lutein monomethyl ether is dissolved in methanol to produce a suitably low concentration ($\sim 10^{-3}$ μg/μl) and stored in a freezer under a nitrogen atmosphere. The concentration is determined spectroscopically using an extinction coefficient of 251 liters/(g·cm) at peak (~ 444 nm), and should be checked periodically.

---

[12] S. L. Jensen and S. Hertzberg, *Acta Chem. Scand.* **20**, 1703 (1966).

Carotenoid Extraction

The macular pigments are extracted from the annuli or disks of retinal tissue by grinding them in a glass tissue homogenizer in 2 to 3 ml of acetone (analytical grade) to which is added an accurately known mass (~ 10 ng) of the internal standard. The acetone extract is centrifuged, filtered through a 0.2-μm nylon-mesh syringe filter, and dried in a 5-ml conical bottom flask under a stream of pure nitrogen. Larger tissue sample extracts are found to contain significant amounts of greasy components. The quantity of these is greatly reduced by taking advantage of the preferential solubility of the carotenoids in cold acetone. The solvent, at $-20°$, is swirled gently around the prechilled 5-ml flask and quickly transferred to a clean flask where it is dried, as before, in preparation for HPLC. The white, greasy components are retained on the walls of the original flask. This procedure may be repeated several times if necessary.

Separation and Quantitation by HPLC

HPLC is performed using a 250 × 4.6 mm, C_{18} reversed-phase column with an efficiency of around 100,000 plates/m [for example, as supplied by Keystone Scientific (Bellefonte, PA), with a 5-μm Spherisorb ODS1 packing]. For the mobile phase, an isocratic eluent composed of 92% methanol and 8% water-acetonitrile (3:1, v/v) at a flow rate of 1 ml/min provides essentially baseline separation of lutein and zeaxanthin. Their retention times are about 16 and 17 min, respectively, and that of the internal standard is about 24 min. The detector should be set to a wavelength of 450 nm, and its sensitivity should impose a lower detection limit of no more than about 0.2 ng for each carotenoid.

The quantities of zeaxanthin and lutein are determined by their peak areas relative to that of the internal standard. In the case of lutein, which has the same extinction coefficient as the standard, no weighting of areas is required. Zeaxanthin, however, has a larger extinction coefficient at 450 nm than that of lutein and the internal standard and, accordingly, the peak area should be multiplied by 1.087 to correct for this.

Stereochemistry of the Macular Pigment Components

The specific diastereoisomers of zeaxanthin and lutein that constitute the macular pigment may be investigated by additional HPLC on a suitable chiral column [for example, a 250 × 4-mm Sumichiral OA-2000 from Regis Chemical Company (Morton Grove, IL)]. Dibenzoate esters of zeaxanthin and lutein are separable into their individual diastereoisomers on such a column. To prepare the esters, samples of accumulated macular zeaxanthin and lutein (~0.25 μg) are dissolved in 125 μl of anhydrous pyridine, to which is added 5 to 10 μl of benzoyl chloride. The mixture is stirred under a dry nitrogen atmosphere for 30 min at 10°. The products are dissolved in hexane and washed with sodium bicarbonate solution followed by water. After evaporating the hexane under a stream of nitrogen, the dibenzoate esters may be purified on the reversed-phase column. Their retention times are about 50% greater than for the parent compounds.

Elution conditions with the chiral column consist of an eluent composed of hexane/methylene chloride/ethanol (90:10:0.1), as used by Maoka et al.,[13] and a flow rate of 1 ml/min. Under these conditions, macular zeaxanthin is resolved into similar amounts of two diastereoisomers, m-zeaxanthin [(3R,3′S)-β,β-carotene-3,3′-diol] (first eluting peak) and zeaxanthin [(3R,3′R)-β,β-carotene-3,3′-diol] (second eluting peak). Macular lutein elutes as the single diastereoisomer, lutein [(3R,3′R,6′R)-β,β-carotene-3,3′-diol]. Retention times are typically 20 to 30 min.

For comparison, standards of the three diastereoisomers of zeaxanthin may be prepared from rhodoxanthin.[13] One hundred micrograms of rhodoxanthin (from yew berries) is reacted in 0.5 ml of methylene chloride with 25 μl of acetic acid and 1 mg of activated zinc powder at 20° for 30 min. The main product, β,β-carotene-3,3′-dione, is reduced with 5 mg of sodium borohydride in 0.5 ml of methanol to give a racemic mixture of zeaxanthin diastereoisomers. Dibenzoate esters of this mixture, and of a lutein standard, may be prepared as described above. On the chiral column, the zeaxanthin diastereoisomers elute in the following order: (3R,3′S), (3R,3′R), and (3S,3′S). For identification of the macular zeaxanthin and lutein diastereoisomers, coinjections with the racemic zeaxanthin and lutein standards are essential.

Comments

A sample set of chromatograms is shown in Fig. 2 (see Ref. 2). Those labeled (a) and (b) in Fig. 2 were obtained with single trephines, whereas (c) and (d) in Fig. 2 are examples of those obtained with the set of trephines shown in Fig. 1. In the majority of cases, zeaxanthin was the dominant component from the center of the fovea out to a radial distance of about 2.5 mm. Beyond this point, lutein was found in greater abundance. The total amount of carotenoids per unit area was found to decrease from

---

[13] T. Maoka, A. Arai, M. Shimizu, and T. Matsuno, *Comp. Biochem. Physiol. B* **83B**, 121 (1986).

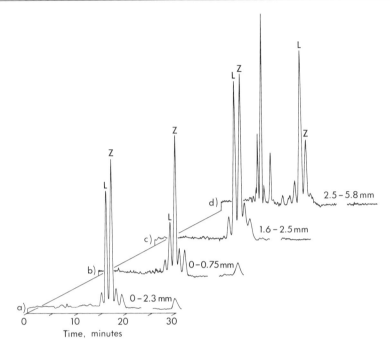

FIG. 2. Representative chromatograms of pigments extracted from retinal tissue. L, Lutein; Z, zeaxanthin; detector wavelength, 450 nm. Chromatogram (a) was obtained from a central disk (0–2.3 mm). Chromatograms (b), (c), and (d) illustrate the dramatic change in the zeaxanthin–lutein ratio with retinal eccentricity. For the sake of clarity, the internal standard peak (retention time ~24 min) has been omitted. (Adapted from Bone et al.,[2] with permission of *Investigative Ophthalmology & Visual Science*.)

approximately 13 ng/mm^2 in the central 0 to 0.25 mm, to around 0.05 ng/mm^2 at eccentricities in the range 8.7 to 12.2 mm.

Chromatograms typically contained a number of small peaks in addition to the two principal macular pigment components. Their exact identity has not been established, although their spectra suggest carotenoids. In Fig. 2a, the two components immediately following zeaxanthin peaked at 444 and 441 nm, respectively, and were further characterized by cis peaks at approximately 335 nm. That preceding lutein peaked at 446 nm, but did not appear to have a cis peak. With larger tissue samples beyond the fovea, other earlier eluting peaks appeared (Fig. 2d), reported to be those of retinoids.[3]

The distribution of individual zeaxanthin diastereoisomers throughout the retina is not known, but is currently under investigation in our laboratory. In the macular region, the average ratio of zeaxanthin to m-zeaxanthin was found to range from 1.08 to 1.34 to 1.

Acknowledgments

This work was supported by NIH Grants EY05452 and RR08205. We gratefully acknowledge the National Disease Research Interchange, Philadelphia, and the Florida Lions Eye Bank for supplying human donor eyes.

[34] Carotenoid Glycoside Ester from *Rhodococcus rhodochrous*

By SHINICHI TAKAICHI and JUN-ICHI ISHIDSU

Introduction

Only eight kinds of fatty acid esters of C_{40}-carotenoid glycosides, including two precursors, have been reported. Kleinig and co-workers have identified an ester of 1'-glucosyloxy-3',4'-didehydro-1',2'-dihydro-β,ψ-carotene, esters of its 3,4-didehydro (myxobactin), 4-keto (myxobactone), and 3-hydroxy derivatives, and esters of its two precursors. Myxobactin ester and myxobactone ester have been isolated only from two species, *Myxococcus fulvus* Mx f2[1,2] and *Stigmatella aurantiaca* Sg a1.[3] The ester of the 3-hydroxy derivative has been isolated only from two species, *Chondromyces apiculatus*[4] and *Sorangium compositum* So c1.[5] These four species belong to three distinct families of fruiting gliding bacteria.[6] Myxobactone ester and its diglucoside type have been isolated from a nonfruiting gliding bacterium, *Herpetosiphon gigantens*.[7] 1'-[(x-O-Palmitoyl-β-D-glucopyranosyl)oxy]-3',4'-didehydro-1',2'-dihydro-β,ψ-caroten-2'-o1 has been isolated from a nocardioform actinomycete, *Nocardia kirovani*.[8] In

---

[1] H. Reichenbach and H. Kleinig, *Arch. Mikrobiol.* **76**, 364 (1971).
[2] H. Kleinig, *Eur. J. Biochem.* **57**, 301 (1975).
[3] H. Kleinig, H. Reichenbach, and H. Achenbach, *Arch. Mikrobiol.* **74**, 223 (1970).
[4] H. Kleinig and H. Reichenbach, *Phytochemistry* **12**, 2483 (1973).
[5] H. Kleinig, H. Reichenbach, H. Achenbach, and J. Stadler, *Arch. Mikrobiol.* **78**, 224 (1971).
[6] H. D. McCurdy, in "Bergey's Manual of Systematic Bacteriology" (J. T. Staley, M. P. Bryant, N. Pfennig, and J. G. Hold, eds.), Vol. 3, p. 2139. Williams & Wilkins, Baltimore, 1989.
[7] H. Kleinig and H. Reichenbach, *Arch. Microbiol.* **112**, 307 (1977).
[8] M.-J. Vacheron, N. Arpin, and G. Michel, *C. R. Hebd. Seances Acad. Sci., Ser. C* **271**, 881 (1970).

all cases, the glycoside is D-glucoside, and the fatty acids are similar to the major ones in cellular lipids. However, whether the glucoside is the α or β type has not yet been reported with sufficient data; furthermore, the bonding position of the fatty acid moiety on the glucose moiety has not yet been reported.

The fatty acid esters of di(β-D-glucopyranosyl)-4,4'-diapocarotene-4,4'-dioate and its precursors from *Methylobacterium rhodinum* B9,[9,10] and [6-*O*-(12-methyltetradecanoate)-α-D-glucopyranosyl]-7',8'-dihydro-4,4'-diapocaroten-4-oate, namely staphyloxanthin, from *Staphylococcus aureus* S41[11] have been isolated. The fatty acids in the esters are also similar to the major ones in cellular lipids. They are esterified with the C-6 hydroxyl group of the glucose moiety, although data on them are still insufficient for structural elucidation. Furthermore, these carotenoid glucosides are a β and an α type, respectively, and their carotenoid moiety is C_{30}-carotenoic acid.

About 10 different carotenoid glycoside esters have been found among 600 structurally identified naturally occurring carotenoids.[12] This may be due to difficulty in isolation, to saponification being used almost routinely during isolation, or to the small amount of the esters. Furthermore, the composition of the fatty acids in the carotenoid glycoside esters is easily influenced by the contamination of a large amount of cellular lipids.

The chemical structure of a carotenoid glucoside ester from a nocardioform actinomycete, *Rhodococcus rhodochrous* RNMS1, has been fully elucidated.[13] It has the 3',4'-saturated structure of the carotenoid moiety in contrast to the myxobactone ester with the 3',4'-unsaturated structure. This chapter describes procedures for identification of the ester, including determination of the structures of its fatty acid moieties.

Biological Material[13,14]

Rhodococcus rhodochrous RNMS1 (IAM 13988) is cultured in a nutrient broth containing 1.0% (w/v) peptone, 0.3% (w/v) meat extract, and 0.2% (w/v) NaCl (pH 7.0) with shaking at 36° for 45–60 hr. As cells form aggregates, they are harvested with a glass microfiber filter (e.g., GF/F;

---

[9] H. Kleinig, R. Schmitt, W. Meister, G. Englert, and H. Thommen, *Z. Naturforsch., C: Biosci.* **34C**, 181 (1979).
[10] P. N. Green and I. J. Bousfield, *Int. J. Syst. Bacteriol.* **33**, 875 (1983).
[11] J. H. Marshall and G. J. Wilmoth, *J. Bacteriol.* **147**, 900 (1981).
[12] O. Straub, in "Key to Carotenoids" (H. Pfander, ed.), 2nd Ed. Birkhäuser, Basel, 1987.
[13] S. Takaichi, J. Ishidsu, T. Seki, and S. Fukuda, *Agric. Biol. Chem.* **54**, 1931 (1990).
[14] S. Takaichi, K. Kinoshita, T. Miyamoto, K. Azegami, J. Ishidsu, T. Seki, and S. Fukuda, *Bull. JFCC* **5**, 96 (1989).

Whatman, Clifton, NJ), washed with 0.9% (w/v) NaCl solution, and stored at −30°.

Isolation and Purification of Carotenoids[13]

After the harvested wet cells (100–200 g) are dehydrated once with methanol by centrifugation, orange-colored material is extracted three or four times with chloroform–methanol (1:1, v/v) by centrifugation. The extract is washed with 50 mM Tris-HCl buffer (pH 8.0) containing 2% (w/v) NaCl and 0.001% (w/v) sodium ascorbate. Floating pellets are removed by centrifugation, followed by filtration through a phase separation filter (e.g., 1PS; Whatman). The solvent, to which ethanol is added for dehydration, is evaporated to dryness under reduced pressure. Purification of the carotenoids is carried out under a dim light without saponification.

The carotenoid extract is dissolved in a small volume of chloroform and diluted with hexane. It is submitted to silica gel 60 (E. Merck AG, Darmstadt, Germany) column chromatography (Fig. 1) and colorless materials are eluted with hexane. A yellow carotenoid (**B**) and then an orange one (**K**) are eluted with chloroform. A carotenoid glucoside (**K-G**) and a carotenoid glucoside ester (**K-G-FA**) are simultaneously eluted with chloroform–methanol (9:1, v/v) as a red eluate. The last eluate is evaporated, and the residue is dissolved in hexane. Then the solution is submitted to DEAE-Sepharose CL-6B (Pharmacia, Uppsala, Sweden) column chromatography, **K-G-FA** being eluted with hexane–chloroform (3:2, v/v) and **K-G** with chloroform–methanol (19:1, v/v). These column chromatographic purifications must be repeated several times for **K-G-FA**, until it is adsorbed as a very narrow and sharp band when submitted to the column. If these purifications are not sufficient, some cellular lipids are present. Finally, these carotenoids are purified by high-performance liquid chromatography (HPLC). A prepacked column of a Radial-PAK μBondapak C_{18} cartridge (8 × 100 mm), installed in an RCM 8 × 10 radial compression separation system, and an absorbance detector model 440 set at 439 nm for carotenoids and at 254 nm for impurities (Waters Associates, Milford, MA) are used. The eluent is methanol (2.0 ml/min).

An elution profile of the crude carotenoid extract described above is shown in Fig. 1. **K-G, K,** and **B** reveal single peaks. **K-G-FA** reveals one multiple peak, which is due to a difference in the fatty acids bonded. Minor components of **B-G** and **B-G-FA** are also found. The relative amounts of the carotenoids are 10% **B,** 8% **K,** 7% **K-G,** and 76% **K-G-FA,** the total carotenoid content being about 0.1 nmol/mg of dried cells.

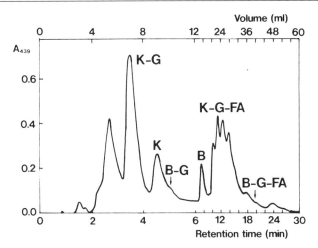

FIG. 1. An elution profile of the crude carotenoid extract from *R. rhodochrous* RNMS1 by reversed-phase HPLC. See text for details of the chromatographic conditions. The chart speed was changed at 6 min.

Chemical Modification of Carotenoids

Reduction. The method of Krinsky and Goldsmith[15] is slightly modified. After a carotenoid is dissolved in 1 ml of 95% (v/v) aqueous ethanol, a few crystals of $NaBH_4$ are added. The mixture should stand for 30–60 min at room temperature. The reduced derivative is dried thoroughly under an N_2 stream, and extracted with hexane or chloroform. The solution is passed through a Teflon filter [pore size around 0.5 μm, e.g., Millipore (Bedford, MA) Columngard or Biofield (Tokyo, Japan) Chromatodisc] equipped with a syringe, and evaporated. The yield is nearly 100%.

Acetylation. The method of Hager and Stransky[16] is slightly modified. After a carotenoid is dissolved in 0.2 ml of dry pyridine [dried over molecular sieve 3A (E. Merck AG)], 0.1 ml of acetic anhydride is added. The mixture should stand for 1 hr at room temperature. The solvent is removed by an N_2 stream. If the reaction time is insufficient, the carotenoid is not fully acetylated.

Trimethylsilylation. The method of McCormick and Liaaen-Jensen[17] is slightly modified. After a carotenoid is dissolved in 0.3 ml of dry pyridine, 0.1 ml of hexamethyldisilazane and 0.05 ml of trimethylchlorosilane are

---

[15] N. I. Krinsky and T. H. Goldsmith, *Arch. Biochem. Biophys.* **91,** 271 (1960).
[16] A. Hager and H. Stransky, *Arch. Mikrobiol.* **71,** 132 (1970).
[17] A. McCormick and S. Liaaen-Jensen, *Acta Chem. Scand.* **20,** 1989 (1966).

added. The mixture is heated at 70° for 5 min. The silyl derivative is dried thoroughly under an N_2 stream and extracted with hexane or chloroform. The solution is passed through the Teflon filter and evaporated. The yield is almost 100%.

Molecular weights of the carotenoids and their chemically modified derivatives are analyzed with a double-focusing gas chromatograph–mass spectrometer (GC-MS) equipped with a field desorption apparatus (FD-MS).[13,18] The molecular weights of the polar carotenoids, **K-G** and **K-G-FA**, and of the high molecular weight derivative, trimethylsilylated **K-G-FA**, can be determined. The number of carbonyl and hydroxyl groups is determined by the increased mass unit of the chemically modified carotenoid described above; i.e., the reduction of one carbonyl group increases the molecular weight by two mass units. By acetylation of one primary or secondary hydroxyl group, it increases by 42 mass units. It increases by 72 mass units by silylation of one hydroxyl group. Although a part of the carotenoids is degraded during the chemical modification, the carotenoid derivatives are distinguishable from the degraded products due to their low intensities on FD-MS spectra.

Identification of B and K[13]

B and **K** are identified as 1′,2′-dihydro-β,ψ-caroten-1′-ol and 1′-hydroxy-1′,2′-dihydro-β,ψ-caroten-4-one, respectively[13] (Fig. 2), from the following evidence: (1) The absorption spectra of **B** and of the reduced derivative of **K** indicate that they are a one-conjugated β ring-type carotenoid and have 11 conjugated double bonds.[19] (2) FD-MS spectra indicate that the molecular weights of **B** and **K** are 554 and 568, respectively, and the presence of one tertiary hydroxyl group in both is indicated by the formation of a monosilyl derivative but no acetyl derivatives. (3) The proton nuclear magnetic resonance (^{1}H NMR) spectra of **B** and **K**, dissolved in $CDCl_3$, are assigned by comparing them with those of authentic β-carotene, canthaxanthin, and rhodopsin.

Structure of K-G and K-G-FA[13]

The absorption spectra of **K-G** and **K-G-FA**[13] are compatible with that of **K**. FD-MS reveals a molecular weight of 730 for **K-G**, and the presence of four primary and/or secondary hydroxyl groups is indicated by the formation of a tetraacetyl or a tetrasilyl derivative. On the other hand,

---

[18] S. Takaichi, K. Shimada, and J. Ishidsu, *Phytochemistry* **27**, 3605 (1988).
[19] S. Takaichi and K. Shimada, this volume [35].

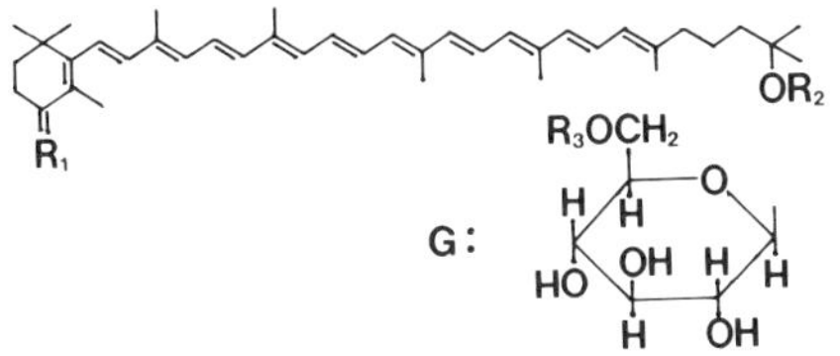

FIG. 2. Structures of the carotenoids from *R. rhodochrous* RNMS1. **B**: $R_1 = H_2$, $R_2 = H$; **K**: $R_1 = O$, $R_2 = H$; **K-G**: $R_1 = O$, $R_2 = G$, $R_3 = H$; **K-G-FA**: $R_1 = O$, $R_2 = G$, $R_3 = $ acyl.

K-G-FA shows several molecular ion peaks at m/z 968–1036, and has three primary and/or secondary hydroxyl groups.

Saponification of K-G-FA

After **K-G-FA** is dissolved in 1 ml of methanol, 0.2 ml of 60% (w/v) KOH solution is added. The mixture is heated at 70° for 5 min. After the addition of 1 ml each of water and chloroform, the carotenoid moiety in the chloroform fraction is collected. It is identified as **K-G** from the following evidence: Its molecular weight of 730, the presence of four primary and/or secondary hydroxyl groups, the same R_f value on silica gel high-performance thin-layer chromatography (HP-TLC) (E. Merck AG) developed with dichloromethane–ethyl acetate (3:1, v/v), and the same retention time on HPLC as described above.

Acidic Methanolysis of K-G-FA

Acidic methanolysis of **K-G-FA** is performed with 0.5–1.0 ml of 5% (w/v) HCl–methanol solution at 60° for 30 min. After evaporation, the carotenoids and the fatty acid methyl esters are extracted with hexane. The esters are further purified by silica gel HP-TLC developed with hexane–diethyl ether–acetic acid (80:30:1, v/v). The TLC plate is sprayed with 0.01% primurine in acetone–water (4:1, v/v) and the esters are detected under UV light of 365 nm as a bluish white fluorescent band.[20] They are extracted from the band with chloroform using the Teflon filter. The esters in cellular lipids are also obtained by the same procedures.

Carotenoid Moiety. Three kinds of carotenoids are detected in the hexane extract described above. They are identified as **K** and as methoxy and dehydrated derivatives of **K** from their molecular weights and retention times on HPLC. These carotenoids are also produced from **K-G** by the same procedures.

Fatty Acid Methyl Esters. The fatty acid composition is analyzed with

---

[20] N. Sato and N. Murata, this series, Vol. 167, p. 251.

gas-liquid chromatography (GLC) fitted with a flame ionization detector and a fused silica capillary column (0.25 mm × 30 m) coated with SP-2330 (Supelco, Bellefonte, PA). The column temperature is 190°, the carrier gas being N_2. The structures of the esters are determined by GC-MS as described below. The fatty acid moieties comprise nine major types: four saturated (C16:0–C19:0), four monounsaturated (C18:1–C21:1), and one branched-chain (10-methyl-C19:0) acid, including the odd-numbered ones. However, the composition is distinct from that in cellular lipids, in which the major ones are C16:0, C16:1, C18:1, and 10-methyl-C19:0.[14]

Sugar Moiety. After the sugar moiety is extracted with water from the residue of the hexane extract described above, it is analyzed with silica gel HP-TLC developed with acetone–2 mM sodium acetate (9:1, v/v). Two blue spots are detected on the basis of a color reaction with 0.2% (w/v) naphthoresorcinol in 10% (v/v) phosphoric acid–ethanol solution. The R_f values are compatible with those of authentic D-glucose and 1-*O*-methyl-D-glucoside. The same sugar moiety is also obtained from **K-G**.

^1H NMR Spectra of **K-G** and **K-G-FA**

The ^{1}H NMR spectra of **K-G** and **K-G-FA** in the carotenoid region show that the signals of H_3-16' and H_3-17' are shifted to 1.25 ppm, the others being compatible with those of **K**. In the glycoside region, **K-G** and **K-G-FA** show the characteristic signals of D-glucose.[21] Doublet signals at 4.45 ppm ($J = 7.8$ Hz) indicate the presence of β-D-glucosyl residue.[22] Although the signals of H-6 and H-6' for **K-G** are at 3.89 and 3.77 ppm (doublet–doublet), those for **K-G-FA** are shifted to 4.40 and 4.28 ppm (doublet–doublet), respectively, indicating that the fatty acid is esterified with the C-6 hydroxyl group of the D-glucose moiety in **K-G-FA**. In addition, **K-G-FA** also shows the characteristic signals of the fatty acids.[23]

In conclusion, the structure of **K-G** is the β-D-glucoside of **K** at C-1', 1'-[(β-D-glucopyranosyl)oxy]-1',2'-dihydro-β,ψ-caroten-4-one (Fig. 2). **K-G-FA** is a mixture of the fatty acid monoesters of **K-G** at C-6 of the D-glucoside moiety, 1'-[(6-*O*-acyl-β-D-glucopyranosyl)oxy]-1',2'-dihydro-β,ψ-caroten-4-one.

Structures of Fatty Acids from **K-G-FA** and Cellular Lipids

The fatty acid methyl esters purified by silica gel HP-TLC as described above are separated into a saturated and an unsaturated fraction by silica gel HP-TLC impregnated with $AgNO_3$ and developed with hexane–diethyl

---

[21] H. J. Koch and A. S. Perlin, *Carbohydr. Res.* **15,** 403 (1970).
[22] J. H. Bradbury and J. G. Collins, *Carbohydr. Res.* **71,** 15 (1979).
[23] K. Kates, "Techniques of Lipidology," 2nd Ed. Elsevier, Amsterdam, 1986.

ether (9:1, v/v).[20] They are located by the fluorescence of primurine under UV light as described above. The R_f value of the saturated esters is higher than that of the unsaturated ones. They are extracted with chloroform using the Teflon filter. The mass spectra of the chemical derivatives of the fatty acids are obtained by GC-MS equipped with a glass column (3 mm × 1 m) packed with 2% OV-1 on a Uniport HP 80/100 (GL Sciences, Tokyo, Japan). The temperature of the ion source is 205°, and that of the column is programmed to increase from 200 to 250° at a rate of 3°/min. The ionization voltage is 70 eV and the carrier gas is helium.

Position of Double Bonds of Unsaturated Fatty Acids[24]

The monounsaturated esters are oxidized to diols with OsO_4, and the resulting dihydroxyl esters are converted into the corresponding trimethylsilyloxy derivatives. The principal fragmentation of the derivatives occurs between the trimethylsilyloxy-substituted carbon atoms, giving rise to a pair of intense fragments.[25,26] In the case of the C20:1 derivatives from **K-G-FA,** two intense fragments at m/z 215 and 287 are derived from the cleavage at $\omega 9$. Furthermore, the C20:1 derivatives show the positional isomers of $\omega 7$, $\omega 8$, and $\omega 10$. The amounts of the $\omega 7$, $\omega 8$, $\omega 9$, and $\omega 10$ isomers are estimated from mass fragmentography,[26] the composition being 32, 25, 32, and 11%, respectively. Similarly, the other monounsaturated esters from **K-G-FA** show a heterogeneous location of the double bond, the major positions being $\omega 7$, $\omega 8$, and $\omega 9$. Such heterogeneity has not been reported previously.

In cellular lipids, C16:1 is composed of 22% Δ^9 and 73% Δ^{10} isomers, while C18:1 is composed only of one Δ^9 species. The presence of 10-C16:1, in addition to 9-C16:1, has also been reported in cellular lipids of *Rhodococcus equi, Rhodococcus fascians,*[26] *Mycobacterium bovis,* and *Mycobacterium smegmatis.*[27]

Structure of Saturated Fatty Acids

The saturated esters are converted into the pyrrolidide derivatives by the method of Andersson and Holman.[28] Because the mass spectrum of the major branched-chain fatty acid indicates the absence of the fragment at m/z 224, it is identified as 10-methyl-C19:0.[29] Similarly, 10-methyl-C17:0 is identified. From the retention time on GLC, a homologous series of

[24] S. Takaichi, *Agric. Biol. Chem.* **54,** 2139 (1990).
[25] P. Capella and C. M. Zorzut, *Anal. Chem.* **40,** 1458 (1968).
[26] K. Suzuki, A. Kawaguchi, K. Saito, S. Okuda, and K. Komagata, *J. Gen. Appl. Microbiol.* **28,** 409 (1982).
[27] J. G. C. Hung and R. W. Walker, *Lipids,* **5,** 720 (1970).
[28] B. A. Andersson and R. T. Holman, *Lipids* **9,** 185 (1974).
[29] B. A. Andersson and R. T. Holman, *Lipids* **10,** 716 (1975).

branched-chain fatty acids (C17:0–C22:0) is also found in **K-G-FA** and cellular lipids. The mass spectra of other derivatives of the saturated fatty acids in **K-G-FA** and cellular lipids indicate that they are even- and odd-numbered, saturated, straight-chain fatty acids. The presence of 10-methyl-C17:0 and 10-methyl-C19:0 is widely known in Actinomycetes.

Carotenoids from *Rhodococcus*

In preliminary experiments, **B, B-G,** and **B-G-FA** were identified in a yellow mutant of *R. rhodochrous* RNMS1, **K, K-G,** and **K-G-FA** being absent. The fatty acid composition of **B-G-FA** was compatible with that of **K-G-FA** from the wild type. *Rhodococcus rhodochrous* JCM 3202 (ATCC 13808, American Type Culture Collection, Rockville, MD) also had **K, K-G,** and **K-G-FA**. The fatty acid species of **K-G-FA** from this strain were similar to those from *R. rhodochrous* RNMS1, but the relative amount was different. From HPLC analysis, *Rhodococcus rhodnii* JCM 3203 and *R. equi* JCM 1311 seemed to have **B, K, K-G,** and **K-G-FA**. However, *Rhodococcus erythropolis* IFO 12320 had only γ-carotene and 4-keto-γ-carotene.

In *R. rhodochrous* RNMS1, the carotenogenesis seems to be as follows: **B** is oxidized to a keto derivative of **K, K-G** is formed by β-D-glucosylation of **K**, and finally **K-G-FA** is formed by acylation of **K-G**. However, the pathway in *M. fulvus* is different. After D-glucosylation and then acylation of the acyclic carotenoid, the β-ionone ring is formed.[2]

Acknowledgments

The authors wish to thank Professor K. Harashima of Yachiyo International University for valuable suggestions and Dr. N. Sato of The University of Tokyo for helpful advice on the fatty acid analysis. We also thank Professor S. Fukuda and co-workers of Nippon Medical School for their generous support.

[35] Characterization of Carotenoids in Photosynthetic Bacteria

By SHINICHI TAKAICHI and KEIZO SHIMADA

Introduction

General methods for the isolation, purification, and identification of carotenoids in phototrophic organisms have been summarized by Liaaen-Jensen and Jensen[1] and Britton.[2] The chemistry of carotenoids in photo-

[1] S. Liaaen-Jensen and A. Jensen, this series, Vol. 23, p. 586.

synthetic bacteria has been reviewed by Liaaen-Jensen.[3] About 80 different carotenoids are synthesized by photosynthetic bacteria. Their distribution in four major families of photosynthetic bacteria, namely the Rhodospirillaceae, Chromatiaceae, Chlorobiaceae, and Chloroflexaceae, has been reviewed by Schmidt.[4] In general, the characteristics of carotenoids in photosynthetic bacteria are as follow: (1) Most carotenoids are an aliphatic type, except for some aromatic or β end group types in the Chlorobiaceae and Chloroflexaceae; (2) the cross-conjugated aldehyde and the tertiary methoxy group are confined to the carotenoids of photosynthetic bacteria; (3) several kinds of carotenoids are found in each species; (4) all the carotenoids are bound to light-harvesting or reaction center complexes; and (5) structural elements such as allenic or acetylenic bonds, epoxides or furanoxides, or C_{45} or C_{50} carotenoids are not encountered.

Since 1978, photosynthetic pigments, namely several carotenoids and bacteriochlorophyll (Bchl), have been found in some aerobic bacteria, such as *Erythrobacter longus*,[5] *Roseobacter denitrificans*[6,7] (previously named *Erythrobacter* species OCh 114),[8] and *Pseudomonas radiora*.[9] Although photosynthetic bacteria synthesize pigments under anaerobic conditions, these aerobic bacteria cannot grow anaerobically even in the light and they synthesize pigments only under high aeration.

The carotenoids of *E. longus* are quite different from those of typical photosynthetic bacteria. More than 20 different carotenoids are found in cells cultured under various conditions; all of them have been identified.[10-12] A major carotenoid group consists of the novel carotenoid sulfates erythroxanthin sulfate and caloxanthin sulfate. *In vivo,* these carotenoid sulfates are not associated with the pigment-protein complex, which is the sole antenna complex of this bacterium, and do not function as light-harvesting pigments.[13] Other carotenoids in *E. longus* are C_{40}

---

[2] G. Britton, this series, Vol. 111, p. 113.
[3] S. Liaaen-Jensen, *in* "The Photosynthetic Bacteria" (R. K. Clayton and W. R. Sistrom, eds.), p. 233. Plenum, New York, 1978.
[4] K. Schmidt, *in* "The Photosynthetic Bacteria" (R. K. Clayton and W. R. Sistrom, eds.), p. 729. Plenum, New York, 1978.
[5] T. Shiba and U. Simidu, *Int. J. Syst. Bacteriol.* **32**, 211 (1982).
[6] K. Harashima and H. Nakada, *Agric. Biol. Chem.* **47**, 1057 (1983).
[7] S. Takaichi, K. Furihata, and K. Harashima, *Arch. Microbiol.* **155**, 473 (1991).
[8] T. Shiba, *Syst. Appl. Microbiol.* **14**, 140 (1991).
[9] Y. Nishimura, M. Shimadzu, and H. Iizuka, *J. Gen. Appl. Microbiol.* **27**, 427 (1981).
[10] S. Takaichi, K. Shimada, and J. Ishidsu, *Phytochemistry* **27**, 3605 (1988).
[11] S. Takaichi, K. Shimada, and J. Ishidsu, *Arch. Microbiol.* **153**, 118 (1990).
[12] S. Takaichi, K. Furihata, J. Ishidsu, and K. Shimada, *Phytochemistry* **30**, 3411 (1991).
[13] K. Shimada, H. Hayashi, and M. Tasumi, *Arch. Microbiol.* **143**, 244 (1985).

carotenoids, which can be classified into three groups: (1) bicyclic carotenoids, consisting of β-carotene and its hydroxyl derivatives β-cryptoxanthin, zeaxanthin, caloxanthin, and nostoxanthin, (2) the monocyclic carotenoids rubixanthin, bacteriorubixanthin, and bacteriorubixanthinal, and (3) the acyclic carotenoids anhydrorhodovibrin and spirilloxanthin. The $(3R)$-3-hydroxy-β end group has rarely been found in the carotenoids of the Rhodospirillaceae and Chromatiaceae, while the acyclic carotenoids described above have been found exclusively in photosynthetic bacteria.

This chapter describes the characterization of the carotenoids from *E. longus* by modern techniques; the procedures are applicable to other carotenoids from photosynthetic bacteria. In particular, the procedures for analysis of the absorption spectra may be useful. It also describes the procedures for the identification of the carotenoid sulfates from this bacterium, including some specific methods.

Biological Materials[11]

Erythrobacter longus OCh 101 (ATCC 33941, American Type Culture Collection, Rockville, MD) is cultured aerobically (by bubbling) or semiaerobically (by stirring) in the dark for 60–70 hr. A medium described by Shioi[14] is used: 2 g yeast extract (Difco, Detroit, MI), 1.0 g polypeptone, 1.0 g casamino acids, 1.0 ml glycerol, 20.0 g NaCl, 5.0 g $MgCl_2 \cdot 6H_2O$, 2.0 g Na_2SO_4, 0.5 g KCl, 0.5 g $CaCl_2 \cdot 2H_2O$, 0.2 g $NaHCO_3$, and 0.1 g ferric citrate per liter (pH 7.5). Cells cultured aerobically in the presence of 130 μM diphenylamine or 10 mM nicotine are also used. They are harvested by centrifugation, washed with 25 mM phosphate buffer (pH 7.5), and stored at $-30°$.

Isolation and Purification of Carotenoids from *Erythrobacter longus*[10-12]

Carotenoids are extracted from four types of wet cells (20–30 g) as described above with chloroform–methanol (1:2, v/v) followed by centrifugation. The extract, to which ethanol is added for dehydration, is evaporated to dryness under reduced pressure, the pigments being dissolved in a small volume of chloroform. Their elution profiles by high-performance liquid chromatography (HPLC), using a μBondapak C_{18} column (8 × 100 mm; Waters Associates, Milford, MA) as described in Chapter [34] of this volume[15] are shown in Fig. 1. Many carotenoid peaks can be detected.

[14] Y. Shioi, *Plant Cell Physiol.* **27**, 567 (1986).
[15] S. Takaichi and J. Ishidsu, this volume [34].

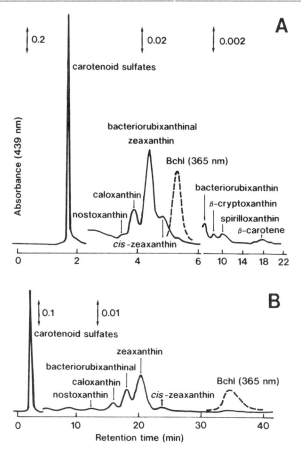

FIG. 1. Elution profiles of the carotenoids from the aerobically grown cells of *E. longus* by reversed-phase HPLC eluted with (A) methanol or (B) methanol–H_2O (9:1, v/v). The flow rate is 2.0 ml/min. See text for details of the chromatographic conditions. Bchl, Bacteriochlorophyll.

For purification, the carotenoids are dissolved in a small volume of chloroform and diluted with hexane. They are submitted to silica gel 60 (E. Merck AG, Darmstadt, Germany) column chromatography and eluted successively with hexane, hexane–chloroform (9:1, 3:1, 3:2, and 2:3, v/v), chloroform, and chloroform–methanol (9:1 and 3:1, v/v). After each fraction is evaporated, this column chromatographic procedure is repeated. Further purification of each carotenoid except the carotenoid sulfates is achieved by DEAE-Sepharose CL-6B (Pharmacia, Uppsala, Sweden) column chromatography eluted successively with hexane–

chloroform (9:1, 3:1, 3:2, and 2:3, v/v) and chloroform, by silica gel high-performance thin-layer chromatography (HP-TLC) (Merck) developed with dichloromethane–ethyl acetate (3:1 or 1:1, v/v), by reversed-phase TLC (KC_{18}; Whatman, Clifton, NJ) developed with chloroform–methanol (1:4, v/v), by HPLC eluted with methanol or methanol–H_2O (9:1, v/v), or by a combination of these procedures.

The carotenoid sulfates are eluted with chloroform–methanol (3:1, v/v) from the first silica gel column chromatography described above. After evaporation and dissolution in chloroform, the purification is repeated once more. Then the carotenoids dissolved in chloroform–methanol (3:1, v/v) are adsorbed on a SEP-PAK NH_2 cartridge (Waters Associates) and eluted with chloroform–methanol (1:2, v/v) containing 0.13 M NH_4OH. After evaporation and dissolution in chloroform–methanol (3:1, v/v) the carotenoids are applied to HPLC equipped with a preparative column of a μBondapak C_{18} cartridge (25 × 100 mm) installed in an RCM 25 × 10 radial compression separation system (Waters Associates) and eluted with methanol–H_2O (7:3, v/v) containing 0.04 M NH_4OH. Erythroxanthin sulfate (all-trans form) is eluted as a red eluate, which is followed by its cis forms, and then caloxanthin sulfate (all-trans form) is eluted as a yellow eluate, which is followed by its cis forms. After each eluate is diluted with a third volume of H_2O, the carotenoid sulfate is again adsorbed on a SEP-PAK C_{18} cartridge (Waters Associates) and eluted with methanol.

Identification of Carotenoids[10,11]

Absorption Spectra

The absorption spectra of carotenoids are obtained by continuous-monitoring HPLC equipped with a photodiode array detector, MCPD-350 PC system II (230–800 nm, 1.4-nm resolution, 1-sec interval; Otsuka Electronics, Osaka, Japan). The μBondapak C_{18} column is used, the eluent being methanol (2.0 ml/min). The characteristics of this system are as follows: (1) The absorption spectra of all-trans forms of carotenoids can be obtained by separation from cis forms and other carotenoids. (2) The absorption spectra in a simple solvent can be obtained because the eluent is methanol only. (3) The absorption spectrum in the ultraviolet region can be measured without interference from impurities. (4) A few micrograms of a carotenoid is enough to make measurements with good resolution.

The absorption parameters obtained from the carotenoids of *E. longus*,[10,11] *R. denitrificans*,[7] *Rhodospirillum rubrum*, *Rhodocyclus gelatinosus*, *Rhodobacter sphaeroides*, *Chromatium vinosum*, *Rhodococcus rho-*

dochrous,[15,16] *Rhodococcus erythropolis*,[15] *Pyramimonas parkeae*, *Bryopsis maxima*, carrot roots, spinach leaves, and synthetic samples are summarized in Table I. All are measured by the instrument after partial or complete purification, or chemical reduction. The wavelengths of two or three absorption maxima (λ_{max}) and the highest cis peak are described. The terms %III/II and %D_B/D_{II} express the spectral fine structures as illustrated in Fig. 2.[17] The low values of %D_B/D_{II} in Table I indicate the high purity of these carotenoids. In the carotenoids containing a conjugated carbonyl group, such as bacteriorubixanthinal and canthaxanthin, the absorption spectrum has only a single and rounded peak, and λ_{max} is not clear. When the carbonyl group is reduced by $NaBH_4$ into the hydroxyl group as described in Chapter [34] of this volume[15] the absorption spectrum shifts to a shorter wavelength and takes on the usual three-peaked appearance.

It is well known that the positions of main λ_{max} are characteristic of the chromophores of the individual carotenoids. λ_{max} is mainly influenced by the number of conjugated C=C double bonds *(N)* in the compounds, and λ_{max} increases as *N* increases (Table I). The main λ_{max} of phytoene (*N* = 3) in methanol is 285 nm and that of spirilloxanthin (*N* = 13) is 492 nm. The spectral fine structure term of %III/II reaches its maximum value at *N* = 8 in the nonconjugated β ring[18] type, and the values decrease with decreasing or increasing *N* values. Thus, *N* can be estimated from λ_{max}.

The conjugated β ring influences the chromophore. A carotenoid in which the conjugation extends into the β ring has its λ_{max} at shorter wavelengths than a nonconjugated β ring-type carotenoid with the same *N* value. Thus, in *N* = 11 carotenoids, lycopene (nonconjugated β ring type) has its main λ_{max} at 468 nm, while rubixanthin (one-conjugated β ring type) and β-carotene (two-conjugated β ring type) have main λ_{max} at 458 and 449 nm, respectively (Table I). The nonconjugated rings do not contribute to the chromophore. For example, the ε end group in α-carotene and lutein and the 5,6-epoxy β end group in violaxanthin do not affect λ_{max}. Furthermore the %III/II values of the nonconjugated β ring type are more than 60, while those of the one-conjugated β ring type are about 50, and those of the two-conjugated β ring type are about 30 (Table I). Thus the number of conjugated β rings is distinguishable from the %III/II value.

---

[16] S. Takaichi, J. Ishidsu, T. Seki, and S. Fukuda, *Agric. Biol. Chem.* **54,** 1931 (1990).
[17] B. Ke, F. Imsgard, H. Kjøsen, and S. Liaaen-Jensen, *Biochim. Biophys. Acta* **210,** 139 (1970).
[18] In this chapter, the term *conjugated β ring* is used for the derivatives of the β end group that maintain the C-5–C-6 double bond conjugated with the polyethylene chain, and not for those of the β end group with the C-5–C-6 single bond, such as a 5,6-epoxy β end group and a 5,6-dihydroxy β end group.

TABLE I
SPECTROSCOPIC PROPERTIES OF SOME NONCONJUGATED AND ONE- AND TWO-CONJUGATED β RING-TYPE CAROTENOIDS IN METHANOL

N^a	N_p	N_r	λ_{max} (nm)^b				%III/II	%D_B/D_{II}	Carotenoids
3	3	0		(276)	284	(294)			Phytoene (cis-)
5	5	0	254	329	346	365	85.0	10.1	Phytofluene
7	7	0	295	377	398	422	95.0	5.2	Symmetrical ζ-carotene
			295	374	395	419	89.9	11.1	Asymmetrical ζ-carotene
8	8	0	309	397	419	446	97.7	5.8	Siphonaxanthin reduced
9	9	0	328	413	436	465	90.2	6.8	Neurosporene
			328	412	437	465	87.1	7.8	Chloroxanthin
			329	413	437	465	87.6	8.7	3,4-Dihydrospheroidene
			328	414	436	465	89.1	7.3	3,4-Dihydrospheroidenone[7]
			326	415	436	466	89.9	5.0	Violaxanthin
10	10	0	344	427	452	482	78.5	10.6	Spheroidene
			344	426	451	482	78.2	11.9	Hydroxyspheroidene
11	11	0	360	442	468	499	68.7	9.2	Lycopene[11]
12	12	0	372	454	480	513	65.2	10.7	Anhydrorhodovibrin
13	13	0	385	464	492	524	63.8	14.1	Spirilloxanthin[11]
9	8	1	313	(402)	423	447	55.7	6.2	β-Apo-8'-carotenol
10	9	1	335	(422)	442	471	61.2	5.6	α-Carotene
			330	(422)	443	470	61.7	5.3	Lutein
11	10	1	346	(436)	458	489	48.1	8.9	Rubixanthin[11]
			348	(436)	458	487	46.3	9.8	Hydroxy-γ-carotene[15]
			346	(436)	458	487	44.3	8.2	Carotenoid B[15,16]
12	11	1	361	(446)	468	499	46.2	17.4	Bacteriorubixanthin[10]
11	9	2	341	(429)	449	475	24.6	5.4	β-Carotene
			341	(427)	448	475	27.4	5.7	Zeaxanthin[11]
			341	(426)	447	474	32.7	5.0	Isozeaxanthin
			339	(427)	448	475	31.6	6.5	Caloxanthin[11]
			341	(425)	448	475	30.9	9.9	Nostoxanthin[11]
			339	(426)	447	474	33.8	6.8	Crustaxanthin

^{a} N, The number of conjugated double bonds; N_p, N in the polyethylene chain; N_r, the number of conjugated β rings that maintain the C-5–C-6 double bond; $N = N_p + N_r$.
^{b} Parentheses indicate the position of the shoulders of the absorption spectra obtained from first derivatives of the spectra.

The nonconjugated β ring-type carotenoids are usually the acyclic carotenoids (derivatives of ψ,ψ-carotene), which are widely found in photosynthetic bacteria. The one-conjugated β ring-type ones are usually the monocyclic carotenoids (derivatives of β,ψ-carotene) or the derivatives of β,ε-carotene. The two-conjugated β ring-type ones are the bicyclic carotenoids (derivatives of β,β-carotene).

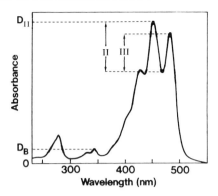

FIG. 2. Explanation of %III/II and %D_B/D_{II} terms used in expressing the shape and fine structure in the absorption spectrum of a carotenoid.[17] THe absorption spectrum of all-transspheroidene in methanol is shown as an example.

Substituents, such as the hydroxyl and methoxy groups, and the presence of the nonconjugated ring or the nonconjugated carbonyl group do not contribute to the chromophore, and have little or no effect on λ_{max}. For example, β-carotene and nostoxanthin, neurosporene and violaxanthin, and 3,4-dihydrospheroidene and 3,4-dihydrospheroidenone have virtually identical absorption spectra (Table I). An exception is symmetrical ζ-carotene and asymmetrical ζ-carotene ($N = 7$), in which the position of the conjugated double bonds is shifted in the molecule.

In summary, Table I shows that the chromophore of the carotenoids can be estimated both by the number of the conjugated double bonds *(N)*, which is determined from the wavelength of the absorption maximum (λ_{max}), and by the number of conjugated β rings, which is determined from the value of %III/II. In the case of the carotenoids containing the conjugated carbonyl group(s), the chromophore can be estimated after the reduction by $NaBH_4$.

Retention Time on HPLC

Polar groups, such as the hydroxyl, methoxy, keto, and aldehyde groups, influence the retention times on reversed-phase HPLC as shown in Fig. 1. The retention time becomes shorter as the number of hydroxyl groups increases. For example, β-carotene, β-cryptoxanthin, zeaxanthin, caloxanthin, and nostoxanthin, which have zero, one, two, three, and four hydroxyl groups, respectively, elute in reverse order. Polarity due to two methoxy groups roughly corresponds to that due to one hydroxyl group as observed in the small difference in retention time between spirilloxanthin and β-cryptoxanthin (Fig. 1A). The addition of H_2O to the eluent methanol facilitates the separation of the polar carotenoids as observed in the separation of caloxanthin and nostoxanthin (Fig. 1B). Thus the kinds and numbers of polar groups can be estimated from the retention times on HPLC. In this system, chlorophyll *a* and neutral lipids elute at about 9 and 18 min, respectively.

Other Spectroscopic Methods

The molecular weights of the carotenoids are analyzed with a double-focusing gas chromatograph-mass spectrometer equipped with a field desorption apparatus (FD-MS). The number of carbonyl and hydroxyl groups is determined by the increase in mass unit on chemical modification, such as reduction, acetylation, and trimethylsilylation (described in [34]).[15]

The structures of the carotenoids are finally determined by proton nuclear magnetic resonance (^{1}H NMR) spectra in $CDCl_3$ at room temperature. It is well known that an ^{1}H NMR spectrum of a carotenoid consists of the spectra of the left and right halves of the molecule. ^{1}H NMR spectrum data of more than 70 different end groups and 136 derivatives have been summarized by Englert.[19,20] Unfortunately, only four sets of data on the derivatives of the ψ end group, which are widely distributed in the photosynthetic bacteria, are included.

Chirality of the carotenoids is determined by circular dichroism (CD) spectra in ether–2-propanol–ethanol (5:5:2, v/v) measured with a spectropolarimeter at room temperature. Some CD spectrum data have been summarized by Sturzenegger *et al.*[21]

From spectroscopic and chemical evidence the chemical structures of the carotenoids can be determined. For example, the structures of bacteriorubixanthin [(3*R*)-1'-methoxy-3',4'-didehydro-1',2'-dihydro-β,ψ-caroten-3-ol] and bacteriorubixanthinal [(3*R*)-9'-*cis*-3-hydroxy-1'-methoxy-3',4'-didehydro-1',2'-dihydro-β,ψ-caroten-19'-al] (Fig. 3) from *E. longus*, which are novel carotenoids, have been determined.[10]

---

[19] G. Englert, *in* "Carotenoid Chemistry and Biochemistry" (G. Britton and T. W. Goodwin, eds.), p. 107. Pergamon, Oxford, 1982.
[20] G. Englert, *Pure Appl. Chem.* **57,** 801 (1985).
[21] V. Sturzenegger, R. Buchecker, and G. Wagniere, *Helv. Chim. Acta* **63,** 1074 (1980).

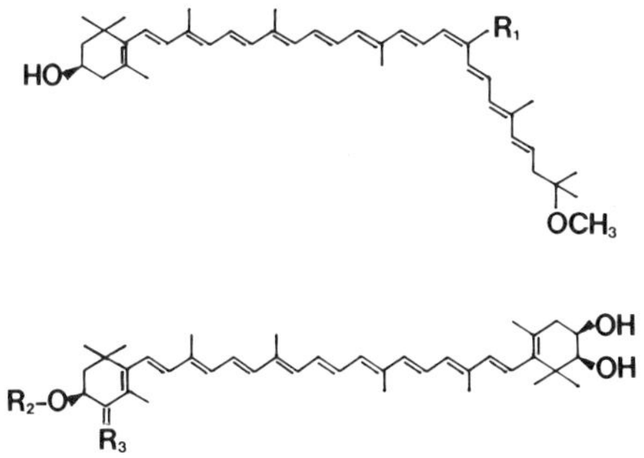

FIG. 3. Structures of novel carotenoids from *E. longus*.[10,12] Bacteriorubixanthin: R_1 = CH_3, all-trans; bacteriorubixanthinal: R_1 = CHO, 9'-cis; caloxanthin: R_2 = H, R_3 = H_2; caloxanthin sulfate: R_2 = SO_2OH, R_3 = H_2; erythroxanthin: R_2 = H, R_3 = O; erythroxanthin sulfate: R_2 = SO_2OH, R_3 = O.

Identification of Carotenoid Sulfates from *Erythrobacter longus*[12]

Absorption Spectra

The absorption spectra of erythroxanthin sulfate and its carotenoid moiety (erythroxanthin)[12] in methanol show only one broad peak at around 470 nm. The reduced derivative of erythroxanthin by $NaBH_4$ shows λ_{max} at 448 and 474 nm and the %III/II value of 36, and caloxanthin sulfate also shows λ_{max} at 450 and 475 nm and the %III/II value of 29. These absorption spectra are both compatible with that of β-carotene (Table I). Therefore the chromophore of these carotenoids is the two-conjugated β ring type (derivatives of β,β-carotene).

Molecular Weight

The molecular ions of the carotenoid sulfates could not be detected by FD-MS, possibly due to thermal decomposition and/or the presence of the acidic group.

First, the molecular weights of the carotenoid moieties are determined by FD-MS as follows: After treatment of erythroxanthin sulfate with 1 ml of 0.6 M HCl in tetrahydrofuran (THF) for 10 min at room temperature, most of the sulfate group is liberated. The solution is diluted with 1 ml of H_2O, and the carotenoid moiety (erythroxanthin) is adsorbed on the SEP-PAK C_{18} cartridge and eluted with methanol. FD-MS reveals its molecular weight to be 598. One carbonyl group is present, as shown by an increase in the molecular weight by 2 mass units on the reduction by $NaBH_4$ (m/z 600). The presence of three primary and/or secondary hydroxyl groups is indicated by the formation of a triacetyl (m/z 724) or a trisilyl (m/z 814) derivative. Similarly, caloxanthin is obtained by the treatment of caloxanthin sulfate with HCl-THF. Its molecular weight is 584 indicating the presence of three primary and/or secondary hydroxyl groups.

Next, after methylation of the sulfate group, the molecular ions of the methylated derivatives can be detected by FD-MS. The carotenoid sulfates, which are obtained as the sulfate forms, are dissolved in chloroform-methanol (3:1, v/v), diluted with 10 vol of methanol-H_2O (1:1, v/v), soaked in Dowex ion exchange (HCR-W2; Dow Chemical, Midland, MI), and extracted by chloroform. By these procedures they are converted into hydrogen carotenoid sulfates, which can be converted into their methyl esters. For methylation,[22] diazomethane in ether, which is made from N-methyl-N-nitro-N-nitrosoguanidine using a micromole diazomethane generator (GL Science, Tokyo, Japan), is added to the carotenoid sulfates; the mixtures should stand for 30 min at room temperature. The methyl carotenoid sulfates are purified by silica gel HP-TLC developed with dichloromethane-ethyl acetate (3:1, v/v). The molecular ions of methyl erythroxanthin sulfate and methyl caloxanthin sulfate can be detected by FD-MS at m/z 692 and 678, respectively, similar to other carotenoids.

Identification of Sulfate

After treatment of erythroxanthin sulfate with HCl-THF and evaporation, the carotenoid moiety is extracted by chloroform. The sulfate group is extracted from the residue using water. The infrared (IR) spectrum of the sample in a KBr disk shows v_{max} cm^{-1}: 618 s (S—O str.), 1120 vs (S=O str.), 1400 s (N—H def.), 3150 vs (N—H), and 3400 vs (O—H str.), which are compatible with those of authentic Na_2SO_4 and $(NH_3)_2SO_4$. Furthermore, the IR spectrum of erythroxanthin sulfate shows the characteristic absorptions of sulfuric ester,[23] that is, v_{max} cm^{-1}: 600 m, 860 m (S—O str.), 980 s (C—O str.), 1020 s (S=O str.), and 1230 vs (S=O str.), in addition to the characteristic absorptions of a carotenoid.

^{1}H NMR and CD Spectra

The ^{1}H NMR spectra, including the 2D-COSY of the carotenoid sulfates, are measured in $CDCl_3$-CD_3OD (3:1, v/v) because they are sparingly soluble in $CDCl_3$. Assignments of the spectra of these carotenoids are

---

[22] S. Hertzberg, T. Ramdahl, J. E. Johansen, and S. Liaaen-Jensen, *Acta Chem. Scand, Ser. B* **B37**, 267 (1983).

[23] M. Kates, "Techniques of Lipidology," 2nd Ed. Elsevier, Amsterdam, 1986.

made by comparison with those of astaxanthin, zeaxanthin, caloxanthin, and nostoxanthin from *E. longus*.[11] It is concluded that the hydroxyl group at C-3 of erythroxanthin is esterified with the sulfuric acid.

The CD spectra of the reduced derivative of erythroxanthin and of caloxanthin obtained from caloxanthin sulfate by the treatment with HCl-THF are compatible with that of $(2R,3R,2'R)$-caloxanthin from *E. longus*.[11] Therefore, the configuration of erythroxanthin sulfate is $(3S,2'R,3'R)$.

In conclusion, the structure of erythroxanthin sulfate is $(3S,2'R,3'R)$-3,2',3'-trihydroxy-β,β-caroten-4-one 3-sulfate, and that of caloxanthin sulfate is $(2R,3R,3'R)$-β,β-carotene-2,3,3'-triol 3'-sulfate (Fig. 3). These carotenoid sulfates are novel ones. Only two types of naturally occurring carotenoid sulfates have been isolated from marine sponges, *Ianthella* species,[22,24] and ophiuroids, *Ophioderma longicaudum*[25] and *Ophioderma nigra*.[26]

Acidic Group

The presence of an acidic group in carotenoids is easily detected by reversed-phase KC_{18} TLC developed with methanol to which about 50 mM HCl or NH_4OH is added. The R_f value of the salt form of the carotenoid in the alkaline development solvent is higher than that of the hydrogen form in the acidic solvent. In the case of a very polar carotenoid, H_2O is added to the solvents to reduce the R_f values. Thus, the presence of the acidic groups of sulfate in erythroxanthin sulfate from *E. longus* and of carboxylate in synthetic bixin, which has a carboxylic acid, can be detected. In the preliminary experiments, the presence of carotenoid carboxylates from *Pseudomonas radiora* MD-1 was indicated. They were not bound to pigment–protein complexes, while acyclic carotenoids were.

Acknowledgments

The authors wish to thank Professor K. Harashima of Yachiyo International University for valuable suggestions. They would also like to thank Drs. T. Sasa, Y. Nishimura, and H. Hayashi for supplying biological materials. Canthaxanthin and astaxanthin were kindly supplied by Nippon Roche. They also thank Professor J. Ishidsu of Nippon Medical School for generous support.

[24] S. Hertzberg, P. Bergquist, and S. Liaaen-Jensen, *Biochem. Syst. Ecol.* **17,** 51 (1989).
[25] M. V. D'Auria, R. Riccio, and L. Minale, *Tetrahedron Lett.* **26,** 1871 (1985).
[26] M. V. D'Auria, L. Minale, R. Riccio, and E. Uriarte, *J. Nat. Prod.* **54,** 606 (1991).

[36] Enhancement and Determination of Astaxanthin Accumulation in Green Alga *Haematococcus pluvialis*

By SAMMY BOUSSIBA, LU FAN, and AVIGAD VONSHAK

Introduction

The ketocarotenoid astaxanthin (3,3'-diketo-4,4'-dihydroxy-β-carotene) was first described in aquatic crustaceans as an oxidized form of β-carotene, which gives the carapace of these animals its pinkish color.[1] It was later found that this pigment is very common in certain species of fish and birds, in which it plays an important role in coloration during the mating season.[1] Astaxanthin is also found in algae such as *Chlamydomonas nivalis* and *Haematococcus pluvialis*,[2] *Euglena rubida*,[3] and *Acetabularia mediterranea*.[4] There has been a growing interest in the use of this pigment as a colorant for egg yolk in the poultry industry and in aquaculture, where it is used as a feed supplement in the production of salmon and shrimp. In addition, the carotenoids are lipophilic oxygen quenchers with potential anticancer activities, and it has been shown that this carotenoid possesses a higher antioxidant activity than β-carotene.[5]

Little research has been done on the conditions favoring accumulation of this ketocarotenoid by the unicellular alga *H. pluvialis* and what has been published is contradictory.[6] It has been suggested that nitrogen deficiency and high light intensity cause massive accumulation of this red pigment in *H. pluvialis*.[2,7,8] Other hypotheses concerning astaxanthin production in *H. pluvialis* argue either that it is favored by agents that prevent cell division without impairing the ability of the alga to assimilate carbon[9] or that the carbon–nitrogen balance in the medium determines the degree of carotene formation.[10] The present study was aimed at defining conditions favoring astaxanthin accumulation in *H. pluvialis*.

---

[1] B. Czeczuga and H. S. Osorio, *Isr. J. Bot.* **38**, 115 (1989).
[2] F.-C. Czygan, *Arch. Mikrobiol.* **74**, 69 (1970).
[3] B. Czeczuga, *Comp. Biochem. Physiol. B* **48B**, 349 (1974).
[4] B. Czeczuga, *Acta Soc. Bot. Pol.* **55**, 601 (1986).
[5] M. Kobayashi, T. Kakizono, and S. Nagai, *J. Ferment. Bioeng.* **71**(5), 335 (1991).
[6] Y. Y. R. Yong and Y.-K. Lee, *Phycologia* **30**(3), 257 (1991).
[7] M. R. Droop, *Arch. Mikrobiol.* **20**, 391 (1954).
[8] W. Goodwin and M. Jamikorn, *Biochem. J.* **57**, 376 (1954).
[9] M. R. Droop, *Nature (London)* **175**, 42 (1955).
[10] F. Chodat, *Arch. Sci.* **20**, 96 (1938).

Growth Conditions and Measurements

Haematococcus pluvialis Flotow (Chlorophyceae, Volvocales) is obtained from the culture collection of algae at the University of Göttingen, Germany.

The algae are cultivated in a modified BG_{11} medium,[11] which contains $NaNO_3$ (1.5 g/liter) in 500-ml sterilized columns placed in a transparent Plexiglas circulating water bath maintained between 25 and 28°. Light is supplied at a photon flux density of 85 to 120 $\mu mol/(m^2 \cdot sec)$. Continuous aeration is provided by bubbling air containing 1.5% CO_2. Under these conditions, the pH is maintained between 6.8 and 7.5 and the algae remain green until they enter the stationary phase. For growth measurements, four parameters are used[12]: (1) cell number is determined with a Thomas blood cell counter; (2) chlorophyll is extracted with dimethyl sulfoxide (DMSO), the absorbance of the extracts is determined at 672 nm, and chlorophyll content is calculated with an $E_{1cm}^{1\%}$ of 898 according to Seely *et al.*[13]; (3) protein is determined according to Lowry *et al.*[14]; and (4) dry weight is measured by heating sample filtrated on a preweighed filter paper at 70° overnight.

Extraction and Measurement of Astaxanthin

The following procedures are used: cells are harvested by centrifugation (3500 rpm, 5 min), the pellet is resuspended in a solution of 5% (w/v) KOH in 30% (v/v) methanol and heated in a 70° water bath for 5 min to destroy the chlorophyll. The mixture is centrifuged and the supernatant is discarded. The remaining pellet is extracted with DMSO after adding 5 drops of acetic acid, and homogenized to recover the astaxanthin. The mixture is then heated for 5 min at 70°. This last step is repeated if necessary until the cell debris is totally white. The absorbance of the combined extracts is determined at 492 nm, and the amount of the pigment is calculated according to Davies[15] ($E_{1cm}^{1\%}$ 2220). All the processes should be conducted in darkness if possible.

The method described above is a simple and rapid way for ketocarotenoid determination. When the cells are green (Fig. 1a and b), the result is

---

[11] R. Y. Stanier, M. M. Kunisawa, and G. Cohen-Bazire, *Bacteriol. Rev.* **35**, 171 (1971).
[12] S. Boussiba and A. Vonshak, *Plant Cell Physiol.* **32**, 1077 (1991).
[13] G. R. Seely, M. J. Duncan, and W. E. Widaver, *Mar. Biol.* **12**, 184 (1972).
[14] O. H. Lowry, N. J. Rosebrough, A. L. Farr, and R. J. Randall, *J. Biol. Chem.* **193**, 265 (1951).
[15] B. H. Davies, *in* "Chemistry and Biochemistry of Plant Pigments" (T. W. Goodwin, ed.), 2nd Ed., Vol. 2, pp. 38–165. Academic Press, London, 1976.

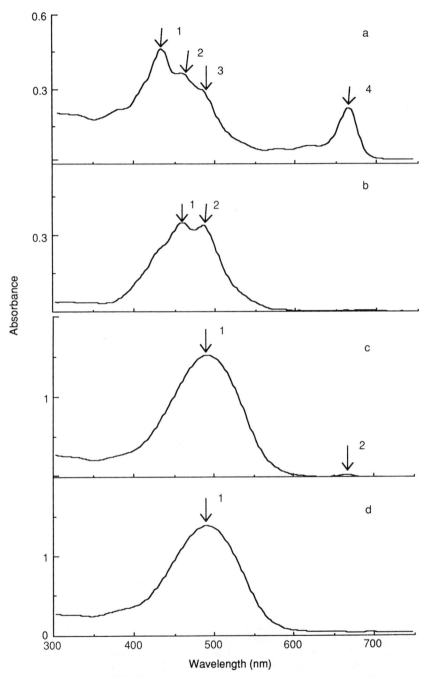

FIG. 1. Absorbance of DMSO extraction of *H. pluvialis*. (a) Chlorophyll (arrows 1 and 4) and total carotenoids (arrows 2 and 3) in green cells. (b) Astaxanthin (arrow 2) and other carotenoids (arrow 1) in green cells. (c) Chlorophyll (arrow 2) and astaxanthin (arrow 1) in red cells. (d) Astaxanthin (arrow 1) in red cells.

the amount of total ketocarotenoids instead of astaxanthin; however, after the cells become red, astaxanthin is the dominant form because of the conversion of intermediates to that pigment (Fig. 1c and d).

Conditions Inducing Astaxanthin Accumulation

Light and Nitrogen

At optimal light intensity for growth [85 μmol/(m$^2 \cdot$sec), L1] the astaxanthin content of logarithmically growing cells remained almost constant. When cultures were exposed to high light intensity [170 μmol/(m$^2 \cdot$sec), L2], a massive accumulation of astaxanthin was observed and the pigment reached a value of 65 pg/cell. Under the same light conditions but a lower nitrogen concentration (0.15 g/liter of NaNO$_3$, N2) this level was somewhat lower, 52 pg/cell (Table I). In high light intensity, the onset and rate of accumulation of astaxanthin was dependent on nitrogen concentration: in high nitrogen the massive accumulation started only on the fifth day, while in lower nitrogen the accumulation started on the second day and increased at a much faster rate.

TABLE I
ENVIRONMENTAL CONDITIONS INDUCING
ASTAXANTHIN ACCUMULATION

	Astaxanthin content (pg/cell)	
Treatments[a]	Start	After 4 days
L1 and N1	2.31	4.5
L2 and N1	2.47	65
L2 and N2	1.89	52
L1 and NP	2.34	20.13
L2 and NP	2.25	58.7
L1 and NS1	1.75	11.29
L1 and NS2	2.73	10.27
L1 and SS	8.75	47.2
L1 and vinblastine	2.18	14.84

[a] L1, 85 μmol/(m$^2 \cdot$ sec); L2, 175 μmol/(m$^2 \cdot$ sec); N1, 1.5 g/liter NaNO$_3$; N2, 0.15 g/liter NaNO$_3$; SS, 0.8% NaCl: NP, no phosphate addition; NS1, without MgSO$_4$; NS2, MgCl$_2$ substituted for MgSO$_4$.

Phosphate Starvation

The influence of phosphate supply in combination with nitrogen on astaxanthin accumulation was studied. Logarithmic cells of *H. pluvialis* grown under optimal conditions are harvested and resuspended in a phosphate-free medium containing either high (N1) or low (N2) concentrations of nitrogen under high light intensity (L2). Phosphate-deprived cells exposed to the high nitrogen concentration do not divide, whereas those grown under the low level of nitrogen grow for 2 days and then stop. Astaxanthin accumulation follows the opposite pattern, being maximal in the phosphate-deprived culture containing a high nitrogen level. It should be noted that the rate and extent of astaxanthin accumulation per cell in phosphate-deprived cells is also dependent on light intensity: under 170 μmol/(m^2·sec) astaxanthin content is 58.7 pg/cell after 4 days, while it was 20.13 pg/cell under 85 μmol/(m^2·sec) (Table I).

Salt Stress

Exposing the logarithmic cells to salt stress by the addition of 0.8% (w/v) NaCl (SS) to the growth medium under optimal light intensity (L1) causes complete cessation of growth. Growth arrest is accompanied by a massive accumulation of astaxanthin, which reaches 47.2 pg/cell after 4 days (Table I).

Sulfate Starvation

Logarithmically growing cells are sulfur starved by suspending them in medium either without MgSO$_4$ (NS1), or in which MgCl$_2$ is substituted for MgSO$_4$ at a concentration of 0.06 g/liter (NS2). Under these conditions, astaxanthin is observed to reach 11.29 and 10.27 pg/cell, respectively, after 4 days (Table I). When MgSO$_4$ is substituted by Na$_2$SO$_4$ at a concentration of 0.043 g/liter, astaxanthin contents remained as constant as that of the control (2.68 and 2.53 pg/cell, respectively).

Cell Division Inhibitor

To evaluate the interaction between cell division and astaxanthin accumulation further, the effect of vinblastine was studied. Cell division in cultures grown under the L1/N1 regime and exposed to 2 or 5 mg/ml of vinblastine is completely inhibited. This inhibition is followed by an increase in the accumulation of astaxanthin in the resting cells (Table I).

In conclusion, astaxanthin accumulation is induced whenever a disturbance in cell division is imposed, and nitrogen is required for this process. Of the five conditions described above, phosphate starvation under high light intensity induced astaxanthin accumulation with the fastest rate and largest cell content. In all cases, the amount of astaxanthin did not exceed 3% of the dry weight.

[37] Simultaneous Quantitation and Separation of Carotenoids and Retinol in Human Milk by High-Performance Liquid Chromatography

By A. R. GIULIANO, E. M. NEILSON, B. E. KELLY, and L. M. CANFIELD

Introduction

The importance of carotenoids in human health has been known for decades.[1] Carotenoids have received the attention of researchers in various fields due to the role carotenoids play as precursors of retinol, as antioxidants, and as effectors of immune function. The effects of β-carotene on the immune system is of particular importance to infants, especially those in less developed countries, as these infants rely almost entirely on breast milk as a source of nutrients. Further, it is now clear that breast milk provides significant immunoprotection to the infant. However, there is essentially nothing known about the quantity or identity of the individual carotenoids in mature breast milk. Previous studies measuring carotenoids in mature breast milk have reported values for total carotene, being unable to quantify individual carotenoids due to the less sensitive and precise methods utilized.[2-6] In a study of carotenoid content of colostrum, Patton et al.[7] reported a method for separating and quantifying carotenoids by high-performance liquid chromatography (HPLC). Due to the significantly lower levels of lipid and higher levels of carotenoids in colostrum compared with mature milk, this method has not been directly applicable

---

[1] A. Bendich and J. A. Olson, *FASEB J.* **3**, 1927 (1989).
[2] Z. A. Ajans, A. Sarif and M. Husbands, *Am. J. Clin. Nutr.* **17**, 139 (1965).
[3] N. F. Butte and D. H. Calloway, *Am. J. Clin. Nutr.* **34**, 2210 (1981).
[4] L. Hussein, A. Drar, H. Allam and B. El Naggar, *Int. J. Nutr. Res.* **57**, 3 (1986).
[5] E. M. Ostrea, J. E. Balun, R. Winkler and T. Porter, *Am. J. Obstet. Gynecol.* **154**, 1014 (1986).
[6] J. E. Chappell, T. Francis and M. T. Clandinin, *Nutr. Res.* **6**, 849 (1986).
[7] S. Patton, L. M. Canfield, A. M. Ferris and R. G. Jensen, *Lipids* **25**, 159 (1990).

to the simultaneous analysis of carotenoids and retinol in mature breast milk.

In this chapter we report methods for the separation, determination, and quantitation of the major carotenoids (and retinol) in mature breast milk. Two methods are presented here: (1) a saponification procedure that converts all the retinyl esters to free retinol, facilitating the simultaneous quantitation of total retinol and β-carotene, and (2) a saponification procedure for the separation and quantitation of the more labile carotenoids α-carotene and lycopene. Although this first procedure is a suitable assay for retinol, quantitation of retinol in mature breast milk is not reported here.

Methods

Milk Pool. Mature milk samples, obtained from a group of urban healthy, nonsmoking mothers,[8] are stored at −80° until the time of pooling. To assure homogeneity of the pooled aliquots, 4-ml aliquots of thawed, pooled milk are obtained while the milk pool is stirring. Aliquots are stored in darkened 20-ml vials at −80° until analysis.

Materials. β-Carotene is obtained from Fluka (Buchs, Switzerland) and lycopene, α-carotene, and lutein are from Sigma (St. Louis, MO). Methanol, hexane, and tetrahydrofuran (THF) are purchased from Baxter (Muskegon, MI) and ethanol is from Quantum Chemical (Tuscola, IL). All solvents used for HPLC are HPLC grade or better. Organic solvents for HPLC are filtered through a 0.45-μm Fluoropore filter (Millipore, Bedford, MA) prior to use.

Sample Preparation. Frozen aliquots of pooled mature milk are thawed at 37° in a shaking water bath at 60 oscillations/min. Figure 1 gives a schematic of the sample preparation scheme. To each 20-ml vial containing 4 ml of thawed pooled milk, 5 ml of absolute ethanol and 3 ml of 50% (w/w) KOH are added. Samples are flushed with argon, capped with Teflon-lined phenolic caps (Titeseal, VMR, San Francisco, CA), and sonicated for 5 min in a water bath sonicator. For the analysis of retinol, β-carotene, and the xanthophylls, samples are saponified for 16 hr at 25° at 130 oscillations/min in an orbital shaker (New Brunswick Scientific, New Brunswick, NJ) in the dark. For the analysis of α-carotene and lycopene, the saponification step is reduced to 0.5 hr.

Following saponification, 4 ml of hexane is added to each sample vial. Samples are vigorously mixed by vortexing and sonicating for 5 min, then

---

[8] L. M. Canfield, J. M. Hopkinson, A. F. Lima, G. S. Martin, K. Sugimoto, J. Burr, L. Clark, and D. L. McGee, *Lipids* **25**, 406 (1990).

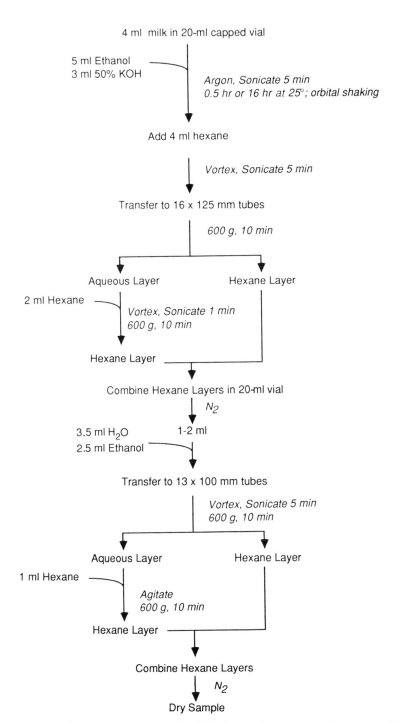

FIG. 1. Sample preparation method for HPLC analysis of carotenoids and retinol in mature breast milk.

transferred to 16 × 125 mm disposable culture tubes and centrifuged for 10 min at 600 g. The resulting hexane layer is retained and the aqueous layer reextracted with 2 ml of hexane as before. The combined hexane layers are evaporated to 1–2 ml under nitrogen.

To remove water-soluble impurities (e.g., KOH and polar lipids) from the combined hexane layers (1–2 ml), samples are extracted with 2.5 ml of absolute ethanol and 3.5 ml of deionized H_2O. This mixture is vortexed and sonicated for 5 min followed by centrifugation for 10 min at 600 g. The resulting hexane layer is removed and saved in a 1.5-ml microcentrifuge tube. The aqueous layer is reextracted with 1 ml of hexane as before. The combined hexane layers are evaporated to dryness with nitrogen and redissolved in 200 μl of THF–methanol (20:80, v/v). A higher percentage of THF was found necessary to adequately solubilize carotenoids. Following a 30-sec sonication, the 200-μl sample is loaded onto the HPLC.

HPLC Analysis. The minimal equipment requirement for the procedure are a single HPLC pump and an absorbance monitor. HPLC analysis in our laboratory is performed using a Waters model 510 pump, a Milton Roy programmable detector model SM 4000, a 50-μl loop, and Waters Maxima 820 version 3.02 system controller (Waters Associates, Milford, MA). A YMC (Morris Plains, NJ) reversed-phase C_{18} column, 5 μm, 120A ODS column (4.6 × 250 mm) is used for all analyses. Carotenoids are isocratically eluted with a running solvent consisting of 10% THF, 90% methanol (v/v), and 0.5 g/liter butylated hydroxytoluene (BHT) at a flow rate of 1.6 ml/min. Carotenoids are detected at 452 nm and retinol at 325 nm. The total time for a single analysis including reequilibration of the column is 20 min.

Standardization. For the quantitation of carotenoids in mature breast milk, external standards are utilized. Authentic standards of retinol, β-carotene, α-carotene, lycopene, and lutein are prepared by serial dilution into HPLC mobile phase from respective stock solutions containing approximately 1 mg/ml solubilized in THF plus 0.5 g/liter BHT.

Comments

Mature breast milk contains an average of 4.5 g/dl fat, approximately twofold more fat than is contained in colostrum.[9] Measurement of carotenoids from mature milk requires saponification to release the carotenoids from this lipid matrix. A saponification step is mandatory for the analysis of the carotenoids in mature breast milk. When samples are extracted

---

[9] R. G. Jensen, "Textbook of Gastroenterology and Nutrition in Infancy," 2nd Ed. pp. 157–208. Raven, New York, 1989.

TABLE I
EFFECT OF TEMPERATURE ON CAROTENOID CONCENTRATION IN MATURE BREAST MILK
FOLLOWING 16-hr SAPONIFICATION

Temperature	Concentration^a (μg/dl)			
	β-Carotene	Lycopene	α-Carotene	Lutein
4°	0.95 (0.36)	1.93 (0.42)	0.19 (0.06)	0.78 (0.15)
25°	1.00 (0.08)	1.45 (0.10)^b	0.19 (0.01)	0.80 (0.09)
37°	0.94 (0.35)	1.28 (0.10)^b	0.20 (0.15)	0.85 (0.20)
45°	0.86 (0.06)	1.00 (0.21)^b	0.19 (0.01)	0.77 (0.02)

^{a} Values represent mean (SD) of three replicates.
^{b} Significantly different than 4° ($p < 0.05$).

without saponification, carotenoids cannot be detected. The choice of conditions for the saponification step are ultimately determined by the carotenoids and retinoids to be quantitated. The saponification and extraction conditions reported here are those optimal to measure β-carotene and retinol simultaneously, after the complete hydrolysis of retinyl esters to retinol. Because substantial degradation of the more labile carotenoids α-carotene and lycopene results from the saponification conditions in this procedure, we report a separate procedure for the analysis of these carotenoids.

The effects of both temperature and time during the saponification step on carotenoid and retinol concentrations were studied. In studies testing the effect of temperature with a 16-hr saponification step, β-carotene, α-carotene, and lutein appeared unaffected by temperature (Table I). No differences in the concentration of these compounds were observed at 4, 25, 37, or 45° with 16 hr of saponification. In contrast, lycopene significantly decreased with increasing temperature, resulting in a 68% loss of lycopene at 45° compared with 4°.

In separate experiments we tested the effect of saponification time (0.5, 2.0, 4.0, and 16 hr) at two temperatures (25 and 45°), on the conversion of retinyl esters (Fig. 2) to retinol and the stability of β-carotene, lutein, α-carotene, and lycopene in breast milk extracts (Fig. 3). Similar to the findings of Thompson and Duval,[10] we found that at 25°, four more hours were needed for the complete conversion of retinyl esters to retinol (Fig. 2). At 45°, complete conversion of retinyl esters to retinol was attained at 2 hr. At both 25° and 45°, the retinol concentration measured in extracts

[10] J. N. Thompson and S. Duval, *J. Micronutr. Anal.* **6**, 147 (1989).

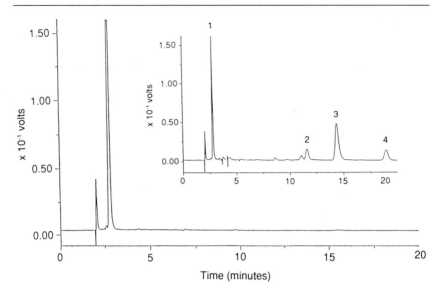

FIG. 2. HPLC chromatogram at 325 nm demonstrating the effects of saponification for 16 and 0.5 hr (inset) at 25° on retinyl ester and retinol profiles. The eluted compounds are (1) retinol, (2) retinyl linoleate, (3) retinyl palmitate, and (4) retinyl stearate.

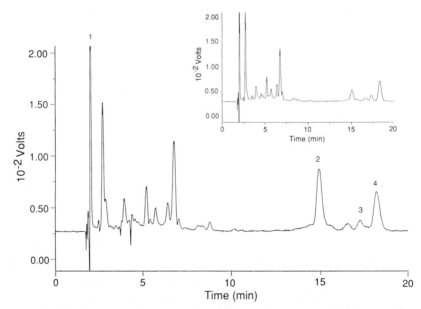

FIG. 3. HPLC chromatogram at 452 nm demonstrating the carotenoid profile of mature breast milk after 0.5 and 16 hr (inset) saponification. Note the significant decrease in lycopene peak after 16 hr of saponification. The carotenoids are (1) lutein, (2) lycopene, (3) α-carotene, and (4) β-carotene.

TABLE II
EFFECT OF TIME AND TEMPERATURE DURING SPONIFICATION ON LYCOPENE
AND α-CAROTENE CONCENTRATIONS

Time (hr)	Degradation over time[a] (% decrease from 0.5 hr)			
	25°		45°	
	Lycopene	α-Carotene	Lycopene	α-Carotene
2.0	0	28	23[b]	47[b]
4.0	0	37	—	—
7.0	—	—	35[b]	47[b]
16.0	25	49[b]	61[b]	49[b]

[a] Each value is the mean of three replicates.
[b] Significantly different than 0.5 hr ($p < 0.05$).

remained constant for up to 16 hr of saponification. Because the concentration of β-carotene and lutein did not significantly change as a function of saponification time at either temperature tested, we chose the less severe temperature (25°) and the length of time of saponification that would allow us to process most conveniently large numbers of samples (16 hr) for the simultaneous analysis of β-carotene and retinol.

Unlike retinol and β-carotene, lycopene was sensitive to both temperature and length of saponification (Tables I and II). Although not significant, there was a 25% reduction in the lycopene measured after 16 hr of saponification compared with 0.5 hr of saponification at 25° (Fig. 2). When the saponification step was conducted at 45°, there was a significant reduction in lycopene measured after both 2 hr (23% decrease) and 16 hr (61% decrease) of saponification. Similarly, the concentration of α-carotene in breast milk extracts significantly decreased (49%) after 16 hr compared with 0.5 hr of saponification at 25°. Under the more severe saponification conditions of 45° an equivalent loss of α-carotene was observed within 2 hr. Therefore, for the analysis of the more labile carotenoids lycopene and α-carotene we recommend that the saponification step be conducted at 25° for 0.5 hr. Under these conditions we found mature breast milk from healthy mothers in the United States to contain the following concentrations of carotenoids (Table III): 1.01 µg/dl β-carotene, 0.32 µg/dl α-carotene, 2.73 µg/dl lycopene, and 1.06 µg/dl lutein. In our pooled samples, β-carotene accounted for 38% of total carotenoid.

Equally important in the quantitation of retinol and β-carotene is the proper mixing of the samples during the saponification step. We found that orbital rather than linear shaking resulted in more efficient saponification

TABLE III
CONCENTRATION OF MAJOR CAROTENOIDS IN
MATURE BREAST MILK AFTER 0.5 hr
SAPONIFICATION[a]

Carotenoid	Concentration μg/dl	Standard deviation
β-Carotene	1.01	0.02
α-Carotene	0.32	0.02
Lycopene	2.73	0.13
Lutein	1.06	0.03

[a] At 25°. Values represent the mean of three determinations on each of 2 days.

and extraction of retinol and the carotenoids. In addition, the speed (130 oscillations/min) of shaking during saponification was an important determinant of saponification efficiency.

Several contaminants were observed at both 325 and 452 nm as a result of the materials used in the saponification step. Use of the urea caps that are sold with standard 20-ml glass scintillation vials led to the appearance of a substantial peak at approximately 2–3 min that interfered with the quantitation of retinol at 325 nm and the quantitation of the xanthophylls at 452 nm. This contamination peak was eliminated when the urea caps were replaced by Teflon-lined phenolic caps (Titeseal, VWR). Similarly, addition of BHT to the saponification mixture resulted in the appearance of a peak between 1.5 and 2.0 min at 325 nm. Removal of BHT from the saponification step eliminated this contamination problem with no significant loss of retinol or β-carotene to oxidation.

Similar to results obtained by others in plasma,[11] we were unable to reproducibly equilibrate exogenous and endogenous carotenoids in mature breast milk. Therefore, quantitation with an internal standard in milk samples to control for extraction efficiency is not meaningful. We tested the efficiency of the extraction procedure by conducting one to five serial hexane extractions following the saponification step. Similar to the findings of other investigators,[10,11] our results indicated that maximal amounts of the carotenoids are extracted with two serial extractions.

The choice of column was found to significantly affect the separation and quantitation of carotenoids from mature breast milk. In our hands, the

---

[11] L. R. Cantilena and D. W. Nierenberg, *J. Micronutr. Anal.* **6,** 127 (1989).

best results were obtained with a YMC (A-303) 5-μm ODS reversed-phase column (YMC, Morris Plains, NJ).

In summary, we have presented two sample preparation methods for the analysis of carotenoids and retinol in mature breast milk by HPLC: (1) a method for the simultaneous analysis of retinol and β-carotene and (2) a method for the analysis of the more labile carotenoids such as lycopene and α-carotene. Using these methods we can reproducibly quantitate β-carotene, α-carotene, lycopene, and lutein. Depending on the HPLC system utilized, total xanthophylls can be further separated, generating values for β-cryptoxanthin and zeaxanthin.

Acknowledgment

The authors wish to acknowledge receipt of NIH Grant R01 HD26715.

Section III

Antioxidation and Singlet Oxygen Quenching

[38] Antioxidant Effects of Carotenoids *in Vivo* and *in Vitro:* An Overview

By PAOLA PALOZZA and NORMAN I. KRINSKY

Introduction

Carotenoid pigments are widely distributed in nature, where they play an important role in protecting cells and organisms against photosensitized oxidations.[1] This chapter deals with their role as biological antioxidants in radical-initiated processes, both *in vitro* and *in vivo*. We define biological antioxidants as "compounds that protect biological systems against the potentially harmful effects of processes or reactions that can cause excessive oxidations."[2]

Several mechanisms have been suggested for carotenoid protection in biological systems, which include the deactivation of electronically activated species such as singlet oxygen ($^1O_2$) and the deactivation of reactive chemical species, such as peroxyl or alkoxyl radicals, that can be generated within cells and might otherwise initiate harmful oxidative reactions.[3]

Antioxidant action is well documented in a number of *in vitro* studies, including lipids in homogeneous solutions, liposomes, isolated membranes, and intact cells. This antioxidant activity is related to the chemical structures of the carotenoids and it has been reported to be particularly effective at low O_2 tension.[4]

Additionally, direct evidence of carotenoids functioning as antioxidants *in vivo* has been reported in animal models. These studies show that carotenoid protection is related either to a direct antioxidant activity or to a modulation of cellular antioxidant levels by carotenoids.

The implications of a role for β-carotene and other carotenoids in human disease in which radical species are thought to be involved, such as cancer, cataract, atherosclerosis, and the process of aging, would have great significance because of the well-known lack of toxicity of β-carotene,[5,6] and the potential use of this compound and other carotenoids as chemopre-

---

[1] N. I. Krinsky, *in* "The Science of Photomedicine" (J. D. Regan and J. A. Parrish, eds.), p. 397. Plenum, New York, 1982.
[2] N. I. Krinsky, *Proc. Soc. Exp. Biol. Med.* **200**, in press (1992).
[3] N. I. Krinsky, *Free Radical Biol. Med.* **7**, 617 (1989).
[4] G. W. Burton and K. U. Ingold, *Science* **224**, 569 (1984).
[5] A. Bendich, *Nutr. Cancer* **11**, 207 (1988).
[6] M. M. Mathews-Roth, *Toxicol. Lett.* **41**, 185 (1988).

ventive agents.[7] Even before all of the evidence is available regarding the chemopreventive activity of carotenoids, it is necessary to offer an explanation for the mechanism of action of these pigments. Antioxidant action of these molecules seems a likely basis for the effects of carotenoids, and this chapter evaluates the available evidence for this explanation of *in vivo* biological activity.

Mechanisms of Carotenoid Action

The mechanisms by which carotenoids can act as protective agents in cells and organisms have been reviewed[3,8] and will be discussed only briefly here. They can be summarized in the abilities of these molecules to act either as photoprotective agents against the harmful effects of light, O_2, and photosensitizing pigments or as compounds reactive against chemical species generated within cells and able to induce oxidative damage.

Photoprotection

The basis for carotenoid (CAR) protection against photosensitized reactions is the ability of these pigments to quench, by an energy transfer process, either triplet sensitizers (^3S) [reaction (1)] or $^1O_2$ [reaction (2)], generated via energy transfer from ^3S to ground-state oxygen ($^3O_2$).

$$^3S + CAR \rightarrow S + {}^3CAR \tag{1}$$

$$^1O_2 + CAR \rightarrow {}^3O_2 + {}^3CAR \tag{2}$$

These reactions limit the potentially dangerous effects that ^3S is able to induce via either a type I process that initiates electron or hydrogen atom transfer from a substrate to ^3S, or a type II process that generates the highly reactive $^1O_2$.[9]

$^1O_2$ can be produced not only by photoexcitation, but by nonphotochemical processes, such as the reaction of ozone with various biological molecules.[10] The generation and possible pathological consequences of $^1O_2$ in biological systems have been well described[9] and $^1O_2$ has been clearly implicated in the inactivation of proteins, in the peroxidation of biological lipids,[11] and in DNA damage.[12]

---

[7] N. I. Krinsky, *in* "Carotenoids: Chemistry and Biology" (N. I. Krinsky, M. M. Mathews-Roth, and R. F. Taylor, eds.), Proc. 8th Int. Symp. Carotenoids, p. 279. Plenum, New York, 1989.
[8] G. W. Burton, *J. Nutr.* **119**, 109 (1989).
[9] C. S. Foote, *in* "Singlet Oxygen" (H. H. Wasserman and R. W. Murray, eds.), p. 139. Academic Press, New York, 1979.
[10] J. R. Kanofsky and P. Sima, *J. Biol. Chem.* **266**, 9039 (1991).
[11] A. W. Girotti, *Photochem. Photobiol.* **51**, 497 (1990).
[12] H. Wefers, D. Schutle-Frohlinde, and H. Sies, *FEBS Lett.* **211**, 49 (1987).

The ability of carotenoids to quench $^1O_2$ is related to the number of conjugated double bonds, with a maximum protection shown by pigments having nine or more conjugated double bonds[13], and to the structure of the pigments. Di Mascio et al.[14] have demonstrated that lycopene was approximately twice as effective as β-carotene in quenching $^1O_2$. Their data suggest that the quenching properties of carotenoids reside not only in the length of the conjugated double bond system, but also in the structure of the pigment.

In both reactions (1) and (2), the carotenoid triplet that is formed can readily lose its energy to the environment and return to its original form, as seen in reaction (3).

$$^3\text{CAR} \rightarrow \text{CAR} + \text{heat} \tag{3}$$

This characteristic makes carotenoids very potent protective agents against photosensitized reactions.

Reaction with Oxygen Radicals

Several forms of reactive oxygen are generated in the cell as a result of various metabolic processes or following exposure to xenobiotics. The products superoxide (O_2^-), hydrogen peroxide (H_2O_2), and the hydroxyl radical ($HO\cdot$) are shown in reaction (4).

$$O_2 \xrightarrow{e^-} O_2^- \xrightarrow{e^-} H_2O_2 \xrightarrow{e^-} OH\cdot \xrightarrow{e^-} H_2O \tag{4}$$

These species have all been shown to cleave DNA, peroxidize lipids, alter enzyme activity, depolymerize polysaccharides, and kill cells.[15,16]

A large number of studies have indicated that carotenoids are able to limit the oxidative damage induced by oxy radical-generating systems. This protection involves both nuclear and lipid molecules. β-Carotene is able to reduce the extent of nuclear damage induced by xanthine oxidase/hypoxanthine or by activated polymorphonuclear leukocytes,[17] as well as to inhibit the lipid peroxidation induced either by enzymatic sources of oxy radicals, such as the xanthine oxidase system[18,19] and NADPH/cytochrome

---

[13] C. S. Foote, R. W. Denny, L. Weaver, Y. Chang, and J. Peters, *Ann N.Y. Acad. Sci.* **171**, 139 (1970).

[14] P. Di Mascio, S. Kaiser, and H. Sies, *Arch. Biochem. Biophys.* **274**, 532 (1989).

[15] T. F. Slater, in "Free Radicals, Lipid Peroxidation and Cancer" (D. C. H. McBrien and T. F. Slater, eds.), p. 243. Academic Press, New York, 1982.

[16] K. Brawn and I. Fridovich, *Arch. Biochem. Biophys.* **206**, 414 (1981).

[17] A. B. Weitberg, S. A. Weitzman, E. P. Clark, and T. P. Stossel, *J. Clin. Invest.* **75**, 1835 (1985).

[18] E. W. Kellogg and I. Fridovich, *J. Biol. Chem.* **250**, 8812 (1975).

[19] W. Miki, *Pure Appl. Chem.* **63**, 141 (1991).

P-450 reductase,[20,21] or by nonenzymatic sources, such as transition metal salts.[19,22-24]

In addition, there is experimental evidence indicating the effectiveness of carotenoids in inhibiting lipid peroxidation induced by paraquat[21] and Adriamycin,[25] both well-known agents implicated in the production of oxy radicals.

Although it is clear that there is an involvement of carotenoids in reactions with oxygen radicals, it is not clear which radical species is primarily involved. Dixit et al.,[20] looking at epidermal skin microsomal lipid peroxidation induced by NADPH, found that β-carotene and other $^1O_2$ quenchers were able to protect the membranes, while hydroxyl radical interceptors were ineffective. It should be pointed out that some of the scavengers used in these experiments as $^1O_2$ interceptors (histidine, dimethylfuran, and β-carotene) are excellent OH· traps as well.[26]

Reaction with Peroxyl Radicals

It is well known that peroxyl radicals, because of the selectivity of their reactions and their ability to diffuse in biological systems, are potentially more dangerous than many other types of radicals.[26] It has been suggested that these species may be implicated in the toxic action of many chemicals and environmental agents,[27] and may be connected with a variety of pathological events, such as heart disease, cancer, and the process of aging.[28]

There is increasing evidence that carotenoids are very effective quenchers of peroxyl radicals,[3] but the mechanism of their antioxidant action has not yet been defined. A hypothesis has been presented by Burton and Ingold[4] and expanded by Burton.[8] Although one might expect that the reaction of β-carotene with a peroxyl radical would form a carotenoid radical species [reaction (5)],

$$\beta\text{-Carotene} + \text{ROO}\cdot \rightarrow \beta\text{-carotene}\cdot + \text{ROOH} \tag{5}$$

[20] R. Dixit, H. Mukhtar, and D. R. Bickers, *J. Invest, Dermatol* **81**, 369 (1983).
[21] H. Kim(Jun), *Korean J. Nutr.* **23**, 434 (1990).
[22] N. I. Krinsky and S. M. Deneke, *JNCI* **69**, 205 (1982).
[23] A. J. F. Searle and R. L. Willson, *Biochem. J.* **212**, 549 (1983).
[24] M. Kurashige, E. Okimasu, M. Inoue, and K. Utsumi, *Physiol. Chem. Phys. Med. NMR* **22**, 27 (1990).
[25] G. F. Vile and C. C. Winterbourn, *FEBS Lett.* **238**, 353 (1988).
[26] M. G. Simic, *Mutat. Res.* **202**, 377 (1988).
[27] G. L. Plaa and H. Witschi, *Annu. Rev. Pharmacol.* **16**, 125 (1976).
[28] B. Halliwell and J. M. C. Gutteridge, "Free Radicals in Biology and Medicine." Clarendon, Oxford, 1985.

these workers proposed that β-carotene could react directly with a peroxyl radical to form a resonance-stabilized, carbon-centered radical [per reaction (6)].

$$\beta\text{-Carotene} + \text{ROO} \cdot \rightarrow \text{ROO-}\beta\text{-Carotene} \cdot \qquad (6)$$

The combination of the carotenoid radical with oxygen would lead to the formation of a carotenoid–peroxyl radical, but this reaction would be dependent on the O_2 tension in the system. If the O_2 tension is sufficiently low, the equilibrium of reaction (7) shifts to the left, reducing the amount of chain-carrying peroxyl radical.

$$\beta\text{-Carotene} \cdot + O_2 \rightleftharpoons \beta\text{-carotene-OO} \cdot \qquad (7)$$

In addition, the β-carotene–peroxyl complex could react with another peroxyl radical, leading to a termination reaction, as shown in reaction (8).

$$\beta\text{-Carotene-OO} \cdot + \text{ROO} \cdot \rightarrow \text{inactive products} \qquad (8)$$

On the other hand, if the O_2 tension is high, the equilibrium of reaction (7) shifts to the right and β-carotene, because of autooxidation, forms a peroxyl radical capable of acting as a prooxidant.[4]

This hypothesis seems to be confirmed by experimental evidence, discussed in the next sections, but unfortunately none of the potential intermediate forms proposed by Ingold and Burton[4] have been isolated. Rather, a variety of products arising from the reactions of radicals with β-carotene have been described in three publications.[29-31]

Analysis of Antioxidant Activity of Carotenoids

Bleaching of Carotenoids

Both spontaneous and radical-initiated oxidation of carotenoids have been followed extensively by observing the loss of color (bleaching) in the visible region of the spectrum (for a review, see Ref. 3). This bleaching, as well as the analysis of carotenoid oxidation products, has been studied in an attempt to evaluate carotenoid antioxidant ability when these compounds react directly with radical species, and also to establish kinetics of radical–carotenoid interactions. Some of the compounds or techniques

---

[29] T. A. Kennedy and D. C. Liebler, *Chem. Res. Toxicol.* **4,** 290 (1991).
[30] G. J. Handelman, F. J. G. M. van Kuijk, A. Chatterjee, and N. I. Krinsky, *Free Radical Biol. Med.* **10,** 427 (1991).
[31] R. C. Mordit, J. C. Walton, G. W. Burton, L. Hughes, K. U. Ingold, and D. A. Lindsay, *Tetrahedron Lett.* **32,** 4203 (1991).

used to study either bleaching or formation of oxidation products of carotenoids induced by nonenzymatic and enzymatic sources of radicals are summarized in Table I.[3,4,29,30,32-41]

The results of these studies, discussed extensively,[3] show that carotenoid molecules effectively trap oxygen and organic free radical intermediates.

Products of Autoxidation and Induced Oxidations

Three publications have appeared regarding the products obtained when β-carotene reacts with radical sources.[29-31] The products identified in these studies, as well as in earlier ones, are presented in Table II.[29-32,42-45] It is clear that the renewed interest in the nature of these products is indicative of an attempt to understand how carotenoids interact with, and therefore quench, radical species.

Antioxidant Effects of Carotenoids *in Vitro*

Inhibition of Lipid Peroxidation

Lipid-Homogeneous Solutions. The possibility of using lipid-soluble azo compounds[46] to generate peroxyl radicals following thermal decomposition has allowed many studies to be carried out evaluating the effectiveness of carotenoids as inhibitors of these reactive species. The use of homogeneous solutions for studying carotenoid protection of lipid substrates has the advantage of simplifying the assay, because both the carotenoids and the lipids are dissolved in organic solvents.

---

[32] A. H. El-Tinay and C. O. Chichester, *J. Org. Chem.* **35**, 2290 (1970).
[33] W. Boguth, R. Patzelt-Wenczler, and R. Repges, *Int. J. Vitam. Res.* **41**, 21 (1971).
[34] M. Elahi and E. R. Cole, *Nature (London)* **203**, 186 (1964).
[35] W. A. Pryor and C. K. Govindan, *J. Org. Chem.* **45**, 4679 (1981).
[36] J. E. Packer, J. S. Mahood, V. O. Mora-Arellano, T. F. Slater, R. L. Willson, and B. S. Wolfenden, *Biochem. Biophys. Res. Commun.* **98**, 901 (1981).
[37] W. Bors, C. Michel, and M. Saran, *Biochim. Biophys. Acta* **796**, 312 (1984).
[38] B. P. Klein, D. King, and S. Grossman, *Adv. Free Radical Biol. Med.* **1**, 309 (1985).
[39] J. Kanner and J. E. Kinsella, *Lipids* **18**, 198 (1983).
[40] J. Kanner and J. E. Kinsella, *Lipids* **18**, 204 (1983).
[41] J. Kanner and J. E. Kinsella, *J. Agric. Food Chem.* **31**, 370 (1983).
[42] R. F. Hunter and R. M. Krakenberger, *J. Chem. Soc.* p. 1 (1947).
[43] J. Friend, *Chem. Ind. (London)* 597 (1958).
[44] C. Zinsou, *Physiol. Veg.* **9**, 149 (1971).
[45] C. O. Ikediobi and H. E. Snyder, *J. Agric. Food Chem.* **25**, 124 (1977).
[46] Y. Yamamoto, S. Haga, E. Niki, and Y. Kamiya, *Bull. Chem. Soc. Jpn.* **57**, 1260 (1984).

TABLE I
BLEACHING OF OR OXIDATION PRODUCTS OF CAROTENOIDS INDUCED
BY RADICAL SOURCES

Radical source[a]	Radical species	Assay Bleaching	Oxidation products	Ref.
Nonenzymatic				
AIBN	Peroxyl radicals	+		32
			+	33
			+	4
			+	30
AMVN	Peroxyl radicals		+	29
NaOCl	Oxygen radicals		+	30
t-BOOH	t-BOO·	+		34
			+	4
TPPO	Singlet oxygen		+	35
Pulse radiolysis of CCl_4	$CCl_3O·$	+		36
Photolysis of t-BOOH or fatty acid hydroperoxides	Alkoxyl radicals	+		37
Enzymatic				
Lipoxidase	No demonstration		For reviews, see 3	
	Enzyme-bound radicals?			38
Peroxidase				
Lactoperoxidase	No demonstration	+		39, 40
Myeloperoxidase	No demonstration	+		41

[a] AIBN, Azobisisobutyl nitrile; AMVN, 2,2-azobis(2,4-dimethylvaleronitrile); t-BOOH, tertbutyl hydroperoxide; and TPPO, triphenylphosphite ozonide.

Experimental evidence, obtained in organic solution from different laboratories, indicates that the antioxidant activity of these pigments is related to the O_2 concentration, to the chemical structures of carotenoids, and to the concomitant presence of other antioxidants. These three aspects will be discussed here briefly.

Oxygen Pressure. A number of interesting observations about the antioxidant activity of β-carotene in homogeneous solutions have been reviewed by Burton,[8] who discusses the effects of varying the concentration of β-carotene (0.05 – 5 mM) and the partial pressure of O_2 (15 – 760 torr) on the azobisisobutyl nitrile (AIBN)-induced oxidation of methyl linoleate. At a low O_2 tension (15 torr) β-carotene is an effective antioxidant, but at high O_2 tensions (760 torr) the initial antioxidant activity is followed by a prooxidant action of β-carotene.[4]

Data supporting the possibility that β-carotene could be more effective

TABLE II
Oxidation Products of β-Carotene When Exposed to Autoxidation, Radical Attack, or Coupled Oxidation

Product	Autoxidation[a]	Radical attack[b]	Coupled oxidation[c]
Hydrocarbons			
cis-β-Carotene	+	+	+
Epoxycarotenoids			
5,6-Epoxy-β-carotene	+	+	+
5,8-Epoxy-β-carotene	+	+	
5,6-5′,6′-Diepoxy-β-carotene	+	+	+
5,6-5′,8′-Diepoxy-β-carotene	+	+	
5,8-5′,8′-Diepoxy-β-carotene	+	+	+/−[d]
15,15′-Epoxy-β-carotene	+		
Cryptoflavine			+
Carbonyls			
Semi-β-carotenone	+	+	
β-Carotenone	+	+	
β-Apo-13-carotenone	+	+	+/−
β-Apo-10′-carotenal	+		+/−
β-Apo-12′-carotenal			+/−
β-Apo-14′-carotenal	+		
β-Apo-15-carotenal (retinal)	+	+	
Retro-dehydro-β-carotenone			+
Conjugated polyene carbonyls	+/−		+/−

[a] Autoxidation was reported by Hunter and Krakenberger[42] and Handelman et al.[30]
[b] Radical attack was reported by Kennedy and Liebler,[29] Handelman et al.,[30] Mordit et al.,[31] and El-Tinay and Chichester.[32]
[c] Coupled oxidation was reported by Friend,[43] Zinsou,[44] and Ikediobi and Snyder.[45]
[d] In some cases, (+/−) only spectral evidence was obtained for these products.

as a chain-breaking antioxidant in a lipid environment at low O_2 pressures have also been presented by Stocker et al.[47] They compared the effect of O_2 at both 20 and 2% on the antioxidant properties of β-carotene, α-tocopherol, and bilirubin. At 20% O_2, α-tocopherol was the most effective antioxidant, with very little difference between β-carotene and bilirubin. However, at 2% O_2, both β-carotene and bilirubin were better antioxidants than under air. The same effect is not observed with α-tocopherol, which in any case remained the most powerful antioxidant.

Similar results have been obtained by Palozza and Krinsky,[48] using a

[47] R. Stocker, Y. Yamamoto, A. F. McDonagh, A. N. Glazer, and B. N. Ames, *Science* **235**, 1043 (1987).
[48] P. Palozza and N. I. Krinsky, *Free Radical Biol. Med.* **11**, 407 (1991).

TABLE III
EFFECT OF β-CAROTENE AND α-TOCOPHEROL ON AIBN-DEPENDENT LIPID PEROXIDATION OF MICROSOMAL LIPIDS[a]

Compound	$pO_2 = 150$ mmHg		$pO_2 = 18.8$ mmHg		Antioxidant effectiveness (vacuum/air) (%)
	MDA[b] (nmol/ml lipid)	Inhibition (%)	MDA[b] (nmol/mg lipid)	Inhibition (%)	
Control	25.0 ± 1.0	0	8.4 ± 0.8	0	
β-Carotene (50 μM)	12.8 ± 0.5	49	3.9 ± 0.1	54	+10
α-Tocopherol (1.1 μM)	13.4 ± 0.6	46	5.7 ± 0.7	32	−30

[a] Under air and under vacuum. (from Ref. 48.)
[b] Malondialdehyde values are means ± SEM of three to six experiments.

mixture of lipids isolated from rat liver microsomal membranes, in hexane solution exposed to AIBN at 37°. Using either malondialdehyde (MDA) or conjugated diene production as measures of lipid peroxidation, they found that both α-tocopherol and β-carotene blocked the lipid peroxidation of microsomal lipids. In air (pO_2 150 mmHg), α-tocopherol is about 40–50 times better than β-carotene as an antioxidant. However, when the pO_2 is reduced to under 20 mmHg, the difference in effectiveness decreases by about 40%, confirming the importance of oxygen tension in the antioxidant activity of β-carotene (Table III). Because the O_2 pressures found in mammalian tissues are low, these observations seem to suggest an important role of β-carotene as an antioxidant in such environments and the possibility of complementary roles between antioxidants *in vivo*.

Chemical Structure. The importance of the carotenoid structure in determining antioxidant activity has been reported by Terao[49] who compared β-carotene, canthaxanthin, astaxanthin, and zeaxanthin with respect to their ability to inhibit the formation of hydroperoxides of methyl linoleate in a radical-initiated system. The antioxidant activities of canthaxanthin and astaxanthin were better and lasted longer than those of either β-carotene or zeaxanthin, suggesting that the presence of a conjugated keto group increases the efficiency of the peroxyl radical-trapping ability of carotenoids.

Miki[19] has also observed the effectiveness of conjugated keto groups in a study of β-carotene, lutein, zeaxanthin, astaxanthin, tunaxanthin, and canthaxanthin, in comparison with α-tocopherol. Using a heme–protein–Fe^{2+} complex as a free radical generator and measuring TBA production, Miki found that astaxanthin is the most efficient scavenger, with an ED_{50}

[49] J. Terao, Lipids **24**, 659 (1989).

of 0.2 μM, followed by zeaxanthin, canthaxanthin, lutein, tunaxanthin, and β-carotene, respectively, showing ED_{50} values in the range of 0.4–1.0 μM. In contrast, the ED_{50} of α-tocopherol was about 3 μM.

Carotenoid–Tocopherol Interactions. Just as the presence of δ-tocopherol enhances the protective effect of β-carotene on $^1O_2$-initiated photooxidation of methyl linoleate,[50] so does β-carotene delay markedly the AIBN-induced loss of endogenous microsomal tocopherols.[48]

Liposomes. The use of liposomes as a model membrane system in the study of inhibition of lipid peroxidation by carotenoids provides many advantages. It allows manipulation of lipid composition, pH, and temperature and avoids complications related to the introduction of the lipid-soluble carotenoids into tissue membranes.

There have been a number of different studies showing the inhibition of lipid peroxidation in liposomes, and these are summarized in Table IV.[22,47,51]

In agreement with the data obtained in homogeneous solutions, Kennedy and Liebler have reported[52] that the antioxidant activity of β-carotene varies with the O_2 concentration in phospholipid liposomes. Although this activity was similar at 160 and 15 torr, it decreased at high pO_2. They suggest that this difference could be due to the rapid formation of autooxidation products of β-carotene at high pO_2 conditions, but this may also be an example similar to the prooxidant effect reported at high pO_2.[4]

Lipoproteins. Two factors make lipoproteins interesting as an *in vitro* system to elucidate the antioxidant activities of carotenoids. These are the increasing evidence that the initiation of atherosclerosis is related to oxidative modifications of low-density lipoproteins (LDLs),[53] and the well-established observation that LDL is the major carrier of β-carotene in humans.[54]

Much of the work on the role of carotenoids as antioxidants in LDL subfractions has been reported by Esterbauer's group. These workers have treated human LDLs with Cu^{2+} as a prooxidant, and have demonstrated that the level of oxidation is highly related to the endogenous level of antioxidants. The presence of the antioxidants prolongs the lag phase that precedes the rapid oxidation of LDLs, and the antioxidants are consumed

---

[50] J. Terao, R. Yamauchi, H. Murakami, and S. Matsushita, *J. Food Process Preserv.* **4,** 79 (1980).
[51] L. Cabrini, P. Pasquali, B. Tadolini, A. M. Sechi, and L. Landi, *Free Radical Res. Commun.* **2,** 85 (1986).
[52] T. A. Kennedy and D. C. Liebler, *J. Biol. Chem.* **267,** 4658 (1992).
[53] D. Steinberg, S. Parthasarathy, T. E. Carew, J. C. Khoo, and J. L. Witztum, *N. Engl. J. Med.* **320,** 915 (1989).
[54] N. I. Krinsky, D. G. Cornwell, and J. L. Oncley, *Arch. Biochem. Biophys.* **73,** 233 (1958).

TABLE IV
CAROTENOID INHIBITION OF LIPID PEROXIDATION IN LIPOSOMES

Substrate	Prooxidant	Carotenoid	Assay of inhibition	Ref.
Phosphatidylcholine	$K_3C_2O_8$	β-Carotene	MDA	22
Phosphatidylcholine	Fe^{2+}	β-Carotene	MDA	22
	Fe^{2+}	Canthaxanthin	MDA	22
Egg lecithin–phosphatidic acid–diarachidonyllecithin	Spontaneous autoxidation	β-Carotene	MDA[a]	51
Phosphatidylcholine	AMVN	β-Carotene	Hydroperoxides[b]	47

[a] No inhibition in the presence of Fe^{2+}.
[b] Increase of inhibition in the presence of 2% O_2 in comparison to 20% O_2.

in the following sequence: α-tocopherol, γ-tocopherol, lycopene, retinyl stearate (now believed to be phytofluene), and β-carotene.[55] Although there was an initial report of a correlation between the content of endogenous tocopherols and resistance to oxidative stress in pig LDLs,[56] they now report that some human LDL preparations, with practically equal amounts of tocopherols, have very different lag phases[57] and they suggest that the presence of carotenoids could explain these differences and could act as a second protective barrier against Cu^{2+}-induced oxidation.

Membranes. A more physiological approach to the study of the antioxidant activity of carotenoids is represented by experiments using isolated membranes. The use of this model is, however, complicated by the fact that the membranes are isolated as aqueous suspensions, and the carotenoids are virtually insoluble in this medium. This fact, in addition to the limited solubility of carotenoids in polar organic solvents such as alcohols, may explain the disparate results obtained by different laboratories, in terms of concentrations and effectiveness of carotenoid pigments in protecting membranes against *in vitro* radical reactions.

The studies on membrane lipid peroxidation in the presence of carotenoids are summarized in Table V[19–21,23,24,58–60] and will be briefly discussed below.

[55] H. Esterbauer, G. Striegl, H. Puhl, and M. Rotheneder, *Free Radical Res. Commun.* **6**, 67 (1989).
[56] G. Knipping, M. Rotheneder, G. Striegl, and H. Esterbauer, *J. Lipid Res.* **31**, 1965 (1990).
[57] H. Esterbauer, M. Dieber-Rotheneder, G. Striegl, and G. Waeg, *Am. J. Clin. Nutr.* **53**, 314 (1991).
[58] G. F. Vile and C. C. Winterbourn, *Biochem. Pharmacol.* **37**, 2893 (1988).
[59] O. Halevy and D. Sklan, *Biochim. Biophys. Acta* **918**, 304 (1987).
[60] P. Palozza, S. Moualla, and N. I. Krinsky, *Free Radical Biol. Med.* **13**, in press (1992).

TABLE V
MEMBRANE LIPID PEROXIDATION IN THE PRESENCE OF CAROTENOIDS

Membrane (mg/ml)	Prooxidant	Carotenoid	MDA inhibition (% of control)	Ref.
Rat skin microsomes (2.3–3 mg/ml)	1 mM NADPH	5 μM β-carotene	58[a]	20
		10 μM β-carotene	69[a]	
Rat liver microsomes (1 mg/ml)	5 μM FeSO$_4$/500 μM cysteine	1mM β-carotene	7[b]	23
Rat liver microsomes (0.5 mg/ml)	100 μM NADPH	10 mM β-carotene	13[b]	58
	1.0 μM FeCl$_3$/30 μM adriamycin	25 μM β-carotene	40[b]	
			70[b,c]	
Rat liver microsomes (0.5–1.0 mg/ml)	25 mM AAPH/AMVN	50 nmol β-carotene/mg prot.	Modest[a]	60
	400 μM NADPH/50 μM FeCl$_3$/1 mM ADP	10 nmol β-carotene/mg prot.	11[b]	
Rat liver microsomes (1 mg/ml)	0.1 mM NADPH/0.1 mM FeCl$_3$/1.7 mM ADP	100 μM β-carotene	20[b]	21
		100 μM α-carotene	37	
		100 μM lycopene	42	
		100 μM lutein	33	
Rat liver microsomes (1 mg/ml)	0.2 mM paraquat/0.5 mM NADPH	100 μM β-carotene	18[a]	21
		100 μM α-carotene	30	
		100 μM lycopene or lutein	33	
Bovine seminal vesicles (0.3–0.5 mg/ml)	Spontaneous	20 μM-β-carotene	34[d]	59
		40 μM β-carotene	56[d]	
Rat liver mitochondria (2 mg/ml)	100 μM Fe^{2+}	4.2 μM astaxanthin	100[a]	24
Rat liver mitochondria (2 mg/ml)	? μM FeSO$_4$	0.1 μM astaxanthin	100	19

[a] Values obtained after 60 min of incubation.
[b] Values obtained after 30 min of incubation.
[c] Incubation at 4 mmHg O$_2$.
[d] Inhibition of prostanoid products.

Dixit et al.[20] studied the effect of β-carotene in a membrane model that consisted of epidermal microsomes undergoing NADPH-dependent lipid peroxidation. At concentrations of 5–10 μM, β-carotene significantly inhibited MDA formation more effectively than α-tocopherol used at the same concentrations. However, to obtain the 10μM concentration used in their experiments, these investigators would have had to have a stock solution of β-carotene in ethanol of 300 μM, which exceeds the solubility of the pigment in this solvent.

Searle and Willson[23] reported a weak inhibition, by 1–10 mM β-carotene, of MDA formation in rat liver microsomes exposed to $FeSO_4$ and cysteine. They suggested that the weak effect was due to the difficulty of dissolving the antioxidant, added as suspension, in the microsomes.

A clearer example of antioxidant activity of β-carotene has been reported by Vile and Winterbourn[58] in Adriamycin-treated rat liver microsomes studied at different O_2 tensions. In the presence of increasing concentrations of either α-tocopherol or β-carotene (1 to 100 nmol/mg protein), β-carotene was a better antioxidant at low pO_2 (4 mmHg) than α-tocopherol in terms of inhibiting MDA formation. At a pO_2 of 8 mmHg and above, α-tocopherol became the more effective antioxidant. These observations support other reports that β-carotene is a strong antioxidant at low pO_2.[4,29,47,48] It is unclear, however, how they[58] incorporated the concentrations of β-carotene indicated in the membranes, using 10 μl of chloroform.

Kim[21] has also looked at rat liver microsomal peroxidations, induced by either Fe^{3+}-ADP/NADPH or paraquat/NADPH, in the presence of several carotenoids including α-carotene, β-carotene, lutein, lycopene, and α-tocopherol. Kim found that lycopene, lutein, and α-carotene were better antioxidants than β-carotene or α-tocopherol. Although the results are suggestive, it is difficult to understand how a 100 μM concentration of carotenoids in absolute ethanol, as described in these experiments, was obtained.

Palozza et al.[60] have added β-carotene and α-tocopherol to rat liver microsomes, and initiated lipid peroxidation with either radical initiators (AAPH/AMVN) or NADPH/ADP/Fe^{3+}. An aliquot of β-carotene in nitrogen-saturated $CHCl_3/CH_3OH$ (2:1, v/v) was evaporated to dryness under nitrogen, microsomes were added in phosphate buffer and homogenized, and the β-carotene was dispersed uniformly in the microsomes. α-Tocopherol was added in an ethanol solution. When AMVN was used as the radical initiator, β-carotene was protective at concentrations as low as 5 nmol/mg protein, but it required 50 nmol/mg protein to protect against AAPH. As reported earlier, β-carotene protection against AMVN-induced peroxidation in solutions of microsomal lipids does not produce a clearly defined lag phase.[48,50] Using NADPH/ADP/Fe^{3+}, 10 nmol of β-carotene/mg protein gives significant protection, which is in agreement with earlier results.[58] Using either of these prooxidant systems, α-tocopherol was a considerably better antioxidant than β-carotene.[60]

A protection by astaxanthin of liver mitochondria from vitamin E-deficient rats, exposed to Fe^{2+} to initiate lipid peroxidation, has been shown by Kurashige et al.[24] The inhibitory effect of astaxanthin, dissolved in dimethyl sulfoxide at concentrations ranging from 0.13 nM to 1.3 mM, on mitochondrial TBA formation seems to be 100–500 times stronger than that of α-tocopherol. Similar results have been reported by Miki[19] in homogenates of rat mitochondria exposed to Fe^{2+}. The author suggests that the strong antioxidant activity of astaxanthin could be related to the high affinity of this molecule for mitochondrial membranes, due to the chemical structure of this pigment.

Cells. The direct demonstration of carotenoids functioning as antioxidants in isolated cells has been reported in only a few cases.

In 1973, Yamane and Lamola[61] found that the hemolysis of red blood cells, induced by cholesterol hydroperoxide, could be inhibited by the addition of small amounts of either β-carotene or α-tocopherol.

The endothelial cells of isolated, perfused rabbit cornea, treated with either β-carotene (1 μg/ml) or α-tocopherol (10 μm/ml), showed doubled survival times over control cells, or cells treated with water-soluble antioxidants such as benzoate, mannitol, or the enzymes superoxide dismutase (SOD) or catalase.[62] An important aspect of their work was the observation that the cells supplemented with α-tocopherol were more resistant to the deterioration than control cells and that the sites of major damage were the mitochondria and endoplasmic reticulum, whereas the plasma and nuclear membranes appeared normal. These data indicate that lipid peroxidation of the membranes may be the first event in the cellular damage induced by free radicals.

Inhibition of Mutagenesis–Carcinogenesis and Immunomodulating Effects

Substantial evidence exists that implicates both oxygen and organic free radical intermediates in the initiation, promotion, and/or progression stages of carcinogenesis (for reviews, see Refs. 63 and 64). Several lines of

---

[61] T. Yamane and A. A. Lamola, *Proc. Annu. Meet., Am. Soc. Photobiol., 1st* abstr. MAM-B10 (1973).
[62] O. Lux-Neuwirth and T. J. Millar, *Curr. Eye Res.* **9,** 103 (1990).
[63] Y. Sun, *Free Radical Biol. Med.* **8,** 583 (1990).
[64] M. Trush and T. W. Kensler, *Free Radical Biol. Med.* **10,** 201(1991).

evidence suggest that the potential chemopreventive and antigenotoxic properties of carotenoids could be related to their antioxidant activity. However, we still have no way of ascertaining whether the observed effects are related to an antioxidant property of these pigments, and therefore must wait for additional evidence before considering the antioxidant hypothesis as affecting mutagenesis, carcinogenesis, or immunomodulation.

Antioxidant Effects of Carotenoids *in Vivo*

Animal Studies

Inhibition of Lipid Peroxidation. Increasing evidence that carotenoids can function as antioxidants *in vivo* has been reported in animal models in which the pigments have been injected or added directly to the diet (Table VI).[19,24,65-70]

From studies in which an inhibition of lipid peroxidation has been observed, two aspects of the antioxidant activity of carotenoids seem to be important. The first is the efficiency of these molecules to decrease peroxidation directly and the second is the ability of these molecules to modulate endogenous levels of other antioxidants.

Direct Antioxidant Activity. A demonstration of the direct antioxidant activity of carotenoid molecules in animal models has been presented by a number of workers.[19,24,65,66]

In these experiments, carotenoids have been either injected[65] or administered orally[19,24,66] and the lipid peroxidation has been induced *in vitro*,[24] *in vivo*,[65] or spontaneously.[66] Although none of these reports have information on the plasma or tissue levels of carotenoids, a decrease of lipid peroxidation products follows the carotenoid administration. In agreement with the *in vitro* experiments, the strong antioxidant activity of astaxanthin seems to be confirmed *in vivo*.[19,24]

Interactions with Other Antioxidants. Although the work presented above seems to prove that carotenoid pigments may provide direct antioxidant activity *in vivo*, it is possible that other mechanisms may be involved in this protection. An alternative mechanism is suggested by Mayne and Parker,[70] who reported that the addition of dietary canthaxanthin to chicks deficient in vitamin E and selenium increased the resistance to lipid perox-

---

[65] K.-J. Kunert and A. L. Tappel, *Lipids* **18**, 271 (1983).
[66] R. Zamora, F. J. Hidalgo, and A. L. Tappel, *J. Nutr.* **121**, 50 (1991).
[67] S. Q. Alam and B. S. Alam, *J. Nutr.* **113**, 2608 (1983).
[68] L. Lomnitski, M. Bergman, I. Schön, and S. Grossman, *Biochim. Biophys. Acta* **1082**, 101 (1991).
[69] B. Leibovitz, M.-L. Hu, and A. L. Tappel, *J. Nutr.* **120**, 97 (1990).
[70] S. T. Mayne and R. S. Parker, *Nutr. Cancer* **12**, 225 (1989).

TABLE IV
EFFECTS OF CAROTENOIDS ON LIPID PEROXIDATION IN ANIMAL MODELS

Species	Prooxidant	Carotenoid	Target	Effect on lipoperoxidation[a]	Ref.
Guinea pigs					
Vitamin C deficient	CCl_4	β-Carotene	Expired air	+	65
Rats					
Sprague-Dawley	Spontaneous	β-Carotene	Plasma	+	66
	$BrCl_3C$	β-Carotene	RBC	+	
Wistar	Fe^{2+}	Astaxanthin	Liver	+	24
(vitamin E deficient)	Xanthine/xanthine oxidase	Astaxanthin	RBC	+	19
Unspecified	Xanthine/xanthine oxidase	Astaxanthin	RBC	+	
(vitamin E deficient)					
Sprague-Dawley	Spontaneous	β-Carotene	Liver	−	67
	Spontaneous	β-Carotene	Plasma	−	68
Charles River	Spontaneous	β-Carotene	Testis	−	
(vitamin E deficient)					
Sprague-Dawley	t-BOOH	β-Carotene	Different organs	±[b]	69
(vitamin E and Se deficient)		β-Carotene	Different organs	±[b]	
Chicks	Fe^{2+}	Canthaxanthin	Liver	+	70
(vitamin E and Se deficient)					

[a] Lipoperoxidation measured *in vitro* as thiobarbituric acid reactive substances (TBARS), except data from Ref. 65 are measured as expired pentane and ethane. +, Inhibition of lipoperoxidation; −, no effect.
[b] ± effect in combination with other antioxidants.

idation, primarily by increasing membrane α-tocopherol levels, and only secondarily providing weak direct antioxidant activity.

The possibility of other interactions between antioxidants *in vivo* seems to be confirmed by Blakely and co-workers.[71] They report that β-carotene is able to modulate the increase of SOD, induced by peroxyl radicals produced by a high-fat diet.

In addition, Leibovitz et al.,[69] looking at the effects of different dietary antioxidants *in vivo*, present data suggesting a clear synergistic action between β-carotene and other antioxidants in the protection of different rat organs from both spontaneous and induced lipid peroxidation.

However, not all investigations are in agreement about the effectiveness of carotenoids functioning as antioxidants. A lack of antioxidant activity and an increase of the peroxide levels in plasma and liver of rats fed with a large dose of β-carotene have been reported by Alam and Alam.[67] Lomnitski et al.[68] reported a significant increase of testis TBA value, as well as an increase of the testis 15-lipoxygenase activity, as a consequence of a β-carotene-supplemented diet.

One of the possible explanations for the lack of antioxidant activity in these experiments may be the poor storage of β-carotene in rat tissues or, in accordance with the hypothesis of Burton and Ingold,[4] the possibility that under certain circumstances (concentration of β-carotene, oxygen tensions in the tissues, modality of administration) the action of β-carotene represents a balance between the anti- and prooxidant properties.

Inhibition of Carcinogenesis. Several studies have examined the relationship between carotenoid administration and cancer in animals. The results of these studies have been discussed.[72] Although the evidence of association is not always positive and many workers have used systems unique to their laboratories, the data seem to suggest a clear anticarcinogenic role of carotenoids in animal models. However, there is no direct evidence that any of these effects can be attributed to an antioxidant action of these carotenoids. The same applies to the studies indicating that carotenoids can enhance immune functions.[73] Again, in all of these studies, there is no direct evidence for an antioxidant action of carotenoids.

Human Studies

Carotenoid Intake and Disease Prevention. There has been increasing interest in the possibility that carotenoid pigments may exert a preventive role in some diseases in which free radicals seem to be involved. Rationales

---

[71] S. R. Blakely, L. Slaughter, J. Adkins, and E. V. Knight, *J. Nutr.* **118**, 152 (1988).

[72] N. I. Krinsky, *Am. J. Clin. Nutr.* **53**, 238 (1991).

[73] A. Bendich, *in* "Carotenoids: Chemistry and Biology" (N. I. Krinsky, M. M. Mathews-Roth, and R. F. Taylor, eds.), Proc. 8th Int. Symp. Carotenoids, p. 323. Plenum, New York, 1989.

for this are either epidemiological studies in humans, showing that carotenoids may reduce the risk of cancer, or the increasing evidence of a clear antioxidant activity of these pigments *in vitro* and in animal models. Notwithstanding this evidence, we cannot as yet claim that β-carotene functions as an antioxidant in disease prevention in humans.

Acknowledgment

Supported by a Fellowship from Associazone Italiana per la Ricera sul Cancro (P.P.) and NCI-CA51506 (N.I.K.).

[39] Efficiency of Singlet Oxygen Quenching by Carotenoids Measured by Near-Infrared Steady-State Luminescence

By ESTHER OLIVEROS, PATRICIA MURASECCO-SUARDI, ANDRÉ M. BRAUN, and HANS-JÜRGEN HANSEN

Introduction

The noxious effects of light on biological systems, in the presence of a dye and molecular oxygen (photodynamic effect), had already been reported at the beginning of the century.[1] In many instances, singlet oxygen $[^1O_2(^1\Delta_g)]$, denoted simply $^1O_2$ below, is the reactive intermediate that induces the damaging oxidation reactions (type II photooxidation).[2] In solution, this activated metastable state of molecular oxygen is most often produced by photosensitization [Eqs. (1)–(3)], which involves energy transfer from an electronically excited state (usually the triplet state) of a sensitizer molecule (e.g., a dye) to molecular oxygen.

$$\text{Sens} \xrightarrow{h\nu} {}^1\text{Sens*} \quad (1)$$

$$^1\text{Sens*} \xrightarrow{k_{isc}} {}^3\text{Sens*} \quad (2)$$

$$^3\text{Sens*} + {}^3O_2 \xrightarrow{k_{et}} \text{Sens} + {}^1O_2 \quad (3)$$

Among the important roles of carotenoid pigments in photosynthesis,

---

[1] O. Raab, *Z. Biol. (Munich)* **39**, 524 (1900).
[2] A. M. Braun, M.-T. Maurette, and E. Oliveros, "Photochemical Technology" (D. G. Ollis and N. Serpone, transl.), Ch. 11. Wiley, Chichester, England, 1991, and references cited therein.

photobiology, and photomedicine,[3] their protective function against photodynamic damage in biological systems[4,5] is a prominent factor explaining the considerable interest in these compounds. Their possible use as therapeutic agents, particularly in dermatology,[5,6] has promoted basic and applied investigations of their mechanisms of action. Carotenoids have been shown to deactivate singlet oxygen mainly by physical quenching [Eq. (4)].[4,7,8]

$$^1O_2 + Q \xrightarrow{k_q} {}^3O_2 + {}^3Q^* \qquad (4)$$

This energy-transfer process produces ground-state molecular oxygen and the carotenoid triplet that, in turn, decays rapidly to its ground state. Thus, harmful chemical reactions of singlet oxygen with surrounding compounds are prevented. The bimolecular rate constants of the physical quenching of singlet oxygen (k_q) by carotenoids provide a measure of their efficiency in deactivating singlet oxygen. These rate constants can be conveniently determined by near-infrared luminescence measurements of singlet oxygen [$^1O_2(^1\Delta_g)$] produced on continuous excitation of the sensitizer.[9-11] In the following paragraphs, we describe the principle of the method of analysis, its limitations, and the equipment used. We then give examples of the determination of k_q in the case of two carotenoids, all-*trans*-β-carotene and canthaxanthin, in a polar and a nonpolar solvent.

Method of Analysis

Singlet oxygen is produced by photosensitization [Eqs. (1)–(3)], using a suitable sensitizer. For example, rose bengal[12,13] and 1*H*-phenalen-1-one[14] are efficient $^1O_2$ sensitizers, the former being soluble in water and polar

---

[3] For a recent review, see T. G. Truscott, *J. Photochem. Photobiol. B: Biol.* **6**, 359 (1990).
[4] C. S. Foote, Y. C. Chang, and R. W. Denny, *J. Am. Chem. Soc.* **92**, 5216 and 5218 (1970).
[5] M. M. Mathews-Roth, *Photochem. Photobiol.* **43**, 91 (1986).
[6] G. Swanbeck and G. Wennersten, *Acta Derm.–Venereol.* **54**, 433 (1974).
[7] A. Farmilo and F. Wilkinson, *Photochem. Photobiol.* **18**, 447 (1973).
[8] G. Speranza, P. Manitto, and D. Monti, *J. Photochem. Photobiol. B: Biol.* **8**, 51 (1990).
[9] A. A. Krasnovsky, Jr., *Chem. Phys. Lett.* **81**, 443 (1981).
[10] R. D. Hall, G. R. Buettner, A. G. Motten, and C. F. Chignell, *Photochem. Photobiol.* **46**, 295 (1987).
[11] P. Murasecco-Suardi, E. Oliveros, A. M. Braun, and H.-J. Hansen, *Helv. Chim. Acta* **71**, 1005 (1988).
[12] J. J. M. Lamberts and D. C. Neckers, *Tetrahedron* **41**, 2183 (1985).
[13] P. Murasecco-Suardi, E. Gassmann, A. M. Braun, and E. Oliveros, *Helv. Chim. Acta* **70**, 1760 (1987), and references cited therein.
[14] E. Oliveros, P. Suardi-Murasecco, T. Aminian-Saghafi, A. M. Braun, and H.-J. Hansen, *Helv. Chim. Acta* **74**, 79 (1991).

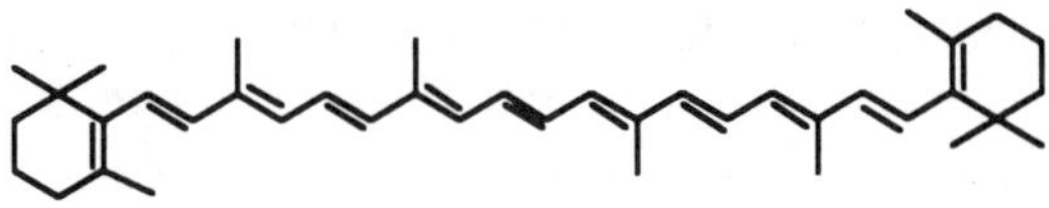

*all-trans-*β Carotene

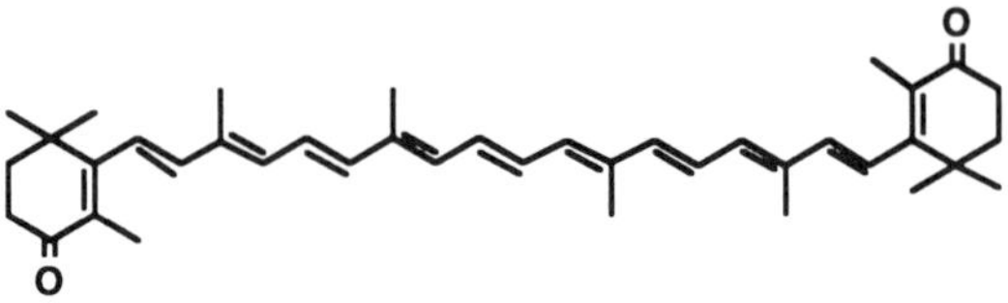

Canthaxanthin

organic media and the latter presenting the advantage of being soluble in polar as well as in nonpolar organic solvents.

In the absence of a physical quencher or a chemical acceptor, $^1O_2$ is deactivated in solution by collision with solvent molecules [main decay pathway, Eq. (5)] and by phosphorescence emission [Eq. (6)].

$$^1O_2 \xrightarrow{k_d} {}^3O_2 \tag{5}$$

$$^1O_2 \xrightarrow{k_e} {}^3O_2 + h\nu'' \tag{6}$$

k_d and k_e are, respectively, the nonradiative and radiative rate constants of $^1O_2$ deactivation in the solvent used.

The $^1O_2$ luminescence can be conveniently monitored in the near-infrared at the emission maximum of 1270 nm,[9] where very few organic compounds absorb or emit. It should be noted that this emission is very weak, the rate constant k_e being negligible compared to that of the deactivation by the solvent (k_d).[9]

The luminescence signal [S(mV)], measured on continuous excitation of the sensitizer, depends on the quantum yield of singlet oxygen luminescence [Φ_e, Eq. (7)], on the photonic flux absorbed by the sensitizer (P_a), and on an apparatus factor C.

$$\Phi_e = P_e/P_a = SC/P_a \tag{7}$$

where P_e is the photonic flux emitted by singlet oxygen. The apparatus factor C is specific to the equipment used. As the $^1O_2$ luminescence is detected at 90° with respect to the axis of the incident light (Fig. 1), C depends on the absorbance of the solution at the wavelength of excitation (for a given geometric setup of the equipment, C is a constant for a given absorbance[15]).

[15] A. M. Braun and E. Oliveros, *Pure Appl. Chem.* **62**, 1467 (1990).

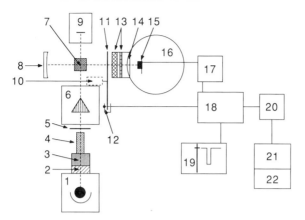

FIG. 1. Scheme of the equipment used for the near-infrared luminescence measurements of singlet oxygen on continuous irradiation of the sensitizer. 1, Light source; 2, condenser; 3, infrared filter; 4, lenses; 5, shutter; 6, monochromator; 7, sample cell; 8, mirror; 9, radiometer; 10, motor; 11, chopper; 12, frequency reference diode; 13, filters; 14, lens; 15, Ge photodiode; 16, Dewar; 17, preamplifier; 18, lock-in amplifier; 19, plotter; 20, digital voltmeter; 21, computer; 22, printer.

In absence of singlet oxygen quencher, the quantum yield of $^1O_2$ luminescence, (Φ_e^0), is related to the quantum yield of $^1O_2$ production by the sensitizer, Φ_Δ^0, by Eq. (8),

$$\Phi_e^0 = \Phi_\Delta^0 \phi_e^0 = \Phi_\Delta^0 k_e \tau_\Delta \qquad (8)$$

where

$$\phi_e^0 = k_e/k_d = k_e \tau_\Delta \qquad (9)$$

is the luminescence efficiency, and τ_Δ (sec) is the lifetime of singlet oxygen in the solvent used, in the absence of quencher. This applies if the quenching of $^1O_2$ by the sensitizer (rate constant: k_q^S) is negligible compared to the deactivation by the solvent, i.e., if $k_q^S[\text{Sens}] \ll k_d$, a condition that is usually met for rose bengal and $1H$-phenalen-1-one at the concentrations currently used.

If a singlet oxygen quencher Q is added to the solution, the quantum yield of $^1O_2$ luminescence is given by:

$$\Phi_e = \Phi_\Delta k_e / [k_d + (k_r + k_q)[Q]] \qquad (10)$$

where k_q and k_r are the bimolecular rate constants of the physical quenching of $^1O_2$ by Q [Eq. (4)] and of the chemical reaction between $^1O_2$ and Q, respectively.

If the rate of quenching of the sensitizer triplet by Q (rate constant: k_q^Q)

is negligible in comparison with the rate of energy transfer from this same triplet state to molecular oxygen [rate constant: k_{et}, Eq. (3)], i.e., if

$$k_{et}[^3O_2] \gg k_S^Q[Q] \tag{11}$$

then

$$\Phi_\Delta = \Phi_\Delta^0 \tag{12}$$

In this case, the variation of the ratio of the luminescence signals in the absence and in the presence of quencher (S_0/S) as a function of the concentration of quencher (Stern–Volmer analysis) is linear, as shown by combining Eqs. (7), (8), and (10):

$$S_0/S = \Phi_e^0/\Phi_e = 1 + (k_r + k_q)\tau_\Delta[Q] \tag{13}$$

provided that the absorbance of the sensitizer is the same in the whole series of measured solutions [C and P_a constant, Eq. (7)].

The value of $(k_r + k_q)$ can be calculated from the slope of $S_0/S = f([Q])$, if τ_Δ is known. If $k_r \ll k_q$, this method allows the direct determination of k_q.

Equipment[11,14]

A fluorescence cell (1 × 1 cm) containing the sample solution is placed on an optical bench and irradiated with a xenon/mercury lamp (Osram, Winterthur, Switzerland, 1 kW) through a condenser, a water filter (10 cm), focalizing optics, and a monochromator (ISA Jobin-Yvon, Paris, France, 6-nm bandwidth). The $^1O_2$ luminescence is measured at 90°, with a germanium photodetector (6 mm^2; Judson, Infrared, Montgomeryville, PA) cooled to $-78°$. A mirror is used to increase the emitted photonic flux at the detector surface. The emitted light is chopped at a frequency of 11 Hz, and the $^1O_2$ luminescence is isolated by a cutoff filter at 1000 nm (Oriel, LOT, Darmstadt, Germany) and an interference filter (51% transmittance at 1270 nm; Oriel). The detector is connected to a preamplifier (30 dB) and to a lock-in amplifier (Princeton Applied Research, Salem, MA, No. 5101). Signals are recorded on a graphic plotter and, via a digital multimeter (HP 3478A; Hewlett-Packard, Palo Alto, CA) on an HP 200 computer. A scheme for the experimental setup is given in Fig. 1.

In a series of measurements, alternation is recommended between reference (solution without quencher) and samples. The fluorescence cells must be previously checked for equivalence. Precisely averaged signals may need irradiation times of up to 5 min. The baseline signals are registered for the same periods of time before and after irradiation of the sample.

The intensity of the luminescence signal depends greatly on the values of Φ_Δ, τ_Δ and k_e [Eq. (8)], i.e., on the sensitizer and on the solvent used. The sensitivity of our equipment allows measurements of luminescence signals in CH_3OH ($\tau_\Delta \approx 10$ μsec,[16,17] $k_e/k_e(C_6H_6) \approx 0.28$[18]) or in ^2H_2O ($\tau_\Delta \approx 60$ μsec,[17,19] $k_e/k_e(C_6H_6) \approx 0.20$[18]) but thermal noise (≈ 20 μV) is too important for measurements in H_2O ($\tau_\Delta \approx 3.6$ μsec,[16,17,20] $k_e/k_e(C_6H_6) \approx 0.12$[18]).

Experimental

Reagents. 1*H*-Phenalen-1-one (EGA Chemie, Steinheim, Germany) and rose bengal (Fluka, Buchs, Switzerland) were used as $^1O_2$ sensitizers. 1*H*-Phenalen-1-one was dissolved in CH_2Cl_2 (Fluka puriss.) and purified on preparative thin-layer chromatography (TLC) silica gel plates (Merck, Darmstadt, Germany) and by subsequent recrystallization from CH_2Cl_2–methanol (2:1, v/v). The purified carotenoid pigments, all-*trans*-β-carotene and canthaxanthin, were provided by Hoffmann-La Roche A.G. (Basel, Switzerland). Carbon tetrachloride (Fluka, puriss.), methanol (Fluka, puriss.), and deuterated methanol (> 99.5% d_4, Dr. Glaser, Basel, Switzerland) were used as solvents.

Near-Infrared Luminescence of Singlet Oxygen. The $^1O_2$ luminescence measurements were made with solutions of identical absorbance (A 0.4) at excitation wavelengths of 366 nm for 1*H*-phenalen-1-one and 546 nm for rose bengal. Absorption spectra were recorded on a UV-260 Shimadzu spectrophotometer (Burkard Instriemente, Zürich, Switzerland). Solutions containing different concentrations of the carotenoids were prepared from standard solutions of the sensitizers. Equivalent 1 × 1 cm fluorescence cells (Hellma, Müllheim, Germany) were used for the experiments. The results are the average of at least two series of measurements with a limit of error of less than 3%.

Carbon tetrachloride and methanol were chosen as solvents. $C^2H_3O^2H$ was used instead of CH_3OH, as $^1O_2$ luminescence signals are stronger in deuterated solvents owing to a large increase in τ_Δ values.[2] Solutions of 1*H*-phenalen-1-one containing up to 3×10^{-7} mol/liter of the carotenoid in CCl_4 and up to 10^{-5} mol/liter in $C^2H_3O^2H$ were irradiated at 366 nm (incident photonic rate of $\sim 7 \times 10^{15}$ photons/sec). In the absence of quencher, a stable signal was obtained for the singlet oxygen luminescence on irradiation of the sensitizer (stationary concentration of $^1O_2$). However,

---

[16] M. A. J. Rodgers, *J. Am. Chem. Soc.* **105,** 6201 (1983).
[17] E. Oliveros, C. Charbonnet, and A. M. Braun, Fall Assembly, Swiss Chemical Society, Bern, 1990, p. 67.
[18] R. Schmidt and E. Afshari, *J. Phys. Chem.* **94,** 4377 (1990).
[19] J. R. Hurst and G. B. Schuster, *J. Am. Chem. Soc.* **105,** 5756 (1983).
[20] S. Y. Egorov, V. F. Kamalov, N. I. Koroteev, A. A. Krasnovsky, Jr., B. N. Toleutaev, and S. V. Zinukov, *Chem. Phys. Lett.* **163,** 421 (1989).

in the presence of β-carotene or canthaxanthin, the signal increased during irradiation. This increase may be explained by a reaction between the sensitizer and the quencher (e.g., hydrogen abstraction), leading to a partial consumption of the quencher and, thus, to an increase of the $^1O_2$ concentration during the irradiation time. For the Stern–Volmer analysis, the intensities of the signals were measured at the beginning of the irradiation. The experiments with β-carotene were repeated in $C^2H_3O^2H$ using rose bengal as a sensitizer (incident photonic rate of $\sim 5 \times 10^{15}$ photons/sec at 546 nm). In this latter case, the luminescence signals were stable in the presence as well as in the absence of β-carotene. The results obtained using the two different sensitizers were, within the limits of experimental error, identical. Under the experimental conditions indicated above (sensitizer, corresponding wavelength of irradiation, absorbance), the signal intensities, in the absence of quencher, were about 20 mV in CCl_4 and 500 μV in $C^2H_3O^2H$.

Solvent Dependence on Rate Constants of Singlet Oxygen Quenching by β-Carotene and Canthaxanthin

Canthaxanthin is used by some cosmetics manufacturers as an orally taken tanning agent, and it has been of interest to determine its bimolecular rate constant of $^1O_2$ physical quenching. all-*trans*-β-Carotene, which otherwise has been more widely investigated,[3] has been taken for comparison. Experiments have been carried out in carbon tetrachloride and in methanol to evaluate medium effects on the efficiency of $^1O_2$ quenching by these two carotenoids.

Stern–Volmer plots of the quenching of $^1O_2$ luminescence by canthaxanthin and β-carotene in CCl_4 and in $C^2H_3O^2H$ are linear (Figs. 2 and 3). In fact, assuming diffusion controlled energy transfers from the sensitizer triplet to both molecular oxygen [Eq. (3)] and quencher [Eq. (4)], the chosen range of carotenoid concentrations complies with condition (11). Reaction (3) must be at least 200 times faster than reaction (4), taking into account the concentrations of oxygen in air-saturated CCl_4 and $C^2H_3O^2H$ of 2.6×10^{-3} mol/liter and 2.1×10^{-3} mol/liter, respectively.[21]

As the slopes of the Stern–Volmer plots represent the product $k_q\tau_\Delta$ [Eq. (13)], τ_Δ in the corresponding solvent must be known for calculating k_q. Singlet oxygen lifetimes can be determined by time-resolved $^1O_2$ luminescence measurements[16,17,19,22] or thermal lensing spectroscopy.[23] Literature

---

[21] S. L. Murov, "Handbook of Photochemistry." Dekker, New York, 1973.
[22] P. R. Ogilby and C. S. Foote, *J. Am. Chem. Soc.* **105**, 3423 (1983).
[23] K. Fuke, M. Ueda, and M. Itoh, *J. Am. Chem. Soc.* **105**, 1091 (1983).

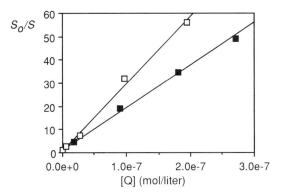

FIG. 2. Stern–Volmer plots of the quenching of singlet oxygen by β-carotene (■) and canthaxanthin (□) in CCl_4 [sensitizer; $1H$-phenalen-1-one ($\lambda_{ex} = 366$ nm)].

values may also be used, but potential errors due to differences in solvent purity or water content must be taken into account. However, τ_Δ need not be known for calculating relative values of k_q in a given solvent for a series of quenchers. For instance, the ratio of the Stern–Volmer slopes obtained for canthaxanthin and β-carotene in CCl_4 is equal to 1.6 (±0.2) (Table I). The discrepancy between published τ_Δ values in CCl_4 is rather large, spanning from 0.9 msec[24] to 26 msec,[9,25] 31 msec,[26] and up to 87 msec.[27] Using a τ_Δ of 28.5 msec,[2] rate constants for $^1O_2$ physical quenching by β-carotene and canthaxanthin of 6.3×10^9 and 1.0×10^{10} liters/(mol·sec), respectively, are obtained (Table I). The value of β-carotene is close to the one reported in the literature [$7(\pm2) \times 10^9$ liters/(mol·sec)].[28]

The reported rate constants for $^1O_2$ quenching by β-carotene and canthaxanthin in benzene (Table I) are about one-half the rate constant of diffusion (k_{diff}) calculated to be 3.0×10^{10} liters/(mol·sec) when oxygen is one of the reaction partners.[29] The k_q in CCl_4 is slightly lower than in benzene; this might be due to the higher viscosity of CCl_4.[21] Canthaxanthin appears to be a more efficient quencher than β-carotene, and some authors attribute this fact to the lower triplet energy of canthaxanthin, which contains 13 conjugated double bonds (11 C=C and 2 C=O) instead of 11 for β-carotene.[30]

[24] J. R. Hurst, J. D. McDonald, and G. B. Schuster, *J. Am. Chem. Soc.* **104**, 2065 (1982).
[25] K. I. Salokhiddinov, I. M. Byteva, and G. P. Gurinovich, *J. Appl. Spectrosc (Engl. Transl.)* **34**, 561 (1981).
[26] T. A. Jenny and N. J. Turro, *Tetrahedron Lett.* **23**, 2923 (1982).
[27] R. Schmidt and H.-D. Brauer, *J. Am. Chem. Soc.* **109**, 6976 (1987).
[28] A. A. Krasnovsky, Jr., *Photochem. Photobiol.* **29**, 29 (1979).
[29] W. R. Ware, *J. Phys. Chem.* **66**, 455 (1962).
[30] F. Wilkinson and W.-T. Ho, *Spectrosc. Lett.* **11**, 455(1978).

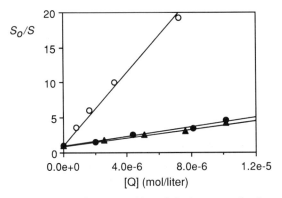

FIG. 3. Stern–Volmer plots of the quenching of singlet oxygen by β-carotene [●: sensitizer, 1H-phenalen-1-one; ▲: sensitizer, rose bengal (λ_{ex} = 546 nm)] and by canthaxanthin (○: sensitizer, 1H-phenalen-1-one).

In $C^2H_3O^2H$, τ_Δ (~ 230 μsec[16,17]) is more than 100 times shorter than in CCl_4 and, consequently, higher quencher concentrations must be used for the Stern–Volmer analysis. The present study shows that k_q of canthaxanthin seems, independent of the solvent, to be about a factor of two-lower than the corresponding rate constants of diffusion. On the other hand, k_q of β-carotene in methanol is 20 times smaller than the corresponding diffusion rate constant and 8 times smaller than k_q of canthaxanthin (Table I).

TABLE I
RATE CONSTANTS OF QUENCHING OF SINGLET OXYGEN BY β-CAROTENE AND CANTHAXANTHIN IN CCl_4 AND IN $C^2H_3O^2H$

Quencher	Solvent	$k_q \tau_\Delta$ (liters/mol)	k_q[b] [liters/(mol · sec)]	Relative efficiency
β-Carotene	CCl_4	1.8 (± 0.1) × 10⁸	6.3 (± 0.6) × 10⁹[c]	1
	$C^2H_3O^2H$	3.4 (± 0.2) × 10⁵	1.5 (± 0.1) × 10⁹	1
		3.2 (± 0.2) × 10⁵	1.4 (± 0.1) × 10⁹[d]	1
Canthaxanthin	CCl_4	2.9 (± 0.1) × 10⁸	1.0 (± 0.1) × 10¹⁰[c]	1.6
	$C^2H_3O^2H$	2.7 (± 0.1) × 10⁶	1.2 (± 0.1) × 10¹⁰	7.8

[a] Sensitizer: 1H-phenalen-1-one, unless otherwise specified.
[b] Calculations of k_q have been made using τ_Δ of 28.5 msec in CCl_4 and 227 μsec in $C^2H_3O^2H$.
[c] For comparison, values of k_q in C_6H_6 are 1.25 (± 0.2) × 10¹⁰ liters/(mol · sec) for β-carotene and 1.45 (± 0.2) × 10¹⁰ liter/(mol · sec) for canthaxanthin. [From F. Wilkinson and W.-T. Ho, *Spectrosc. Lett.* **11**, 455 (1978).]
[d] Sensitizer: rose bengal.

The lower quenching efficiency of β-carotene in methanol is probably related to its low solubility and aggregation in this solvent. In fact, within the concentration range used for the Stern-Volmer analysis, a net deviation from the Beer-Lambert law is found.[11] No aggregational effect of canthaxanthin could be detected in any of the solvents used. This compound, exhibiting a higher polarity due to its two carbonyl groups, is a better reference for kinetic measurements in methanol, as aggregational effects may be avoided.

[40] Assay of Lycopene and Other Carotenoids as Singlet Oxygen Quenchers

By PAOLO DI MASCIO, ALFRED R. SUNDQUIST, THOMAS P. A. DEVASAGAYAM, and HELMUT SIES

Introduction

There is increasing interest in the role of nutrition in the prevention and pathogenesis of cancer.[1] Epidemiological studies in humans have suggested that β-carotene aids in cancer prevention,[2] and β-carotene may exert this effect independent of its role as a precursor of vitamin A.[3] Other carotenoids such as lycopene and oxycarotenoids are present at significant levels in cells and plasma[4] but their possible importance in disease prevention has received far less attention than β-carotene. Furthermore, the physiological functions of the carotenoids may be highly specialized; for example, the oxycarotenoids zeaxanthin and lutein are present in the macular region of the retina in the virtual absence of β-carotene.[5] Significant variation in organ carotenoid patterns has also been reported.[6]

The discovery[7] that carotenoids deactivate singlet molecular oxygen ($^1O_2$) was an important advance in understanding the biological effects of carotenoids. The mechanism by which carotenoids protect biological systems against $^1O_2$-mediated damage appears to depend largely on physical quenching and to a much lesser extent on chemical reaction. Physical

---

[1] B. N. Ames, *Science* **221**, 1256 (1983).
[2] K. F. Gey, G. B. Brubacher, and H. B. Staehelin, *Am. J. Clin. Nutr.* **45**, 1368 (1987).
[3] R. Peto, R. Doll, J. D. Buckley, and M. B. Sporn, *Nature (London)* **290**, 201 (1981).
[4] R. S. Packer, *J. Nutr.* **119**, 101 (1989).
[5] G. J. Handelman, E. A. Dratz, C. C. Reay, and F. J. G. M. van Kuijk, *Invest. Ophthalmol. Visual Sci.* **29**, 850 (1988).
[6] L. A. Kaplan, J. M. Lau, and E. A. Stein, *Clin. Physiol. Biochem.* **8**, 1 (1990).
[7] C. S. Foote and R. W. Denny, *J. Am. Chem. Soc.* **90**, 6233 (1968).

quenching involves transfer of excitation energy from $^1O_2$ to the carotenoid, resulting in the formation of ground state oxygen ($^3O_2$) and triplet excited carotenoid ($^3C^*$); subsequently, the excitation energy is dissipated through rotational and vibrational interactions between $^3C^*$ and the solvent to recover ground state carotenoid [reaction (1)]. In this way, the carotenoid acts as a

$$^1O_2 + C \rightarrow {}^3O_2 + {}^3C^* \rightarrow C + \text{thermal energy} \tag{1}$$

catalyst to deactivate the potentially harmful $^1O_2$.[8,9]

A variety of spectroscopic techniques have been developed to evaluate the $^1O_2$-quenching capacity of carotenoids. Near-infrared photoemission accompanies the spontaneous decay of $^1O_2$ to the triplet ground state [reaction (2)], and a decrease in either

$$^1O_2 \rightarrow {}^3O_2 + h\nu \quad (1270 \text{ nm}) \tag{2}$$

the lifetime or the level of this emission is a measure of $^1O_2$ quenching. Quenching assays based on measurements of $^1O_2$ photoemission have been described in which (1) a pulse of $^1O_2$ is generated with a photochemical source and the lifetime of the photoemission in the absence and presence of carotenoid is measured by time-resolved spectroscopy,[10] and (2) $^1O_2$ is generated continuously either chemically[11] or photochemically[12] and the effect of the carotenoid on the steady state level of photoemission is evaluated. A third technique takes advantage of the unique absorption pattern of the triplet excited carotenoid formed during the quenching process [see reaction (1)]. Triplet carotenoids typically display an absorption maximum at wavelengths greater than that of ground state carotenoid, and the rate of decay of this absorption band after a pulse of $^1O_2$ is a measure of the $^1O_2$ lifetime in the presence of the carotenoid.[13-15] An additional, indirect method of measuring quenching is to determine the extent to which a quencher inhibits $^1O_2$-dependent chemical reactions.

---

[8] C. S. Foote, R. W. Denny, L. Weaver, Y. Chang, and J. Peters, *Ann. N.Y. Acad. Sci.* **171**, 139 (1970).

[9] N. I. Krinsky, in "Singlet Oxygen" (H. H. Wasserman and R. W. Murray, eds.), p. 597. Academic Press, New York, 1979.

[10] P. F. Conn, W. Schalch, and T. G. Truscott, *J. Photochem. Photobiol. B: Biol.* **11**, 41 (1991).

[11] P. Di Mascio, S. Kaiser, and H. Sies, *Arch. Biochem. Biophys.* **274**, 532 (1989).

[12] E. Oliveros, P. Murasecco-Suardi, A. M. Braun, and H.-J. Hansen, this volume [39].

[13] A. Farmilo and F. Wilkinson, *Photochem. Photobiol.* **18**, 447 (1973).

[14] F. Wilkinson and W.-T. Ho, *Spectrosc. Lett.* **11**, 455 (1978).

[15] M. A. J. Rodgers and A. L. Bates, *Photochem. Photobiol.* **31**, 533 (1980).

Two such reactions that have been used to measure the quenching activity of carotenoids are the photooxidation of rubrene, which can be monitored as a decrease in rubrene absorption at 520 or 525 nm,[16-18] and the chlorophyll-sensitized photooxidation of soybean oil, monitored as oxygen consumption.[19]

To determine overall quenching rate constants for carotenoids and related compounds we have used the thermodissociation of the endoperoxide of 3,3'-(1,4-naphthylidene) dipropionate ($NDPO_2$), as $^1O_2$ source and tested the effect of the compounds on the steady state level of $^1O_2$ photoemission. The synthesis of $NDPO_2$, the apparatus for measuring 1270 nm emission, and the quenching assay are described below.

Principle

$NDPO_2$ dissociates at 37° to the parent compound, 3,3'-(1,4-naphthylidene) dipropionate (NDP), and the triplet ground and singlet excited states of molecular oxygen [reaction (3)]. Light

$$2NDPO_2 \xrightarrow{37°} 2NDP + {}^3O_2 + {}^1O_2 \qquad (3)$$

emission due to the phosphorescent transition of $^1O_2$ to ground state oxygen [see reaction (2)] is monitored as monomol emission at 1270 nm with a germanium photodiode detector. In this system $^1O_2$ quenching is observed as a reduction in the steady state level of photoemission on introduction of the quencher.

Photoemission resulting from the thermodissociation of $NDPO_2$ can also be monitored as dimol emission at 634 or 703 nm using a single-photon counting system equipped with a red-sensitive photomultiplier as described in detail elsewhere.[20,21] However, due to the greater intensity of monomol emission and the first-order relationship between $NDPO_2$ concentration and monomol emission, detection at 1270 nm is better suited to determining $^1O_2$-quenching rate constants.

---

[16] M. M. Mathews-Roth, T. Wilson, E. Fujimori, and N. I. Krinsky, *Photochem. Photobiol.* **19**, 217 (1974).
[17] S. R. Fahrenholtz, F. H. Doleiden, A. M. Trozzolo, and A. A. Lamola, *Photochem. Photobiol.* **20**, 505 (1974).
[18] D. J. Carlsson, T. Suprunchuk, and D. M. Wiles, *J. Polym. Sci.* **B11**, 61 (1973).
[19] S.-H. Lee and D. B. Min, *J. Agric. Food Chem.* **38**, 1630 (1990).
[20] E. Cadenas and H. Sies, this series, Vol. 105, p. 221.
[21] M. E. Murphy and H. Sies, this series, Vol. 186, p. 595.

Method

Assay Reagents

Assay solvent: chloroform–ethanol (1:1, v/v; spectroscopic grade) purged with nitrogen for 1 hr on ice (ethanol is present to enhance the solubility of $NDPO_2$)

$NDPO_2$ stock solution: 0.3 M in ^2H_2O (see below) Carotenoid stock solutions: 0.01, 0.1, and 1 mM solutions prepared daily in deaerated chloroform and stored on ice away from light

Preparation of NDP

NDP is synthesized by modification of the method of Saint-Jean and Cannone[22] as described.[23] 1,4-Bis(bromomethyl)naphthalene is first prepared by bromination of 1,4-dimethylnaphthalene. In a dry 2-liter flask a solution of 70 g of 1,4-dimethylnaphthalene is placed in 800 ml of carbon tetrachloride; 2 g of benzoyl peroxide, 2 g of azodiisobutyronitrile, and 175 g of N-bromosuccinimide are powdered, mixed well, and added to the flask. The solution is refluxed and irradiated with an ultraviolet source for 1 hr and stored overnight at room temperature. The solvent is then removed under vacuum and the resulting residue warmed in chloroform on a steam cone and filtered hot. The resulting residue is digested several times with hot chloroform to remove all remaining product, followed by repeated recrystallization in benzene. Diethyl α,α'-dicarbethoxy-1,4-naphthalene dipropionate is prepared from the above product by malonic synthesis[24] and subjected to saponification and decarboxylation[25] to yield NDP. NDP can either be purified as the free acid by column chromatography (activated alumina, 3.5 × 21 cm, prepared in water) and used directly in the synthesis of $NDPO_2$, or it can be first neutralized with sodium methylate in methanol, purified by column chromatography, and used as the disodium salt.

Preparation of $NDPO_2$

$NDPO_2$ is prepared from NDP by the method of Aubry,[26] in which $^1O_2$ produced from the decomposition of H_2O_2 by Na_2MoO_4 [reaction (4)] adds to NDP to yield the endoperoxide. NDP (1.0 g), 2.4 g of

$$H_2O_2 \xrightarrow{Na_2MoO_4} {}^1O_2 + {}^3O_2 \qquad (4)$$

[22] R. Saint-Jean and P. Cannone, *Bull. Soc. Chim. Fr.* No. 9, p. 3330 (1971).
[23] P. Di Mascio and H. Sies, *J. Am. Chem. Soc.* **111**, 2909 (1989).
[24] C. S. Marvel and B. D. Wilson, *J. Org. Chem.* **23**, 1483 (1958).
[25] G. Lock and E. Walter, *Ber. Gesamte Biol., Abt. B* **75B**, 1158 (1942).
[26] J. M. Aubry, *J. Am. Chem. Soc.* **107**, 5844 (1985).

Na$_2$MoO$_4$, 170 mg of NaHCO$_3$, 210 mg of Na$_2$CO$_3$, and 10 ml of distilled water are placed in a 50-ml plastic centrifuge tube along with a magnetic stirring bar and equilibrated at 20° by means of a water jacket. With continuous stirring, 6 ml of 30% H$_2$O$_2$ is added to the mixture as 1-ml aliquots at 14-min intervals. The initial red-brown color that develops on addition of H$_2$O$_2$ fades to golden yellow during the 14-min incubation period. After the last addition of H$_2$O$_2$ the solution is stirred for 25 min. The water bath is then replaced with an ice bath, and 10 ml of ice-cold 2 M H$_3$PO$_4$ is slowly added to precipitate the free acid of NDPO$_2$. The mixture is centrifuged at 4000 rpm for 10 min to sediment the off-white precipitate, and then the yellow supernatant is discarded and the precipitate washed with 20 ml of ice-cold water; this procedure is repeated twice. With the tube on ice, the sediment from the final centrifugation is carefully dissolved with a minimum amount of ice-cold 2 M NaOH in D$_2$O. The final pH of the solution should be ~7. The solution is distributed into small aliquots in cryotubes and frozen in liquid nitrogen for storage at $-70°$.

The concentration of the above NDPO$_2$ stock solution can be determined as follows: 10 μl of the solution is diluted to 50 ml with ice-cold water in a volumetric flask and the absorbance measured at 288 nm (Abs$_1$); the flask is then incubated at 50° for 1 hr to convert NDPO$_2$ to NDP and again the absorbance is measured (Abs$_2$). Estimates of the purity and overall concentration of the preparation are calculated using the extinction coefficients of NDP (7780 M^{-1} cm^{-1}) and NDPO$_2$ (140 M^{-1} cm^{-1}) and the following equations:

$$\text{NDPO}_2(\%) = [1 - (\text{Abs}_1/\text{Abs}_2)] [1 - (\varepsilon_{\text{NDPO}_2}/\varepsilon_{\text{NDP}})]^{-1}$$

$$[\text{NDP} + \text{NDPO}_2] = (\text{Abs}_2) (5000) (\varepsilon_{\text{NDP}})^{-1}$$

Germanium Diode Setup

The photoemission of ^{1}O$_2$ at 1270 nm is measured using a liquid nitrogen-cooled germanium photodiode detect (model EO-817L; North Coast Scientific Co., Santa Rosa, CA) fitted to a light-proof sample chamber (Fig. 1A). A bias power supply (model 823; North Coast Scientific Co.) is used to provide ±12 VDC to the detector, and the signal from the detector is processed with a lock-in amplifier (model 5205; EG & G, Brookdeal Electronics, Princeton Applied Research, Bracknell, Berkshire, England) and followed with a strip chart recorder. The optical chopper control unit (model OC 4000; Photon Technology, Inc., Princeton, NJ) drives a 5-slot chopper disc (100-mm diameter) fitted between the cuvette holder and the shutter at an exposure frequency of 30/sec. The signals from the detector and chopper control unit are monitored by oscilloscope.

The photodiode is prepared for measurements by putting it in a vertical

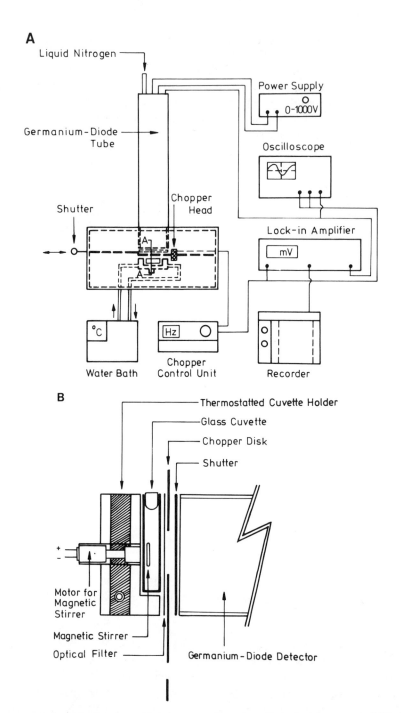

FIG. 1. Schematic diagrams of (A) the germanium photodiode detector setup and (B) the sample chamber cross-section (A–A).

position and filling the cooling jacket through a funnel with liquid nitrogen. Liquid nitrogen is added periodically until the jacket is full, with the entire cooling process requiring 2 hr.

A cross-section (A–A) of the sample chamber is diagrammed in Fig. 1B. The detector is operated in the horizontal position. The cuvette holder is maintained at temperature with an external circulating water bath connected through the walls of the chamber with black tubing. An optical glass cuvette (35 × 6 × 55 mm; Welabo Laboratory Suppliers, Düsseldorf, Germany) with a Teflon cover is used for the quenching assays. The solution in the cuvette can be stirred by means of the magnetic stirrer and a small stirring bar or with a small spatula. Care should be taken to minimize stray radiation and vibration reaching the detector during operation to maintain a smooth baseline signal.

Procedure

Three milliliters of assay solvent is placed in the cuvette and the cuvette is covered and inserted into the sample chamber. After a 3-min incubation to bring the cuvette to ~37°, $NDPO_2$ is added at 5 mM and the photoemission monitored on the chart recorder until a maximum level of emission (designated S_0) is achieved (5–6 min). A signal of 2 mV is typically observed with 5 mM $NDPO_2$. Ten to 50 μl of a carotenoid stock solution is then added and the resulting, decreased level of photoemission (designated S) is recorded. The overall quenching constant ($k_q + k_r$) is determined by assaying the carotenoid over a range of concentrations and then analyzing the data on a Stern–Volmer plot, which is based on the equation

$$S_0/S = (k_q + k_r)\tau[Q] + 1$$

where k_q is the physical quenching rate constant, k_r is the chemical reaction rate constant, τ is the $^1O_2$ lifetime, and [Q] is the quencher (i.e., carotenoid) concentration. With carotenoids, $k_q \gg k_r$ and so the contribution of chemical reaction to the quenching process can be neglected. The lifetime of $^1O_2$ in the assay solvent was determined to be 33 ± 1 μsec according to Ref. 27. Examples of Stern–Volmer plots used for the calculation of rate constants for lycopene, β-carotene, and lutein are given in Fig. 2.

Comments

$^1O_2$ quenching constants have been determined for carotenoids, oxycarotenoids, and related compounds using the technique of steady state $^1O_2$ production and monomol photoemission detection described above,[11] and

---

[27] G. Valduga, S. Nonell, E. Reddi, G. Jori, and S. E. Braslavsky, *Photochem. Photobiol.* **48**, 1 (1988).

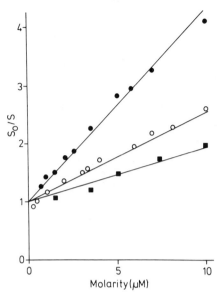

FIG. 2. Stern–Volmer plots of data on the quenching of $NDPO_2$-generated $^1O_2$ by (●) lycopene, (○) β-carotene, and (■) lutein in chloroform–ethanol (1:1, v/v), 37°, collected from measurements made with a germanium photodiode detector.

a partial listing of the results are given in Table I. At the time of these measurements the best estimate of the $^1O_2$ lifetime in the mixed solvent employed was 10 μsec,[28] which is the $^1O_2$ lifetime in ethanol (the lifetime in chloroform is ~200 μsec and the contribution of the small amount of water introduced with $NDPO_2$ is negligible). Direct measurements of the $^1O_2$ lifetime in our assay solvent have been made according to Ref. 27 since these results were reported, and quenching constants recalculated from the original data and with a $^1O_2$ lifetime of 33 μsec are listed in parentheses in Table I.

Also included in Table I are quenching rate constants taken from the literature and arranged according to method of measurement. From the range of values reported for a specific compound it is clear that significant variations can arise when comparing rate constants determined by different techniques. For example, with β-carotene the literature values vary by as much as fourfold (see Table I). A second point to note is that the ordering of carotenoids according to quenching ability can differ with

[28] B. M. Monroe, in "Singlet O_2" (A. A. Frimer, ed.), Vol. I, p. 177. CRC Press, Boca Raton, Florida, 1985.

TABLE I
SINGLET OXYGEN QUENCHING CONSTANTS OF CAROTENOIDS AND RELATED COMPOUNDS

	k_q (10^9 M^{-1} sec^{-1}), as determined from:				
	$^1O_2$ Photoemission		Triplet excited carotenoid extinction	$^1O_2$-dependent decomposition of probe	Other method
Quencher	Steady state	Pulse			
Carotenoids					
Lycopene	31 (9)[a]	14–19[b]			7[c]
γ-Carotene	25 (7)[a]	0.8[b]			
α-Carotene	19 (6)[a]	8[b]			
β-Carotene	6[d]; 14 (4)[a]	7[e]; 10–14[b]	13[f]	6[g]; 8[h]; 23[i]	13[j]
Oxycarotenoids (xanthophylls)					
Astaxanthin	24 (7)[a]	14[b]			10[c]
Canthaxanthin	10[d]; 21 (6)[a]	13[b]	14[k]; 18[l]		
Zeaxanthin	10 (3)[a]	12[b]	3[l]		7[c]
Lutein	8 (2)[a]	16[b]		21[i]	6[c]
Cryptoxanthin	6 (2)[a]				
Other polyenes					
Bixin	14 (4)[a]	9[b]		18–23[m]	
Crocin	1 (0.3)[a]			2[n]	
Retinoic acid	<0.1[a]				

[a] P. Di Mascio, S. Kaiser, and H. Sies, *Arch. Biochem. Biophys.* **274**, 532 (1989). Presented are values as originally calculated with the singlet oxygen lifetime of 10 μsec for ethanol (see Ref. 28); values in parentheses are calculated with the lifetime of 33 μsec, determined as described in Ref. 27 for the mixed solvent employed.

[b] P. F. Conn, W. Schalch, and T. G. Truscott, *J. Photochem. Photobiol. B: Biol.* **11**, 41 (1991). A range of values determined in different solvents is given.

[c] S.-H. Lee and D. B. Min, *J. Agric. Food Chem.* **38**, 1630 (1990). Quenching activity was measured as the inhibition of chlorophyll-sensitized soybean oil photooxidation.

[d] P. Murasecco-Suardi, E. Oliveros, A. M. Braun, and H.-J. Hansen, *Helv. Chim. Acta* **71**, 1005 (1988).

[e] A. A. Krasnovsky, Jr., *Photochem. Photobiol.* **29**, 29 (1979).

[f] A. Farmilo and F. Wilkinson, *Photochem. Photobiol.* **18**, 447 (1973).

[g] S. R. Fahrenholtz, F. H. Doleiden, A. M. Trozzolo, and A. A. Lamola, *Photochem. Photobiol.* **20**, 505 (1974).

[h] D. J. Carlsson, T. Suprunchuk, and D. M. Wiles, *J. Polym. Sci.,* **B11**, 61 (1973).

[i] M. M. Mathews-Roth, T. Wilson, E. Fujimori, and N. I. Krinsky, *Photochem. Photobiol.* **19**, 217 (1974).

[j] C. S. Foote, in "Singlet Oxygen" (H. H. Wasserman and R. W. Murray, eds.), p. 139. Academic Press, New York, 1979.

[k] F. Wilkinson and W.-T. Ho, *Spectrosc. Lett.* **11**, 455 (1978).

[l] M. A. J. Rodgers and A. L. Bates, *Photochem. Photobiol.* **31**, 533 (1980).

[m] G. Speranza, P. Manitto, and D. Monti, *J. Photochem. Photobiol. B: Biol.* **8**, 51 (1990). Bleaching of the quencher was taken as a measure of quenching activity; a range of values determined in different solvents is given.

[n] P. Manitto, G. Speranza, D. Monti, and P. Gramatica, *Tetrahedron Lett.* **28**, 4221 (1987). Bleaching of the quencher was taken as a measure of quenching activity.

technique as well; for lycopene (Ly), astaxanthin (A), zeaxanthin (Z), and lutein (Lu) the following orders have been reported: Ly > A > Z > Lu,[11] Ly > Lu > A > Z,[10] and A > Ly = Z > Lu.[19] Although different solvents were employed in the studies represented in Table I, choice of solvent alone cannot explain these discrepancies; rate constants determined for β-carotene,[10] lycopene,[10] and bixin[29] in different solvents but by the same technique varied by less than 40% for each compound, which is slight compared to the range of values determined, for example, for β-carotene by various techniques (i.e., 400%). While it may not be possible to critically compare individual rate constants reported in the literature, it is nevertheless clear from these studies that carotenoids such as lycopene and several oxycarotenoids are highy efficient quenchers of $^1O_2$.

In addition to carotenoids, we have determined $^1O_2$ quenching rate constants for a wide range of biomolecules and drugs with the $NDPO_2$/ germanium photodiode detection system and suitable solvents. For less lipophilic compounds such as bile pigments[11] and tocopherols[30] an ethanol–D_2O (1:1, v/v) solvent was employed, while for hydrophilic compounds, such as thiols,[31] amino acids,[31] and nucleosides,[32] 50 mM sodium phosphate buffer in D_2O, pD 7.4, could be used. The intensity of $^1O_2$ photoemission generated from $NDPO_2$ (i.e., S_0) in these solvents is comparable.

Acknowledgments

We gratefully acknowledge the generous supply of carotenoids by Dr. Bausch (Hoffmann-La Roche, Basel, Switzerland), and helpful discussions with him as well as with Prof. S. E. Braslavsky (Mülheim) and Prof. A. M. Braun (Lausanne).

Our own studies were supported by the National Foundation for Cancer Research, Bethesda, MD.

---

[29] G. Speranza, P. Manitto, and D. Monti, *J. Photochem. Photobiol. B: Biol.* **8,** 51 (1990).

[30] S. Kaiser, P. Di Mascio, M. E. Murphy, and H. Sies, *Arch. Biochem. Biophys.* **277,** 101 (1990).

[31] T. P. A. Devasagayam, A. R. Sundquist, P. Di Mascio, S. Kaiser, and H. Sies, *J. Photochem. Photobiol. B: Biol.* **9,** 105 (1991).

[32] T. P. A. Devasagayam, S. Steenken, M. S. W. Obendorf, W. A. Schulz, and H. Sies, *Biochemistry* **30,** 6283 (1991).

[41] Biosynthesis of β-Carotene in *Dunaliella*

By AVIV SHAISH, AMI BEN-AMOTZ, and MORDHAY AVRON

Introduction

Unicellular algae of the genus *Dunaliella* are classified in the order Volvocales, of the class Chlorophyta (green algae). *Dunaliella* are ovoid in shape, contain one large cup-shaped chloroplast, and are motile with two equal long flagellae. They lack a rigid polysaccharide cell wall and the cell is enclosed by an elastic plasma membrane.[1] *Dunaliella* are halotolerant, and the most tolerant species can be grown in a wide range of salt concentrations, from 50 mM to saturated (around 5.5 M) NaCl. The algae osmoregulate by accumulating intracellularly essentially a single substance, glycerol.[2] Among the many strains of the genus *Dunaliella* described, only two — *D. bardawil* and *D. salina* Teod. — have been shown capable of producing large amounts of β-carotene,[3,4] under proper inductive conditions. Under optimal conditions more than 10% of the dry weight of *D. bardawil* is β-carotene while under noninductive conditions it contains less than 0.3%. The conditions optimal for carotene biosynthesis include high light intensity, coupled with a growth-limiting parameter such as high sodium chloride concentration, low temperature, or nitrate or sulfate deficiency.[5] The β-carotene is accumulated within oily globules in the interthylakoid spaces of the chloroplast and is composed mainly of two stereoisomers: 9-cis and all-trans.[6] The ratio between the 9-cis and all-trans isomers of β-carotene is higher, the higher the light intensity.[6] The massively accumulated β-carotene in *D. bardawil* protects the alga against damage by high irradiation by its absorption of the blue region of the spectrum.[7] Utilizing inhibitors of β-carotene biosynthesis it was shown that the intermediates phytoene, phytofluene, ζ-carotene, β-zeacarotene, lycopene, and γ-carotene are all composed of two stereoisomers.[8] By using the protective action of β-carotene as a selection criterion several carotene-rich mutants of *D. bardawil* were obtained.[9]

[1] M. A. Borowitzka and L. J. Borowitzka, in "Microalgal Biotechnology" (M. A. Borowitzka and L. J. Borowitzka, eds.), p. 27. Cambridge University Press, Cambridge, England, 1988.
[2] M. Avron, *Trends Biochem. Sci.* **11,** 5 (1986).
[3] A. Ben-Amotz, A. Katz, and M. Avron, *J. Phycol.* **18,** 529 (1982).
[4] L. Loeblich, *J. Mar. Biol. Assoc. U.K.* **62,** 493 (1982).
[5] A. Ben-Amotz and M. Avron, *Plant Physiol.* **12,** 593 (1983).
[6] A. Ben-Amotz, A. Lers, and M. Avron, *Plant Physiol.* **86,** 1286 (1988).
[7] A. Ben-Amotz, A. Shaish, and M. Avron, *Plant Physiol.* **91,** 1040 (1989).
[8] A. Shaish, M. Avron, and A. Ben-Amotz, *Plant Cell Physiol.* **31,** 689 (1990).
[9] A. Shaish, A. Ben-Amotz, and M. Avron, *J. Phycol.* **27,** 652 (1991).

Induction of β-Carotene Biosynthesis

Alga. Dunaliella bardawil Ben-Amotz and Avron, a local isolated species, is available from the American Type Culture Collection (Rockville, MD) as No. 30861; and from the Culture Collection of Algae and Protozoa (The Ferry House, Ambleside, Cumbria, UK) as CCAP 19/30.

Growth Conditions. The algae are cultivated with continuous slow shaking, at 26° in 250-ml Erlenmeyer flasks containing 50 ml of the following growth medium: 1–3 M NaCl, 5 mM KCl, 5 mM MgCl$_2$, 5 mM Na$_2$SO$_4$, 5 mM NaNO$_3$, 0.2 mM CaCl$_2$, 0.2 mM KH$_2$PO$_4$, 2 μM FeCl$_3$, 5 μM NaEDTA, 7 μM MnCl$_2$, 1 μM CuCl$_2$, 1 μM ZnCl$_2$, 1 μM CoCl$_2$, 1 μM (NH$_4$)Mo$_7$O$_{24}$, and 50 mM NaHCO$_3$ (pH ~ 8.0). Continuous light at about 10 W/m^2 is supplied by cool white fluorescent lamps. Lower light intensity (5 W/m^2) is achieved by covering the flasks with cloth. When high light intensities (500 W/m^2) are desired the algae are placed in a temperature-controlled water bath and illuminated from below with high-intensity halogen lamps.

Analysis. Cell number is determined in a Coulter counter (Luton, UK) with a 100 μm orifice. For carotene analysis a sample of about 2 ml of algae is centrifuged at 1000 g for 5 min and the pellet extracted by 3 ml of ethanol–hexane (2:1, v/v). Two milliliters of water and 4 ml of hexane are added and the mixture thoroughly mixed and centrifuged at 1000 g for 5 min. The hexane layer is separated and its absorption at 450 nm determined: A_{450} (19.6) equals the micrograms of carotene in sample. For HPLC analysis the hexane layer is dried completely under N$_2$ and redissolved in a minimal volume of hexane.

Induction by Nitrate or Sulfate Starvation. Cells are pregrown with a complete medium at very low light intensity (5 W/m^2) to start with relatively low β-carotene-containing cells. Cultures containing about 2–4 × 10^5 cells/ml are washed twice and resuspended in a nitrate or sulfate free medium (at 2 × 10^5 cells/ml) and incubated under normal growth conditions (light intensity 10 W/m^2). Within 3–5 days the cells triple to quadruple their β-carotene content with no more than a doubling in the cell number.

Induction by High Light Intensity. Cells pregrown at low light intensity (5 W/m^2) are centrifuged (1000 g, 5 min), resuspended in complete medium at 2 × 10^5 cells/ml and incubated with slow shaking in a temperature-controlled water bath at 26°, equipped with halogen lamps that provide 500 W/m^2. β-Carotene begins to accumulate after a few hours.[10]

---

[10] A. Lers, Y. Biener, and A. Zamier, *Plant Physiol.* **93**, 389 (1990).

Induction by High Light and Limiting Nutrients. Cells pregrown at low light intensity (5 W/m^2) are centrifuged (1000 g, 5 min), resuspended in a complete medium that contains only 0.5 mM NaNO$_3$ at 2 × 10^5 cells/ml, and incubated as described in the preceding paragraph. The highest final β-carotene content per cell is obtained under this procedure.

Pathway for β-Carotene Biosynthesis with Inhibitors

The routine procedures to induce high β-carotene biosynthesis in *D. bardawil*, i.e., exposure to high-intensity irradiation (described above) cannot be employed for examining the effects of inhibitors in the carotene biosynthesis pathway because in the presence of the inhibitor, when no β-carotene is synthesized, *D. bardawil* is highly sensitive to photodestruction, which results in rapid cell death.[8] Therefore only the procedure described in "Induction by Nitrate or Sulfate Starvation" (above) can be employed. The inhibitors are added during transfer to the nitrate-free medium and the pigments are extracted for analysis following 3–5 days of incubation. The following inhibitors of β-carotene formation were shown to be effective in *Dunaliella*.

Norflurazon (SAN 9789; Sandoz, Basel, Switzerland) is an inhibitor of the enzyme phytoene desaturase. At higher concentrations (~0.3 μM) it fully inhibits β-carotene biosynthesis, while an equivalent amount of phytoene is accumulated.[11] At lower concentrations (~0.1 μM) phytofluene also accumulates.[8]

J-334 (R112334, ICI Agrochemicals) is also a desaturation inhibitor. At higher concentrations (~ μM) it fully inhibits β-carotene biosynthesis, while phytoene and ζ-carotene accumulate. At lower concentrations (~0.04 μM) β-zeacarotene is the major intermediate that accumulates.[8]

CPTA and MPTA (Amchem Products, Amber, PA) are compounds that inhibit cyclization in carotene biosynthesis. When used directly they severely inhibit growth and survival with no significant accumulation of intermediates. However, when the algae are preloaded with phytoene (by preincubation in the presence of 0.3 μM norflurazon), the inhibitor removed by washing, the cyclization inhibitors added in fresh nitrate-free medium, and the cells incubated for another day, several intermediates can be observed. At concentrations of 50 μM CPTA or MPTA, lycopene is the major accumulated intermediate, indicating that the cyclization of lycopene to γ-carotene is blocked.

Nicotine is a cyclization inhibitor. At 100 μM, when used following the

---

[11] A. Ben-Amotz, J. Gressel, and M. Avron, *J. Phycol.* **23,** 176 (1987).

procedure outlined above for CPTA or MPTA, it causes the accumulation of lycopene. However, at lower concentrations (~1 μM) γ-carotene is the major accumulated intermediate.

Stereoisomers

All intermediates identified in the carotenogenesis pathway, including β-carotene, are always composed of two major stereoisomers that can be separated by HPLC (see below). In the case of β-carotene the two major stereoisomers are identified as the all-trans and the 9-cis isomers.

HPLC Analysis

The identification of the intermediates is best done by separation on HPLC coupled with spectral analysis of the eluted peaks by on-line diode array spectrophotometry. A stainless steel column, 25 cm × 4.6 cm i.d., packed with C_{18} reversed-phase material of 5-μm particle size [Vydac (Hesperia, CA) TP201 54] has proved appropriate for pigment analysis. This column is particularly efficient in separating the various stereoisomers.[8] Elution is performed isocratically at 1 ml/min with methanol–acetonitrile (9:1, v/v). A better but slower separation can be obtained by using methanol as the mobile phase at 0.5 ml/min.

The following wavelengths and the extinction coefficients, $E_{1cm}^{1\%}$, have been employed for detection and analysis of the various pigments: lycopene, 504 nm (2950); β-carotene, 450 nm (2550); β-zeacarotene, 427 nm (1940); ζ-carotene, 402 nm (2550); phytofluene, 347 nm (1540); phytoene, 287 nm (915); and γ-carotene, 489 nm (2720). The areas under each relevant peak with the appropriate extinction coefficient can be used for quantitative estimations.

Protection against Photoinhibition

It is generally believed that the interthylakoid β-carotene protects plants by quenching excited chlorophyll and damaging reagents (e.g., singlet oxygen) produced by excessive excitation of chlorophyll.[12,13] However, because of its location outside the thylakoids and the short lifetime of these compounds, the massively accumulated β-carotene in *D. bardawil* cannot be effective by this mechanism. Nevertheless it clearly protects the algae

---

[12] C. S. Foote, *in* "Free Radicals in Biology" (M. A. Pryor, ed.), Vol. 2, p. 85. Academic Press, San Diego, 1990.
[13] N. I. Krinsky, *Free Radical Biol. Med.* **7**, 617 (1989).

from damage by excessive irradiation. It was demonstrated that its protection was due to its absorption properties, reducing the light intensity that reaches the thylakoids. Thus, the massively accumulated β-carotene protects the algae only when the photoinhibitory light is composed of wavelengths absorbed by β-carotene (i.e., blue light). Under such conditions the high β-carotene alga is highly resistant to photoinhibition. However, when red light (not absorbed by β-carotene) is used as the photoinhibitory light, no protection is afforded by the massively accumulated β-carotene.[7]

Photoinhibition can be monitored by placing the algae (5×10^5 cells/ml) in a glass vessel equipped with continuous slow stirring and a water jacket for circulating water to maintain the temperature at 25°. High light intensity can be provided by several high-intensity slide projectors located in an appropriate distance from the reaction vessel to provide the desired light intensity (around 2000 W/m^2). The light must be filtered through a heat filter (cutoff around 730 nm) and at least 2 cm of water. Red light is produced by insertion of a Schott (Mainz, Germany) OG 580 filter and blue light with a Corning (Corning, NY) 9788 filter. Samples are removed and assayed for photosynthetic oxygen evolution, with a Clark-type oxygen electrode, or for survival. Low-β-carotene algae lose all ability to evolve oxygen photosynthetically following 3–5 hr of illumination. By that time most of the algae are killed by the intense illumination.

Selection of High β-Carotene Mutants

The ability of β-carotene in *Dunaliella* to protect the algae from killing by high blue irradiation can be utilized to select *D. bardawil* mutants that accumulate a higher content of β-carotene.[9]

At the logarithmic growth phase *D. bardawil* are incubated under inductive conditions for β-carotene biosynthesis (see "Induction by High Light Intensity," above) for 5 hr. They are then exposed to mutagenesis conditions by UV irradiation (0.68 J/sec) for 10 min (1–5% survival) and kept in the dark for 24 hr to prevent photoactivation repair. The algae are then incubated under low light intensity (10 W/m^2) for 9 days. Under these low-light conditions, the wild type contains about 5 pg of β-carotene per cell, a low level that does not protect against killing by strong blue light. The algae are exposed to high-intensity blue light (2000 W/m^2) for 8 hr, to select mutants that contain high β-carotene levels, and allowed to recover under normal growth conditions (10 W/m^2) for 3 days. This selection procedure is repeated three times. The surviving algae are plated on agar plates or grown in wells following infinite dilution.

Plating on agar is performed by mixing the algae with 3 ml of melted 0.4% (w/v) agarose at 37° and spreading the mixture on a layer of 5 ml of pregelled 1.5% (w/v) Bactoagar (Dijco, Detroit, MI) in a petri dish. The agarose and the agar are dissolved in a growth medium containing 10 mM NaHCO$_3$. The petri dishes or wells are placed under normal growth conditions (10 W/m^2). Colonies that phenomenologically appear reddest under these noninductive conditions are selected, and individually tested for their ability to synthesize β-carotene under a variety of conditions.

[42] Carotenoid Free Radicals

By MICHAEL G. SIMIC

Introduction

The unusual conjugated structure of carotenoids gives them exceptional physicochemical properties that may explain their presence in most biochemical systems, from chloroplasts in photosynthesis to human physiology, as well as in biosensors and biomimetic devices. The mechanisms of carotenoid action are not fully understood, and new biochemical and physiological roles are continuously uncovered. Despite their widespread presence in biosystems, the redox and free radical chemistry of carotenoids is still in its infancy. The need for greater understanding of these compounds is even more critical in view of discoveries of free radical processes *in vivo*.

The electronic and free radical mechanisms studied by radiation techniques and methodologies will be reviewed in this chapter. Consideration of radiation technology in the overall review of carotenoids is important because of the unique contributions of this field toward understanding of the fast, time-resolved processes and short-lived, transient intermediates in chemistry and biochemistry.[1-3]

Although many water-soluble biochemical components and their models have been comprehensively studied by pulse radiolysis[4] of aqueous solutions, information on carotenoids is scant. This is a consequence of their insolubility in water and the less complete understanding of radiation

[1] M. G. Simic, this series, Vol. 186, p. 89.
[2] C. von Sonntag, "The Chemical Basis of Radiation Biology." Taylor & Francis, London, 1987.
[3] Farhataziz and M. A. J. Rodgers (eds.), "Radiation Chemistry." Verlag Chemie, New York, 1987.
[4] Pulse radiolysis is a technique best described as generation of free radicals by pulsed radiation and time-resolved monitoring of their reactions. Time resolutions are usually on the order of microseconds, but occasionally, if required, may be even less than nanoseconds.

chemistry of nonaqueous media in which studies of carotenoids could be conducted. Strong indications that carotenoids may play a role in the prevention of degenerative diseases, such as cancer and cardiovascular disorders, may be expected to stimulate new studies of the carotenoid defense mechanisms against oxidative processes, i.e., their free radical chemistry.

Generation of Free Radicals

Radiation techniques are an invaluable source of primary free radicals, which in turn are used to generate secondary radicals of carotenoids and their model compounds. In general, there are four basic reactions between initial solvent radicals and carotenoids: oxidation, reduction, addition, and abstraction. Carotenoids can be oxidized by oxidizing radicals such as solvent positive ions, Br_2^-, and many other radicals with high redox potentials. Reduction occurs via electrons (hydrated in water and nonsolvated in nonpolar media) and reducing free radicals with low redox potentials, such as α- hydroxy radicals. Some radicals such as ·OH and ·H add readily to unsaturated bonds and aromatic systems. Abstraction of H usually occurs from C–H and S–H bonds and is determined by energetics; the rates are faster for greater differences of bond strength between the participating bonds of the reactants. The prevalence of a reaction type is governed by the energetics, activation energies, and entropic factors.

Solvent Radicals

Ionizing radiation excites or ionizes the solvent molecule. Excitation, however, does not occur in every solvent.

In water three initial water radicals are formed by ionization of H_2O,

$$H_2O \rightsquigarrow \cdot OH, e_{aq}^-, \cdot H \tag{1}$$

In organic nonpolar media,[3] such as aliphatic hydrocarbons, RH,

$$RH \rightsquigarrow \cdot RH^+ + e^- \tag{2}$$

In contrast to aqueous media, the electrons, e^-, are not solvated in nonpolar media and are consequently more mobile than hydrated electrons, e_{aq}^-. The very high diffusion-controlled rate constants are therefore even higher in nonpolar media than in aqueous solutions (see Tables I and II).

In media consisting of halogenated hydrocarbons, the electrons are scavenged by the solvent itself before they have a chance to react with the solute,

$$e^- + RHX_n \rightarrow \cdot RHX_{n-1} + X^- \tag{3}$$

TABLE I
REACTION RATE CONSTANTS FOR SELECTED UNSATURATED
ALIPHATIC COMPOUNDS AND WATER RADICALS[a]

Solute, S	$k(S + radical), M^{-1} sec^{-1}$		
	e_{aq}^-	·H	·OH
H_3-CH_3	$<10^4$	2.3×10^6	1.8×10^9
$H_2C=CH_2$	$<3 \times 10^5$	3.0×10^9	4.0×10^9
Cyclohexene	$<10^6$	4.1×10^9	8.8×10^9
$HC\equiv CH$	2×10^7	2.2×10^9	4.7×10^9
$H_2C=CHCH_2OH$	7.2×10^7	1.1×10^9	6.0×10^9
$H_2C=CHCONH_2$	2.2×10^{10}	—	—
$H_2C=CHCH=CH_2$	8.0×10^9	1.0×10^{10}	7.0×10^9
1,3-Cyclohexadiene	1.0×10^9	9.7×10^9	9.9×10^9
1,4-Cyclohexadiene	$<10^6$	4.7×10^9	7.7×10^9
Crocetin[2−]	7.1×10^{10}	—	2.3×10^{10}
Crocin	1.1×10^{11}	—	3.1×10^{10}
Hematoporphyrin	1.5×10^{10}	$\sim 10^{10}$	$>10^{10}$

[a] In aqueous solutions at 25°, measured by pulse radiolysis.[5]

Reactivity of Model Compounds

Reaction rate constants[5] of water free radicals with selected saturated, monounsaturated, and polyene model compounds can be obtained by pulse radiolysis and are shown in Table I. The reactivity of hydrated electrons increases with unsaturation and conjugation until it reaches diffusion-controlled rates, $k \sim 10^{11} M^{-1} sec^{-1}$. Rate constants with nonsolvated electrons are about an order of magnitude higher. The abstraction rate constant for ·H from ethane, for example, is relatively low, whereas the addition of ·H to double bonds is much faster. Abstraction rate constants for ·OH are only slightly lower than those for addition to unsaturated bonds. Abstraction of H atoms by either ·H or ·OH does not take place from C—H bonds associated with double bonds because of the high bond strength of $>C=C-H$ bonds. The intermediates of model compounds in Table I are poorly characterized spectroscopically because of their weak absorption.

[5] G. V. Buxton, C. L. Greenstock, W. P. Helman, and A. B. Ross, *J. Phys. Chem. Ref. Data* **17**, 513 (1988).

TABLE II
Spectral Properties of Carotenoid Radicals and Electron Transfer Rate Constants for Reactions of Carotenoid Radicals[a]

Carotenoid (C)	Number of Conjugated double bonds	·C+		k(·C+ + Chl) (M^{-1} sec^{-1})		·C−		k(·C− + Chl) (M^{-1} sec^{-1})	
		λ_{max} (nm)	$\varepsilon \times 10^{-5}$ (M^{-1} cm^{-1})	Chl a	Chl b	λ_{max} (nm)	$\varepsilon \times 10^{-5}$ (M^{-1} cm^{-1})	Chl a	Chl b
7,7′-Dihydro-β-carotene	8	830		5.4×10^9	10^{10}	785		8.0×10^9	2.5×10^{10}
Septapreno-β-carotene	9	915		1.1×10^{10}	6×10^9	785		7.0×10^9	8.0×10^{10}
15,15′-cis-β-Carotene	11	1050	1.54	1.2×10^{10}	$<10^8$	900	2.51	8.7×10^9	1.4×10^{10}
all-trans-β-Carotene	11	1040	2.18	6.0×10^9	$<10^8$	880	4.42	8.5×10^9	1.7×10^{10}
all-trans-Lycopene	11	1070		1.7×10^9	—	950		7.0×10^9	—
Decapreno-β-carotene	15	1250	3.25	4.7×10^9	$<10^8$	1130	4.1	5.4×10^9	10^9

[a] Radical cation (·C+) and anion (·C−) with chlorophyll a (Chl a) and chlorophyll b (Chl b) in hexane as a function of the number of conjugated double bonds.[6,7,13]

Carotenoid Radicals

The reactions of β-carotene (Car) in organic solvents, with both electrons and positive radical ions, are very fast and lead to formation of strongly absorbing intermediates. In hexane[6,7]

$$\text{Car} + e^- \xrightarrow{(k = 1.4 \times 10^{12} M^{-1} \text{sec}^{-1})} \cdot \text{Car}^- \qquad (4)$$

$$\text{Car} + \cdot \text{RH}^+ \xrightarrow{(k = 2.5 \times 10^{10} M^{-1} \text{sec}^{-1})} \cdot \text{Car}^+ + \text{RH} \qquad (5)$$

The spectroscopic properties of these and other carotenoid (C) radicals are shown in Table II. A simple rule is evident: an increasing number of conjugated double bonds in C shifts absorption bands for both $\cdot C^+$ and $\cdot C^-$ to longer wavelengths. Very high ε values with λ_{max} in the near infrared allow detection and monitoring of these radicals, even in complex biosystems.[8]

In aqueous solutions β-carotene radicals can be generated in micelles (Mic),[9,10] e.g., in Triton X-100 micelles,[9]

$$\text{Car(Mic)} + \text{Br}_2^{-} \xrightarrow{(k = 1 \times 10^8 M^{-1} \text{sec}^{-1})} \cdot \text{Car}^+(\text{Mic}) + 2\text{Br}^- \qquad (6)$$

The relatively slower rate of oxidation of β-carotene results from inaccessibility of the nonpolar inside portion of the micelle, in which β-carotene is lodged, to hydrophilic Br_2^- radicals.

Properties of Carotenoid Radicals

The unpaired electron in both $\cdot C^+$ and $\cdot C^-$ is highly delocalized due to the resonant structure of the intermediates. The $\cdot C^-$ radicals are strongly reducing (low redox potential) and react rapidly with oxygen,[6]

$$\cdot C^- + O_2 \xrightarrow{(k \sim 10^9 M^{-1} \text{sec}^{-1})} C + \cdot O_2^- \qquad (7)$$

The $\cdot C^+$ radicals, owing to their higher redox potential[8] and resonant structure,[11] apparently do not react with oxygen,

$$\cdot C^+ + O_2 \rightarrow \text{not observed} \qquad (8)$$

---

[6] E. A. Dawe and E. J. Land, *J. Chem. Soc., Faraday Trans. 1* **71**, 2162 (1975).

[7] J. Lafferty, T. G. Truscott, and E. J. Land, *J. Chem. Soc., Faraday Trans. 1* **74**, 2760 (1978).

[8] L. J. Land, D. Lexa, R. V. Bensasson, D. Gust, T. A. Moore, A. L. Moore, P. A. Liddell, and G. A. Nemeth, *J. Phys. Chem.* **91**, 4831 (1987).

[9] J. P. Chauvet, R. Vlovy, E. J. Land, R. Santus, and T. G. Truscott, *J. Phys. Chem.* **87**, 592 (1983).

[10] M. Almgren and J. K. Thomas, *Photochem. Photobiol.* **31**, 3291 (1980).

[11] M. G. Simic, S. V. Jovanovic, and M. Al-Sheikhly, *Free Radical Res. Commun.* **6**, 113 (1989).

The redox potential of carotenoids, $E(\cdot C^+/C)$, decreases with an increasing number of conjugated bonds. For example, radical cations of septapreno-β-carotene (Sept), a carotenoid with 9 double bonds, readily oxidize β-carotene with 11 double bonds,[7]

$$\text{Car} + \cdot\text{Sept}^+ \xrightarrow[(k = 5.2 \times 10^9 \, M^{-1} \, \text{sec}^{-1})]{} \cdot\text{Car}^+ + \text{Sept} \qquad (9)$$

According to the general rule described for the correlation between the redox potential and the number of conjugated double bonds, retinoid radicals would also be expected to oxidize β-carotene and its analogs.

An interesting reaction offers a mechanism for the regeneration of β-carotene in the presence of chlorophyll. The $\cdot\text{Car}^+$ radical rapidly oxidizes chlorophyll a,[7]

$$\text{Chl } a + \cdot\text{Car}^+ \xrightarrow[(k = 6 \times 10^9 \, M^{-1} \, \text{sec}^{-1})]{} \cdot\text{Chl } a^+ + \text{Car} \qquad (10)$$

The reaction of $\cdot\text{Car}^+$ with chlorophyll b, a metalloporphyrin with a slightly higher redox potential (0.65 V for chlorophyll b and 0.54 V for chlorophyll a vs SCE),[12] was not observed.

$$\text{Chl } b + \cdot\text{Car}^+ \xrightarrow[(k < 6 \times 10^8 \, M^{-1} \, \text{sec}^{-1})]{} \cdot\text{Chl } b^+ + \text{Car} \qquad (11)$$

Whether reaction (11) occurs is not clear because it may proceed at rates below the detection limits for this particular system and conditions. From comparison of the kinetics of reactions (10) and (11), one can conclude that the redox potential of β-carotene is higher than that of chlorophyll a but close to the value of chlorophyll b. This is in agreement with a value of 0.72 vs SCE, measured directly with cyclic voltametry.[8]

The reaction of carotenoid radicals with porphyrins (P) can be summarized by two equations. Electron adducts of carotenoids rapidly transfer an electron to most porphyrins,[13]

$$\text{P} + \cdot\text{C}^- \xrightarrow[(k \sim 10^{10} \, M^{-1} \, \text{sec}^{-1})]{} \cdot\text{P}^- + \text{C} \qquad (12)$$

whereas a carotenoid cation radical is usually unreactive,

$$\text{P} + \cdot\text{C}^+ \rightarrow \cdot\text{P}^+ + \text{C} \qquad (13)$$

In both reactions the specific behavior would depend on relative one-electron redox potentials of the reactants.

---

[12] G. R. Seely, *Photochem. Photobiol.* **27**, 639 (1978).
[13] J. McVie, R. S. Sinclair, D. Tait, T. G. Truscott, and E. J. Land, *J. Chem. Soc., Faraday Trans. 1* **75**, 2869 (1979).

Carotenoids as Inhibitors of Free Radicals

As discussed in the section on generation of free radicals, carotenoids have fairly high reactivities with a variety of free radicals. For convenience and because of the differences in applicability, the reactions of carotenoids with free radicals are classified according to the type of free radical: nonoxyl (C centered, S centered) or oxyl (hydroxyl, peroxyl, alkoxyl) radicals.

Nonoxyl Radicals

Addition of C-centered radicals to double bonds is a well-known reaction in polymerization of unsaturated monomers. Unfortunately, there is very little information on the reactivity of $\cdot R$ with carotenoids.

An interesting reaction between glutathione radicals and retinol (Ret) has been reported.[14] The experiments were conducted in aqueous solutions of methanol, glutathione (GSH), and retinol by pulse radiolysis. A high methanol concentration (60%) in water allowed solubilization of retinol. Methanol radicals were generated either by direct ionization of methanol or by $\cdot OH$ radicals. In the subsequent reaction S-centered radicals were generated,

$$GSH + \cdot CH_2OH \rightarrow GS \cdot + CH_3OH \tag{14}$$

The reaction of the glutathione radical was fast,

$$Ret + GS \cdot \longrightarrow \cdot Ret^+ + GS^- \tag{15a}$$
$$\longrightarrow GS - \dot{R}et \tag{15b}$$
$$(k = 1.4 \times 10^9 \, M^{-1} s^{-1})$$

The relative contributions of the two reactions were not resolved, although the transient was well characterized ($\lambda_{max} = 380$ nm, $\varepsilon = 4 \times 10^4 \, M^{-1} cm^{-1}$). Reaction (15) was suggested to be a possible mode for inactivation of potentially damaging S-centered radicals.

Oxyl Radicals

Because of their relatively low concentration in biosystems, carotenoids would be inefficient as $\cdot OH$ scavengers despite their very high reactivity (Table I).

The less reactive peroxyl radicals, however, have apparently been suc-

---

[14] M. D'Aquino, C. Dunster, and R. L. Willson, *Biochem. Biophys. Res. Commun.* **161**, 1199 (1989).

cessfully scavenged by carotenoids, which behave as antioxidants.[15,16] The reaction with $Cl_3COO\cdot$, the most reactive of all peroxyl radicals, was very fast.[17]

$$Car + Cl_3COO\cdot \xrightarrow[(k = 1.5 \times 10^9 \, M^{-1} \, sec^{-1})]{} \cdot Car^+ + Cl_3COO^- \qquad (16)$$

The reactivities of the nonhalogenated peroxyl radicals with carotenoids normally encountered in biosystems, which are formed via reaction (17),

$$\cdot R + O_2 \xrightarrow[(k \sim 10^9 \, M^{-1} \, sec^{-1})]{} ROO\cdot \qquad (17)$$

have not been reported, probably due to their much lower reactivities. In general, the rate constants of $ROO\cdot$ radicals with antioxidants are two to three orders of magnitude lower than those of the $Cl_3COO\cdot$ radical.[1]

The addition of $ROO\cdot$ to the double bond of β-carotene adjacent to the ring has been suggested to be the antioxidant mechanism of β-carotene.[16] Addition of oxyl radicals in the middle of the conjugated double bonds in crocin[18] (a compound with seven conjugated double bonds), as deduced from the bleaching of the crocin absorption, contrasts with the previous suggestion that addition occurs at the beginning of the conjugated chain.[16] Rapid bleaching of the β-carotene absorption bands at 450–480 nm in cyclohexane by radiolytically generated $ROO\cdot$[19] would favor the hypothesis of addition in the middle of the chain.[18] Further mechanistic studies are required to resolve this issue.

The efficient antioxidant activity of β-carotene[15] can be demonstrated in cylindrical micelles of linoleic acid (H_2L)[20,21] in which β-carotene reacts with linoleic peroxyl radicals, $HLOO\cdot$. Consequently, the chain length of radiation-induced peroxidation of linoleic acid is reduced in the presence of β-carotene, as reflected in lower observed oxygen uptake. The inhibition of peroxidation by β-carotene is comparable to that observed for α-tocopherol and bilirubin,[22] although it is not as efficient under these conditions.

It is interesting to note that $RO\cdot$ radicals, like $\cdot OH$ and $ROO\cdot$, also

---

[15] N. I. Krinsky, *Free Radical Biol. Med.* **7**, 617 (1989).
[16] G. W. Burton and K. U. Ingold, *Science* **224**, 569 (1984).
[17] J. E. Packer, J. S. Mahood, V. O. Mora-Arellano, T. F. Slater, R. L. Willson, and B. S. Wolfenden, *Biochem. Biophys. Res. Commun.* **98**, 901 (1981).
[18] W. Bors, C. Michel, and M. Saran, *Biochim. Biophys. Acta* **796**, 312 (1984); W. Bors, C. Michel, and M. Saran, *Int. J. Radiat. Biol.* **41**, 493 (1982).
[19] M. G. Simic, unpublished observation (1978).
[20] M. Al-Sheikhly and M. G. Simic, *Abstr. 4-ICOR* p. 50 (1987).
[21] M. Al-Sheikhly and M. G. Simic, *J. Phys. Chem.* **93**, 3103 (1988).
[22] M. Al-Sheikhly and M. G. Simic, in "Anticarcinogenesis and Radiation Protection" (P. Cerutti, O. F. Nygaard, and M. G. Simic, eds.), p. 47. Plenum, New York, 1987.

bleach carotenoid absorption, i.e., they add to the double-bond system. Superoxide radical, on the other hand, does not seem to react with carotenoids with seven conjugated double bonds,[18]

$$C + \cdot O_2^- \rightarrow \text{not observed} \quad (18)$$

In contrast, the protonated form of superoxide, $\cdot O_2H$ ($pK_a = 4.8$), should be as reactive with carotenoids as $ROO\cdot$ radicals are.

Molecular Electronic Devices

Conjugated polyenes and carotenoids, in particular, have been used in the development of models of the biophysical stages of photosynthesis to elucidate the mechanisms of photoinduced charge separation. The experiments have been extended to the design of simple molecular devices for utilization of solar energy. The high conductivity of polyenes has also been used in developing subcomponents for ultrafast molecular computers. These uses of carotenoid are pertinent to this chapter because the devices deal with excess electrons and positive holes, which may be defined as the carotenoid free radicals $\cdot C^-$ and $\cdot C^+$.

Photoinduced Charge Separation

A combination of carotenoids, porphyrins, metalloporphyrins (MP), and quinones (Q) has been used to mimic multistep intramolecular charge separation and transfer in photosynthesis as well as to achieve long-lived, high-energy, charge-separated states. The original triad arrangement (C-P-Q)[8] has been expanded to pentads (C-ZnP-P-diQ).[23] The simplified mechanism of photoinduced charge separation in a pentad is as follows:

$$\text{C-ZnP-P-diQ} \xrightarrow[\text{(560 nm)}]{h\nu} \text{C-}^1\text{ZnP-P-diQ} \xrightarrow[(k_1 = 2.3 \times 10^{10} \text{ sec}^{-1})]{k_1} \text{C-ZnP-}^1\text{P-diQ}$$
$$\xrightarrow[(k_2 = 7 \times 10^8 \text{ sec}^{-1})]{k_2} \text{C-ZnP-}\dot{P}^+\text{-di}\dot{Q}^- \longrightarrow \text{C-Zn}\dot{P}^+\text{-P-di}\dot{Q}^- \quad (19)$$
$$\xrightarrow[(k_3 = 55-340 \ \mu\text{sec})]{k_3} \dot{C}^+\text{-ZnP-P-di}\dot{Q}^- \longrightarrow \text{products}$$

Much faster intramolecular electron transfer than in the original observation in cobalt complexes[24] is achieved by resonant coupling between the components. The electron-donating ability of a carotenoid with 10 conju-

---

[23] D. Gust, T. A. Moore, A. L. Moore, S.-J. Lee, E. Bittersmann, D. K. Luttrull, A. A. Rehms, J. M. DeGraziano, X. C. Ma, F. Gao, R. E. Belford, and T. T. Trier, *Science* **248**, 199 (1990).
[24] M. Z. Hoffman and M. G. Simic, *J. Am. Chem. Soc.* **94**, 1757 (1972).

gated double bonds coupled to a benzene ring, for example, is sufficiently high to make the carotenoid a positive charge trap.

Molecular Electronics

Conjugated polyene components (CPE) were coupled to chromophores A, B, C, and D through conjugated fused rings (RR) in order to fit within a vesicle and test soliton switching (AB to A^+B^-; DC to D^-C^+; B^-C^+ to BC; and the final return to the initial state via D^-A^+ to DA) with the following arrangement:

$$\begin{array}{c} A - CPE \\ \diagdown \\ \text{RR-RR} \\ \diagup \\ D - CPE \end{array} \begin{array}{c} CPE - C \\ \diagup \\ \\ \diagdown \\ CPE - B \end{array} \qquad (20)$$

in which the groups A, B, C, and D on the surface of the vesicle are photoactivated, while PE and RR act as electronic conductors.[25]

To construct such electronic devices for testing of soliton mechanisms, the compatibility of long conjugated polyenes with Langmuir–Blodgett film was ascertained.[25] The pressure-vs-area curves for β-carotene showed a sharp rise. A detrimental, less sharp rise in successive runs indicated a gradual oxidation of β-carotene, a problem that must be overcome before such devices make significant inroads as high-technology products.

Conclusion

A comprehensive understanding of redox and free radical mechanisms of carotenoids is critical not only for prevention of degenerative diseases through appropriate diets but also in the high-technology field of molecular electronics, in which inhibition of autoxidation processes is crucial for the development and practical application of devices for solar-energy utilization and ultrafast computing.

Acknowledgments

Partial support of this work was provided by DNA and the ILSI Risk Science Institute. Special thanks are due to Karen A. Taylor for numerous suggestions and discussions and Gloria Wiersma for technical assistance.

[25] F. L. Carter, P. C. Berg, and W. R. Barger, *in* "Molecular Electronic Devices" (F. L. Carter, R. E. Siatkowski, and H. Wohltjen, eds.), p. 465. Elsevier/North-Holland, Amsterdam, 1988.

[43] Lipid Hydroperoxide Assay for Antioxidant Activity of Carotenoids

By JUNJI TERAO, AKIHIKO NAGAO, DONG-KI PARK, and BOEY PENG LIM

Introduction

In general, carotenoids exert an antioxidant activity by quenching excited triplet sensitizer and singlet molecular oxygen in photosensitized oxidation or by trapping oxygen radicals responsible for autocatalytic oxidation. Polyunsaturated lipids are highly susceptible to attack by such activated species and the resulting lipid peroxidation is undoubtedly of importance in oxygen toxicity to biological systems. Thus, the measurement of the inhibition of lipid peroxidation by carotenoids is an effective method for evaluating their ability as biological antioxidants. Lipid peroxidation generally consists of complex processes resulting in a wide variety of oxidation products. However, direct analysis of lipid hydroperoxides, primary products of lipid peroxidation, is surely valuable.[1,2] This analysis will give a kinetic approach to the antioxidant activity of carotenoids if hydroperoxides accumulate quantitatively through lipid peroxidation.

Here we describe the application of the method to the measurement of the peroxyl radical-trapping ability of carotenoids. The method is also applicable to the measurement of the singlet oxygen-quenching ability and the antioxidant activity of other compounds.[3]

The principle for the measurement of antioxidant activity is based on hydroperoxidation of polyunsaturated fatty acids (LH) via radical chain reaction, in which the thermal decomposition of an azo compound (A—N=N—A) gives chain-initiating peroxyl radical (AOO·) at a known and constant rate[4] (see Fig. 1). Peroxyl radicals (AOO·, LOO·) mediate chain reaction by abstracting a hydrogen to produce lipid radical (L·). The extent of oxidation can be measured by monitoring the production of hydroperoxides (LOOH).[5,6] The radical-trapping activity of carotenoids (Car) competes with LH for the chain-propagating peroxyl radical and hence inhibits the formation of hydroperoxides.

---

[1] J. Terao, I. Asano, and S. Matsushita, *Arch. Biochem. Biophys.* **238**, 326 (1984).
[2] J. Terao and S. Matsushita, *Free Radical Biol. Med.* **3**, 345 (1987).
[3] R. Stocker, A. F. McDonagh, A. N. Glazer, and B. N. Ames, this series, Vol. 186, p. 325.
[4] E. Niki, this series, Vol. 186, p. 100.
[5] Y. Yamamoto, E. Niki, and Y. Kamiya, *Bull. Chem. Soc. Jpn.* **55**, 1548 (1982).
[6] Y. Yamamoto, E. Niki, Y. Kamiya, and H. Shimasaki, *Biochim. Biophys. Acta* **795**, 332 (1984).

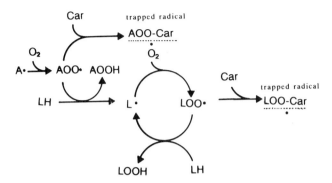

FIG. 1. Proposed mechanism of inhibition by carotenoids of peroxyl radical-mediated lipid peroxidation.

This competitive reaction with methyl linoleate in homogeneous solution (methyl linoleate hydroperoxide method) is designed for evaluating the inherent trapping ability of carotenoids. The reaction with phosphatidylcholine liposomes (phospholipid hydroperoxide method), a model of biomembrane lipids, however, is planned to understand how the structural property of carotenoids affects their trapping ability in the heterogeneous system consisting of aqueous and lipid phases.

Reagents and Solutions

Methyl Linoleate Solution. Commercial methyl linoleate (99% grade) is purified by column chromatography to remove preformed hydroperoxides. Methyl linoleate (10 g) is charged onto a glass column (20-mm i.d. × 200 mm) packed with Florisil (60–100 mesh; Floridin, NY) and eluted with n-hexane (200 ml). The eluent containing purified methyl linoleate is collected and the solvent is evaporated under reduced pressure. The amount recovered is determined by weighing. The purified methyl linoleate is dissolved in n-hexane–2-propanol (1:1, v/v) at 0.1 M concentration and stored at $-20°$.

Phosphatidylcholine Solution. Commercially available egg yolk phosphatidylcholine should be purified to remove contaminating hydroperoxides. Phosphatidylcholine (100 mg) is dissolved in chloroform–methanol–water (1:10:0.5, v/v/v, 1.0 ml) and then applied to a Lichroprep RP-8 column (40- to 63-μm, 10-mm i.d. × 240 mm; E. Merck AG, Darmstadt, Germany), which is connected to a high-performance liquid chromatography (HPLC) pump system. Phosphatidylcholine is eluted with the same solvent at a flow rate of 1.8 ml/min and fractions of 1.0 ml are collected. To verify the presence of phosphatidylcholine, an aliquot of each fraction is developed on HPTLC plates (RP-8F254; E. Merck AG) by the same solvent as above and exposed to iodine vapor (R_f values: 0.30 for phosphatidylcholine and 0.45 for phosphatidylcholine hydroperoxides). Phosphatidylcholine fractions are collected and evaporated under reduced pressure. After weighing, purified phosphatidylcholine is dissolved in chloroform and stored at $-20°$ until use.

Carotenoid Solution. A carotenoid for testing the antioxidant activity is dissolved in tetrahydrofuran or chloroform (5 mM) and stored at $-20°$ in the dark.

Radical Initiator. Lipid-soluble 2,2'-azobis(2,4-dimethylvaleronitrile) (AMVN) (Wako Pure Chemical Industries, Osaka, Japan) is dissolved in n-hexane (100 mM). Water-soluble 2,2-azobis (2-amidinopropane) hydrochloride (AAPH) (Wako Pure Chemical Industries) is dissolved in 0.01 M Tris-HCl buffer (pH 7.4, 110 mM). These initiator solutions are prepared freshly just before the oxidation is started.

Buffer Solution. Tris-HCl buffer (0.01 M, pH 7.4) containing 0.5 mM diethylenetriaminepentaacetic acid is prepared. This chelator is needed to prevent decomposition of the hydroperoxide due to contaminant metals from the buffer and other reagents.

Hydroperoxide Standards. Commercial methyl linoleate (0.5 g) is spread out at the bottom of glass vial (15-cm i.d.) and oxidized at 50° in the dark. It takes several days to accumulate hydroperoxides at a yield suitable for preparation.[7] Accumulation of methyl linoleate hydroperoxides is verified by spotting on silica thin-layer chromatography (TLC) plates and developing with n-hexane–diethyl ether–acetic acid (8:7:0.1, v/v/v). When the spot of hydroperoxides is clearly visualized at the lower position (R_f value, 0.55) by exposure to iodine vapor, the hydroperoxides are then purified from oxidized sample by silica gel column chromatography (silica gel 60, 60–230 mesh; column size, 20-mm i.d. × 250 mm.[8] Unoxidized methyl linoleate is eluted with the first solvent [n-hexane–diethyl ether, 95:5 (v/v), 200 ml] and methyl linoleate hydroperoxides with the second solvent [n-hexane–diethyl ether, 50:50 (v/v), 200 ml]. The second eluent is evaporated under reduced pressure and dissolved in hexane after weighing. This solution is stored at $-20°$ and used as the standard sample of methyl linoleate hydroperoxides. For the preparation of phosphatidylcholine hydroperoxides, commercial phosphatidylcholine from egg yolk is

---

[7] J. Terao and S. Matsushita, *Lipids* **21**, 255 (1986).
[8] J. Terao and S. Matsushita, *J. Am. Oil Chem. Soc.* **54**, 234 (1977).

subjected to photosensitized oxidation using methylene blue as a photosensitizer and the resulting hydroperoxides are purified on a Licroprep RP-8 column similar to the purification of phosphatidylcholine as mentioned above. The oxidation and purification procedures were described previously.[9] Hydroperoxides obtained are dissolved in chloroform and stored at $-20°$. These standard solutions are stable for at least 1 month.

Methyl Linoleate Hydroperoxide Method[10]

Oxidation Procedure. To 1.0 ml of methyl linoleate solution (0.1 M), carotenoid solution (tetrahydrofuran, 0.04–0.2 ml) is added and then preincubated at 37° for 5 min in the dark. The oxidation is initiated by adding AMVN solution (0.1 ml). The mixture is incubated with continuous shaking under air in the dark at 37°.

HPLC Procedure for the Determination of Hydroperoxides. At each 6- to 10-min interval, an aliquot of the reaction solution (10 μl) is withdrawn by a microsyringe and then injected into a normal-phase column immediately. A column of YMC-pack SIL (unbonded silica, 5-μm particle size, 6.0-mm i.d. × 150 mm; Yamamura Chemical Laboratories Co., Kyoto, Japan) is eluted with n-hexane–2-propanol (99:1, v/v) at a flow rate of 2.0 ml/min. The eluent is monitored by UV absorption at 235 nm for conjugated diene structure of the hydroperoxides. A typical HPLC chromatogram is shown in Fig. 2. The peaks are identified as (1) methyl 13-hydroperoxy-9Z,11E-octadecadienoate, (13-Z,E), (2) methyl 13-hydroperoxy-9E,11E-octadecadienoate (13-E,E), and (3) methyl 9-hydroperoxy-10E,12Z-octadecadienoate (9-E,Z) and methyl 9-hydroperoxy-10E,12E-octadecadienoate (9-E,E). These four conjugated diene hydroperoxides are formed by the peroxyl radical-mediated peroxidation of methyl linoleate.[11] The sum of these peak areas is obtained and the concentration of hydroperoxides in the reaction mixture is determined from the calibration curve of the hydroperoxide standard.

Phospholipid Hydroperoxide Method

Oxidation Procedure. Purified phosphatidylcholine solution (5.5 μmol) is introduced into a 10-ml screw-capped test tube and a known amount of carotenoid (0.011–0.055 μmol in chloroform) is then added. When the oxidation is designed to initiate within the liposomes, AMVN solution

---

[9] J. Terao, I. Asano, and S. Matsushita, *Lipids* **20**, 312 (1985).
[10] J. Terao, *Lipids* **24**, 659 (1989).
[11] E. N. Frankel, W. E. Neff, W. K. Rohwedder, B. P. S. Khambay, R. F. Garwood, and B. C. L. Weedon, *Lipids* **12**, 908 (1977).

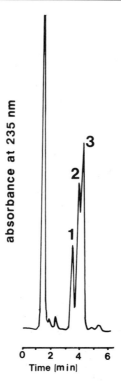

FIG. 2. Typical chromatogram of the reaction mixture for methyl linoleate hydroperoxide method. (1) 13-Z,E isomer; (2) 13-E,E isomer; (3) the mixture of 9-E,Z isomer and 9-E,E isomer.

(0.55 μmol) is also added in the test tube. The solvent is removed with a stream of nitrogen and finally under partial vacuum (10 min). Tris-HCl buffer (0.5 ml) is added at once, and the contents are vortexed for 1.0 min followed by ultrasonic irradiation with a B-12 ultrasonifier (Branson Sonic Power Co., Danbury, CT) (55 kHz) for 30 sec at ambient temperature. Tris-HCl buffer (0.6 ml) is added and this final mixture is incubated at 37° with continuous shaking. When the oxidation is designed to initiate on the surface of liposomes, AAPH is used instead of AMVN as radical initiator. In this case, AAPH is not included in the sonicated suspension. After sonication, Tris-HCl buffer (0.5 ml) is added and is preincubated in the dark at 37° for 5 min. Then the oxidation is initiated by the addition of AAPH solution (100 μl). Incubation of the liposomal suspension is carried out at 37° with continuous shaking.

HPLC Procedure for the Determination of Hydroperoxides. At each 8- to 10-min interval, an aliquot sample (10 μl) is withdrawn by microsyringe and immediately injected into a reversed-phase column (YMC-pack C_8,

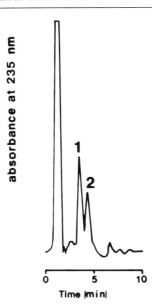

FIG. 3. Typical chromatogram of the reaction mixture for phospholipid hydroperoxide method. (1) Palmitoyl group-containing species; (2) stearoyl group-containing species.

octane-bonded silica, 5-μm particle size, 6.0-mm i.d. × 150 mm; Yamamura Chemical Laboratories Co.) that is eluted with methanol–water (95:5, v/v) at a flow rate of 2.0 ml/min. A UV detector is used to monitor the absorption at 235 nm for eluent. Figure 3 shows the chromatogram obtained from the oxidized phosphatidylcholine. Phosphatidylcholine hydroperoxides are separated into two peaks due to the difference between the molecular species containing a palmitoyl group and those containing a stearoyl group.[12] The two peak areas are summed up and the concentration of phosphatidylcholine hydroperoxides in the reaction mixture is determined by the calibration curve of the hydroperoxide standard.

Comments

First, some carotenoids give interfering peaks in the chromatogram of methyl linoleate hydroperoxides by the reaction with peroxyl radicals. However, the yields of (9-E,Z + 9-E,E) and (13-Z,E + 13-E,E) are always equal[10] and thus the peak areas of the hydroperoxides can be corrected from the uninterfered peak.

[12] J. Terao, S. S. Shibata, K. Yamada, and S. Matsushita, in "Medical, Biochemical and Chemical Aspects of Free Radicals" (O. Hayaishi, E. Niki, M. Kondo, and T. Yoshikawa, eds.), p. 781. Elsevier, Amsterdam, 1989.

Second, attention should be paid to the isomeric composition of hydroperoxides when the singlet oxygen-mediated photosensitized oxidation is applied to the preparation of standard phosphatidylcholine hydroperoxides. This reaction gives isomers containing both conjugated diene structures and nonconjugated diene structures,[8] although peroxyl radical-mediated peroxidation gives only conjugated diene structure-containing isomers. In addition, egg yolk phosphatidylcholine can be replaced by other commercially available products. In the case of soybean phosphatidylcholine, the species whose two acyl groups are both hydroperoxidized (dihydroperoxides) also appears together with hydroperoxides because of the abundance of the dilinoleoyl species.[13]

By using these methods, we have clarified the peroxyl radical-trapping ability of some carotenoids in solution[10] and in liposomal suspension.[14] This leads us to suggest that oxycarotenoids, so-called xanthophylls, are of importance as biological antioxidants.

[13] J. Terao, Y. Hirota, M. Kawakatsu, and S. Matsushita, *Lipids* **16**, 427 (1981).
[14] J. Terao, A. Nagao, J. H. Song, and D. K. Park, unpublished observation (1990).

[44] Antioxidant Radical-Scavenging Activity of Carotenoids and Retinoids Compared to α-Tocopherol

By MASAHIKO TSUCHIYA, GIORGIO SCITA, HANS-JOACHIM FREISLEBEN, VALERIAN E. KAGAN, and LESTER PACKER

Introduction

Free radicals are potentially dangerous for the cell,[1-8] although *in vivo* they may be generated by metabolism.[9] These radicals can be classified into two groups: (1) oxygen-derived radicals and active oxygen species,

[1] M. R. Clemens and H. D. Waller, *Chem. Phys. Lipids* **45**, 251 (1987).
[2] G. W. Teebor, R. J. Boorstein, and J. Cadet, *Int. J. Radiat. Biol. Relat. Stud. Phys., Chem. Med.* **54**, 131 (1988).
[3] A. Bindoli, *Free Radical Biol. Med.* **5**, 247 (1988).
[4] P. Hochstein and A. S. Atallah, *Mutat. Res.* **202**, 363 (1988).
[5] C. E. Vaca, J. Wilhelm, and M. Harms-Ringdahl, *Mutat. Res.* **195**, 137 (1988).
[6] T. L. Dormandy, *J. R. Coll. Physicians London* **23**, 221 (1989).
[7] M. Comporti, *Chem. Biol. Interact.* **72**, 1 (1989).
[8] D. Chiu, F. Kuypers, and B. Lubin, *Semin. Hematol.* **26**, 257 (1989).
[9] R. J. Youngman, *Trends Biochem. Sci.* **9**, 280 (1984).

including superoxide anion radical, hydroxyl radical, hydrogen peroxide, singlet molecular oxygen, and (2) organic radicals that usually originate in the course of lipid peroxidation. According to the location in which they are generated, oxygen-derived radicals are believed to react mainly with components in the aqueous phases such as extra- and intracellular fluids, while organic radicals interact mainly with constituents of the hydrophobic membranous phase.

The steady state concentrations of free radicals are known to be controlled by specific antioxidant enzymes (predominantly cytosolic) and by radical scavengers, antioxidants.[10] These antioxidants are also classified into two major groups according to their location and radical scavenging sites: water-soluble antioxidants (such as ascorbic acid, uric acid, and thiols) and lipid-soluble antioxidants (such as vitamin E, flavonoids, and ubiquinols).

Carotenoids and retinoids have been reported to act as peroxyl radical-scavenging antioxidants,[11-13] in addition to their recognized singlet oxygen-quenching ability.[14,15] However, no quantitative data on the antioxidant potency of carotenoids compared to other lipid-soluble antioxidants have been reported so far. This is probably due to the lack of a reliable methodology for quantitation of peroxyl radical-scavenging activity for lipid-soluble antioxidants.

The goal of the present research was to study the antioxidant activity of carotenoids and retinoids compared to α-tocopherol and its homologs. For the assay in the aqueous phase, a modified method of a well-known phycoerythrin fluorescence-based procedure developed by A. N. Glazer[16] was used, whereas for the assay in the hydrophobic phase a new *cis*-parinaric acid fluorescence-based assay was developed.

Experimental Procedures

Chemicals. 2,2'-Azobis (2-amidinopropane) dihydrochloride (AAPH) and 2,2'-azobis(2,4-dimethylvaleronitrile) (AMVN) were purchased from Polysciences, Inc. (Warrington, PA). *cis*-Parinaric acid was purchased from Molecular Probes (Junction City, OR). Carotenoids (β-carotene,

---

[10] B. Halliwell, *Free Radical Res. Commun.* **9**, 1 (1990).
[11] G. W. Burton and K. U. Ingold, *Science* **224**, 569 (1984).
[12] N. I. Krinsky, *Free Radical Biol. Med.* **7**, 617 (1989).
[13] M. Hiramatsu and L. Packer, this series, Vol. 190, p. 273.
[14] C. S. Foote and R. W. Denny, *J. Am. Chem. Soc.* **90**, 6233 (1968).
[15] C. S. Foote, *in* "Free Radicals in Biology" (W. A. Pryor, eds.), Vol. 2, p. 85. Academic Press, New York, 1976.
[16] A. N. Glazer, this series, Vol. 186, p. 161.

cryptoxanthin, lutein, canthaxanthin, lycopene, and zeaxanthin) and retinoids (retinyl palmitate, retinoic acid, and retinol) were purchased from Sigma Chemical Company (St. Louis, MO). Trolox, an α-tocopherol water-soluble homolog, was purchased from Aldrich Chemical Company (Milwaukee, WI). Other reagents were commercial products of analytical grade. B-Phycoerythrin was a gift from Prof. A. N. Glazer (Department of Microbiology, University of California, Berkeley). d-α-Tocopherol was a gift from Henkel Corp. (LaGrange, IL).

Phycoerythrin Fluorescence-Based Assay in Aqueous System

This assay is a modification of the method of A. N. Glazer.[16] A hydrophilic azo initiator, [HCl·NH=C(NH$_2$)—C(CH$_3$)$_2$—N=N—C(CH$_3$)$_2$—C(NH$_2$)=NH·HCl] (AAPH), thermally decomposes to produce the reactive peroxyl radicals at a constant rate:[17]

$$R-N=N-R \rightarrow [R \cdot N_2 \cdot R] \rightarrow 2eR \cdot + (1-e)R-R + N_2$$
$$R \cdot + O_2 \rightarrow ROO \cdot$$

The AAPH-induced peroxyl radicals oxidatively modify B-phycoerythrin. This is monitored by the loss of its characteristic fluorescence (excitation at 540 nm and emission at 575 nm). The reaction mixture (2 ml) consists of 5×10^{-10} M β-phycoerythrin, 25 mM AAPH (added in methanolic solution) in 75 mM sodium phosphate buffer (pH 7.4). Antioxidants are dissolved in absolute methanol or water (according to their solubility) and added to the reaction mixture after the initiation of peroxidation. Carotenoids and retinoids are first solubilized in taurodeoxycholic acid (20 mM) and then added to the incubation medium in the micellar form.[18]

cis-Parinaric Acid Fluorescence-Based Assay in Hexane

The assay is performed using a hydrophobic azo initiator of radicals, 2,2′-azobis(2,4-dimethylvaleronitrile) [CH(CH$_3$)$_2$—CH$_2$—C(CH$_3$)CN—N=N—C(CH$_3$)CN—CH$_2$—CH(CH$_3$)$_2$] (AMVN), which produces peroxyl radicals by its thermal decomposition in the same way as AAPH,[19,20]

---

[17] E. Niki, M. Saito, Y. Yoshikawa, Y. Yamamoto, and Y. Kamiya, *Bull. Chem. Soc. Jpn.* **59**, 471 (1986).
[18] H. Westergaard and J. N. Dietschy, *J. Clin. Invest.* **58**, 97 (1976).
[19] J. M. Braughler and J. F. Pregenzer, *Free Radical Biol. Med.* **7**, 125 (1989).
[20] E. Niki, this series, Vol. 186, p. 100.

and a hydrophobic reporting molecule, cis-parinaric acid, in a hydrophobic solvent, hexane. AMVN-induced peroxyl radicals oxidizes cis-parinaric acid, which is monitored by a decay of its characteristic fluorescence (excitation at 304 nm and emission at 421 nm). The reaction mixture (3 ml) contains 100 mM AMVN, which is added to the mixture as chloroform solution, and 30 μM cis-parinaric acid. Antioxidants dissolved in absolute methanol (α-tocopherol and Trolox) or chloroform (carotenoids and retinoids) are added to the incubation medium during the course of AMVN-induced fluorescence loss of cis-parinaric acid.

cis-Parinaric Acid Fluorescence-Based Assay in DOPC Liposomes

Dioleoylphosphatidylcholine (DOPC) liposomes with incorporated cis-parinaric acid and antioxidant are made by sonication of DOPC dispersion under nitrogen (1 min with a 30-sec interval, repeated three times). The reaction mixture (2 ml) contains 5 mM AMVN and 2.5 mM DOPC liposomes (containing 45 μM cis-parinaric acid and antioxidants at different concentrations) in 20 mM Tris-HCl buffer (pH 7.4). The addition of hydrophobic diazo initiator, AMVN, induces the peroxidation of liposome-incorporated cis-parinaric acid, which is monitored by its fluorescence decay as described above. The optimal excitation and emission wavelengths (excitation at 323 nm and emission at 421 nm) are slightly different from the values in hexane.

All the reactions are carried out at 40° in a thermostatted cuvette equipped with a magnetic stirring device. The data are expressed as mean ± SD.

Results

Phycoerythrin Fluorescence-Based Assay in Aqueous System

Exposure of the phycoerythrin to AAPH-derived peroxyl radicals induced an immediate and almost linear decrease in its fluorescence emission (Fig. 1). When water-soluble antioxidants (e.g., ascorbic acid) were added to the system, a complete protection of phycoerythrin against the peroxidative stress was achieved, and the fluorescence intensity remained constant until the antioxidant added had been consumed.[21] Low concentrations of absolute methanol, a solvent for antioxidants, had little effect on this reaction.

---

[21] A. N. Glazer, *FASEB J.* **2**, 2487 (1988).

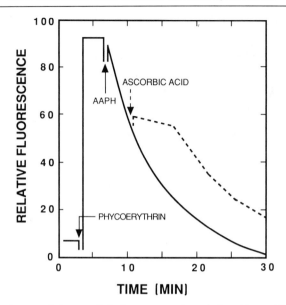

FIG. 1. Time course of phycoerythrin fluorescence decay induced by AAPH (excitation at 540 nm and emission at 575 nm with 5-nm slits, at 40°). The incubation system contained 5×10^{-10} M B-phycoerythrin and 25 mM AAPH in 75 mM sodium phosphate buffer (pH 7.4). The dashed line indicates the effect of ascorbic acid (10 μM).

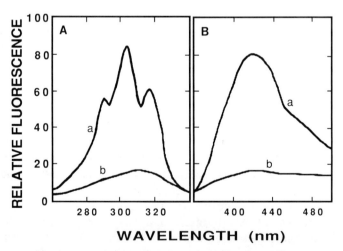

FIG. 2. Fluorescence excitation spectra (A) and emission spectra (B) of *cis*-parinaric acid before (a) and after (b) AMVN-induced peroxidation. The reaction mixture contained *cis*-parinaric acid (30 μM) and AMVN (100 mM) in hexane. Excitation spectra were recorded with a 5-nm slit and an emission wavelength of 421 nm, whereas the emission spectra were recorded with a 5-nm slit and an excitation wavelength of 304 nm, at 40°.

cis-Parinaric Acid Fluorescence-Based Assay in Hexane

In hexane, cis-parinaric acid has maxima at 304 and 421 nm in its fluorescence excitation and emission spectra, respectively (Fig. 2). The reaction with AMVN-derived peroxyl radicals results in disappearance of the fluorescence. Following the addition of AMVN, the fluorescence intensity of cis-parinaric acid decreased linearly until the end of the reaction (Fig. 3). This peroxidation process can be efficiently inhibited by hydrophobic antioxidants (e.g., α-tocopherol). Unless the antioxidant was not consumed only a small decrease in cis-parinaric acid fluorescence intensity was observed. Low concentrations of absolute methanol or chloroform, solvents for antioxidants, had little effect on this reaction.

Comparison of Radical-Scavenging Effects of α-Tocopherol and Trolox in Aqueous System and in Hexane

Figure 4 shows the comparison of radical-scavenging activity of α-tocopherol and Trolox, as measured in aqueous medium with phycoerythrin and in hexane with cis-parinaric acid. A hydrophobic antioxidant, α-tocopherol, did not protect against the AAPH-induced fluorescence loss of phycoerythrin. However, Trolox, a hydrophylic homolog of α-tocopherol, taken at the same concentrations, showed a potent protective effect against the oxidative damage of phycorythrin. The converse effects

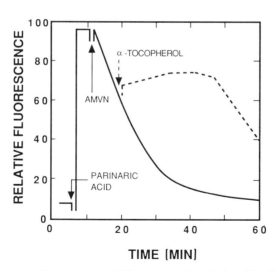

FIG. 3. Time course of cis-parinaric acid fluorescence decay induced by AMVN in hexane. The peroxidation of cis-parinaric acid was stopped by the addition of 250 μM α-tocopherol (dashed line). Other conditions were as in Fig. 2.

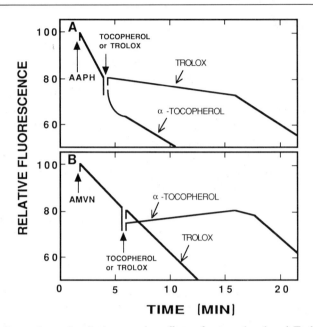

FIG. 4. Comparison of radical-scavenging effects of α-tocopherol and Trolox in (A) a phycoerythrin fluorescence-based assay and (B) a cis-parinaric acid fluorescence-based assay. In the phycoerythrin-based assay 10 μM Trolox or α-tocopherol was used; in the cis-parinaric acid-based assay 100 μM Trolox or α-tocopherol was used. Other conditions were as in Figs. 1 and 3.

were observed in the cis-parinaric acid fluorescence-based assay: α-tocopherol showed a pronounced protection of cis-parinaric acid against AMVN-induced damage while Trolox was inefficient.

Radical-Scavenging Effect of Carotenoids and Retinoids in Aqueous System and in Hexane

No delay in the AAPH-induced phycoerythrin fluorescence decay indicative of radical scavenging was observed on addition of carotenoids or retinoids to the aqueous assay system (Fig. 5A). However, these compounds slowed down the rate of fluorescence decay in a concentration-dependent manner (Fig. 5A and B). Retinyl palmitate had the greatest effect.

In contrast, carotenoids caused pronounced delay in the AMVN-induced cis-parinaric acid fluorescence decay in hexane (158 ± 6 sec by 1.7 μM β-carotene; Fig. 6). The same lag period was observed in other carotenoids. Retinoids showed no protective effect. In particular, retinyl palmitate did not affect the fluorescence decay at all.

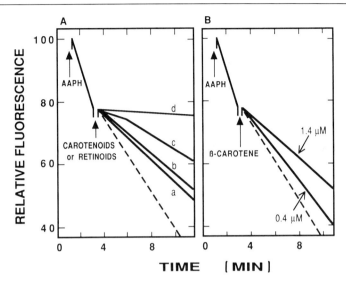

FIG. 5. The effect of carotenoids and retinoids (0.9 μM each) on phycoerythrin fluorescence decay (A) and the dose-dependent effect of β-carotene (B). The measurement with micelles that did not contain carotenoids and retinoids was performed as a control (dashed line). Other conditions were as described in Fig. 1. (a) Cryptoxanthin, β-carotene, zeaxanthin, retinol, lutein, and lycopene; (b) retinoic acid; (c) canthaxanthin; (d) retinyl palmitate.

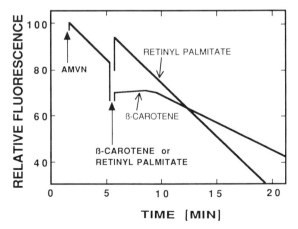

FIG. 6. The effect of β-carotene and retinyl palmitate (1.7 μM each) on cis-parinaric acid fluorescence decay in hexane. Other conditions were as in Fig. 4.

Radical-Scavenging Effects in Liposomal Membrane

Figure 7 shows the effects of AMVN and antioxidants on the fluorescence of cis-parinaric acid in the DOPC liposome system. Antioxidants were initially incorporated into liposomes containing cis-parinaric acid. The addition of AMVN induced peroxidation of cis-parinaric acid in liposomal membrane, resulting in its fluorescence decay (Fig. 7; control). Both Trolox and α-tocopherol produced a delay in the cis-parinaric acid fluorescence loss. α-Tocopherol was significantly more efficient than Trolox.

β-Carotene was very efficient in radical scavenging in liposomal system, whereas the effect of retinyl palmitate was much less pronounced.

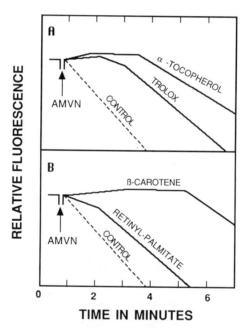

FIG. 7. The effect of antioxidants on cis-parinaric acid fluorescence decay in DOPC liposomes (excitation at 323 nm and emission at 421 nm with 5-nm slit, at 40°). The reaction mixture contained the following: 2.5 mM DOPC liposome [containing 45 μM cis-parinaric acid and (A) 100 μM Trolox or α-tocopherol; (B) 10 μM β-carotene or retinyl palmitate] and 5 mM AMVN in 20 mM Tris-HCl buffer (pH 7.4). Control measurement was made with DOPC liposome that did not contain antioxidants (dashed lines).

Discussion

The generators of free radicals may be located in the two different cell compartments: membranes and cytosol, i.e., in a hydrophobic or an aqueous phase, respectively. In accordance with this, the efficiency of free radical scavenging should be evaluated in these two phases for the hydrophobic and hydrophilic antioxidants. The phycoerythrin assay proved to be appropriate for the quantitations of the antioxidant efficiency in the aqueous phase.[16,21] However, in this assay both the source of peroxyl radicals (AAPH) and the reporting fluorescent detector (phycoerythrin) are water-soluble molecules. These features of the system may cause its inapplicability to quantitate the efficiency of free radical scavenging by hydrophobic antioxidants.

α-Tocopherol[22-25] and Trolox[26-28] have the same H-donating hydroxy group in their chromanol nucleus (which is directly involved in the radical scavenging) but differ significantly in their polarity, Trolox being a more hydrophylic homolog of α-tocopherol. In agreement with this, Trolox had a prominent radical-scavenging effect in the phycoerythrin system whereas α-tocopherol was inefficient. Similarly, carotenoids and retinoids did not show any reasonable structure–function-related radical-scavenging efficiency in the phycoerythrin-based assay. This was the impetus to develop an assay for quantitating the radical-scavenging activity of hydrophobic substances. Thus, we took advantage of a hydrophobic azo initiator of peroxyl radicals, AMVN, and a hydrophobic fluorescent reporting molecule, cis-parinaric acid,[29-33] to elaborate a hydrophobic assay system. cis-Parinaric acid with its four conjugated double bonds is extremely sensitive to peroxyl radical-induced oxidation (Fig. 8). The fluorescence decay of cis-parinaric acid in hexane due to its peroxidative destruction by AMVN-derived radicals was linear in time and could be continuously measured in this system. Antioxidant-dependent peroxyl radical scavenging prevented

---

[22] G. W. Burton and K. U. Ingold, *J. Am. Chem. Soc.* **103**, 6472 (1981).
[23] E. Niki, T. Saito, A. Kawakami, and Y. Kamiya, *J. Biol. Chem.* **259**, 4177 (1984).
[24] K. Mukai, Y. Watanabe, Y. Uemoto, and K. Ishizu, *Bull. Chem. Soc. Jpn.* **59**, 3113 (1986).
[25] K. Mukai, K. Fukuda, and K. Ishizu, *Chem. Phys. Lipids* **46**, 31 (1988).
[26] G. W. Burton, L. Hughers, and K. U. Ingold, *J. Am. Chem. Soc.* **105**, 5950 (1983).
[27] L. R. C. Barclay, S. T. Locke, J. M. MacNeil, J. VanKessel, G. W. Burton, and K. U. Ingold, *J. Am. Chem. Soc.* **106**, 2479 (1984).
[28] Y. Yamamoto, S. Haga, E. Niki, and Y. Kamiya, *Bull. Chem. Soc. Jpn.* **57**, 1260 (1984).
[29] L. A. Sklar, B. S. Hudson, and R. D. Simoni, *Proc. Natl. Acad. Sci. U.S.A.* **72**, 1649 (1975).
[30] L. A. Sklar, B. S. Hudson, M. Petersen, and J. Diamond, *Biochemistry* **16**, 813 (1977).
[31] L. A. Sklar, B. S. Hudson, and R. D. Simoni, *Biochemistry* **16**, 819 (1977).
[32] H. Matuo, T. Mitsui, K. Motomura, and R. Matuura, *Chem. Phys. Lipids* **30**, 55 (1981).
[33] W. J. Pjura, A. M. Kleinfeld, and M. J. Karnovsky, *Biochemistry* **23**, 2039 (1984).

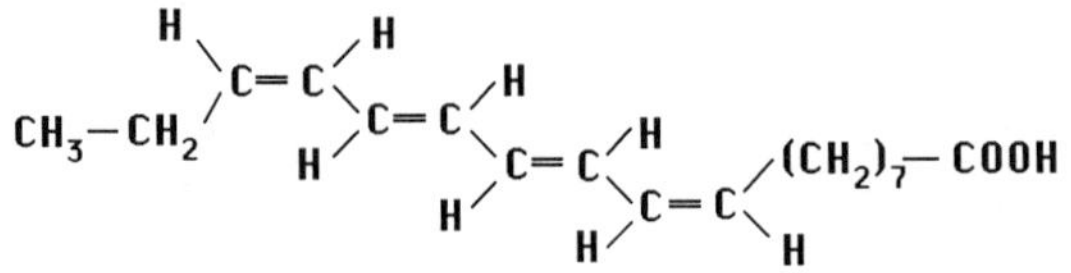

FIG. 8. Structure of cis-parinaric acid.

cis-parinaric acid fluorescence decay, thus providing an index for the antioxidant efficiency.

In this system α-tocopherol produced a delay in cis-parinaric acid fluorescence decay indicative of its efficient competitive scavenging peroxyl radicals. Based on the reported stoichiometry of peroxyl radical scavenging by α-tocopherol (2:1)[22-25] and the delay in fluorescence decay (1543 ± 15 sec by 250 μM α-tocopherol; Fig. 3) we can calculate the rate of peroxyl radical generation by AMVN. This estimation gives the rate of radical production a value of $(3.2 \pm 0.2) \times 10^{-6}$ M/[AMVN]·sec for the reaction in hexane at 40°, which is in good agreement with the data in the literature for AMVN decomposition in methanol at 37° (4.3×10^{-6} M/[AMVN]·sec).[19] The assumption as to α-tocopherol also gives the reaction stoichiometric value of carotenoids and retinoids with peroxyl radicals in this system by the following equation.

$$\text{Stoichiometry} = [\text{AMVN}][\text{R}][\text{time}]/[\text{antioxidant}]$$

where [AMVN] is the concentration of AMVN *(M)*, [R] is the rate of peroxyl radical production by AMVN, [time] is the duration of consumption of carotenoids or retinoids (sec), and [antioxidant] is the concentration of carotenoids or retinoids *(M)*. Consequently the value for β-carotene was 30.8 ± 0.8 mol peroxyl radicals consumed/mol, and the others are listed in Table I.

The water-soluble homolog of α-tocopherol, Trolox, which was efficient in the phycoerythrin system, produced no delay in AMVN-induced cis-parinaric acid fluorescence decay in hexane. Thus, the (AMVN + cis-parinaric acid) assay system is not applicable to water-soluble antioxidants, but gives adequate estimations of free radical-scavenging efficiency for hydrophobic antioxidants.

Carotenoids and retinoids differed qualitatively in their ability to produce a delay in cis-parinaric acid fluorescence decay: carotenoids showed very high radical-scavenging efficiency, while retinoids were inefficient. There was no substantial difference in the radical-scavenging efficiency between the carotenoids studied. We suggest that carotenoids with their 11 conjugated double bonds (isoprenoid chain + β-ionone rings) are unequiv-

TABLE I
REACTION STOICHIOMETRY OF CAROTENOIDS AND RETINOIDS
WITH PEROXYL RADICALS IN HEXANE

Compound	Reaction stoichiometry (mol peroxyl radicals consumed/mol)	Number of conjugated double bond
β-Carotene	30.8 ± 0.8	11
Cryptoxanthin	29.4 ± 1.2	11
Lutein	29.6 ± 1.6	11
Canthaxanthin	30.2 ± 0.8	11
Lycopene	27.2 ± 1.0	11
Zeaxanthin	31.4 ± 0.8	11
Retinyl palmitate	0.0	5
Retinoic acid	0.0	5
Retinol	0.0	5

ocal winners in their competition with cis-parinaric acid (4 conjugated double bonds) for AMVN-derived peroxyl radicals. In contrast, retinoids, having only 5 conjugated double bonds in their molecules, are not overwhelming competitors with cis-parinaric acid for peroxyl radicals, thus producing no quenching of peroxyl radicals (Table I). The extremely high stoichiometric value of β-carotene may reflect the possible involvement of different β-carotene oxidation products (with a still very high polyunsaturation index) in interactions with peroxyl radicals that dominate the reactvity of cis-parinaric acid.

The determination of the radical-scavenging efficiency in hexane gives an estimation of the chemical reactivity but does not report on the structural hindrances imposed by a possibly limited mutual accessibility of peroxyl radicals and scavenger molecules in the membrane lipid bilayer. To address this question, DOPC liposomes containing cis-parinaric acid were employed as a model membrane system for evaluation of the scavenging of peroxyl radicals induced by a hydrophobic azo initiator, AMVN. cis-Parinaric acid fluorescence was earlier reported to be sensitive to lipid peroxidation induced in red blood cells.[34,35] Due to the concentration of water-insoluble AMVN in a small volume of membranous liposomal phase we were able to reduce its concentration in this system without a significant decrease in its fluorescence decay rate.

[34] J. J. M. Van den Berg, F. A. Kuypers, J. H. Qju, D. Chiu, B. Lubin, B. Roelofsen, and J. A. F. Op den Kamp, *Biochim, Biophys. Acta* **944,** 29 (1988).
[35] J. J. M. Van den Berg, F. A. Kuypers, B. Roelofsen, and J. A. F. Op den Kamp, *Chem. Phys. Lipids* **53,** 309 (1990).

In the liposome system α-tocopherol was much more efficient than Trolox, which is in accord with our findings in hexane. However, Trolox did not show any free radical scavenging in hexane but produced some delay in cis-parinaric acid fluorescence decay in DOPC liposomes. This may be due to the partitioning of both Trolox and AMVN into polar surface regions of the lipid bilayer, where they could react to cause a delay in fluorescence decay.

β-Carotene demonstrated a free radical-scavenging potency in liposomes that was far beyond that of α-tocopherol. However, the mechanism of action of β-carotene as an antioxidant is considered to be quite different from that of phenolic or SH-containing antioxidants,[36] because β-carotene does not have these groups. Repetitive attacks by peroxyl radicals on β-carotene result in the fragmentation of the molecule and the formation of products that may impair membrane integrity and inhibit cell function. Moreover, once β-carotene is oxidized it cannot be regenerated by other antioxidants, like the recycling of vitamin E.[37] Therefore, β-carotene may not be as important as vitamin E in peroxyl radical scavenging in the cell.

[36] P. Di Mascio, M. E. Murphy, and H. Sies, *Am. J. Clin. Nutr.* **53**, 194S (1991).
[37] V. E. Kagan, E. A. Sherbinova, and L. Packer, *Arch. Biochem. Biophys.* **282**, 221 (1990).

[45] Epoxide Products of β-Carotene Antioxidant Reactions

By DANIEL C. LIEBLER and TODD A. KENNEDY

Carotenoids display antioxidant properties in several chemical and biochemical model systems.[1,2] Expression of these antioxidant effects in living organisms is thought to contribute to the photoprotective and anticarcinogenic properties of carotenoids.[3,4] However, because a variety of radical and nonradical oxidants may cause biological oxidative damage, carotenoids could inhibit oxidations by a variety of mechanisms. Carotenoids efficiently quench singlet oxygen,[5] yet also react rapidly with hydroxyl

[1] N. I. Krinsky, *Free Radical Biol. Med.* **7**, 617 (1989).
[2] N. I. Krinsky, *Pure Appl. Chem.* **51**, 649 (1979).
[3] W. F. Malone, *Am. J. Clin. Nutr.* **53**, S305 (1991).
[4] C. W. Boone, G. J. Kelloff, and W. F. Malone, *Cancer Res.* **50**, 2 (1990).
[5] C. S. Foote and R. W. Denny, *J. Am. Chem. Soc.* **90**, 6233 (1968).

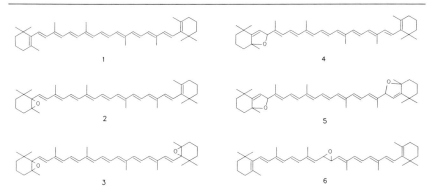

FIG. 1. Chemical structures of β-carotene and its epoxide oxidation products.

radicals[6] and peroxyl radicals.[7] By scavenging peroxyl radicals, which are the principal chain-carrying species in biological lipid peroxidation, carotenoids may function as chain-breaking antioxidants in membranes and other lipid-rich environments.[8]

β-Carotene (**1**, Fig. 1), the most widely studied carotenoid, has served as a prototype for examining the interactions of carotenoids with free radical oxidants. Previous studies of β-carotene autooxidation or cooxidation in organic solvents and plant oils indicate that multiple products result. Among those identified (see Fig. 1) were 5,6-epoxy-β,β-carotene (**2**), 5,6,5′,6′-diepoxy-β,β-carotene (**3**), and their corresponding furanoid rearrangement products, 5,8-epoxy-β,β-carotene (**4**) and 5,8,5′,8′-diepoxy-β,β-carotene (**5**).[9,10] Compounds **2** and **3** also result from the ferrous phthalocyanine-catalyzed oxidation and the lipoxygenase-catalyzed cooxidation of **1**.[11] In a recent study, peroxyl radicals generated in hexane by the thermolysis of azobis(2,4-dimethylvaleronitrile) (AMVN) oxidized **1** to epoxides **2** and **3** and 15,15′-epoxy-β,β-carotene (**6**), Fig. 1, a unique product not previously reported.[12] The reaction also oxidized **1** to a mixture of polar products that were not identified. This chapter describes methods for the production, isolation, and analysis of the major epoxide products that result from peroxyl radical trapping reactions of β-carotene.

---

[6] W. Bors, M. Saran, and C. Michel, *Int. J. Radiat. Biol.* **41**, 493 (1982).
[7] J. E. Packer, J. S. Mahood, V. O. Mora-Arellano, T. F. Slater, R. L. Wilson, and B. S. Wolfenden, *Biochem. Biophys. Res. Commun.* **98**, 901 (1981).
[8] G. W. Burton and K. U. Ingold, *Science* **224**, 569 (1984).
[9] A. H. El-Tinay and C. O. Chichester, *J. Org. Chem.* **35**, 2290 (1970).
[10] R. F. Hunter and R. M. Krakenberger, *J. Chem. Soc.* p. 1 (1947).
[11] J. Friend, *Chem. Ind. (London)* p. 597 (1958).
[12] T. A. Kennedy and D. C. Liebler, *Chem. Res. Toxicol.* **4**, 290 (1991).

AMVN-Dependent Oxidation of β-Carotene in Homogeneous Solution

Reagents. Compound **1** was obtained from Fluka Chemical Corp. (Ronkonkoma, NY) and was found by reversed-phase high-performance liquid chromatography (HPLC) analysis (see below) to contain > 95% all-*trans*-β-carotene. One or more cis isomers comprised the remaining material, which eluted in a single peak immediately following all *trans*-β-carotene. AMVN was from Polysciences, Inc. (Warrington, PA) and was used without further purification. Hexane for these experiments was HPLC grade (Baxter/Burdick & Jackson, Muskegon, MI).

Procedure

All procedures involving compound **1** and its oxidation products are carried out under reduced light. Compound **1** (50 mg, 93 μmol) and AMVN (23 mg, 93 μmol) are dissolved in 100 ml air-saturated hexane. The mixture is then incubated in the dark at 37° with gentle shaking. Aliquots of the reaction mixture are taken at intervals and evaporated to dryness under a stream of nitrogen, dissolved in HPLC mobile phase, and analyzed by reversed-phase HPLC (see below).

Epoxides **2** and **6** begin to accumulate immediately, along with polar products of compound **1** oxidation. The epoxides are further oxidized to polar products under the reaction conditions, however.[12] Under air at 37°, maximum epoxide yield occurs at about 12 hr, when total epoxide yield approaches 20% of compound **1** consumed. Epoxides are then gradually consumed by further oxidation over the following 12–24 hr. An oxygen atmosphere accelerates both compound **1** oxidation and further oxidation of the epoxides. In addition to oxidation, **1** also isomerizes to one or more *cis*-β-carotene isomers. Whether this process is free radical dependent is unclear.

HPLC of β-Carotene and Its Epoxides

Reversed-phase HPLC may be used to separate **1** and its oxidation products into at least three product fractions, each of which can be further resolved into individual components by cyano column HPLC. Compound **1** and its epoxide products are separated on a Whatman Partisil ODS-2, 10-μm, 4.6 × 250 mm analytical column (Alltech, Deerfield, IL). A Spherisorb ODS-2, 5-μm, 4.6 × 250 mm analytical column (Alltech) was also found to be satisfactory. Compounds were eluted from either column with methanol–hexane (85:15, v/v) at 1.5 ml/min. This reversed-phase HPLC system easily resolves **1** from the polar products (fraction A), the ring epoxides **2–5** (fraction B), and the 15,15′-epoxide **6** (fraction C) (Fig. 2).

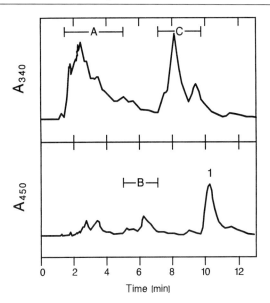

FIG. 2. HPLC chromatograms of 1 and its oxidation products with absorbance detection at 340 nm (top) and 450 nm (bottom). Elution of 1 and product fractions A, B, and C are indicated. [From T. A. Kennedy and D. C. Liebler, *Chem. Res. Toxicol.* **4**, 290 (1991). Used with permission.]

Epoxides 2-6 can be detected by absorbance, although their maxima differ. Table I lists the absorbance maxima for 1 and epoxides 2-6 in the reversed-phase HPLC mobile phase described above. In addition to the all-*trans*-carotenoids listed in Table I, *cis* isomers of carotenoids and their epoxides display an additional absorbance band at about 110 nm below their absorbance maxima.[13] Simultaneous detection of 1 and its epoxides requires a diode array detector, multiple-wavelength detector, or a programmable variable-wavelength detector. Of these detection methods, only diode array detection with continuous ultraviolet–visible (UV–VIS) spectral acquisition provides the spectroscopic information needed to distinguish closely related and isomeric products (see below).

Resolution of Epoxides 2 through 5

In the oxidation system described above, epoxide 2 is the principal product in fraction B (Fig. 2). However, lesser amounts of epoxide 5 appear to be formed during compound 1 oxidation and elute just prior to

---

[13] L. Zechmeister, A. L. LeRosen, W. A. Schroeder, A. Polgar, and L. Pauling, *J. Am. Chem. Soc.* **65**, 1940 (1943).

TABLE I
UV–VIS SPECTRAL DATA FOR COMPOUND 1 AND
ITS EPOXIDE PRODUCTS

Compound	Absorbance maxima[a,b]
1	275, 431, **450**, 475
2	267, 425, **445**, 471
4[c]	252, 404, **427**, 452
5	237, 385, **403**, 425
6[d]	**331**
6[e]	257, **329**

[a] Absolute spectral maxima are listed in boldface.
[b] All spectral data are in reversed-phase HPLC mobile phase (methanol–hexane, 85:15, v/v).
[c] Recorded in cyano column HPLC mobile phase (hexane–ethyl acetate, 99.5:0.5, v/v).
[d] trans-15,15'-Epoxy-β,β-carotene.
[e] cis-15,15'-Epoxy-β,β-carotene.

epoxide 2. Trace amounts of epoxides 3 and 4 may be formed by compound 1 oxidation, although we have not unequivocally detected these products in our studies. Diode array scans of the tailing portion of the epoxide 2 peak indicate the elution of one or more cis isomers of 2, which display an absorbance band at about 335 nm. Without the continuous UV–VIS scanning capability that diode array detection provides, it would not be possible to distinguish elution of *cis* isomers of 2 from simple peak tailing.

Improved resolution of epoxides 2–5 is achieved by HPLC on a Spherisorb CN, 5-μm, 4.6 × 250 mm analytical column (Alltech). The 5,6-epoxide 2 is resolved from several minor products (possibly *cis* isomers of 2 or 4) by elution with hexane–ethyl acetate (99.5:0.5, v/v) at 1.5 ml/min.

Resolution of Isomers of Epoxide 6

The 15,15'-*cis* and 15,15'-*trans* isomers of epoxide 6 are incompletely resolved on reversed-phase HPLC. These may be resolved by HPLC on a Spherisorb CN column with hexane–ethyl acetate (99.99:0.01, v/v) at 1 ml/min. The first isomer of epoxide 6 to elute (at about 3–4 min) is tentatively identified at the 15,15'-*cis*-epoxide, based on its relatively weak molar absorptivity at 330 nm (ϵ_{330} 6200 M^{-1} cm^{-1}) and the presence of a second, weak absorbance band at 258 nm. *cis*-Carotenoids typically display weaker molar absorptivities than their all-*trans* isomers.[13] The 15,15'-*trans* isomer elutes at about 6 min (ϵ_{330} 25,000 M^{-1} cm^{-1}).

Electrochemical Detection of β-Carotene and Its Epoxides

Sensitive HPLC detection of **1** and its epoxides can be achieved with an electrochemical detector. We have used an ESA model 5100A coulometric detector, equipped with a standard analytical cell (porous graphite electrode) and a standard guard cell (ESA, Inc., Bedford, MA). Amperometric detection with a thin-layer electrode has also been used to analyze **1** and retinol[14] and presumably would also be suitable for analyzing epoxides **2-6**. Compound **1** and its epoxides are analyzed on a Whatman Partisil ODS-2, 10-μm, 4.6 × 250 mm analytical column eluted with methanol–hexane (85:15, v/v) containing 20 mM sodium acetate. Hydrodynamic voltammograms for **1** and epoxides **2** and **6** are shown in Fig. 3. The half-wave potential for the series decreases with increasing length of polyene conjugation. Consequently, an electrode potential of approximately 1.0 V is required for optimum sensitivity in detecting **6**, whereas **2** produces maximum detector response at about 0.75 V. Approximate limits of detection for **2** and **6** in this system at a detector potential of 0.8 V are approximately 50 pg and 1 ng, respectively.

Storage and Stability of β-Carotene Epoxides

Like β-carotene, epoxides **2-6** all have extended conjugation and should be considered light and oxygen sensitive. All may undergo isomerization from all-*trans* to *cis* isomers. Isomerization reactions may also be catalyzed by trace acid. For this reason, these products should not be stored in chloroform, which usually contains traces of HCl. Epoxides **2** and **3** easily undergo acid-catalyzed rearrangement to their furanoid isomers **4** and **5**, respectively.[15] Epoxides **2-5** may be stored for about 1 week under argon at −20° without significant degradation, whereas epoxide **6** appears stable for up to 2-3 weeks under these conditions. HPLC analysis is a more reliable indicator of epoxide purity than UV–VIS spectroscopy. Older epoxide samples may contain significant amounts of polar decomposition products that are not readily detected by UV–VIS spectroscopy, particularly when only the spectral region above 350 nm is scanned.

[14] W. A. MacCrehan and E. Schonberger, *Clin. Chem.* **33**, 1585 (1987).
[15] K. Tsukida and L. Zechmeister, *Arch. Biochem. Biophys.* **74**, 408 (1958).

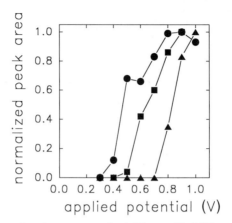

FIG. 3. Hydrodynamic voltammograms for **1** (●), epoxide **2** (■) and epoxide **6** (▲).

Mass Spectrometry of β-Carotene Epoxides

Mass spectrometric analysis of carotenoids is complicated by the facile thermal decomposition of these compounds under typical analysis conditions.[16] Nevertheless, epoxides **2–6** may be analyzed by direct probe mass spectrometry in the electron-impact mode. Gradual heating of the probe to cause sample desorption causes considerable sample decomposition and a substantial increase in the low mass background, although molecular ions and some major fragments may be discerned. Most of the sample desorbs at about 200°. To minimize thermal decomposition, the probe is heated rapidly (approximately 2–3 sec) from ambient temperature to 200°. Rapid sample desorption results and the first several scans at 200° produce acceptable spectra. High-resolution mass spectra of both the *cis* and *trans* isomers of epoxide **6** were obtained with this procedure.[12] Electron impact (EI) spectra for epoxides **2–6** are summarized in Table II. Although the base peak for all of these epoxides is the molecular ion (m/z 552), ring epoxides **2/4** and 15,15′-epoxide **6** produce distinctly different spectra. Ring epoxides **2** and **4** both produce a characteristic M − 80 loss, whereas epoxide **6** yields strong ions at m/z 378 and 349. Spectra for 5,6-epoxide **2** and its furanoid isomer **4** are virtually identical, which suggests that rearrangement may occur under the conditions employed.

[16] R. F. Taylor and M. Ikawa, this series, Vol. 67, p. 233.

TABLE II
MASS SPECTRAL DATA FOR β-CAROTENE
EPOXIDES

Compound	Major fragment ions, m/z (relative intensity)[a]
2	552 (M+, 100%), 550 (71), 472 (38), 406 (11), 336 (24), 205 (68), 177 (43), 165 (23)
4	552 (M+, 100%), 550 (76), 472 (43), 406 (14), 336 (27), 205 (49), 177 (60), 165 (9)
6	552 (M+, 100%), 550 (23), 534 (13), 378 (58), 349 (32), 255 (13)

[a] All compounds were analyzed in the electron impact mode at an ionizing voltage of 70 kV.

Acknowledgments

We thank Mr. Peter F. Baker for obtaining mass spectra. This work was supported by a Basic Research Support Grant from the College of Pharmacy, University of Arizona, and by the Small Grants Program, University of Arizona. T.A.K. was supported by NIH predoctoral training Grant ES07091-11.

[46] Techniques for Studying Photoprotective Function of Carotenoid Pigments

By MICHELINE M. MATHEWS-ROTH

An important function of carotenoid pigments is their ability to protect organisms that contain them against photosensitization. This was first demonstrated by Sistrom *et al.*,[1] who found that the carotenoids of photosynthetic bacteria protected the organisms against lethal photosensitization by their own chlorophyll. This photoprotective ability was soon found to apply to nonphotosynthetic bacteria as well,[2,3] and since then has been

---

[1] W. R. Sistrom, M. Griffiths, and R. Y. Stanier, *J. Gen. Physiol.* **48**, 473 (1956).
[2] R. Kunisawa and R. Y. Stanier, *Arch. Mikrobiol.* **31**, 146 (1958).
[3] M. M. Mathews and W. R. Sistrom, *Arch. Mikrobiol.* **35**, 139 (1960).

demonstrated in other organisms[4,5] and made use of in the treatment of human photosensitivity diseases.[6] This chapter will describe the techniques used to study carotenoid photoprotection in various classes of organisms. These techniques can be used to determine if carotenoids of interest can give protection, as well as to study the mechanisms of protection. The references cited are not meant to be all inclusive, but will give workers enough information to start experiments, and suggest where to begin a more thorough literature search.

Bacteria

Carotenoid protection against photosensitization by either endogenous or exogenous photosensitizers can be easily studied in bacteria. Examples of the former are the bacteriochlorophyll of photosynthetic bacteria and the flavins and porphyrins in nonphotosynthetic bacteria. An exogenous photosensitizer is any compound that can be added that will cause photodamage, such as various dyes (e.g., toluidine blue), and porphyrins.

It is first necessary to obtain colored carotenoidless ("white") mutants of the organism to be studied. It should be pointed out that the colorless carotenoid precursor, phytoene, does not confer significant photoprotection; thus it is not necessary to wipe out carotenoid production completely, just production of colored pigments. Mutants can be induced by either exposing suspensions of bacteria in buffer to sufficient levels of germicidal (UV-C) ultraviolet light to attain 99.99% kill, or exposing them to various chemical mutagens such as ethyl methane sulfonate.[7] Some pigmented bacteria will produce spontaneous pigment mutants. Standard bacteriological techniques are used to purify and store the cultures of the spontaneous or induced pigment mutants.[7]

It is important to make sure that the mutants contain no colored carotenoids. The pigments are detected by growing 1–2 liters of culture in the appropriate culture medium, centrifuging the cells, washing the cell mass with water, and extracting the carotenoids by adding absolute methanol to the cell paste and placing the mixture in a 50° water bath for 30 min and then centrifuging, repeating this step until the methanol layer is colorless. The absorption spectrum of the methanol is then taken in a sensitive

---

[4] D. D. Goldstrohm and V. G. Lilly, *Mycologia* **57,** 612 (1965).
[5] V. M. Koski and J. H. C. Smith, *Arch. Biochem. Biophys.* **34,** 189 (1951).
[6] M. M. Mathews-Roth, *Perspect. Biol. Med.* **28,** 127 (1984).
[7] P. Gerhart, R. G. E. Murray, R. N. Costilow, E. E. Nester, W. A. Wood, N. R. Krieg, and G. B. Phillips, "Manual of Methods for General Bacteriology." American Society for Microbiology, Washington, D.C., 1981.

spectrophotometer. For the mutant to be classified as a colorless mutant, there should be no absorption peaks above 350 nm. One can also study the photoprotective ability of pigments having absorption maxima between 350 and 400 nm: one often finds mutants with pigments absorbing in this range, with no absorption above 400 nm. However, if these are to be studied, an effort should be made to separate the individual carotenoids present and to identify them by the absorption spectra. Techniques for the separation and identification of carotenoids are described elsewhere.[8]

An alternative method to achieve depletion of colored carotenoids is to grow the bacteria in the presence of diphenylamine (DPA), which interrupts the carotenoid biosynthetic pathway just after phytoene. A final concentration of 10^{-4} M is used in the culture medium. A stock solution of 10^{-2} M DPA is made in 95% ethanol, and a 1:100 dilution is made into the culture medium to attain the proper final concentration. Various organisms may differ in the amount of DPA that is effective in inhibiting pigment synthesis: it is wise to opt for the least amount of DPA necessary to accomplish blockage. Here also, DPA-treated cells should be extracted to verify that no significant amount of pigment is produced.

For photobiological experiments to determine the protective ability of carotenoids, cultures of the wild type and mutant(s) to be studied are harvested in the logarithmic phase of growth, centrifuged, and resuspended in phosphate or Tris buffer to contain about 10^6 to 10^7 cells/ml. If an exogenous photosensitizer is to be used, such as toluidine blue or rose bengal (2.5×10^{-6} M final concentration), it is added at this time and the suspension is mixed thoroughly. Solutions of photosensitizer are made to a concentration of 10^{-4} M in water, autoclaved or filtered to sterilize, and the appropriate volume added to the medium to get the proper final concentration. The suspension of cells and photosensitizer (if added) is divided equally between two flat-sided bottles (such as Cat. No. K882750-6360; Kontes Glass Co., Vineland, NJ), one to be exposed to light and the other to be covered to serve as a dark control. They are placed in a glass-sided aquarium fitted with a constant-temperature control. Each bottle is fitted with a tube and air, filtered through a cotton plug, is bubbled into the cell suspension throughout the period of light exposure. If exogenous photosensitizers are to be studied, the amount of light used is about 10,700 lx (17.4 J/sec/m^2): if endogenous photosensitizers are studied, the amount of light is much greater, at about 645,600 lz (1.04 kJ/sec/m^2). Here also, the amount of light needed may differ with the organism studied: preliminary experiments are needed to determine the correct level

---

[8] B. H. Davies, in "Chemistry and Biochemistry of Plant Pigments" (T. W. Goodwin, ed.), p. 38. Academic Press, New York, 1976.

of light exposure to use. The temperature of the water bath is either the optimum temperature for the organism under study, or 0–4°, if studies of the initial carotenoid light-induction reaction are to be done. To determine the level of photokilling, samples are taken from both dark control and light-exposed bottles just before the start of exposure, and at intervals during light exposure diluted appropriately, plated on culture media appropriate for the bacteria studied, and incubated for 4 days at the optimal growth temperature for the organism. The plates are then counted, and the decrease over time of the logarithm of the viable count is calculated and plotted.

Other Microorganisms

Unicellular algae are easily grown in liquid culture and many mutants can be induced by UV exposure[9]: there is probably no need to resort to mutagens, but they could certainly be used. Algal carotenoids could also be inhibited by diphenylamine. Carotenoidless mutants are photosensitive, the endogenous photosensitizer being their chlorophyll.

Carotenoid-containing yeast and fungi (e.g., *Ustilago* and *Neurospora*) can also be studied[10,11]: these pigments also protect the cells against photokilling by exogenous and endogenous photosensitizers. In both cases mutants can be easily induced by UV exposure, or pigments inhibited by diphenylamine. The cells contain endogenous photosensitizers, or exogenous ones can be added.

Some studies have been performed with ciliates, adding carotenoids dissolved in ethanol to the water in which the organisms are grown to obtain photoprotection.[12]

The techniques described above for photobiological experiments on bacteria can also be used with these groups of organisms.

Invertebrates

Flies, mosquitoes, and *Daphnia* have been used to study photosensitization and photoprotection. Endogenous carotenoids in the insect eye have been studied, and carotenoids have been added to the diet.[13–15] For flies,

[9] M. B. Allen, T. W. Goodwin, and S. Phagpolngarm, *J. Gen. Microbiol.* **23**, 93 (1960).
[10] O. H. Will, N. A. Newland, and C. R. Reppe, *Curr. Microbiol.* **10**, 295 (1984).
[11] P. L. Black, R. W. Tuveson, and M. L. Sargent, *J. Bacteriol.* **125**, 616 (1976).
[12] K. C. Yang, R. K. Prusti, E. B. Walker, P. S. Song, M. Watanabe, and M. Furuya, *Photochem. Photobiol.* **43**, 305 (1986).
[13] J. R. Robinson and E. P. Beatson, *Pestic. Biochem. Biophys.* **24**, 375 (1985).
[14] W. E. Bennett, J. I. Maas, S. A. Sweeney, and J. Kagan, *Chemosphere* **15**, 781 (1986).
[15] H. Zhu and K. Kirschfeld, *J. Comp. Physiol.* **154**, 153 (1984).

feeding beef muscle results in carotenoid-depleted insects, and feeding beef liver leads to carotenoid enrichment. Carotenoids dissolved in dimethyl sulfoxide (DMSO) can also be added to the water in which *Daphnia* and other such animals are grown. One could conceivably also add carotenoid-containing "beadlets" to the water (see the section, Mammals, below for a discussion of methods for using beadlets). The advantage of using insects and other small invertebrates is that they can be grown in large numbers, easily exposed to light, and apparently can accumulate through their diet enough carotenoids to obtain photoprotection.

Mammals

Carotenoid photoprotection has been studied in mammalian cells and in whole animals (rodents). Patients with photosensitive skin diseases such as erythropoietic protoporphyria and polymorphic light eruption have also been studied.

Various kinds of fibroblasts, including mammary cells and erythrocytes from protoporphyric patients have been studied; the former have been used to determine the ability of carotenoids to prevent various kinds of cellular and nuclear damage, and the latter to see if carotenoids can prevent photohemolysis. The main difficulty in these systems is to make sure that adequate amounts of carotenoids are absorbed by the cells. Dissolving the pigments to be studied in a solvent such as *n*-hexane[16] or DMSO and adding an appropriate dilution (10^{-5} to 10^{-6} M) seems to be effective in either tissue culture cells or erythrocytes. An additional method that has also proved effective in photohemolysis studies, and should also work with other cells, is to use serum from individuals who have ingested large amounts of β-carotene, and who have serum levels of at least 600 μg/dl, in the cell culture medium in the place of calf serum.[17] One can also dissolve β-carotene beadlets (Hoffmann-La Roche, Nutley, NJ) in a water emulsion[18] and add appropriate dilutions to the cell culture medium. This may be sterilized by filtering through a 0.8-μm filter and then through a 0.22-μm filter (Millipore, Bedford, MA). The carotenoid concentration is calculated from the absorption spectrum after filtering. As the beadlets contain antioxidants that may have some actions similar to the carotenoids under study, parallel experiments must also be performed with placebo beadlets, also available from Hoffmann-La Roche. Measure the spectra of both preparations at 280 nm (at which wavelength carotenoids do not absorb) after filtration to equalize the amounts of placebo and carotenoid-containing beadlets.

[16] S. Som, M. Chatterjee, and M. H. Banerjee, *Carcinogenesis* **5**, 937 (1984).
[17] G. Swanbeck and G. Wennersten, *Acta Derm.-Venereol.* **53**, 283 (1973).
[18] A. Pung, J. Rundhaug, C. H. Yoshizawa, and J. S. Bertram, *Carcinogenesis* **9**, 1533 (1988).

For animal work, the best method is to add carotenoids to the diet. Although intraperitoneal injections of emulsified carotenoids have worked in some cases, the trauma to the animals from repeated injections, which would be necessary to achieve sustained elevated blood or tissue levels of carotenoids, is too great, thus the method cannot be recommended. Carotenoid-containing beadlets (Hoffmann-La Roche), added to powdered rodent chow to given final concentrations of carotenoids from 0.1 to 3% of the diet, have been successful in attaining significant blood and tissue levels of the pigments, and preventing UV light-induced cancer development.[19] Here also placebo beadlets must be included in the chow of the control animals. Either kind of beadlet is mixed thoroughly with the powdered food and placed in the cages in containers made specially for powdered food. The same final concentration of placebo beadlets as carotenoid-containing beadlets is used. We have found that up to 3% final concentration of carotenoids (30% beadlets) does not cause toxic side effects in mice. The beadlets contain carbohydrate and protein, and thus have nutritional value.

Patients with light-sensitive skin diseases react to light because of the endogenous photosensitizers characteristically present in their conditions. The presence of photosensitivity can be measured by exposing them in the laboratory to measured amounts of light (phototesting).[20] Carotenoids, specifically β-carotene, the only pigment approved by the U.S. Food and Drug Administration for human use at this time, are administered by mouth, the usual adult dose being 60 mg, three times a day. Improvement in light tolerance is usually measured by phototesting, or by carefully designed questionnaires administered before treatment starts and during carotenoid ingestion.

Conclusions

In this chapter we have given descriptions of how to perform experiments in various kinds of organisms to study the ability of carotenoid pigments to protect against photosensitization. There is still much to be studied about this function, especially its likely mechanism, the ability of the pigments to quench excited species. Because excited species may also be involved in cellular phenomena other than photosensitization (one possible example being cancer induction), the importance of carotenoids to cellular well-being may indeed be widespread. Thus continued studies of this most important function of carotenoids are well warranted.

[19] M. M. Mathews-Roth, *Pure Appl. Chem.* **57,** 717 (1985).
[20] M. M. Mathews-Roth, E. H. Kass, T. B. Fitzpatrick, M. A. Pathak, and L. C. Harber, *Arch. Dermatol.* **115,** 1381 (1979).

Author Index

Numbers in parentheses are footnote reference numbers and indicate that an author's work is referred to although the name is not cited in the text.

A

Aakermann, T., 244
Abd, M., 166
Achenbach, H., 366
Achiwa, K., 78
Adams, A. J., 360
Adams, A., 230
Adkins, J., 419
Adler, A. D., 90
Adler, A. J., 221, 224(11, 12), 226(11), 227(11), 228(11), 229(12), 230(12), 337, 338(12), 339, 341, 342(12), 360
Afshari, E., 425
Agnew, M. P., 189
Agroud, M., 162
Aitzetmüler, K., 14
Ajans, Z. A., 391
Al-Sheikhly, M., 448, 451
Alam, B. S., 417, 418(67), 419(67)
Alam, S. Q., 417, 418(67), 419(67)
Albert, E. P., 159
Alberts, D. S., 198, 280
Albrecht, M. G., 32
Allam, H., 391
Allen, J. P., 39
Allen, M. B., 482
Almgren, M., 448
Ames, B. N., 410, 412(47), 415(47), 429, 454
Aminin-Saghafi, T., 421
Amstrong, R. J., 78, 83(14)
Andersson, B. A., 373
Andreoni, L., 266
Andrews, A. G., 75
Anno, T., 155
Anton, J. A., 98, 99
Arai, A., 364
Arbers, I., 104
Armarego, W. L. F., 80
Armitt, G. M., 100

Arnaboldi, A, 266
Arpin, N., 366
Asano, I., 454, 457
Ashikawa, I., 296, 299
Assour, J., 90
Atallah, A. S., 460
Aubry, J. M., 432
Auran, J. D., 221, 360
Avron, M., 337, 439, 441, 443(7, 9)
Azegami, K., 367

B

Babler, J. B., 78
Bailey, C. A., 189
Baker, A., 166
Baker, B. R., 79
Baker, J. T., 189
Bakhtiari, M., 39
Balun, J. E., 391
Banerjee, M. H., 483
Barber, M. S., 61
Barbón, P. G., 101, 104, 107
Barclay, L. R. C., 469
Bardat, F., 62
Barger, W. R., 453
Barrera, M. R., 208, 209(29), 349, 350(14)
Barrett, D., 87, 91(1)
Bartlett, D. L., 64
Barua, A. B., 186, 189, 197(4), 199(4), 265, 273, 274, 275(5), 276(3, 5)
Bashor, M. M., 206
Bashor, M., 185, 189
Basu, T. K., 143
Bates, A. L., 430, 437
Batres, R. O., 186, 189, 197(4), 199(4), 265, 274, 275(5), 276(5)
Bauernfeind, J. C., 3, 143, 146(3), 147(3), 160

Bayanyai, M., 251
Beatson, E. P., 482
Beaudry, J., 198
Beecher, G. R., 113, 114, 123, 149, 151(32), 152, 153(47), 158(41), 192, 197(18), 194, 203, 206, 208, 209(13, 14, 19, 22, 28, 29, 30), 211, 212(14), 214(13), 218, 252, 273, 276(2), 294, 323, 336, 337(1), 338(1), 347, 349, 350, 351(16), 352, 353, 354(9), 355(9), 357(8)
Beinayme, A., 160
Belasco, J. G., 101
Belford, R. E., 452
Ben, K. E., 160
Ben-Amotz, A., 337, 439, 441, 443(7, 9)
Ben-Aziz, A, 258
Benasson, R. V., 87, 91(1)
Bendich, A., 175, 266, 322, 323(1), 391, 403, 419
Bennett, W. E., 482
Bensasson, R. V., 87, 91(1), 448, 449(8)
Bensasson, R., 306
Berg, P. C., 453
Bergen, H. R. III, 284
Bergen, H. R., 49, 52(1a), 53(1b)
Bergman, M., 417, 418(68), 419(68)
Bergquist, P., 385
Bernhard, K., 42, 232
Bernstein, H. J., 104
Berry, R. A., 252, 258(19)
Berset, C., 129, 132, 134, 135(8), 136, 137, 139(4, 11), 140, 141(12), 195
Berthet, D., 54
Bertram, J. S., 176, 266, 483
Beyer, P., 62, 63, 64, 73
Bhiwapurkaar, S., 54
Biacs, P. A., 152, 154
Bianchi, A., 266
Bickers, D. R., 406, 413(20), 414(20), 415(20)
Bieman, K., 111
Biener, Y., 440
Bieri, J. G., 206, 219(12)
Bindoli, A., 460
Bittersmann, E., 87, 91(1), 452
Bjørnland, T., 170, 232, 239
Black, P. P., 482
Blaizot, P. R., 165
Blaizot, P., 160
Blakely, S. R., 419

Boekenoogen, H. A., 159
Boey, P. L., 165
Bogoni, P., 159
Boguth, W., 408, 409(33)
Boland, W., 63
Bolt, J. D., 306, 308(5)
Bondi, A., 182
Bone, R. A., 220, 229(4), 230(4), 336, 360, 361(2)
Bonnett, R., 233, 237(16)
Boone, C. W., 472
Boorstein, R. J., 460
Booth, V. H., 147, 314
Borch, G., 323
Bornstein, B., 143, 147(1)
Borowitzka, L. J., 439
Borowitzka, M. A., 439
Bors, W., 408, 409(37), 451, 473
Botnen, J., 186, 189, 265
Bousfield, I. J., 367
Boussiba, S., 387
Boutin, D., 161
Bowen, P. E., 189
Bownds, D., 274
Bradbury, J. H., 158, 159(54), 372
Brandon, C., 85
Braslavsky, S. E., 435
Brauer, H.-D., 427
Braughler, J. M., 462, 470(19)
Braumann, T., 151
Braun, A. M., 420, 421, 421, 422, 425, 426(17), 427(2), 429(11), 430, 437
Brawn, K., 405
Brett, M., 101, 172
Brgon, P. G., 106
Britton, G., 3, 25, 75, 100, 101, 104(13, 14), 107, 167, 168(3), 169(3), 170(3), 171(3), 208, 209(21), 246, 258, 266, 314, 375
Brixius, L., 160
Brockmann, H., 315, 317(10), 318(10)
Brody, S. S., 266
Broger, E., 208, 209(25)
Brown, E. D., 206, 219(12), 267
Brown, P. K., 221, 274
Brown, P. R., 154, 199
Brown-Thomas, J. M., 187, 191(11)
Brubacher, G. B., 429
Brush, A. H., 313, 314, 315
Buchecker, R., 31, 208, 209(23, 24), 382
Buchwald, M., 101, 104(9), 108(9), 110(9)

Buckley, J. D., 143, 205, 347, 429
Budil, D. E., 306
Budowski, P., 182
Budzikiewicz, H., 43
Buettner, G. R., 421
Bunnell, R. H., 143, 147(1)
Bureau, J. L., 149, 152(34), 153(34), 157(34), 159(34)
Burger, W., 45, 46(4)
Burgess, M., 189
Burr, J., 392
Burton, G. W., 143, 175, 403, 404, 406(8), 407, 408(4, 31), 409(4, 8), 410(31), 412(4), 415(4), 419(4), 451, 461, 469, 470(22), 473
Busch, K. L., 251
Bushway, R. J., 149, 152, 153(34, 45), 157(34), 159(34), 189, 193, 194(23)
Butrum, R., 145
Butte, N. F., 391
Buttner, W. J., 78
Buxton, G. V., 446
Byteva, I. M., 427

C

Cabrini, L., 412, 413(51)
Caccamese, S., 323, 326(7), 327(7), 329(7), 333(7)
Cadenas, E., 431
Cadet, J., 460
Callender, R., 101
Calloway, D. H., 391
Camara, B., 62
Campbell, T. C., 266
Canfield, L. M., 391, 392
Cannone, P., 432
Cantilena, L. R., 206, 280, 398
Capella, P., 373
Carevale, J., 120
Carew, T. E., 412
Carey, P. R., 104
Carlsson, D. J., 431, 437
Carnevale, J., 175, 178(8)
Carter, F. L., 453
Cavina, G., 189
Ceccaldi, M. J., 101, 104(11)
Chachaty, C., 87, 91(1)
Chadwick, B. W., 306, 308, 310(6), 311(6)

Chandler, L. A., 148, 296
Chang, Y. C., 421
Chang, Y., 405, 430
Chapman, J. R., 112
Chappell, J. E., 391
Charbonnet, C., 425, 426(17)
Chatterjee, A., 407, 408(30), 409(30), 410(30)
Chatterjee, M., 483
Chauvet, J. P., 448
Chavez, M., 145
Cheah, K. Y., 166
Cheesman, D. F., 100, 101, 104(2, 11)
Chen, B. H., 189
Chen, T. S., 152, 157
Chen, T., 203
Chichester, C. O., 132, 144, 175, 208, 209(21), 324, 408, 409(32), 410(32), 473
Chignell, C. F., 421
Chirno, A., 39
Chiu, D., 460, 471
Cho, S. Y., 251
Chodat, F., 386
Choo, Y. M., 160, 161, 162(75), 164(74), 165(74), 166, 167
Chopra, A. K., 75
Choudhry, S. C., 45, 46(4)
Christensen, R. G., 187, 191(11)
Chumanov, G. D., 32, 41(9)
Clandinin, D. R., 282
Clandinin, M. T., 391
Clark, E. P., 405
Clark, L., 392
Clemens, M. R., 460
Clifford, A. J., 282, 283(10, 11), 284, 290(11)
Coenen, J. W. E., 162
Cogdell, R. J., 305, 306, 308, 310(6), 311(6), 312(2)
Cogdell, R., 168
Cohen-Bazire, G., 387
Cole, E. R., 54, 120, 175, 178(8), 408, 409(34)
Collins, J. G., 372
Comporti, M., 460
Conn, P. F., 430, 437
Conner, M. W., 205
Connolly, J. S., 87, 91(1)
Cooney, J. J., 252, 258
Cooney, R. V., 176
Corey, E. J., 78

Cornwell, D. G., 412
Costa, S. M. De B., 306, 308(5)
Costes, C., 133
Costilow, R. N., 480
Cotter, R. G., 323, 336(8)
Cotter, R. J., 113
Cotton, T. M., 32, 36, 37, 38(15), 39, 41
Craft, N. E., 187, 189, 190, 191(11), 193(14), 194, 196, 197(29), 198(25), 199, 200(37), 201, 203, 204(25), 267
Craft, N., 330
Crank, G., 175, 178(8)
Creighton, J. A., 32
Criel, G. R., 253, 262(20)
Critchley, P., 135
Cronin, S. E., 251
Crouzet, J., 57, 58, 59(12), 61(13), 62(12, 13)
Crow, F. W., 112
Cullum, M. E., 282
Curl, A. L., 318
Cuvier, P., 160
Czeczuga, B., 316, 386
Czygsan, F.-C., 386

D

D'Aquino, M., 450
D'Auria, M. V., 385
Daitch, C. E., 218
Daood, H. G., 152, 154
Daumas, R., 101
David, J. B., 61
Davies, B. H., 190, 246, 387, 481
Davies, D. I., 78
Davis, T. P., 198
Dawe, E. A., 448
Day, W. C., 54
De Graziano, J. M., 87, 91(1)
de las Rivas, J., 101, 104(16)
De Leenheer, A. P., 162, 189, 192, 206, 252, 253, 262(20), 265(17), 282, 318
De Ritter, E., 190, 314, 352
De Ruyter, M. G. M., 253, 282
De Schryver, F. C., 87, 91(1)
DeGraziano, J. M., 452
Delori, F. C., 221, 360, 362(4)
Delva, B. R. A., 162
Demchenko, A. I., 160
Demole, E., 54

Deneke, S. M., 266, 406, 412(22), 413(22)
Denisenko, Y. I., 159
Denny, R. W., 405, 421, 429, 430, 461, 472
Des Marais, D. J., 251
Desty, D. H., 341
Devasagayam, T. P. A., 438
Dexter, D. L., 309
Di Mascio, P., 266, 323, 405, 430, 432, 435(11), 437, 438, 472
Diamond, J., 469
Dickerson, J. W., 144
Dieber-Rotheneder, M., 413
Dieringer, S. M., 258
Dietschy, J. M., 177
Dietschy, J. N., 462
Diment, J., 208, 209(27)
Dixit, R., 406, 413(20), 415(20)
Doi, Y., 55
Doleiden, F. H., 431, 437
Doll, R., 143, 205, 347, 429
Donnelly, S. K., 230, 360
Dormandy, T. L., 460
Downs, J., 254
Drar, A., 391
Dratz, E. A., 220, 225(5), 227(5), 228(5), 229(5), 230(5), 269, 336, 339, 341(3), 360, 362(3), 365(3), 429
Drawert, F., 54
Driskell, W. J., 189, 206
Driskell, W., 185, 189
Droop, M. R., 386
Dunagin, P. E., 281
Duncan, M. J., 387
Dunster, C., 450
Duval, S., 395

E

Eckey, E. W., 166
Efremov, E. S., 32
Efremov, R. G., 32, 41(9)
Egawa, M., 166
Egger, H.-J., 282
Egorov, S. Y., 425
El Naggar, B., 391
El-Tinay, A. H., 132, 175, 408, 409(32), 410(32), 473
Elahi, M., 408, 409(34)
Endo, Y., 166

F

Englert, G., 31, 61, 75, 111, 114(6), 115(6), 119(6), 132, 133(2), 218, 232, 239, 294, 301, 323, 367, 382
Enzell, C. R., 111
Epler, C. S., 330
Epler, K. S., 194, 198(25), 201, 203, 204(25)
Erdman, J. G., 54
Esterbauer, H., 413
Esteves, W., 162
Eugster, C. H., 31, 208, 209(23, 24)

F

Fahrenholtz, S. R., 431, 437
Falk, H., 63
Farhataziz, 444
Farmilo, A., 421, 430, 437
Farr, A. L., 387
Farrow, P. W., 322, 323(3), 326(3)
Feher, G., 39
Felber, M., 308
Fenselau, C., 323, 336(8)
Fenton, T. W., 282
Férézou, J. P., 75, 83(7), 86(7)
Fernandez, L., 220, 229(4), 230(4), 336, 360, 361(2)
Ferris, A. M., 391
Ferris, F. L. III, 220
Field, F. H., 112
Finarelli, F. D., 90
Findlay, J. B. C., 101
Findlay, J. C. B., 172
Fine, S. L., 220
Firdovich, I., 405
Fisher, C., 152, 203
Fisher, J. F., 252
Fitch, K., 269
Fitzpatrick, T. B., 484
Fleischmann, M., 31
Foote, C. S., 404, 405, 421, 426, 429, 430, 437, 442, 461, 472
Francis, G. W., 111, 246, 248
Francis, T., 391
Frank, H. A., 305, 306, 307, 308, 311(7, 8), 312(2)
Frankel, E. N., 457
Fresno, J. A., 106
Frick, J., 312

Fridovich, I., 405
Friend, J., 135, 408, 410(43), 473
Froehling, P. E., 159
Frohring, P. R., 165
Frye, C. L., 78
Fujimori, E., 431, 437
Fujiwara, T., 17
Fukasawa, C., 206
Fuke, K., 426
Fukuda, K., 469, 470(25)
Fukuda, S., 367, 370(13), 379
Furihata, K., 375, 376(12), 378(7)
Furr, H. C., 49, 52(1a), 186, 189, 197(4), 199(4), 265, 273, 274, 275(5), 276(3, 5), 277, 282, 283, 284, 290(11)
Furrow, G., 211, 350, 351(16)
Furuya, M., 482

G

Gallinella, B., 189
Gao, F., 87, 91(1), 452
Gárate, A. M., 101, 104, 107
Garozzo, D., 323, 326(7), 327(7), 329(7), 333(7)
Garrido-Fernandez, J., 159
Garwood, R. F., 457
Gassmann, E., 421
Gast, P., 306
Gawienowski, A. M., 162
George, S. A., 252, 262(7)
Gerhard, F. P., 162
Gerhart, P., 480
Gerspacher, M., 75
Gey, K. F., 429
Gilmore, A. M., 198, 346
Girotti, A. W., 404
Glazer, A. N., 410, 412(47), 415(47), 454, 461, 462(16), 463, 469(16, 21)
Glinka, C. J., 192
Glinz, E., 232
Goh, S. H., 160, 161, 162(75), 164(74), 165(74), 166
Goldmacher, J., 90
Goldsmith, T. H., 318, 369
Goldstrohm, D. D., 480
Goli, M. B., 206, 208, 209(13, 14, 19, 29), 211(14), 212(14), 214(13), 218(13, 19),

273, 276(2), 336, 337(1), 338(1), 347, 349, 350(5, 14)
Gómez, R., 101, 104, 107
Goodwin, T. W., 3, 25, 133, 146, 147, 167, 168, 169(1, 3), 170(1, 3), 171(3), 175, 231, 232(4), 258, 265, 266(2), 482
Goodwin, W., 386
Gore, R. T., Jr., 78
Govindan, C. K., 408, 409(35)
Grady, C. M., 162
Granatica, P., 437
Grandolfo, M. C., 14
Grass, H., 208, 209(18)
Green, P. N., 367
Greenstock, C. L., 446
Gregory, G. K., 152, 203
Gressel, J., 441
Griffiths, M., 479
Grimme, L. H., 151
Grossman, S., 408, 409(38), 417, 418(68), 419(68)
Grundmann, H. P., 266
Gunson, H. H., 266
Gunter, E. W., 189, 206
Gurinovich, G. P., 427
Gust, D., 87, 91(1), 306, 308(6), 310(6), 311(6), 448, 449(8), 452
Gutteridge, J. M. C., 406

H

Hadorn, M., 208, 209(31), 210(31)
Haga, S., 408, 469
Hager, A., 369
Hajdu, F., 152, 154
Halevy, O., 413, 414(59)
Hall, R. D., 421
Halliwell, B., 406, 461
Hama, I., 165, 166, 167
Hamada, K., 17
Handelman, G. J., 198, 220, 221, 224(11, 12), 225, 226(6, 11, 15), 227(5, 11), 228(5, 11), 229(5, 12), 230(4, 5, 11, 12), 265, 269, 273, 276(1), 336, 337, 338(12), 339, 341, 342(12), 360, 362(3), 365(3), 407, 408(30), 409(30), 410(30), 429
Handelman, G., 198, 203
Hankin, J., 145

Hansen, H.-J., 421, 429(11), 430, 437
Hara, N., 166, 167
Harashima, K., 31, 375, 378(7)
Harber, L. C., 484
Harding, L. O., 87, 91(1)
Harkay-Vinkler, M., 154
Harms-Ringdahl, M., 460
Harold, E. P., 159
Harris, J. C., 322, 323(3), 326(3)
Harrison, D. E. F., 254
Harshall, J. H., 367
Hartley, C. W. S., 160, 161(69)
Hasegawa, T., 337
Hashimoto, H., 246, 253, 296, 300, 301, 305
Haugan, J. A., 232, 234(10)
Haxo, F. T., 14, 231, 232, 233, 242(2)
Hayashi, H., 85, 305
He, Y., 266
Heald, R., 36, 39(13), 41(13)
Heasley, V. L., 78
Heathcock, C. H., 83
Heilbron, I. M., 232, 233(13)
Heindze, I., 54
Heinonen, M. I., 150, 154, 155(35), 157(35), 158(35), 159(35)
Helfenstein, A., 232
Heller, S. R., 121
Helman, W. P., 446
Hendra, P. F., 31
Hertzberg, S., 75, 79(3), 86(3), 208, 209(20), 362, 384, 385
Hervé Du Penhoat, C., 79
Hidaka, T., 155
Hidalgo, F. J., 417, 418(66)
Higosaki, N., 281
Hildebrandt, P., 38
Hino, R., 17
Hiramatsu, M., 461
Hirata, J. A., 232
Hirata, Y., 13, 15, 17, 23, 25, 460
Hites, R. A., 112
Ho, W.-T., 427, 428, 430, 437
Hochstein, P., 460
Hoff, A. J., 308
Hoffman, M. Z., 452
Holden, J., 208, 209(29), 349, 350(14)
Holman, R. T., 182, 373
Holt, R. E., 36, 37, 38(15), 39(13, 15), 41
Holzwarth, A. R., 87, 91(1)
Honig, B., 101

Hopkinson, J. M., 392
Horgiguchi, M., 281
Hoschke, A., 154
Hoshiga, I., 166
Hosokawa, N., 337
Hosomi, M., 296, 300, 301
Hostettman, K., 314
Hostettman, M., 314
Hsu, H. C., 90
Hu, M.-L., 417, 418(69), 419(69)
Hubbard, R., 274
Hudon, J., 315
Hudson, B. S., 469
Huff, D. L., 189, 206
Hughers, L., 469
Hughes, L., 407, 408(31), 410(31)
Hung, J. G. C., 373
Hünig, S., 79, 86(19)
Hunt, D. F., 112, 121
Hunter, D. J., 205
Hunter, R. F., 408, 473
Hurst, J. R., 425, 426(19), 427
Husbands, M., 391
Hussein, L., 391
Hyeon, S. B., 54
Hyman, L., 220

I

Ibers, J. A., 99
Iizuka, H., 375
Ikawa, M., 251, 282, 478
Ikediobi, C. O., 408, 410(45)
Imai, T., 55
Imanishi, J., 337
Imfeld, M., 208, 209(25)
Imsgard, F., 244, 379, 381(17)
Ingold, K. U., 175, 403, 407, 408(4, 31), 409(4), 410(31), 412(4), 415(4), 419(4), 451, 461, 469, 470(22), 473
Inoue, M., 406, 413(24), 414(24), 416(24), 417(24), 418(24)
Inoue, Y., 296, 299
Isaksen, M., 246, 248
Ishidsu, J., 367, 370, 375, 376, 378(10, 11), 379, 382(10), 385(11)
Ishizu, K., 469, 470(24, 25)
Isler, O., 3, 13, 44, 51, 146, 208, 280, 349
Isoe, S., 54

Ito, M., 13, 15, 17, 232
Itoh, M., 426
Iwata, K., 305
Iwata, T., 15
Izumimoto, H., 166

J

Jackman, L. M., 61
Jacobs, N. J., 254
Jacobs, P. B., 22
Jamikorn, M., 386
Javanovic, S. V., 448
Jeanmaire, D. L., 32
Jeffrey, S. W., 233
Jenks, W. P., 101, 104(9), 108(9), 110(9)
Jennings, W., 55
Jenny, T. A., 427
Jensen, A., 25, 233, 374
Jensen, R. G., 391, 394
Jensen, S. L., 362
Jermstad, S. F., 282
Johansen, J. E., 14, 49, 384, 385(22)
Johnson, E. L., 186, 187(7)
Jonckheere, J. A., 253, 262(20)
Jones, A. D., 282, 283(10), 284
Jori, G., 435
Jos, J. S., 158, 159(56)
Jose, A. T. Q., 162
Jucker, E., 3, 54, 61(9), 233
Judd, D. H., 221
Julia, M., 75, 79, 81(5), 83(7), 86(7)
Junga, I. C., 79

K

Kaegi, H., 45
Kagan, J., 482
Kagan, V. E., 472
Kaiser, S., 266, 323, 405, 430, 435(11), 437, 438
Kakizono, T., 386
Kam, T. S., 166
Kamalov, V. F., 425
Kamiya, Y., 408, 454, 462, 469, 470(23)
Kanaji, M., 296, 299, 300(1, 2), 301(1), 305(1, 2)
Kanasawud, P., 57, 58, 59(12), 61(13), 62(12, 13)

Kanner, J., 408, 409(39, 40, 41)
Kanofsky, J. R., 404
Kaplan, L. A., 206, 429
Kaplan, L., 186, 189, 199(6)
Kapur, N. S., 54
Karnovsky, M. J., 469
Karrer, P., 3, 54, 61(9), 232, 233
Kass, E. H., 484
Katayama, N., 246, 300, 301(6)
Kates, K., 372
Kates, M., 384
Kato, G., 31
Kato, H., 55
Kato-Jippo, T., 296, 300
Katrangi, N., 206
Katsuyama, M., 23, 25
Katz, A., 439
Katz, J. J., 14
Katzenellenbogen, J. A., 78
Kawaguchi, A., 373
Kawakami, A., 469, 470(23)
Kawakatsu, M., 460
Ke, B., 244, 379, 381(17)
Kelloff, G. J., 472
Kellogg, E. W., 405
Kennedy, T. A., 407, 408(29), 409(29), 410(29), 412, 415(29), 473, 474(12), 478(12)
Kenner, G. W., 90
Kensler, T. W., 416
Khachik, F., 113, 114, 123, 149, 152, 151(32), 153(47), 158(41), 192, 194, 197(18), 203, 206, 208, 209(13, 14, 19, 22, 28, 29, 30), 211, 212(14), 214(13), 218, 252, 273, 276(2), 294, 323, 336, 337(1), 338(1), 347, 349, 350, 351(16), 352, 353, 354(9), 355(9), 357(8)
Khambay, B. P. S., 75, 457
Khoo, J. C., 412
Kidokoro, K., 281
Kienzle, F., 13
Kikendall, J. W., 189
Kim(Jun), H., 406, 413(21), 414(21), 415(21)
Kimura, O., 337
King, D., 408, 409(38)
Kinoshita, K., 367
Kinsella, J. E., 408, 409(39, 40, 41)
Kirkland, J. J., 186, 187(8)
Kirschfeld, K., 221, 482

Kito, M., 296, 299, 300(2), 305(2)
Kjøsen, H., 14, 231, 233(2), 242(2), 244, 379, 381(17)
Klaus, M., 282
Klein, B. P., 408, 409(38)
Kleinfeld, A. M., 469
Kleinig, H., 62, 63, 64, 366, 367, 374(2)
Kleo, J., 306
Kliegl, R., 230, 360
Kline, M. C., 187, 191(11)
Knafo, G., 165
Knight, E. V., 419
Knipping, G., 413
Knowles, J. R., 101
Kobayashi, M., 386
Koch, H. J., 372
Kocis, J. A., 152, 203
Koelle, E. U., 282
Koglin, E., 32
Kohmura, E., 337
Koike, H., 296, 299
Koivistoinen, P. E., 150, 155(35), 157(35), 158(35), 159(35)
Komagata, K., 373
Komiya, H., 39
Komori, K., 25
Komori, T., 23, 25
Konishi, R., 337
Korhals, H. J., 251
Koroteev, N. I., 425
Korsakoff, L., 90
Koski, V. M., 480
Kothe, G., 312
Koyama, Y., 246, 253, 291, 294(5), 296, 299, 300, 301, 305, 357
Krakenberger, R. M., 408, 473
Krane, J., 244
Krasnovsky, A. A., Jr., 421, 422(9), 425, 427, 437
Kreuz, K., 62
Krieg, N. R., 480
Krinsky, G. J., 221, 224(11)
Krinsky, N. I., 3, 143, 144(8), 175, 186, 198, 220, 221, 226(6, 11, 15), 227(5, 11), 228(5, 11), 230(5, 11), 265, 266, 267(1), 268, 269(1, 24), 272(1), 273, 276(1), 304, 305, 318, 336, 337, 338(12), 340, 341, 342(12), 360, 403, 404, 406, 407, 408(3, 30), 409(3, 30), 410, 411(48), 412, 413(22), 415, 416(48, 60), 419,

430, 431, 437, 442, 451, 461, 472
Krinsky, N., 198, 203, 369
Kröncke, U., 73
Krstulovic, A., 154, 199
Kubota, W., 165
Kubota, Y., 165
Kunert, K.-J., 417, 418(65)
Kunisawa, M. M., 387
Kunisawa, R., 479
Kurashige, M., 406, 413(24), 414(24), 416(24), 417(24), 418(24)
Kurata, T., 55
Kusin, J. A., 145
Kuypers, F. A., 471
Kuypers, F., 460

L

La Roe, E. G., 54
Lachenmeier, A., 208, 209(31), 210(31)
Lafferty, J., 448, 449(7)
Lambert, W. E., 253, 282
Lamberts, J. J. M., 421
Lamola, A. A., 416, 431, 437
Land, E. J., 306, 448, 449
Land, L. J., 448, 449(8)
Land, R. I., 337
Landi, L., 412, 413(51)
Landrum, J. T., 220, 229(4), 230(4), 336, 360, 361(2)
Lasorenko-Manevich, R. M., 32
Lau, J. M., 429
Lau, S. Y. M., 100
Lauren, D. R., 189
Lausen, N. C. G., 265, 267(1), 268, 269(1, 24), 272(1)
Lavens, P., 253, 262(20)
LeBoef, R. D., 22
Lee, M.-J., 199
Lee, S.-H., 431, 437
Lee, S.-J., 87, 91(1), 452
Lee, T. C., 208, 209(21)
Lee, W. L., 100, 104(2)
Lee, Y.-K., 386
Leibovitz, B., 417, 418(69), 419(69)
Lenz, F., 203
Lepper, M., 205, 347
LeRosen, A. L., 475, 477(13)
Lers, A., 337, 439, 440

Lesellier, E., 140, 195
Lexa, D., 448, 449(8)
Liaaen-Jensen, S., 11, 14, 25, 49, 75, 79(3), 86(3), 111, 208, 209(20), 231, 232, 233(2), 234(10), 239, 242(2), 244, 246, 319, 323, 342, 349, 351(11), 369, 374, 375, 379, 381(17), 384, 385
Liddell, P. A., 87, 91(1), 306, 308(6), 310(6), 311(6), 448, 449(8)
Liebler, D. C., 407, 408(29), 409(29), 410(29), 412, 415(29), 473, 474(12), 478(12)
Liebman, A. A., 45, 46(4)
Liedvogel, B., 63
Lilienfeld, A. M., 220
Lilly, V. G., 480
Lim, C. L., 145, 147(15), 150, 154(36), 155(36), 156(36), 157(36), 158(36), 159(36)
Lim, H. T., 161
Lima, A. F., 392
Lindsay, D. A., 407, 408(31), 410(31)
Lindsey, J. S., 90
Linkola, E. K., 150, 155(35), 157(35), 158(35), 159(35)
Little, R. G., 99
Liu, S., 205, 347
Llewellyn, C. A., 252
Loach, P. A., 98, 99
Lock, G., 432
Locke, S. T., 469
Loeber, D. E., 208, 209(27)
Loeblich, L., 439
Lomnitski, L., 417, 418(68), 419(68)
Longo, F. R., 90
Lowry, O. H., 387
Lu, T., 36, 38(15), 39(15), 41
Lubin, B., 460, 471
Lukac, T., 208, 209(26)
Lunde, K., 291
Lusby, W. R., 113, 152, 158(41), 194, 206, 208, 209(13, 14, 19, 22, 28, 29), 211(14), 212(14), 214(13), 218, 252, 273, 276(2), 323, 336, 337(1), 338(1), 347, 349, 350(5, 8, 13, 14), 352(8), 357(8)
Luttrull, D. K., 452
Luttrull, D., 87, 91(1)
Lux-Neuwirth, O., 416
Lyubushkin, V. T. I. V., 159

M

Ma, X. C., 87, 91(1), 452
Maas, J. I., 482
Maathews-Roth, M. M., 302
Maatsushita, S., 454
Macarulla, J. M., 101, 104
MacCrehan, W. A., 187, 191(11), 198, 203, 477
Machlin, L. J., 269
Machnicki, J., 306, 308, 311(7)
MacNeil, J. M., 469
Madden, H., 75
Mahood, J. S., 408, 409(36), 451, 473
Makings, L. R., 87, 91(1), 306, 308(6), 310(6), 311(6)
Malarek, D. H., 45, 46(4)
Mallans, A. K., 233, 237(16)
Malone, W. F., 472
Mamuro, H., 165
Mamuro, Y., 165
Manitto, P., 421, 437, 438
Mantoura, R. F. C., 252
Manz, U., 314, 317(4)
Maoka, T., 23, 25, 364
Marbet, R., 48, 208, 209(18)
Marenchic, I. G., 322, 323(3), 326(3)
Marinyuk, V. V., 32
Markmen, A. L., 160
Marks, H. W., 258
Marletta, G. P., 203
Marmur, J., 85
Marsh, A. C., 147, 148(25), 185
Marston, A., 314
Martin, G. S., 392
Marty, C., 129, 132, 134, 135(8), 136, 137, 139(4, 11), 140, 141(12), 195
Marvel, C. S., 432
Mathews, M. M., 479
Mathews-Roth, M. M., 3, 144, 265, 266, 267(1), 268, 269(1, 24), 272(1), 403, 431, 437, 480, 484
Mathews-Roth, Y. C., 421
Mathis, P., 306
Matsuno, T., 22, 23, 25, 31, 172, 266, 364
Matsushita, S., 412, 416(50), 456, 457, 459, 460
Matuo, H., 469
Matus, Z., 192, 203, 251
Matuura, R., 469
Maudinas, B., 306
Maurer, R., 132, 133(2), 294
Maurette, M.-T., 420, 427(2)
May, W. E., 187, 191(11), 330
Mayer, H., 13, 51, 208, 280, 349
Mayer, M. P., 64
Mayer, M., 63
Mayne, S. T., 271, 417, 418(70)
McCombie, S. W., 90
McCommams, S. A., 22
McCormick, A., 233, 237(16), 369
McCurdy, H. D., 366
McDonagh, A. F., 454
McDonald, J. D., 427
McDonough, A. F., 410, 412(47), 415(47)
McEwen, C. N., 121
McGann, W. J., 306, 307, 311(8)
McGee, D. L., 392
McMurry, J. E., 21
McNaughton, D. E., 189
McQuillan, A. J., 31
McVie, J., 449
Meduna, V., 132, 133(2), 294
Meister, W., 367
Merck, E., 189
Merritt, K., 254
Merry, A. H., 266
Metiu, H., 32
Mettal, U., 63
Metz, J. G., 41
Meunier, J., 161
Meyer, J. C., 266
Michel, C., 408, 409(37), 451, 473, 366
Mikhitaryan, S. S., 159
Miki, W., 405, 406(19), 411(19), 413(19), 414(19), 417(19), 418(19)
Milborrow, B. V., 75
Milicua, J. C. G., 101, 104, 107
Millar, T. J., 416
Miller, J. C., 252, 262(7)
Miller, J., 186, 189, 199(6)
Miller, K. W., 189
Milne, D. B., 265
Milne, D., 186, 189
Milne, G. W. A., 121
Min, D. B., 431, 437
Minale, L., 385
Minguez-Mosquera, M. J., 159
Miozzi, M. S., 336
Mitsui, T., 469

Miyamoto, T., 367
Miyata, A., 296, 299, 300
Möbius, D., 32
Moens, L., 253, 262(20)
Moissonnier, J. P., 140
Monéger, R., 62
Monger, T. G., 308
Monroe, B. M., 436
Monti, D., 421, 437, 438
Monties, B., 133
Moore, A. L., 87, 91(1), 306, 308(6), 310(6), 311(6), 448, 449(8), 452
Moore, T. A., 87, 91(1), 306, 308(6), 310(6), 311(6), 448, 449(8), 452
Moorthy, S. N., 158, 159(56)
Mora-Arellano, V. O., 408, 409(36), 451, 473
Mordit, R. C., 407, 408(31), 410(31)
Morf, R., 232
Morrison, W. R., 342
Moss, G. P., 75
Motomura, K., 469
Mott, D. J., 269
Motten, A. G., 421
Moualla, S., 415, 416(60)
Muccino, R. R., 45, 46(4)
Mukai, K., 469, 470(24, 25)
Mukhtar, H., 406, 413(20), 414(20), 415(20)
Mundy, A. P., 75
Munson, M. B. S., 112
Murakami, H., 412, 416(50)
Murakoshi, M., 337
Murasecco-Suardi, P., 421, 429(11), 430, 437
Murata, N., 371, 373(20)
Murov, S. L., 426
Murphy, M. E., 431, 438, 472
Murray, K. F., 54
Murray, R. G. E., 480

N

Nabiev, I. R., 32, 41(9)
Nagai, S., 386
Nagao, A., 460
Nagasawa, H., 337
Nagayama, K., 296, 300
Nair, R. B., 158, 159(56)
Nakada, H., 375
Nakahara, J., 31
Nakamura, A., 166
Nakamura, M., 167
Nakanishi, K., 85
Nakasato, S. J., 165
Nakata, M., 305
Nakatsu, S., 155
Nam, H. S., 251
Nechaev, A. P., 159
Neckers, D. C., 421
Neese, J. W., 206
Neese, J., 185, 189
Neff, W. E., 457
Nelis, H. J. C. F., 162, 189, 206, 253, 262(20), 318
Nelis, H. J. C., 192
Nelis, H. J., 252, 253, 265(17)
Nelson, D., 120
Nemeth, G. A., 87, 91(1), 448, 449(8)
Nester, E. E., 480
Newberne, P. M., 205
Newland, N. A., 482
Ng, J. H., 162, 252
Nierenberg, D. W., 206, 280, 398
Nievelstein, V., 73
Niki, E., 408, 454, 462, 469, 470(23)
Ninomiya, T., 281
Nishino, H., 337
Nishizawa, T., 22
Noga, G., 203
Nonell, S., 435
Norden, D. A., 101
Norgård, S., 14, 231, 232, 233(2), 242(2)
Norris, J. R., 306
Nuccio, E., 230, 360
Nussbaum, J. J., 360, 362(4)

O

Obendorf, M. S. W., 438
Ogilby, P. R., 426
Oglesby, P., 205, 347
Oh, J. J., 306, 308(6), 310(6), 311(6)
Ohata, S., 145
Ohmacht, R., 192, 203
Oishida, A., 165
Okabe, A., 165
Okabe, T., 166

Okimasu, E., 406, 413(24), 414(24), 416(24), 417(24), 418(24)
Okuda, S., 373
Okuzumi, J., 337
Olie, J. J., 162
Oliveros, E., 420, 421, 422, 425, 426(17), 427(2), 429(11), 430, 437
Ollilainen, V., 150, 155(35), 157(35), 158(35), 159(35)
Olson, J. A., 49, 52(1a), 53(1b), 143, 186, 189, 197(4), 199(4), 265, 273, 274, 275(5), 276(3, 5), 281, 282, 283(11), 284, 290(11), 322, 323(1), 391
Omololu, A., 145
Oncley, J. L., 412
Ong, A. S. H., 160, 161, 162(75), 164(74), 165, 166, 167
Ooi, C. K., 161, 162(75), 164(74), 165(74), 166, 167
Ooi, T. L., 165
Ookubo, M., 22, 31
Op den Kamp, J. A. F., 471
Ortiz de Montellano, P., 37
Osorio, H. S., 386
Ostrea, E. M., 391

P

Packer, J. E., 408, 409(36), 451, 473
Packer, L., 461, 472
Packer, R. S., 429
Page, H., 232
Palcic, B., 266
Palmisano, A. C., 251
Palozza, P., 410, 411(48), 412(48), 415, 416(48, 60)
Pappin, D. J. C., 101, 172
Park, D. K., 460
Parker, R. S., 271, 417, 418(70)
Parrott, M. J., 78
Parson, W. W., 308
Parthasarathy, S., 412
Pasquali, P., 412, 413(51)
Patel, A. K., 100
Pathak, M. A., 484
Patroni-Killam, M., 148
Patte, H. E., 159
Pattenden, G., 20
Patton, S., 391

Patzelt-Wenczler, R., 408, 409(33)
Pauling, L., 132, 475, 477(13)
Pavisa, A., 152
Pease, P. L., 360
Pease, P., 230
Pecora, P., 189
Peiper, B., 232
Peiser, G. D., 175, 182(9)
Peña, M. R., 21
Peng, Y. M., 198
Peng, Y., 280
Peredi, J., 159, 160(61)
Perlin, A. S., 372
Perrin, D. D., 80
Perry, C. W., 45, 46(4)
Pessiki, P. J., 87, 91(1)
Peters, J., 405, 430
Peterson, M., 469
Peto, R., 143, 205, 347, 429
Pfander, H., 7, 75, 206, 208, 209(31), 210(31), 218(15), 347, 349(4)
Phagpolngarm, S., 482
Philip, T., 152, 157, 203
Phillips, G. B., 480
Phipers, R. F., 232, 233(13)
Picorel, R., 36, 38(15), 39, 41
Pierson, H. F., 190, 193(14)
Pjura, W. J., 469
Plaa, G. L., 406
Polgar, A., 475, 477(13)
Porra, R., 189
Porter, T., 391
Powls, R., 107
Pregenzer, J. F., 462, 470(19)
Pruett, R. C., 360, 362(4)
Prusti, R. K., 482
Pryor, W. A., 408, 409(35)
Puhl, H., 413
Pullman, A., 136
Pung, A. O., 266
Pung, A., 483
Purcell, A. E., 159, 190, 314, 352

Q

Qiu, J. H., 471
Quackenbush, F. W., 189, 190, 191(13), 193, 194(22)
Quarmby, R., 101

R

Raab, O., 420
Ramdahl, T., 384, 385(22)
Ranalder, U. B., 282
Randall, R. J., 387
Rapoport, H., 14, 231, 232, 233(2), 242(2)
Reames, S. A., 296, 300
Reay, C. C., 220, 225(5), 227(5), 228(5), 229(5), 230(5), 269, 336, 339(3), 341(3), 360, 362(3), 365(3), 429
Reddi, E., 435
Rees, D. C., 39
Rehms, A. A., 452
Reichenbach, H., 366
Reist, E. J., 79
Renstrom, B., 323
Repges, R., 408, 409(33)
Reppe, C. R., 482
Rhee, J. S., 251
Riccio, R., 385
Rich, M. R., 266
Rigassi, N., 61, 111, 114(6), 115(6), 119(6), 323
Ritacco, R. P., 208, 209(21)
Robinson, J. R., 482
Robson, D. C., 20
Rodgers, M. A. J., 425, 426(16), 430, 437, 444
Rodriguez, D. B., 208, 209(21)
Roelofsen, B., 471
Rohwedder, W. K., 457
Ronchi-Proja, F., 145
Rosebrough, N. J., 387
Rosen, T., 83
Ross, A. B., 446
Ross, A. C., 277
Rothemund, P., 90
Rotheneder, M., 413
Rougée, M., 87, 91(1)
Rouseff, R. L., 252
Ruddat, M., 186, 294, 314
Rüedi, P., 252
Ruegg, R., 146
Rundhaug, J. E., 266
Rundhaug, J., 483
Russel, S. W., 208, 209(27)
Russett, M. D., 220, 221, 224(11), 226(6, 11), 228(11), 230(11), 273, 276(1), 336, 337, 338(12), 339, 341, 342(12), 360

Russett, M., 198, 203, 265, 267(1), 268, 269(1, 24), 272(1)
Russett, R. O., 265

S

Saiki, K., 291, 292(4), 293(6), 294(5), 296, 297, 298(6), 299, 300, 305, 357
Saint Martin, P., 140
Saint-Jean, R., 432
Saito, K., 373
Saito, M., 462
Saito, T., 469, 470(23)
Sakai, T., 337
Sakan, T., 54
Salares, V. R., 104
Salokhiddinov, K. I., 427
Sander, L. C., 187, 190, 192, 193, 194, 195, 196, 197(29), 198(25), 204(25), 330
Santamaria, L., 266
Santus, R., 448
Saran, M., 408, 409(37), 451, 473
Sargent, M. L., 482
Sarif, A., 391
Sato, A., 15
Sato, N., 371, 373(20)
Satoh, K., 296
Sauer, K., 306, 308(5)
Scalia, S., 246
Schaffer, R., 187, 191(11)
Schalch, W., 430, 437
Scheidt, K., 266
Schgal, P. K., 265, 267(1), 269(1, 24), 272(1)
Schiedt, K., 170, 172(9)
Schimmer, B. P., 318
Schlegel, V., 37
Schmidt, K., 258, 375
Schmidt, R., 425, 427
Schmitt, R., 367
Schmitz, C., 75, 81(5)
Schnyder, U. W., 266
Schön, I., 417, 418(68), 419(68)
Schönberger, E., 198, 203
Schonberger, E., 477
Schrager, T. F., 205
Schreier, P., 54
Schreiman, I. C., 90
Schreurs, W. H. P., 148
Schrijver, J., 148, 149(31), 156(31), 157(31)

Schroeder, W. A., 475, 477(13)
Schudel, P., 280
Schulz, W. A., 438
Schuster, G. B., 425, 426(19), 427
Schutle-Frohlinde, D., 404
Schutt, G.-W., 282
Schwartz, J. L., 266
Schwartz, S. J., 148, 296
Schweeter, V., 61
Schwieter, U., 75, 111, 114(6), 115(6), 119(6), 146, 323
Schwrtzel, E. H., 258
Scott, W. J., 21
Searle, A. J. F., 406, 413(23), 414(23), 415(23)
Sechi, A. M., 412, 413(51)
Seeger, B., 266
Seely, G. R., 87, 91(1), 387
Seely, G., 87, 91(1)
Segers, L., 252
Sehgal, P. K., 268
Seibert, H. F., 165
Seibert, M., 32, 36, 38(15), 39, 41
Seki, T., 367, 370(13), 379
Selly, G. R., 449
Sensui, N., 337
Sequaris, J.-M., 32
Serico, L., 45, 46(4)
Sethi, S. K., 112
Shaish, A., 439, 441(8), 443(7, 9)
Shannon, J. S., 120
Shapiro, D., 266
Shapiro, S. S., 269
Sharpless, K. B., 78
Shekelle, R. B., 205, 347
Shen, B., 198, 225, 226(15)
Sherbinova, E. A., 472
Shiba, T., 375
Shibata, S. S., 459
Shibata, Y., 13, 15, 232
Shiina, H., 165
Shimada, K., 370, 375, 376(10, 11, 12), 378(10, 11), 382(10), 385(11)
Shimamura, T., 246, 253, 296, 299, 300, 301, 305(1, 2)
Shimasaki, H., 454
Shimizu, I., 22
Shimizu, M., 364
Shimura, T., 296
Shinna, H., 165

Shioi, Y., 376, 379(14)
Shipley, P. A., 54
Shipton, J., 54
Shklar, G., 266
Shone, C. C., 100
Shryock, A. M., 205, 347
Sidek, B., 161
Siefermann-Harms, D., 346
Sies, H., 266, 323, 404, 405, 430, 431, 432, 435(11), 437, 438, 472
Sim, P., 404
Simic, M. G., 406, 444, 448, 451, 452
Simidu, U., 375
Simon, P. W., 148, 153(29)
Simoni, R. D., 469
Simpson, K. L., 144, 175, 208, 209(21), 324
Simpson, K., 154, 199
Sinclair, R. S., 449
Singer, J. T., 258
Singh, R. K., 258
Singh, U., 158, 159(54)
Sistrom, W. R., 479
Sitte, P., 63
Sklan, D., 413, 414(59)
Sklar, L. A., 469
Slater, T. F., 405, 408, 409(36), 451, 473
Sloane, D., 266
Slughter, L., 419
Smallidge, R. L., 189, 190, 191(13)
Smidt, C. R., 282, 283(10)
Smith, A. M., 258
Smith, J. C. J., 267
Smith, J. C., 206, 208, 209(14, 19), 211(14), 212(14), 218(19)
Smith, J. C., Jr., 206, 219(12)
Smith, J. H. C., 480
Smith, K. M., 90
Smith, L. M., 342
Snodderly, D. M., 220, 221, 224, 226(6, 11), 227(11), 228(11), 229(12), 230(11, 12), 265, 273, 276(1), 336, 337, 339, 341, 360
Snodderly, M., 198, 203
Snyder, H. E., 408, 410(45)
Snyder, L. R., 186, 187(8)
Soares, J. H., Jr., 189, 199, 200(37), 201(37)
Som, S., 483
Song, J. H., 460
Song, P. S., 482
Sonntag, N. O. V., 166

Sorgeloos, P., 253, 262(20)
Soukup, M., 208, 209(25, 26)
Sowell, A. L., 189, 206
Spark, A. A., 233, 237(16)
Sparrock, J. M. B., 360
Speek, A. J., 148, 149(31), 156(31), 157(31)
Speek-Saichua, S., 148
Speranza, G., 421, 437, 438
Sporn, M. B., 143, 205, 347, 429
Sreekumari, M. T., 158, 159(56)
Stacewicz-Sap8untzakis, M., 189
Stachelin, H. B., 429
Stadler, J., 366
Stainer, R. Y., 387, 479
Stamler, J., 205, 347
Stancher, B., 159, 203
Steenbergen, C. L. M., 251
Steenken, S., 438
Stein, E. A., 206, 429
Stein, E., 186, 189, 199(6)
Steinberg, D., 412
Stemmler, E. A., 112
Stemmler, I., 79, 86(19)
Stevenson, R., 186, 187(7)
Stewart, I., 148, 185
Stich, H. F., 266
Stille, J. K., 21
Stockburger, M., 38
Stocker, R., 410, 412(47), 415(47), 454
Stoessel, S. J., 21
Stone, J., 221
Stossel, T. P., 405
Strain, H. H., 14, 231, 233(2), 242(2)
Stransky, H., 369
Stratton, F., 266
Straub, O., 146, 231, 232(3), 322, 367
Striegl, G., 413
Sturzenegger, V., 382
Sugawa, N., 337
Sugimoto, K., 392
Sugimoto, T., 337
Sugiura, M., 291, 292(4), 296(4)
Sukuki, K., 373
Sun, Y., 416
Sundquist, A. R., 438
Suprunchik, T., 431, 437
Suraci, C., 189
Svec, W. A., 14, 231, 232, 233(2), 242(2)
Swain, T., 135
Swanbeck, G., 421, 483

Swanson, B., 37
Swärd, K., 21
Sweeney, J. P., 147, 148(25), 185
Sweeney, S. A., 482
Swift, I. E., 75
Szabolcs, J., 225, 251

T

Tabor, J. M., 165
Tadolini, B., 412, 413(51)
Tait, D., 449
Takaichi, S., 367, 370, 373, 375, 376, 378(7, 10, 11), 379, 382(10), 385(11)
Takatsuka, I., 296, 299, 300(2), 305
Takayasu, J., 337
Takii, T., 291, 294(5), 296, 300, 305, 357
Tan, B., 155, 160, 162, 252
Tanaka, N., 17
Tanaka, Y., 165, 166, 167
Tangney, C. C., 189
Tappel, A. L., 417, 418(65, 66, 69), 419(69)
Taremi, S. S., 306, 307
Tarsis, S. L., 220, 229(4), 230(4), 336, 360, 361(2)
Taschner, M. T., 83
Tasumi, M., 305
Tauber, J. D., 22
Taylor, R. F., 3, 251, 254, 282, 302, 322, 323(3), 326(3), 478
Tchapla, A., 195
Tee, E. S., 143, 145, 147(15), 150, 154(36), 155(36), 156(36), 157(36), 158, 159(36, 55)
Tee, J. L., 233, 237(16)
Teebor, G. W., 460
Temalilwa, C. R., 148, 149(31), 156(31), 157(31)
Temple, N. J., 143
Tenorio, M. D., 208, 209(29), 349, 350(14)
Terao, J., 266, 411, 412, 416(50), 454, 456, 457, 459, 460
Thomas, J. K., 448
Thomas, T. A., 83
Thommen, H., 367
Thompson, J. N., 395
Thurnham, D. I., 336
Tidmarsh, M. L., 101
Tjang, T. D., 162

Toleutaev, B. N., 425
Tomimoto, K., 296, 299, 300(2), 305(2)
Tondeur, Y., 284
Tonucci, L. H., 218
Toppo, P., 306, 311(7)
Tóth, G., 251
Toube, T. B., 208, 209(27)
Trakhanov, S. D., 32
Tressl, R., 55
Trier, T. T., 87, 91(1), 452
Trouiller, J., 140
Trozzolo, A. M., 431, 437
Truscott, T. G., 420, 426(3), 430, 437, 448, 449
Trush, M., 416
Tsang, S. S., 266
Tsou, S. C. S., 324
Tsou, S. T. C., 144
Tsugeta, T., 55
Tsukida, K., 13, 15, 17, 232, 291, 292(4), 293(6), 294(5), 296, 297, 298(6), 299, 300, 305, 357, 477
Turro, N. J., 427
Tuveson, R. W., 482

U

Ueda, M., 426
Uemoto, Y., 469, 470(24)
Umarav, A. U., 160
Umbreit, M. A., 78
Underwood, B. A., 145
Ungers, G. E., 258
Upham, R. A., 121
Uphaus, R. A., 32
Urréchaga, E., 101, 104(14)
Utsumi, K., 406, 413(24), 414(24), 416(24), 417(24), 418(24)

V

Vaca, C. E., 460
Vacheron, M.-J., 366
Valduga, G., 435
Valleja, G., 161
Van den Berg, J. J. M., 471
Van Den Bosch, G., 159

Van der Auweraer, M., 87, 91(1)
Van Duyne, R. P., 32, 36
van Kuijk, F. J. G. M., 269, 220, 225(5), 227(5), 228(5), 229(5), 230(5), 336,339(3), 341(3), 360, 362(3), 365(3), 407, 408(30), 409(30), 410(30), 429
Vanderslice, J. T., 211, 350, 351(16)
VanKessel, J., 469
Varo, P. T., 150, 155(35), 157(35), 158(35), 159(35)
Vecchi, M., 132, 133(2), 282, 294
Verstraete, W., 252
Vetter, W., 61, 111, 114(6), 115(6), 119(6), 282, 323
Veysey, S. W., 282
Vile, G. F., 406, 413, 414(58)
Villarroel, A., 101, 104(15)
Violette, C. A., 307
Vlovy, R., 448
Vogtmann, H., 282
Völker, O., 315, 317(10), 318(10)
Von Schütz, J. U., 312
von Sonntag, C., 444
Vonshak, A., 387

W

Wagner, H. P., 208, 209(25)
Wagniere, G., 382
Wahlberg, I., 111
Wald, G., 220
Walker, E. B., 482
Walker, R. W., 373
Waller, H. D., 460
Walls, G. J., 221
Walter, E., 432
Walther, W., 282
Walton, J. C., 407, 408(31), 410(31)
Ware, W. R., 427
Watanabe, M., 482
Watanabe, Y., 469, 470(24)
Weaver, L., 405, 430
Weber, A., 31, 208, 209(23)
Weber, E. J., 226
Weedon, B. C. L., 3, 61, 75, 143, 146(2), 147(2), 208, 209(27), 233, 237(16), 457
Weeks, O. B., 75
Wefers, H., 404

Wegfahrt, P., 14, 231, 232, 233(2), 242(2)
Wehrli, H., 232
Weiler, L., 78, 83(14)
Weinhaus, R. S., 224
Weitberg, A. B., 405
Weiter, J. J., 269
Weitzman, S. A., 405
Welankiwar, S., 186, 265, 267(1), 268, 269(1, 24), 272(1), 340
Wennersten, G., 421, 483
Werner, J. S., 230, 360
Westergaard, H., 177, 462
Wheaton, T. A., 148
Wheaton, T., 185
Whitfield, F. B., 54
Whittaker, N. F., 114, 149, 151(32), 192, 197(18), 203, 252, 349, 350(12), 354(12)
Widaver, W. E., 387
Widmer, E., 21, 208, 209(18, 25, 26)
Wilday, P. S., 78
Wiles, D. M., 431, 437
Wilhelm, J., 460
Wilkinson, F., 421, 427, 428, 430, 437
Will, O. H., 482
Will, O. H., III, 186, 294, 314
Willett, W. C., 205
Williams, C. M., 144
Willis, B. G., 252, 262(7)
Willson, R. L., 406, 408, 409(36), 413(23), 414(23), 415(23), 450, 451
Willstätter, R., 232
Wilmoth, G. J., 367
Wilson, A. M., 152, 153(45)
Wilson, B. D., 432
Wilson, R. L., 473
Wilson, T., 431, 437
Winkler, R., 391
Winterbourn, C. C., 406, 413, 413(58)
Wise, S. A., 187, 189, 192, 193, 194, 195, 196(27), 198(25), 199, 200(37), 201(37), 204(25), 330
Witschi, H., 406
Witztum, J. L., 412
Wolf, H. C., 312
Wolfenden, B. S., 408, 409(36), 451, 473
Wolff, X. Y., 148, 153(29)
Wolleb, H., 75
Wood, W. A., 480
Woodward, S., 87, 91(1)

Y

Yamada, K., 459
Yamaguchi, T., 342
Yamamoto, H. Y., 198, 346
Yamamoto, K., 337
Yamamoto, Y., 408, 410, 412(47), 415(47), 454, 462, 469
Yamane, T., 416
Yamashita, J., 296, 299, 300(2), 305(2)
Yamauchi, R., 412, 416(50)
Yang, C. S., 189, 199
Yang, K. C., 482
Yang, S. F., 175, 182(9)
Yap, S. C., 161, 162(75), 164(74), 165(74), 166(74)
Yeates, T. O., 39
Yelle, L. M., 322, 323(3), 326(3)
Yogo, Y., 166
Yong, Y. Y. R., 386
Yoshikawa, Y., 462
Yoshizawa, C. H., 483
Yoshizawa, C. N., 266
Young, D. S., 212
Young, N. M., 104
Youngman, R. J., 460

Z

Zagalsky, P. F., 100, 101, 104, 172
Zakaria, M., 154, 199
Zamier, A., 440
Zamora, R., 417, 418(66)
Zechmeister, L., 7, 291, 292(1), 475, 477
Zell, R., 208, 209(18, 25)
Zeng, S., 282, 283(11), 290(11)
Zhu, H., 482
Ziegler, R. G., 144, 194, 198(25), 201, 203, 204(25)
Zinsou, C., 408, 410(44)
Zinukov, S. V., 425
Zonta, F., 159, 203
Zorzut, C. M., 373

Subject Index

A

AAPH. *See* 2,2'-Azobis(2-amidinopropane) dihydrochloride
Abstraction, generation of carotenoid free radical by, 445
Acetabularia mediterranea, 386
5-(4-Acetamidophenyl)-10,15,20-tris(4-methylphenyl)porphyrin, 88, 96–97
Acetic anhydride, 318
Acetone. *See also* Hexane-acetone extraction
   as solvent in carotenoid separation, 22–25, 199, 235, 299, 325, 350
Acetone-methanol extraction, 233–234, 241
Acetone-petroleum ether extraction, 60
Acetonitrile. *See also* Petroleum ether-acetonitrile-methanol extraction
   acidity of, 346
   as solvent in carotenoid separation, 197–198, 204
Acetonitrile-dichloromethane-methanol extraction, 206, 254, 278, 318
Acetonitrile-methanol-ethyl acetate extraction, 325–326
Acetonitrile-methanol extraction, 226, 340, 342
4-(N-Acetylamino)benzyltriphenylphosphonium bromide, 92, 94
Acetylated polyamide, 234, 236
Acetylation, in characterization of carotenoids, 25
Acetylene, 45, 47
   ^{13}C-labeled, 46
Acetylide(s), 13
Acyclic carotenoid(s), 376, 385
Acyclic $C_{40}H_{56}$ structure, 4
Addition, generation of carotenoid free radical by, 445
Adonirubin
   reduction of, 109–110
   structure of, 109
Adriamycin, lipid peroxidation induced by, 406, 415
Adsorbosphere-HS material. *See* Alltech C_{18} Adsorbosphere-HS column
Adsorption chromatography, 292–293, 300
Age-related macular degeneration, 220, 336
AIBN. *See* Azobisisobutyl nitrile
C_{20} Alcohol
   preparation of, 84–85
   transformation into phosphonium salt, 85–86
γ-*cis*-C_{15} Alcohol, 78
   obtention of exocyclic olefin, 82–83
Aldehyde derivative(s), formed during thermal degradation of β-carotene, 135–136
Aldehyde fixative(s), 224
Aldol condensation, 13, 22
Aldrich Chemical Company, 51, 462
Alfalfa meal, 147
Algal carotenoid(s), 8, 13, 146, 151, 169–171, 252, 254–255. *See also* Dunaliella; *Haematococcus pluvialis*
   chemosystematic markers for, 232
   classification of, 171
   HPLC analysis of, 241–242
   from unicellular algae, photoprotective activity of, 482
Alkaline ethanol, 315
Alkaline hydrolysis, 232, 351
Alkaline (KOH) methanol, 318, 342, 352
Alkaline plates, for thin-layer chromatography, 236
Alkoxyl radical(s), carotenoid deactivation of, 403
Alkylamine, 186
Alkyl glycol, 186
Alkylidenebutenolide(s), synthesis of, 14–17
   general procedure for, 15–17
4-Alkylidenebutenolide, 14
Alkylnitrile, 186
C_{22}-Allenic sulfone
   preparation of, 19–20
   structure of, 14
   synthesis of, 14, 17–20

SUBJECT INDEX

Alloxanthin
  isolated from *Paralithodes brevipes*, 27–28
  structure of, 28
Alltech C_{18} Adsorbosphere-HS column, 226, 337, 340–341, 474, 476
Allylic hydroxy group, methylation of, 25
Alumina column, 186, 233, 253, 294
Aluminum oxide thin-layer chromatography plates, 60, 318, 320
AMD. *See* Age-related macular degeneration
American Organization of Analytical Chemists, 158–159
American Type Culture Collection, 440
Aminocarotenoid(s), 88, 90, 92, 100
  structure of, 88–89
  synthesis of, 93–95
5-(4-Aminophenyl)-10,15,20-tris(4-methylphenyl)porphyrin, 88, 97
Ammonium acetate buffer, in carotenoid separation, 198, 204, 344
Amphidinium. *See* Dinoflagellate(s)
AMVN. *See* 2,2′-Azobis(2,4-dimethylvaleronitrile)
2′,3′-Anhydrolutein
  in human plasma, chromatographic profile of, 215, 218
  standards for, plant sources of, 350
  structure of, 216
Anhydroretinol
  packed-column gas chromatography of, 281–282
  retention indices of, 288–289
Anhydrorhodovibrin, 260, 376
  in bacteria, 380
Animal carotenoid(s), 171–172. *See also* Avian carotenoid(s); Invertebrate carotenoid(s); Mammalian carotenoid(s)
  formation of, 168
  in marine species, 172
  source of, 168
Antenna complexes, 39
Antheraxanthin, 6–7
  in algae, 257, 259
  commercial sources of, 253
Anthocyanin(s), 142
Antioxidant(s)
  biological, definition of, 403
  carotenoids as, 143, 266, 403–420. *See also* Lipid hydroperoxide assay; Lipid peroxidation
  analysis of, 407–408
  direct activity of, 417
  effects of oxygen pressure on, 409–411
  in vitro studies of, 403, 408–420
    in animals, 417–419
    in humans, 419–420
    interactions with other antioxidants, 417–419
    interactions with tocopherol, 412
    in isolated cells, 416
    in isolated membranes, 413–416
    in lipoproteins, 412–413
    in liposomes, 412–413, 455
    mechanism of, 454
    structure of, role of, 411–412, 444
  classification of, 461
  in light degradation of β-carotene, 177–182, 185
  lipid-soluble, 461
  in thermal degradation of β-carotene
    methods for application of, 139–140
    percentage of loss during extrusion cooking with, 140–142
    protective effects of, 137–142
  water-soluble, 461
Antioxidant radical scavenger(s), 460–472
  cis-parinaric acid fluorescence-based assay
    in DOPC liposomes
      procedures for, 463
      results of, 468, 471–472
    in hexane
      procedures for, 462–463
      results of, 465–467
  phycoerythrin fluorescence-based assay in aqueous system
    carotenoids versus retinoids, 470–471
    merits of, 469
    procedures for, 462
    results of, 463–467
  study of
    discussion of, 469–472
    experimental procedures for, 461–463
    results of, 463–468
AOAC. *See* American Organization of Analytical Chemists
7′-Apo-7′-(4-aminomethylphenyl)-β-carotene, 91, 95
7′-Apo-7′-(4-aminophenyl)-β-carotene, 88, 92, 94
7′-Apo-7′-(4-carbamylphenyl)-β-carotene, 88, 91, 94–95

SUBJECT INDEX

7′-Apo-7′-(4-carbomethoxyphenyl)-β-carotene, 88, 90–91, 93
7′-Apo-7′-(4-carboxyphenyl)-β-carotene, 88, 91, 93
Apocarotenal(s), C_{22}-allenic, preparation of, for carotenoid synthesis, 19
β-Apo-8′-carotenal, 9, 126, 264, 349, 354
   commercial sources of, 114, 274, 349
   molecular weight of, 118
   spectrometric and chemical characterization of, 136
β-Apo-10′-carotenal, 225
   spectrometric and chemical characterization of, 136
β-Apo-12′-carotenal, spectrometric and chemical characterization of, 136
β-Apo-14′-carotenal, spectrometric and chemical characterization of, 136
β-Apo-15-carotenal, spectrometric and chemical characterization of, 136
β-Apo-12′-carotenal-O-t-butyl-oxime, synthesis of, 339–340
8′-Apo-β-carotene, 88, 90–91, 93
(3R)-8′-Apo-β-carotene-3,8′-diol, 209–210
   chromatographic profile of, 215
   structure of, 217
β-Apo-8′-carotenoic acid ethyl ester, 9–10, 264
Apocarotenoid(s), 7, 126
   formed during thermal degradation of β-carotene
      oxidative break of carotene double bonds corresponding to, 136, 138
      spectrometric and chemical characterization of, 136
   gas chromatography-mass spectrometry of, 281–290. *See also* Gas chromatography-mass spectrometry
      limitations on, 289
   Kovats retention indices, 284, 288–290
β-Apo-8′-carotenol, in bacteria, 380
β-Apo-13-carotenone, oxidation products of, 408, 410
β-Apo-8′-carotenyl decanoate, 274, 280
7′-Apo-7′-(4-hydroxymethylphenyl)-β-carotene, 91, 95
7′-Apo-7′-(4-iodomethylphenyl)-β-carotene, 91, 95–96
Aporetinoid(s), 288–290
Applied Biosystems, 326
Apricot(s), 158, 350, 354

carotenoid HPLC profile of extract from, 357
Ara macao, 320–321
Argon, 477
Artemia cysts, 252, 254, 262
   chromatogram of, 262–264
   preparation of, 256
Artocarpus heterophyllus. *See* Jackfruit
Ascorbic acid, 461
Astacus leptodactylus, 101
   carotenoproteins from, 101, 104, 106. *See also* Carotenoprotein(s)
Astaxanthin, 6–7, 101–102
   accumulation of, in *Haematococcus pluvialis*, 386–391
      conditions inducing, 389–391
      effects of cell division inhibitor on, 390
      effects of light intensity on, 389
      effects of nitrogen concentration on, 389
      effects of phosphate starvation on, 390
      effects of salt stress, 390
      effects of sulfate starvation on, 390
      extraction and measurement of, 387–389
      growth conditions and measurements, 387
   in animals, 171–172
   antioxidant activity of, 386, 411–412, 416
   in Arthropoda, 23
   in birds, 320
   commercial uses of, 386
   in fungi, 171
   in *Homarus americanus*, separation of standards of, 319
   in *Paralithodes brevipes*, 26–28
   in *Procambarus clarkii*, 108
   racemic, 10
   reduction of, 109–110
   singlet oxygen quenching constant of, 437
   stereoisomers of, 7
      HPLC separation of, 25–26
   structure of, 28, 109
Attex Ultrasphere TH Cyano, 236
Aurochrome, 61
   standards for, plant sources of, 350
Auroxanthin, in pumpkin, 156
Autoxidation, products of, 408, 410
Averrhoa carambola. *See* Starfruit
Avian carotenoid(s)
   extraction of, 314

from feathers, 315–317
from soft parts, 316–317
identification of, 312–321
historical development of, 313–314
procedures for, 317–319
isolation of, 317–318
locations of, 313–314
pigment structure of, and absorption spectrum, 319
quantitation of, 317
types of, 319–321
2,2′-Azobis(2-amidinopropane) dihydrochloride, commercial sources of, 461–462
2,2′-Azobis(2,4-dimethylvaleronitrile), 468–469
commercial sources of, 461–462
Azobisisobutyl nitrile, methyl linoleate oxidation induced by, 409

B

Bacillariophyceae, 231
Bacteria
carotenoidless (white) mutants
classification of, 481
induction of, 480
fruiting gliding, 366
nonfruiting gliding, 366
Bacterial carotenoid(s), 170
depletion of, methods for, 480–481
detection of, 480–481
extract, preparation of, for HPLC-photodiode array, 254–255
from nonphotosynthetic bacteria, 170
photoprotective function of, 480–482
analysis of
procedures for, 481–482
sample preparation for, 481
from photosynthetic bacteria, 170, 480.
See also Erythrobacter longus
characterization of, 374–385
distribution of, 375
number of, 375
zero-field splitting parameters of complexes from, 311
Bacteriochlorophyll(s), 37, 299, 306, 311, 480
in aerobic bacteria, 375
Bacteriorhodopsin, surface-enhanced Raman scattering spectroscopy of, 41

Bacterioruberin, 7
Bacteriorubixanthin, 376, 380
structure of, 382
Bacteriorubixanthinal, 376
structure of, 382
Banana, 157
Barley oil, 159
BC. *See* β-Carotene
BChl. *See* Bacteriochlorophyll(s)
Beans, carotenoids in, 148–149
Beckman/Altex HPLC system, 176
Beckman model LS 7500 spectrometer, 46
Beer-Lambert law, 429
Benzene, 235, 301
Berry(ies), 157. *See also* Blueberry(ies)
BHA 100, 140
BHT. *See* Butylated hydroxytoluene
BHT 50, 140, 142
Bicyclic carotenoid(s), 75–86, 376
cyclization of, 75
immediate precursor of, 75
synthesis of
biomimetic approach to, 75
procedures for, 79–86
three known types of, 75
two C_{50} isoprene units of, 75
Bilirubin, antioxidant effects of, 410, 451
Biochemistry of Carotenoids (Goodwin), 3
Biochemistry of Natural Pigments (Britton), 3
5,15-Bis(4-acetamidophenyl)-10,20-bis(4-methylphenyl)porphyrin, 88, 96–97
5,15-Bis(4-aminophenyl)-10,20-bis(4-methylphenyl)porphyrin, 88, 96–97
Bixin, 147
conversion to bixindiol, 114
singlet oxygen quenching constant of, 437
Bixin powder, 114
Bleaching, of oxidation products of carotenoids, 407–409, 451–452
Blueberry(ies), carotenoids in
absorbance spectra of, 326, 328, 330
analysis of, 325–327
extraction of, 324
Borohydride reduction method, carotenoid-protein binding studies with, 100–110
Brassica [genus], 149, 151
Brassica oleracea. See Kale
Brevibacterium KY-4313, 252, 254
chromatogram of, 256–257, 264

preparation of, 254-255
Broccoli, 147
Bryopsis maxima, 379
Büchner funnel, 324
Buffer(s)
 in HPLC carotenoid separation, 198, 204, 344
 for lipid hydroperoxide assay, 456
 in surface-enhanced resonance Raman scattering spectroscopy, 38
Butenolide carotenoid(s), 13
Butylated hydroxytoluene, 176, 225, 280
 protective effects on β-carotene, 176, 178-182, 185
Butyllithium, 45, 47, 51

C

C_{18} Adsorbosphere-HS material. *See* Alltech C_{18} Adsorbosphere-HS column
Calcium, 147
Calcium carbonate columns, 233
Calcium hydroxide, 253, 299-300
γ-*cis*-C_{15} alcohol. *See* Alcohol
C_{20} alcohol. *See* Alcohol
California poppy, 169
C_{22}-allenic sulfone. *See* Allenic sulfone
Caloxanthin, 376, 380
Caloxanthin sulfate, 375
 absorption spectra of, 383
 structure of, 385
Calthaxanthin, structure of, 216
Canary xanthophyll(s), 315, 320
Cantaloupe(s), 157-158, 354
Canthaxanthin, 6-7, 10, 259
 in animals, 171-172
 antioxidant activity of, 411-412, 417
 versus β-carotene, 427-429
 near-infrared steady-state luminescence study of, 426
 reported rate constants for, 427-428
 Stern-Volmer plots of, 426-429
 in bacteria, 379-380
 benzene solution of, isomerization shifts of, 301-302
 in birds, 320
 chemical ionization of, 125-126
 chromatogram of, 257-258, 262-264
 commercial sources of, 114, 252, 425, 462
 commercial uses of, 426

distribution of, in pooled extracts of mammalian organs, 270-271
electron-capture negative-ion production of, 126
electron ionization of, 125
high-performance liquid chromatography of, eluent for, 300
in *Homarus americanus*, standards of, separation of, 319
in human diet, 265-266
in human plasma, 266
isomeric, HPLC lime column analysis of, 296
mammalian metabolism studies of, 265-272
 animals used in, 267-268
 HPLC analysis of, 268-269
 materials used in, 266-267
 results of, 269-272
mass spectrometric analysis of, 125-127
in *Paralithodes brevipes*, 27-28
quantitation of, 264
singlet oxygen quenching constant of, 437
structure of, 28, 116, 422
all-*trans*-Canthaxanthin, in bacteria, 254, 256
Capsanthin, 147
4-Carbomethoxybenzyltriphenylphosphonium bromide, 92-93
5-(4-Carbomethoxyphenyl)-10,15,20-tris(pentafluorophenyl)porphyrin, 88, 97
Carbon, isotopic variants of, 43
Carbon-11, 43
Carbon-12, 43
Carbon-13, 43, 48
Carbon-14, 43
Carbon-carbon double bond, formation of, 13
Carbon disulfide, 235
Carbon isotope(s)
 carotenoids labeled with, 42-49
 retinoids labeled with, 45
Carbon tetrachloride, 425-426
Carbonyl(s)
 oxidation products of, 408, 410
 reduction of, 318
Carcinogenesis, carotenoid inhibition of, 13, 143-144, 175, 322-323, 386, 403-404, 416-417, 445, 472

in animal studies, 419
and dietary considerations, 347, 429
in human plasma, 205-206, 336-337
in mammalian metabolism, 265
Carcinus maenas, 101
Cardiovascular disease(s), early prophylaxis in, 13
Cardueline(s), red pigments in, 320
Carlo Erba gas chromatograph, 283
Carotenal oxime, synthesis of, 338-340
Carotene(s), 4, 186
  biosynthesis of, in higher plants, 62
  15-*cis*-, identification of, 73
  cis-and poly-*cis*-, 63
  cyclic, HPLC system II analysis of, 73
  desaturation and cyclization of, 62-74
  from natural sources, highest concentration of, 147
  purification of, 71-73
  separation of
    from polar lipids, 66-67
    from xanthophylls, 66-67
  standards for, plant sources of, 350
  structural isomers of, FAB-MS of, 330
Carotene [stem name]
  prefixes to, 4-5
  structure and numbering of, 4-5
α-Carotene, 228
  antioxidant activity of, 415-416
    demonstration of, 451
  in bacteria, 380
  in blueberries, 328
  in carrots, 152-154, 274, 326, 328, 333-334
  chromatographic separation of, 326
  in chloroplasts of higher plants, 168
  commercial source of, 323, 392
  cyclization of, from lycopene, 62
  FAB mass spectrum of, 330, 332
  in fruits, 157
  in green vegetables, 150-151
  high-performance liquid chromatography of, composition of solvent mixtures, flow rate, and retention time for, 275
  in human milk, 395-398
  in human plasma, 206, 213-214, 273
    chromatographic characteristics of, 278-281
    quantification of, 276
    spectral characteristics of, 276
  identification of, 323, 326

LC-MS spectra of, 333-335
mammalian metabolism of, 337
molecular weight of, 118
in palm oil, 162, 164-165
in pumpkin, 156, 354-356
resolution of, 335
singlet oxygen quenching constant of, 437
standards for
  plant sources of, 350
  preparation of, 394
  structure of, 217
in tomatoes, 326, 328
in vegetables and fruits, 149-151
β-Carotene, 4, 6-7, 10, 380
  absorption spectra of, 380, 383
  in algae, 170, 257, 259
  all-*trans*-, 192
  commercial sources of, 425
  extrusion cooking of, 129-137
    assay of residual products of, with liquid chromatography, 137-142
    chemical reactions revealed by, 142
    degradation products of, 131-132
    fractionation by liquid-phase chromatography, 131-132
    epoxy derivatives formed during, fractionation and identification of, 133-134
    protective effects of antioxidants during, 137-142. *See also* Antioxidant(s)
  iodine-catalyzed photoisomerization of, 292, 294
  stereoisomerization of, 291-292, 294
    chemicals for, 291
    typical results for, 293-294
  structure of, 292, 422
  synthetic, absorption spectra of, 263-264
  thermal isomerization of, 291, 294-295
  AMVN-dependent oxidation of, in homogeneous solution, 474
    procedures for, 474
    reagents for, 474
  analogs of, 49-53, 264
  in animals, 171-172
  anticarcinogenic properties of, 175, 184, 336-337, 403-404, 429. *See also* Carcinogenesis
  antioxidant activity of, 411-413

in Adriamycin-induced oxidation, 415
in AMVN-induced oxidation, 415–416
versus canthaxanthin, 427–429
effects of oxygen pressure on, 409–411
mechanism of, 442–443, 472
near-infrared steady-state luminescence study of, 426
reported rate constants for, 427–428
Stern-Volmer plots of, 426–429
suppression of, 419
in apricots, 357
in bacteria, 254, 256, 376
in benzene solution, isomerization shifts of, 301–302
biosynthesis of, in *Dunaliella*, 439–444
HPLC analysis of, 442
induction of, 440–441
by high light intensity, 440
by limiting nutrients, 441
by nitrate or sulfate starvation, 440
inhibitors of, 441
optimal conditions for, 439
pathway for, with inhibitors, 441
protection against photoinhibition during, 442–443
role of, 439
selection of carotene-rich mutants, 443–444
in blueberries, 326–330
in carrots, 152–154, 274, 326, 328, 330–331, 333–334
chromatographic separation of, 326
chemical ionization of, 118–120
versus electron ionization, 119
in chloroplasts of higher plants, 168
chromatographic separation of, 192, 196–197, 204, 226, 326
effects of pore diameter on, 192
cis-
isomers of, 132, 338
analysis of, scope and limitation of method for, 293–298
classical adsorption chromatography of, 292–293
identification of, procedures for, 293–297
preparative TLC of, 293
resolution of, 292–297
separation of, 291–298, 356–357
structural assignment of, 293

NMR structural determination study of, 291
separation of, 291
HPLC methods for, 294–295
ultraviolet/visible spectrum characteristics of, 296–297
cis-parinaric acid fluorescence assay of, 466–467
commercial sources of, 114, 252, 274, 323, 392, 462, 474
control of oxygen radical reactions by, 405–406
crystalline, triplet state EPR of, 311
cyclization of, from lycopene, 62
degradation of, factors in, 178
deuterated, 50, 53
synthesis of, 49–53
materials for, 51
procedures for, 51–53
6e-electrocyclized, formation of, 297–298
electron-capture negative-ion production of, 120
electronic spectra of, 9
electron ionization of, 115–118
FAB mass spectrum of, 330, 332
free radical scavenging activity of, in liposomal system, 468, 472
in fruits, 156–158
in fungi, 171
gas chromatography-mass spectrometry of, 290
geometric isomers of
cis forms of, 291
number of, 291
gradual oxidation of, 453
in green vegetables, 149–151
leafy versus nonleafy, 150–151
high-performance liquid chromatography of
composition of solvent mixtures, flow rate, and retention time for, 275
eluent for, 300
in human milk, 395–398
in human plasma, 206, 213–214, 220, 273
chromatographic profile of, 215, 279–281
quantification of, 276
spectral characteristics of, 276
hydrophobic nature of, 333
identification of, 323, 326
immunoprotective effects of, 391, 403

inhibition of MDA formation by, 415
inhibition of red blood cell hemolysis by, 415
isolated from *Paralithodes brevipes*, 27–28
isomeric, 301–302
  HPLC lime column analysis of, 296
  Raman spectra of, 304–305
  Kovats retention index of, 290
labeled with radioisotopes, synthesis of, 44, 46–49
lack of toxicity of, 403
LC-MS spectra of, 333–335
light degradation of, 177–182, 185
  under fluorescent light, 178–183
  under ultraviolet light, 178–182
major carrier of, in humans, 412
mammalian metabolism of, 271, 337
mass spectrometric analysis of, 115–120
  mode of introduction for, 126–128
modulation of SOD increase by, 419
molecular weight of, 118
in mustard green leaves, 274
in *Narcissus majalis*, 147
natural occurrence of, 147
oxidative products of, 408, 410, 473
in palm oil, 162, 164–165
in papaya, 358–359
in primate retina, 228–229
in pumpkin, 156, 354, 356
purification of, 129–130
quantitation of, 264
reactions of, in organic solvents, 448
reactions with radical sources, products of, 408
resolution of, 335
role of, 13
in roots and tubers, 158–159
separation of, 253
singlet oxygen quenching by, 405. See also Singlet oxygen quenching
singlet oxygen quenching constant of, 437
stability of, 175–185
  optimal laboratory conditions for, 184–185
  under simulated incubation conditions, 176–179
  under storage conditions, 177, 183–184
standards for
  plant sources of, 350

preparation of, 394
stereoisomers of, 442
stoichiometric value of, 471
structure of, 28, 116, 217, 473
synthesis of, 9
  historical development of, 9
synthetic forms of, 354
thermal degradation of, 54–55, 58
  aldehyde derivatives formed during, 135–136
  apocarotenals formed during
    oxidative breaks of carotene double bonds corresponding to, 136, 138
    spectrometric and chemical characterization of, 136
  components of, 54
  by extrusion cooking, 129–137
  hydroxyl derivatives formed during, 136–137
    HPLC of, 136, 139
  ketone derivatives formed during, 135–136
  long-term protective agent for, 142
  nonvolatile compounds formed by, 129–142
    identification of, 61–62
  oxidation reactions during, 132–137, 142
  pigment extraction from, 131
  protective effects of antioxidants during, 137–142. See also Antioxidant(s)
  residual pigments after, 141
  volatile compounds formed by
    kinetics of, 58–59
    main types of, 57
in thylakoid membranes, 40–41
in tomatoes, 154–155, 326, 328, 330–331
5,10-*trans*-, structure of, 292
in tuberous vegetables, 151
in vegetable oils, 159
in vegetables and fruits, 149–150
vitamin A activity of, 144, 175, 184, 429
γ-Carotene, 150–151, 153
in apricots, 357
in fruit, 158
in human plasma, 213–214
  chromatographic profile of, 215
  intermediate accumulation of, 442

in palm oil, 162, 164
singlet oxygen quenching constant of, 437
stereoisomers of, 439
structure of, 217
β,β-Carotene
  in algae, 242
  derivatives of, 383
  in dinoflagellates, 245
δ-Carotene, in palm oil, 162, 164–165
ζ-Carotene, 153
  in apricots, 357
  in bacteria, 380
  15-cis-, ^{14}C-labeled, biosynthesis of, 64
  in fruit, 158
  in human plasma, 213–214
    chromatographic profile of, 215
  identification of, 73
  intermediate accumulation of, 441
  in palm oil, 162, 164–165
  in papaya, 358–359
  photoisomerization of, 64–65
  in pumpkin, 156, 354–356
  purification of, 71–73
  stereoisomers of, 439
  storage of, 73
  structure of, 217
  15-trans-, ^{14}C-labeled, as substrate, 64
Carotene alcohol
  structure of, 88–89
  synthesis of, 95–96
β,ε-Carotene chromophore, 337–338
ε,ε-Carotene chromophore, 337–338
ε,ε-Carotene-3,3'-dione, 214–216, 218
α-Carotene 5,8-epoxide, in palm oil, 162
β-Carotene 5,6,5',6'-epoxide, 126
  commercial sources of, 114
β-Carotene 5,6-epoxide, 126
  commercial sources of, 114
  in palm oil, 162
  in pumpkin, 156
β-Carotene epoxides, 133
  absorbance maxima for, 475–476
  analysis of, 472–479
  chemical structures of, 473
  electron impact spectra for, 478–479
  formed during extrusion cooking, 132–135, 141
    high-performance liquid chromatogram of, 134
    reaction sequence of formation, 135

generation of, with m-chloroperbenzoic acid, 114
high-performance liquid chromatography of, 474–477
  for resolution, 475–476
  for separation, 474
  for simultaneous detection, 475, 477
  for verification of purity, 477
  hydrodynamic voltammograms for, 477–478
mass spectrometry of, 478–479
storage and stability of, 477
Carotene fractions, 66–67
  HPLC formation of, 67
  obtained from Tangerine tomato, 73
Carotene intermediates, biosynthetically active, configuration of, 63
Carotene iodide
  structure of, 88–89
  synthesis of, 95–96
β-Carotene radical(s)
  generation of, 448
  regeneration of, in presence of chlorophyll, 449
Carotene substrates, 65–66. See also Reference carotenes
  incubation assays
    analysis of, 67–71
    quantitation of, 74
    spectral identification of, 74
  purification of, 67–71
  radiolabeled, biosynthesis of, 63–65
  source of, 65–66
Carotenogenic flowers, three main groups of, 169
Carotenoic acid
  structure of, 88–89
  synthesis of, 92–93
β,β-Carotenoid(s), primate metabolism of, 337
β,ε-Carotenoid(s), primate metabolism of, 337
Carotenoid(s)
  ability of organisms to produce, 146
  absorption spectra of, 379–381
    acidogenic, 315–316, 320
    oxidation of, 318
  in aerobic photosynthesis, 168
  antioxidant effects of. See Antioxidant(s); Antioxidant radical scavenger(s)

asymmetric molecules of, 39
in bacteria. *See* Bacterial carotenoid(s)
biosynthesis of, 11–12
  largest quantity of, 231
biosynthetic pathway of, interruption of, 481
in birds. *See* Avian carotenoid(s)
butenolide, 13
C_{40}, classification of, 376
characterization of, 13–22, 25–29, 322
  instruments for, 27
chemical properties of, 11
cis-, biomolecular functions of, 291
cis-trans isomerism of, 7–9
commercial sources of. *See individual species*
complex biological mixtures of, analysis of, 322
concentrations of, 8
configurations of, determination of, 298–305
coupling of, with cyclic tetrapyrroles, 91
degradation products of, removal of, 303
de novo synthesis of, 168
derivation of name for, 142
dietary, available for utilization by human body, 206
and disease prevention, 419–420. *See also* Carcinogenesis; Immuno response; Mutagenesis
distribution of, 167–172
esterified, enzymatic hydrolysis of, 25
extraction of, 23–25
  avoidance of isomerization during, 299
  chromatographic methods for. *See* High-performance liquid chromatography
  inadvertent isomerization during, 148
  induction of isomerization during, 148
four major types of, 8–9, 146
functions of, 13
high-performance liquid chromatography of, composition of solvent mixtures, flow rate, and retention time for, 275
and human disease. *See* Carcinogenesis; Immuno response; Mutagenesis
human use of, FDA approval for, 484
hydrophobic, 333
industrial production of, 3, 9–11
as inhibitors of free radicals, 450–452
  classification of, 450
  nonoxyl, 450
  oxyl, 450–451
interactions with tocopherol, 412
isocratic separation of, 201, 204, 326
isomeric
  configuration of, determination of, 298–305
  extraction of, 298–300
    procedures for, 299
  labeling of, 43–46
    with radioisotopes, synthesis of, 42–49
    with tritium, 42–49
  mammalian metabolism of, other than β-carotene, 265–272
and molecular electronic devices, 452–453
molecular geometry of, relative to other chromophores, 306, 311–312
molecular weights of, determination of, 118, 128, 382–384
natural occurrence of, 8–9
from natural sources, 7, 142–167
  characterization of, 22–31
  concentrations of, 147
  known number of, 146
  methods for isolation of, 11, 146–167
  most abundant type of, 146, 175
  overall pattern of, 147
  richest concentration of, 156
  structure of, 22–31
  total production of, 8, 142, 146, 175
nomenclature of, 4–8, 206–207, 347–348
  rules for, 4
oxidation products of, bleaching of, 407–409, 451–452
oxygenated. *See* Xanthophyll(s)
photoprotective mechanisms of. *See* Photoprotection
phycoerythrin fluorescence assay of, 466–467
physiological functions of, specialized nature of, 429
purified sample of
  identification of, 23
  structural determination of, 23
reaction stoichiometry of, with peroxyl radicals in hexane, 470–471
redox potential of, 449

research on
  history of, 3
  interdisciplinary approach to, 3
 roles of, 143
 sample introduction of
  by HPLC/photodiode array detector/
   particle beam mass spectrometer
   interface, 126–128
  for mass spectrometric analysis, 112–
   113
 singlet oxygen quenching ability of. See
   Singlet oxygen quenching
 in solid state, prolonged thermal exposure
   of, 119
 stereochemistry of, 7–8
 storage of, 11
 structure of, 4–8, 88–89, 109, 116. See
   also individual species
  electron paramagnetic resonance studies of, 306, 310–311
   zero-field splitting parameters, 311
  in vivo determination of, 312
  physicochemical properties related to,
   411–412, 444
 synthesis of, 13–22
  from isopentenyl pyrophosphate, 11–
   12
  sulfone method for, 15–16
 synthetic, 9–10
  commercial uses of, 10
  structures of, 10
  total sales of, 11
 technical and nutritional applications of, 3
 that lack provitamin A activity, 265
 as therapeutic agents, 421
 thermal degradation of, 54–55
  nonvolatile compounds formed by, 60–
   62
  procedure for, 55
  reagents in, 54
  volatile compounds formed by, 54–62
   kinetics of, 58–60
   quantitation of, 58
   recovery of, 55–56
 trans-cis isomeric, large-scale separation
   of, 293
 triplet states of, 305–306
  detection of, 306
  formation of, mechanism of, 306, 309–
   310
  polarization pattern of EPR signals
   from, 308–310
 and singlet oxygen quenching, 430
 in vegetables. See specific vegetable; Vege-
   table(s)
Carotenoid action, mechanisms of, 404–
   407
Carotenoid-chlorophyll-protein complexes,
   232
Carotenoid-containing beadlets, 483–484
 commercial sources of, 484
Carotenoid epoxides, 126
Carotenoid free radical(s), 444–453
 electron transfer rate constants for, 446–
   448
 generation of, 445–448
  four basic reactions involved in, 445
 properties of, 448–450
 reactions of, with porphyrins, 449
 reactivity of model compounds, 446
 solvent, 445
 spectral properties of, 447–448
 water free, 446
Carotenoid glycoside ester(s)
 from *Rhodococcus rhodochrous*, 366–374
 types of, 366
Carotenoid-protein complex(es)
 borohydride reduction method studies of,
   100–110
 carotenoid triplet states in, 310
 classification of, 101
 from photosynthetic systems, 299
 stoichiometry of, 101
Carotenoids (Karrer and Jucker), 3
Carotenoid sulfates, 375
 from *Erythrobacter longus*
  acidic groups in, 385
  circular dichroic spectra of, 384–385
  identification of, 383–385
  molecular weight of, 383–384
  nuclear magnetic resonance of, 384–
   385
Carotenol acyl bisesters, mass spectrometric
   analysis of, 123–125
Carotenol acyl esters, conversion into hydroxycarotenoids, 352
Carotenol fatty acid esters, in fruits and vegetables, 158, 353, 357–359
Carotenol fatty acyl monoester, 126
Carotenolipoprotein(s), 101

red and orange, borohydride reduction of, 110
β-Caroten-4-one, spectrometric and chemical characterization of, 136
Carotenoporphyrin(s)
  functions of, 87
  structure of, 88–89, 91
  synthesis of, 87–100
    methods for, 92–100
Carotenoporphyrin molecules, synthetic, carotenoid triplet states in, 308
Carotenoprotein(s)
  bathochromic shifts of, 101
  blue water-soluble, 101, 104
    from *Homarus americanus*, 108
    from *Procambarus clarkii*
      accessibility of astaxanthin in, 108
      borohydride reduction of, 107–110
      extraction of, 104–105
      KBH_4 reduction of, 108–110
      purification of, 105–106
  with carotenoid-carotenoid interactions, 103–104, 106, 110
  with carotenoid-lipid interactions, 102, 104, 106–107, 110. *See also* Carotenolipoprotein(s)
  with carotenoid-protein interactions, 101–102, 105–106
  classification of, 101–104
  with mixed interactions, 103–104, 106, 110
    borohydride reduction of, 110
  purification of, 105–107
  purple, from *Pachygrapsus marmoratus* eggs, 104
    purification of, 106
  red insoluble, 104
    from *Procambarus clarkii*, 104–105
  red or orange, from *Procambarus clarkii*, 106–107
  reduction of, 100–110
    borohydride method for, 104, 107–108
    KBH_4 method for, 104, 107–108
    structure of carotenoids involved in, 109
  spectral characteristics of, 101
  yellow
    from *Astacus leptodactylus*, 104, 106
    from *Homarus americanus*, borohydride reduction of, 110

Carotenopyropheophorbide, 89, 92, 99–100
Carpodacus roseus, 320
Carrot(s), carotenoids in, 142, 147, 149–152, 274–275, 350, 354, 356
  absorbance spectra of, 326, 328, 330
  analysis of, 325
  concentrations of, 152–154
  extraction of, 324, 326
  identification of, 330–331
    by LC-MS, 333–334
Cassava, 158–159
Caulerpales, 170
CC. *See* Column chromatography
CCD detector, 304
CD. *See* Circular dichroic spectra
Chemical ionization mass spectrometry, 112, 284, 323
  analyte ions in, rationalization of, 112
  characteristics of, 112
  definition of, 112
  disadvantage of, 323
  versus electron-capture negative-ion production, 112
  parameters of, 112
  sensitivity for determination of molecular weight, 118
Chicks, 418
Chlamydomonas nivalis, 386
Chlorobiaceae, 375
Chloroflexaceae, 375
Chloroform, 199, 300, 463
Chloroform-ethanol, as singlet oxygen quenching assay solvent, 432
Chloroform-methanol extraction, 376–378
Chlorohydric acid, 61
Chloromonadophyta, 171
m-Chloroperbenzoic acid, 114
Chlorophyceae, 439
Chlorophyll(s), 8, 232, 303
  in fruits, 354
  natural sources of, 142
  in pigment-protein complexes, 299
  triplet sensitizer function of, 299
  triplet states of, carotenoid control of, 87
  in vegetables, 149, 151, 354
    chromatographic profiles of, 353–354
Chlorophyll derivative(s)
  structure of, 88–89
  synthesis of, 90

Chlorophyta, 169–171
Chloroplast(s), 13
  of higher plants, carotenoids in, 168–169
Chloroxanthin, in bacteria, 380
Cholesterol, 346
Cholesteryl esters, elimination from plasma sample, 342
Cholestryl oleate, in human plasma, 213–214
Chondromyces apiculatus, 366
Chromatiaceae, 375–376
Chromatium vinosum, 378
Chromatographic retention, with FAB-MS, 335
Chromatography. See Column chromatography; Gas chromatography; High-performance liquid chromatography; Medium-pressure liquid chromatography; Thin-layer chromatography
Chromatophore(s)
  carotenoids in membranes of, location of, 38–39
  carotenoid triplet states in, 308
  isolated from *Rhodobacter sphaeroides*, spectra of, 39–40
  surface-enhanced Raman scattering spectroscopy studies of, 36–41
Chromobacterium viscosum, 339, 342
Chromophores, 32, 37
  of carotenoids, estimation of, 379–381
  coupled with polyene components, in molecular electronics, 453
  molecule geometry of, relative to carotenoids, 306
  surface-enhanced resonance Raman scattering spectroscopy studies of, 42
Chromophyta, 231
Chromoplast membranes
  formation of carotene chromophore and ionone rings in, 62–74
  phytoene incubation of, 67–68
Chrysophyceae, 231
Chrysophyta, 171
CI. See Chemical ionization mass spectrometry
Ciliate(s), 482
Circular dichroic spectra, 23
  of carotenoid sulfates, 384–385
  for characterization of carotenoids, 25–27, 382

for identification of fucoxanthin, 234, 239
for identification of peridinin, 244
instruments for, 27
9-cis isomers, spectra of, 9
15-cis isomers, spectra of, 9
Cis-trans isomerism, of carotenoids, 7–9
Citoscynt, 176
Citral, 54
Citranaxanthin, 10
Citrus microcarpa. See Musk lime
Citrus products. See Fruit, carotenoids
Clextral BC 45 extruder cooker, assembly of screws in, 130
CN column, 236
Coccothraustes verpertina, 320
Colaptes auratus, 315, 320
Colostrum, 391, 394
Columba livia, 320
Column chromatography, 23–25
  equipment for, 79, 235–236, 278
  in isolation of fucoxanthin and peridinin, 233–236
  for purification of carotenoids from bacteria, 377–378
  for separation of carotenoids from plants, 146–147, 323–324
  silica gel, 76
Cone photoreceptors, 220–221
Continuous-flow fast-atom bombardment. See Fast-atom bombardment mass spectrometry
Continuous-wave gas laser(s), in surface-enhanced Raman scattering spectroscopy, 33
Coriander, 148
Corn oil, 159
Corynebacterium poinsettiae, 75
Cottonseed oil, 159
CPTA, 441
Crocetin, 169
Crocin, 437
Crocus, 169
α-Crustacyanin, 101, 108
Crustaxanthin, 380
  structure of, 109
Cryptoflavine, 408, 410
Cryptophyta, 171
α-Cryptoxanthin
  mammalian metabolism of, 271, 337
  structure of, 216

β-Cryptoxanthin, 358
  in bacteria, 376
  in chloroplasts of higher plants, 168
  high-performance liquid chromatography of, composition of solvent mixtures, flow rate, and retention time for, 275
  in human milk, 399
  in human plasma
    quantification of, 276
    spectral characteristics of, 276
  mammalian metabolism of, 271, 337
  in papaya, 358–359
  in primate retina, 228
  in pumpkin, 156
  in red bell pepper, 274
  structure of, 216
Cryptoxanthin(s), 150–151
  in fruits, 157–158
  in human plasma, 206, 273
    chromatographic profile of, 215, 218, 278–279, 281
  resolution of, 338
  singlet oxygen quenching constant of, 437
  standards for, plant sources of, 350
  in tuberous vegetables, 151
C_5 terpenoid precursor, 11
Cucurbita maxima. See Pumpkin
Cucurbita moschata. See Pumpkin
Culture Collection of Algae and Protozoa, 440
Cyanophyta, 170
Cyanopropylphenylmethylsilicone column, 283, 289
β-Cyclo-citral, 57
Cytochrome b-559, 41

D

Daffodil. See *Narcissus pseudonarcissus*
Daphnia, 482–483
Daucus carota. See Carrot(s)
DEAE-Sepharose column chromatography, 377–378
n-Decanoic anhydride, 274
Decanoyl chloride, 114
Decaprenoxanthin, 75
  structure of, 76
  synthesis of, 75
DEP. See Direct exposure probe
Deuterium (D)-labeled analog, of β-carotene, synthesis of, 49–53

scheme for, 50
Deuterium labeling, 43
Deuterium oxide, 51
2-Devinyl-2-carboxy-methylpyropheophorbide, 89, 92, 97
DHA. See Dihydroactinidiolide
1-(N,N-Diacetylamino)-4-methylbenzene, 92–93
Diadinoxanthin, in dinoflagellates, 245
Dialdehyde, 44, 48
Diapocarotenoid(s), 7
Diatoxanthin
  in dinoflagellates, 245
  in *Paralithodes brevipes*, 27–28
  structure of, 28
Diazomethane, 282
Dichloroethane, 278
Dichloromethane. See also Acetonitrile-dichloromethane-methanol extraction
  as solvent in carotenoid separation, 55–56, 199, 342, 352
Didehydroastaxanthin
  in *Paralithodes brevipes*, 27–28
  structure of, 28
Diepoxy-β-carotene, 61
5,6,5′,6′-Diepoxy-β,β-carotene, structure of, 473
5,8,5′,8′-Diepoxy-β,β-carotene, structure of, 473
Diethyl ether, 232, 235, 326
Dihydroactinidiolide, 57, 59
Dihydrodiactinidiolide, 54
5,6-Dihydroxy-5,6-dihydrolycopene, 213, 215–216, 218–219
Dimethylfuran, 406
Dinoflagellate(s), 232, 234
  carotenoids in, high-performance liquid chromatography of, 245
Dinophyceae, 232
Diode array detector. See Photodiode array detector
Diol acetate, 76
  preparation of, from 2-methyl-3-buten-2-ol, 81–82
  primary allylic hydroxyl group of, protection of, 82
Dioleoylphosphatidylcholine liposomes, 463
  as model membrane system, 461–472
Diphenylamine, 481–482
Direct exposure probe, 113–114, 117
DNA damage, 404–405

DOPC. *See* Dioleoylphosphatidylcholine liposomes
Doradexanthin, in birds, 320
α-Doradexanthin, 31
 circular dichroism spectra of, 30–31
 structure of, 29
Dorotheanthus bellidiformis pollen, 252, 254
 chromatogram of, 261–262
 preparation of, 255
DPA. *See* Diphenylamine
Dunaliella
 β-carotene biosynthesis in, 439–444. *See also* β-Carotene, biosynthesis of
 carotene-rich mutants of, 439, 443–444
 characteristics of, 439
 commercial sources of, 440
 growth conditions for, 440
 preparation of, for carotene analysis, 440
 sensitivity to photodestruction of, 440
Dunaliella bardawil, 170, 439
Dunaliella salina, 439
Dunaliella tertiolecta, 252, 254
 chromatogram of, 258–259, 261–263
 preparation of, 255
 principal carotenoids in, 257, 259
Dyads, 98–100

E

Echinenone
 in bacteria, 254, 256
 in birds, 320
 commercial sources of, 252
 in *Homarus americanus*, 319
 in *Paralithodes brevipes*, 27–28
 structure of, 28
ECNI. *See* Electron-capture negative-ion production
EDTA. *See* Ethylenediamine tetraacetic acid
EI. *See* Electron impact mass spectrometry
Elaeis guineensis. *See* Palm oil
Elaeis oleifera. *See* Palm oil
Electron-capture negative-ion production, 112, 114–115
 versus electron ionization or chemical ionization, 112
 sensitivity for determination of molecular weight, 118
Electron impact mass spectrometry, 111, 114, 323

 disadvantage of, 323
 versus electron-capture negative-ion production, 112
 sensitivity for determination of molecular weight, 118
Electron paramagnetic resonance spectroscopy
 for detection of carotenoid triplet states, 305–312
 development of, 308
 materials and methods for, 307
 results and discussion of, 308–312
 double-modulation, for detection of carotenoid triplet states, 307
 triplet state signals
  radical-pair spin polarization pattern of, 308–310
  temperature dependence of, 308
  zero-field splitting parameters detected by, 311
Electron transfer, 90
 study of, 87
Emmons-Horner reaction, 13
End capping, 192
Enzymatic hydrolysis, 22–23
 of esterified carotenoids, 25
Enzyme activity, alteration of, 405
3'-Epilutein, 7
 in human plasma, chromatographic profile of, 214–215, 218
 structure of, 216
5,6-Epoxy-β-carotene, 61
5,6-Epoxy-β,β-carotene, structure of, 473
5,8-Epoxy-β,β-carotene, structure of, 473
15,15'-Epoxy-β,β-carotene
 resolution of isomers of, by HPLC, 476–477
 structure of, 473
Epoxycarotenoid(s), 358
 oxidation products of, 408, 410
Epoxy compounds
 analysis of, 132–135
 functions of, characterization of, 133
C_{15}-Epoxyformyl ester, 20–21
5,6-Epoxy-β-ionone, 57, 59
EPR. *See* Electron paramagnetic resonance spectroscopy
Erythrobacter longus, 375
 carotenoids from, 375–376
  absorbance spectra of, 378–381
  elution profiles of, 377

HPLC analysis of, retention time on, 381–382
identification of, 378–382
isolation and purification of, 376–378
structures of, determination of, 382
carotenoid sulfates from, 383–385. *See also* Carotenoid sulfates
preparation of, for carotenoid extraction, 376
Erythropoietic protoporphyria, 483
Erythroxanthin, derivatives of, 383
Erythroxanthin sulfate, 375, 384
absorption spectra of, 383
structure of, 385
ESA model 5100A coulometric detector, 477
Eschscholzanthin, 169
Esterified carotenoid(s), enzymatic hydrolysis of, 25
Ethanol, 316, 337, 364. *See also* Chloroform-ethanol
commercial sources of, 392
Ether, 350, 362
Ethyl acetate, 199, 325–326
Ethyl-β-apo-8′-carotenoate, 209, 349
chromatographic profile of, 215
commercial sources of, 349
gas chromatography of, 289–290
structure of, 217
Ethylenediaminetetraacetic acid, 104, 222, 339
Ethylhydroxylamine, 225
Ethyl-β-ionylidene acetate, 50, 52
Ethynyl β-ionol, 45
Euglena rubida, 386
Euglenophyta, 171
Extrasynthese, 252
Extrudates, bleaching kinetics of, 141–142
Extrusion cooking
of all-*trans*-β-carotene, 129–137
conditions of, 130
definition of, 129
equipment for, 130
extrudates used in, bleaching kinetics of, 141–142
Eye(s)
human
commercial source of, 361
cross-sectional diagram of, 222
primate, anatomy of, 222

F

FAO. *See* Food and Agricultural Organization
Fast-atom bombardment mass spectrometry
advantages of, in carotenoid analysis, 323
of carotene structural isomers, 330
in carotenoid analysis, 322–336
equipment for, 325
procedures for, 325
purification of fractions for, 324
results and discussion of, 326–335
limits of detection for, 335
selection of matrix for, 333
Fatty acid esters, of carotenoids, 366–367
structure of, 372–374
Fatty acids, 358
Feathers. *See* Avian carotenoid(s)
Fennel, 148
Fibroblast(s), 483
Field desorption apparatus (FD-MS), for determination of molecular weights, 382–384
Finnigan-MAT model 4500 spectrometer, 114
Flame-ionization detection, 283, 372
Flash photolysis, 306
Flavin(s), 480
Flavobacteria, 170
Flavobacterium dehydrogenans, 75
Flavonoid(s), 461
Flavoxanthin, 350
Flow method, for avoidance of isomerization during Raman spectroscopy, 303–304
Fluka Biochemica, 176, 349, 392
Fluka Chemical Corporation, 274, 474
Fluorescence spectroscopy, 32
Fly, 482–483
Food and Agricultural Organization, 151
Foods
carotenoids in. *See also* Fruit(s), carotenoids; Vegetable carotenoid(s)
chromatographic profiles of, 353–354
high-performance liquid chromatography of, 348–353
identification of, by HPLC separation in order of elution, 355
number of, 347

separation and quantitation of, 347–359
  carotenoid standards for, 349
  chromatographic conditions for, 348–349
  experimental procedures for, 348–353
  by HPLC, 353–359
  instrumentation for, 348
  internal standards for, 349–350
  preparation of food extracts for, 350–351
  preparation of standard curves for, 352–353
  saponification, 351–352
 types and concentrations of carotenoids in, 144–145
Fourier transform infrared absorption
 for identification of fucoxanthin, 234, 239
 for identification of peridinin, 244
Fovea, carotenoids in, 360–361, 364–366
Fractionation, of palm oils. *See* Palm oil, extracts from
Free radical(s)
 carotenoid inhibition of, 450–452
 classification of, 450, 460–461
 generation of, 445–448
 location of generators of, 469
 steady state concentrations of, antioxidant enzyme control of, 461
Free-radical scavenger(s), 175
 efficiency of, 469
French marigold, carotenoids in, 249–250
Fritschiellaxanthin, 31
 circular dichroism spectra of, 30–31
 in *Paralithodes brevipes*, 27–28
 structure of, 28–29
Fruit. *See also individual species*
 carotenoids, 156–158
  chromatographic profiles of, 353
  classification of, 169
  extraction of, 350–351
  high-performance liquid chromatography of
   for identification of in order of elution, 355
   for separation, 353–359
  standards for, source of, 350
  synthesis of, 169
  three classes of, 158
  green, carotenoids in, 353–354
  ripe versus unripe, carotenoids in, 157
  vitamin A activity of, 158
  yellow-orange, carotenoids in, 357–359
  yellow-red, carotenoids in, 354–357
FT-IR. *See* Fourier transform infrared absorption
Fucoxanthin, 6–9, 146, 231
 in algae
  isolation of, 237
  production of, 231
 allenic bond in, chirality and isomerization of, 232
 all-*trans*-
  circular dichroic spectra of, 239
  Fourier transform infrared absorption spectra of, 240
  identification criteria for, 234, 238–239
 de novo production of, 231
 high-performance liquid chromatography of
  for isolation, 237–238
  for qualitative detection, 234
  for semipreparative isolation, 234
 isolation of, 231–245
  flow diagram for, 233
  general precautions with, 236
  procedures for, 233, 237–238
  reagents and instrumentation for, 235–236
  strategy for, 232–234
 partition data for, in hexane-aqueous methanol, 234
 purification of, 237–238
 structure of, 231
Fucus serratus. *See* Algal carotenoid(s)
Fungal carotenoid(s), 171
 photoprotective activity of, 482
Furanoid oxide, functions of, characterization of, 133

G

Gas chromatography, of carotenoids, 251
Gas chromatography-mass spectrometry, 129
 for analysis of molecular weights, 370, 382–384

of apocarotenoids and retinoids, 281–290
  equipment for, 283
  results of, 283–290
  direct interfacing of equipment, 284–385
  limitations of, 289
  packed columns, versus bonded-phase capillary columns, 282
Gas-liquid chromatography, 372
Gas-phase reactions, 112
Geranial, 58
  kinetic comportment of, 59–60
  trans-cis isomerization of, 60
Geranyl acetate, 75
  electrophilic alkylation-cyclization of, 76, 80–81
Geranylgeranyl pyrophosphate, 11–12
Germanium photodetector, 424, 431
  setup of, 433–435
GLC. *See* Gas-liquid chromatography
D-Glucoside, 367
Glutathione radical(s), and retinol, reaction between, 450
Glycerol, 254, 325, 333–334
Gradient elution, 186, 199–204, 273–274, 353–354
Grape(s), 157
Grapefruit(s), 158, 354
Green beans, 353–354
Green leaves, carotenoids in, 147–156. *See also* Vegetable(s)
  three main types of, 146
Green plant photosynthetic systems, surface-enhanced resonance Raman scattering spectroscopy studies of, 39–41
Guinea pig(s), 418
Gymnosperm(s), 168

H

Haematococcus pluvialis, 171
  absorbance of DMSO extraction of, 388
  astaxanthin accumulation in, 386–391. *See also* Astaxanthin
  commercial sources of, 387
Hanasakigani. *See Paralithodes brevipes*
Helium, 283
Helium sparging, 187
1-Heptanol, 58
Hermit crab. *See Paralithodes brevipes*

Herpetosiphon gigantens, 366
Hewlett-Packard gas chromatograph, 283
Hewlett-Packard model 1090L liquid chromatograph, 115
Hewlett-Packard 5996 quadrupole mass spectrometer, 57
Hexadecanoate, 290
Hexane
  commercial sources of, 392
  as solvent in carotenoid separation, 61, 225, 233, 235, 448
Hexane-acetone extraction, of avian carotenoids, 315–318
n-Hexane layer, purification of, 299
Hexanoic anhydride, 274
High-performance liquid chromatography, 8, 23, 131, 139, 322–323
  for carotenoid isolation
    equipment for, 324
    procedures for, 324–325
  for carotenoid separation, 253–254, 263
  historical development of, 185–186
  with lime column, 291, 294–296
  of stereoisomers, 26, 293–296
  coupled with mass spectrometry. *See* Liquid chromatography-mass spectrometry
  for determination of hydroperoxides, 457–459
  eluent solutions for, 300
  equipment for, 27, 176, 191, 246–247, 278, 300, 324, 340, 394, 474
  high-resolution, of plasma carotenoids, 336–346. *See also* Plasma carotenoid(s)
  of macular pigments, 360
    internal standards for, 362
  with photodiode array detector, 251–265. *See also* Photodiode array detector
  preparative, 300
    of carotenoids, 246–250
    choice of system for, 247–248
    general procedures for, 246–250
    operation of, 248–249
    stationary phase of, 248–249
  retention time on, 381–382
  reversed-phase, 206, 247
    for carotenoid separation, 185–205, 253, 330–331
    column packing for, 187

regeneration of, 249
column parameters for, 191–204
  effects of particle size and shapes on, 191
  effects of pore diameter/surface coverage, 191–192
  effects of temperature on, 195–196, 204
  end capping, 192, 194
  limitations of, 262
  mobile phase of, 196–200, 278, 340, 475
  procedures for, 186–191
  stationary phase synthesis, 192–196, 204
drawbacks of, 355–356
gradient elution (System A)
  carotene fraction formed on, 67
  for carotenoid separation, 199–204, 207, 211, 273–274, 348–349, 353–354
isocratic elution (System B), for carotenoid separation, 187–189, 199, 201, 207, 211–212, 219, 273–274, 279, 349, 355
nonaqueous, 253, 279
  composition of mobile phases used in, 254
versus normal-phase, 186
retention times on, 381–382
sample introduction to mass spectrometry with, 113, 126–128
solvent systems for. *See* Solvent system(s)
system I, 67–68
system II, 68–70
  calibration of, with carotenes from tomato, 68–70
  separation of substrate and products from cyclization assay using, 71–73
system III, 69, 71
  separation of *cis*-lycopene standards using, 69, 71
Histidine, 406
Hitachi M-80 instrument, 27
Hoffmann-LaRoche, Inc., 51, 176, 252–253, 314, 319, 323, 425, 483
Homarus americanus. See Lobster
Homarus gammarus. See Lobster
Homocarotenoid(s), 7

HPLC. *See* High-performance liquid chromatography
Human milk
  carotenoids and retinol in
    effects of sample mixing during saponification, 397–398
    effects of saponification time on, 395–398
    effects of temperature on, following saponification, 395, 397
    HPLC quantitation and separation of, 391–399
      choice of column for, 398–399
      equipment for, 394
      external standards for, 394
      materials for, 392
      procedures for, 392–394
      results of, 394–399
      sample preparation for, 392–394
      saponification methods for, 394–398
    composition of, versus colostrum, 391, 394
Human plasma
  carotenoids in. *See* Plasma carotenoid(s)
  sample preparation of, 210–211, 339
  vitamin A in, 206
  vitamin E in, 206
Human retina(s). *See* Retina(s), primate
Hydrocarbon carotenoid(s), 4, 186, 228, 329
  chromatographic separation of, 349, 353–357
  FAB ionization of, 327
  in fruits, 158, 354–357
  gas chromatography of, 282
  in human plasma, 213–214
  mass spectrometric analysis of, 115–120
  oxidation products of, 408, 410
  in primate retina, 229
  in vegetables, 149, 151, 354–357
Hydrodynamic voltammograms, for β-carotene epoxides, 477–478
Hydrogen, 283
Hydrogen isotopes, retinoids labeled with, 45
Hydrogen peroxide, 405, 461
Hydrolysis, enzymatic. *See* Enzymatic hydrolysis
Hydroperoxides
  isomeric composition of, 460
  standard preparation of, 456–457
  storage of, in solution, 457

Hydrophobic assay system(s), 469–470
Hydrophobic carotenoid(s), 333
Hydroxycarotenoid(s), 329, 352
  isomeric, 360
  mass spectrometric analysis of, 120–123
3-Hydroxy-β,ε-caroten-3'-one, 215–216
3'-Hydroxy-β,ε-caroten-3'-one, 214, 218
3'-Hydroxy-ε,ε-caroten-3-one, 214, 216, 218–219
3-Hydroxy-4-ketocarotenoid(s), identification of, 318
Hydroxyl derivative(s), formed during thermal degradation of β-carotene, 136–137
Hydroxyl group(s), acetylation of, 318
Hydroxyl radical(s), 405, 461
Hydroxyspheroidene, 380
2-Hydroxy-2,6,6-trimethylcyclohexane-1-carboxaldehyde, 57
2-Hydroxy-2,6,6-trimethylcyclohexanone, 57

I

Ice plant. *See Dorotheanthus bellidiformis* pollen
ICN, 176
Idoxanthin, structure of, 109
Immuno response, carotenoid enhancement of, 175, 266, 391, 416–417
INEPT experiments, 291
Infrared light absorption spectra, 23, 32, 291
  for characterization of carotenoids, 25–27
  instruments for, 27, 79, 236
Inhoffen synthesis, 9
International Symposium on Carotenoids, 3
International Union of Biochemists, 4
International Union of Pure and Applied Chemistry, 4
International Vitamin A Consultative Group, 144–145
Invertebrate carotenoid(s), photoprotective activity of, 482–483
β-Ionone, 45, 51, 54, 57
  deuterated, 50–51
β-Ionylidene ethanol, deuterated, 50, 52
β-Ionylidene ethyltriphenylphosphonium bromide, deuterated, 50, 52–53
IR. *See* Infrared light absorption spectra
ISCO model V^4 detectors, 278
Isocratic elution, 186–189, 273–274, 278–279, 326, 355

Isocryptoxanthin, commercial sources of, 252
Isomerization
  during nuclear magnetic resonance spectroscopy, methods for avoidance of, 300–302
  during Raman spectroscopy, methods for avoidance of, 303–304
  trans to *cis*, 293
Isomers, all-*trans*, spectra of, 9
Isopentenyl diphosphate, 63
  phytoene synthesized from, 62
[1-^{14}C]Isopentenyl diphosphate, 64, 71
Isopentenyl pyrophosphate, carotenoids synthesized from, 11–12
Isoprene epoxide
  alkylation of geranyl acetate with, 76, 80–81
  synthesis of, 79
Isopropyl acetate, 235
Isorenieratene, 170
Isozeaxanthin, 380
  absorption spectra of, 258
  commercial sources of, 114, 252
  electron-capture negative-ion production of, versus lutein, 123
  molecular weight of, 118
  structure of, 116
Isozeaxanthin bispelargonate
  chemical ionization of, 124
  commercial sources of, 114
  electron-capture negative-ion production of, 124–125
  electron ionization of, 123–124
  mass spectrometric analysis of, 123–125
  molecular weight of, 118
  relative abundance of ions versus rate of probe heating current for, 125
  structure of, 116
IUB. *See* International Union of Biochemists
IUPAC. *See* International Union of Pure and Applied Chemistry
IVACG. *See* International Vitamin A Consultative Group

J

J-334, 441
Jackfruit, 157

Jasco 500-C spectropolarimeter, 27
Jobin-Yvon Miniprep LC system, 246–247, 249
Julia's methods, in carotenoid synthesis, 13

K

Kale, 114, 147
Ketoalloxanthin, structure of, 28
Ketocarotenoid(s), 25, 386
 in Arthropoda, 23, 386
 HPLC analysis of, 337–338
 mass spectrometric analysis of, 125–126
 production of, 336–337
 reduction of, 256
4-Ketocarotenoid(s), in birds, 320
Ketone derivatives, formed during thermal degradation of β-carotene, 135–136
3-Ketoretrodehydrocarotenoid, 320
Ketozeaxanthin
 in *Paralithodes brevipes*, 27–28
 structure of, 28
Keystone Scientific, 363
Key to Carotenoids (Pfander), 7
Kishida Chemical Co., Ltd., 291, 300
Kovats retention indices, 283
 of aporetinoids and retinoids, 284, 288
 relationship of carbon number to, 289

L

Langmuir-Blodgett film, 453
Laser(s). *See* Continuous-wave gas laser(s); Pulsed laser(s); Titanium:sapphire laser(s)
LC-MS. *See* Liquid chromatography-mass spectrometry
LDLs. *See* Low-density lipoproteins
Leprotene. *See* Isorenieratene
Leukocytes, activated polymorphonuclear, 405
LiBr. *See* Lithium bromide
LiChrosorb CN-5 column, 300
Light absorption characteristics, with FAB-MS, 335
Light-sensitive skin disease(s), 484
Lilium, 169
Lilixanthin, 169
Lime, commercial sources of, 291
Lime column, 291, 294–296

Linalool, for β-carotene degradation, 58
Linseed oil, 159
Lipid-homogeneous solution(s), carotenoid inhibition of lipid substrates in, 408–409
Lipid hydroperoxide assay, for antioxidant activity of carotenoids, 454–460
 buffer solution for, 456
 carotenoid solution for, 456
 comments on, 459–460
 hydroperoxide standards for, 456–457
 methyl linoleate hydroperoxide method for, 457–458
 drawbacks of, 459
 methyl linoleate solution for, 455
 phosphatidylcholine solution for, 455–456
 phospholipid hydroperoxide method for, 457–459
 drawbacks of, 460
 radical initiator for, 456
 reagents and solutions for, 455–457
Lipid hydroperoxides, direct analysis of, 454
Lipid peroxidation, 404–405, 454, 461
 carotenoid inhibition of, 408–416, 451. *See also* Antioxidant(s)
 in vitro animal studies of, 417–418
 in liposomes, 412–413
 measurement of, 454
 mechanism of, 454–455
 causes of, 406
 measurement of, 411
 in membranes, in presence of carotenoids, 414
Lipids. *See also* Polar lipids
 in human tissues, 346
Lipoproteins, lipid peroxidation in, carotenoid inhibition of, 412–413
Liposomes, lipid peroxidation in, carotenoid inhibition of, 412–413
Liquid chromatography-mass spectrometry
 in carotenoid analysis, 332–336
 equipment for, 325–326
 isocratic solution for, 326
 procedures for, 326
 results and discussion of, 326–335
 identification of carotenoids by, 333–334
 limits of detection for, 335
 selection of matrix to mobile phase of, 334–335

Lithium aluminum hydride, 51
  conversion of bixin to bixindiol with, 114
Lithium bromide, 315
Lobster, 101
  carotenoids in, standards for, separation of, 319
  carotenoproteins from, 104, 108. *See also* Carotenoprotein(s)
Low-density lipoproteins, 412
Luminescence signal. *See* Near-infrared steady-state luminescence
Lutein, 4, 6–7, 9, 146, 329, 358
  in algae, 257, 259
  antioxidant activity of, 411–412, 415
  in bacteria, 380
  in birds, 318, 320
  in blueberries, 325–329
  in carrots, 326, 328, 333–334
    chromatographic separation of, 326
  chemical ionization of, 121–123
  chemopreventive effects of, 336
  in chloroplasts of higher plants, 168
  9′-*cis*-, 210
  commercial sources of, 252, 274, 462
  electron-capture negative-ion production of, 123
    versus isozeaxanthin, 123
  electron ionization of, 120–121
  in fruits, 157–158
  in green vegetables, 149–151
  high-performance liquid chromatography of
    composition of solvent mixtures, flow rate, and retention time for, 275
    effects of column temperature on, 343
  in *Homarus americanus*, standards for, separation of, 319
  in human milk, 395–398
  in human plasma, 206, 273, 336
    chromatographic profile of, 215, 278–279, 281, 338
    quantification of, 276
    spectral characteristics of, 276
  in human retina, 360–366, 429
    chromatographic profiles of, 364–366
    separation and quantitation of, by HPLC, 363
    stereochemistry of, 363–364
  LC-MS spectra of, 333–335
  mammalian metabolism of, 271, 337

  mass spectrometric analysis of, 120–123
  metabolic conversion of, to ketocarotenoids, 336–337
  molecular weight of, 118
  natural sources of, 274
  in papaya, 358–359
  in *Paralithodes brevipes*, 27–28
  in primate retina, 220, 225, 227–228
  in pumpkin, 156, 354, 356
  (3R,3′S,6′R)-, 214, 218
  separation of, 192, 196–197, 208, 226
  singlet oxygen quenching constant of, 437
  standards for
    plant sources of, 350
    preparation of, 394
  structure of, 28, 116, 216
  in tomatoes, 326, 328
  in tuberous vegetables, 151
  in vegetable oils, 159
Lutein bisdecanoate, commercial sources of, 114
Lutein epoxide, 126
  commercial sources of, 114
  molecular weight of, 118
Lutein-5,6-epoxide, plant sources of standards for, 350
Lutein monopalmitate, 126
Luteochrome, plant sources of standards for, 350
Luteoxanthin
  in algae, 242
  in pumpkin, 156
Lycopene, 4, 6–8, 11–12, 147, 228, 260, 265
  7,9,9′,7′-tetra-cis-[^{14}C]-. *See* Prolycopene
  all-*trans*-, 62
  antioxidant effects of, 413, 415
  in apricots, 357
  in bacteria, 380
  bicyclic carotenoids synthesized from, 75
  in blueberries, 328
  in carrots, 153, 326, 328
  chromatographic separation of, 204
  cis,trans-, 62
  commercial sources of, 252, 274, 392, 462
  cyclization of, 64–65
    to α/β-carotene, 62
  decomposition of, and experimental considerations, 271–272
  distribution of, in pooled extracts of

mammalian organs, 270, 272
FAB mass spectrum of, 330, 332
in fruits, 157–158
in green vegetables, 150–151
high-performance liquid chromatography of, composition of solvent mixtures, flow rate, and retention time for, 275
in human diet, 265
in human milk, 395–398
in human plasma, 206, 213–214, 265, 273
  chromatographic profile of, 215, 278–279, 281
  quantification of, 276, 280
  spectral characteristics of, 276
intermediate accumulation of, 441–442
isolated from plant extracts, identification of, 326
mammalian metabolism studies of, 265–272
  animals used in, 267–268
  HPLC analysis in, 268–269
  materials used in, 266–267
  results of, 269–272
in palm oil, 162, 164–165
in primate retina, 229
red-colored, 62
resolution of, 338
in roots and tubers, 158–159
singlet oxygen quenching activity of, 405
singlet oxygen quenching assay of, 429–438. *See also* Singlet oxygen quenching
singlet oxygen quenching constant of, 437
standards for
  plant sources of, 350
  preparation of, 394
stereoisomers of, 439
storage of, 271–272
structure of, 217
thermal degradation of, 54–55, 58
  components of, 54
  nonvolatile compounds formed during, 61
  volatile compounds formed during
    kinetics of, 59–60
    main types of, 58–59
in tomatoes, 154–155, 274, 326, 328, 330–331
  isolation of, 323–324
in tuberous vegetables, 151

M

Macaque(s), 227
Macropipus puber, red pigments from, 104
Macula lutea, 220. *See also* Retina(s)
  definition of, 221
  human, versus monkey, 227
  tissue specimens of, storage of, 229
Macular degeneration, age-related, 220, 336
Macular pigments
  analysis of, 227–230
  extraction of, 363
  function of, 220–221
  human, 227, 336
    distribution of, 360–366
    versus monkey, 230
    stereochemistry of, 363–364
  yellow, 220
Magnesium hydroxide, 147
Magnetophotoselection, 306
Maize. *See* Corn oil
Malondialdehyde, 411
Mammalian carotenoid(s), photoprotective activity of, 483–484
  analysis of
    procedures for, 484
    tissue cultures for, 483
Manganese, 41
Mango(es), 157, 350, 358
Mass spectrometry, 23. *See also* Chemical ionization mass spectrometry; Electron impact mass spectrometry; Fast-atom bombardment mass spectrometry; Gas chromatography-mass spectrometry
  of canthaxanthin, 125–126
  of β-carotene, 115–120
  of β-carotene epoxides, 478–479
  of carotenoids, 111–128
    for characterization, 25–27
    instrumentation for, 114–115
    methods for, 113–115, 325
    for rationalization, 43
    sample handling for, 113–115
    sample introduction to, 112–113, 126–128
    techniques for, 115–126
  of carotenol acyl bisesters, 123–125
  classical methods of, 111
  coupled with high-performance liquid chromatography. *See* Liquid chromatography-mass spectrometry

equipment for, 27, 79, 236, 284, 324–325
gas-phase reaction in, 111
of hydrocarbons, 115–120
of hydroxycarotenoids, 120–123
ionization voltage of, 284
of isozeaxanthin bispelargonate, 123–125
of ketocarotenoids, 125–126
of lutein, 120–123
quadrupole, 283
techniques of, 111, 323
MDA. *See* Malondialdehyde
Medium-pressure liquid chromatography, 131
equipment for, 79
Melon leaves, 148
Membrane integral enzymes, 62–63
Metalloporphyrin(s), in photoinduced charge separation, 452–453
Methanol. *See also* Acetone-methanol extraction; Acetonitrile-dichloromethane-methanol extraction; Acetonitrile-methanol-ethyl acetate extraction; Acetonitrile-methanol extraction; Chloroform-methanol extraction; Petroleum ether-acetonitrile-methanol extraction
commercial sources of, 392
as solvent in carotenoid separation, 326, 337, 350, 463
as solvent in singlet oxygen quenching study, 425–426
Methanol radical(s), generation of, 450
2-Methylbutanal, 60
2-Methyl-3-buten-2-ol, 76, 78
preparation of C_{15} cyclic diol acetate from, 81–82
Methylene chloride, 364
6-Methyl-3,5-heptadien-2-one, 58, 60
2-Methyl-2-hepten-6-one, 54, 58–59
Methyl linoleate, 455–456
AIBN-induced oxidation of, 409
purification of, 455
Methyl linoleate hydroperoxide method, for lipid hydroperoxide assay, 457–458
Methylobacterium rhodinum, 367
Methylpyropheophorbide *a*, 90
Methyl retinoate
packed-column gas chromatography of, 282
preparation of, from retinoic acid, 282
retention indices of, 289

stability of, under gas chromatography conditions, 282
Methyl retinyl ether, packed-column gas chromatography of, 282
Methylsilicone columns, 283
Micrococcus roseus, 252, 254
chromatogram of, 258–259, 261
preparation of, 255
Milk. *See* Human milk
Molecular electronic devices, carotenoids in, 452–453
Molecular Probes, 461
Molecular weights, of carotenoids, determination of, 118, 128, 382–384
Monkey retina(s). *See* Retina(s), primate
Mono-β-carotene, 61
9-Mono-*cis*-carotene, 291
structure of, 292
13-Mono-*cis*-β-carotene, 291
structure of, 292
17,7-Mono-*cis*-β-carotene, structure of, 292
Monochromator, 424
Monochromator/spectrograph combination, in surface-enhanced Raman scattering spectroscopy, 33, 35
Monocyclic carotenoid(s), 376
Monomeric synthesis, 192–195
Mosquito(es), 482–483
MP. *See* Metalloporphyrin(s)
MPLC. *See* Medium-pressure liquid chromatography
MPTA, 441
MS. *See* Mass spectrometry
Mushroom(s), 266
Musk lime(s), 157
Mustard green leaves, 66, 148, 274
Mutagenesis, carotenoid inhibition of, 265–266, 416–417, 445
Mutatochrome, 61
Mycobacteria, 170
Mycobacterium bovis, 373
Mycobacterium smegmatis, 373
Myxobactin ester, 366
Myxobactone ester, 366
Myxococcus fulvus, 366

N

NADPH, lipid peroxidation induced by, 406
NADPH/cytochrome *P*-450 reductase, 405–406

Nakarai Chemical Industries, 291
Narcissus majalis, 147
Narcissus pseudonarcissus, chromoplast membranes, formation of carotene chromophore and ionone rings in, 62–74
NARP. *See* High-performance liquid chromatography, reversed-phase
National Disease Research Interchange, 361
National Institute of Standards and Technology, 187
NBA. *See* 3-Nitrobenzyl alcohol
NDP, synthesis of, 432
NDPO$_2$
 preparation of, from NDP, 432–433
 thermodissociation of, 431
Near-infrared steady-state luminescence, measurement of singlet oxygen quenching by, 419–420, 430
 experimental procedures for, 425–426
 luminescence signal, 422, 424
  intensity of, 424–425
 procedures for, 421–424
 reagents for, 425
Neo-β-carotene, U and B, 291
 configurations of, 291
Neochrome
 in algae, 242
 standards for, plant sources of, 350
Neoxanthin, 6–7, 9, 146
 in algae, 242, 257, 259
 in chloroplasts of higher plants, 168
 commercial sources of, 253
 high-performance liquid chromatography of, composition of solvent mixtures, flow rate, and retention time for, 275
 standards for, plant sources of, 350
Neral, 58
Nermag R10-10 apparatus, 79
Neurospora, 482
Neurosporene(s), 11–12, 260
 in bacteria, 380
 in benzene solution, isomerization shifts of, 301–302
 high-performance liquid chromatography of
  eluent for, 300
  with lime column, 296
 in human plasma, 213–214
  chromatographic profile of, 215
 in palm oil, 162, 164

 structure of, 217
7,9,9'-tri-*cis*-Neurosporene. *See* Proneurosporene
Nicotine, 441–442
NIST. *See* National Institute of Standards and Technology
Nitrile. *See* Acetonitrile
Nitrile-bonded columns, 212–219
3-Nitrobenzyl alcohol, 325, 333, 335
Nitrogen, 316, 325
 purified, 55–57
NMR. *See* Nuclear magnetic resonance spectroscopy
Nocardia kirovani, 366
Nocardioform actinomycete, 366
Nonapreno-β-carotene, 349, 354
 preparation of, 349–350
Nonoxyl radical(s), 450
Nonvolatile compounds
 formation of, by thermal degradation of carotenoids, 60–61
 identification of, 61–62
 isolation of, 60–61
Norflurazon, 441
Nostoxanthin, 376, 380
Nuclear magnetic resonance spectroscopy, 7, 246, 291
 carbon-13
  for identification of fucoxanthin, 234, 239
  for identification of peridinin, 244
 for carotenoid separation, 151–152
 instruments for, 236
 proton
  for carotenoid identification, 234, 239, 244
  of carotenoid sulfates, 384–385
  for determination of carotenoid configuration, 298–305, 372, 382
   assignment of olefinic signals, 301
   avoidance of isomerization during, 300–301
   procedures for, 300–301

O

$^1O_2$. *See* Singlet oxygen
Octadecanoate, 290
1-Octanol, 278
Octyl (C$_8$), 186

Octyldecyl (C_{18}), 186–187, 212–219
Octyldecylsilane (ODS), 187
Oil palm(s). *See* Palm oil
Olefination, palladium-catalyzed, 20–22
Olefinic ^1H signals
  absolute values of chemical shifts of, 301
  assignment of, 301
Olive oil, 159
Onion(s), 148
Orange(s), 157, 350, 358
Organic radical(s), 460–461
Oxidation(s)
  generation of carotenoid free radical by, 445
  induced, 408
  in lipid hydroperoxide assay, 457–458
Oxycarotenoid(s). *See* Xanthophyll(s)
Oxygenated carotenoid(s). *See* Xanthophyll(s)
Oxygen radical(s), 460–461
  carotenoid quenching of, 405
Oxyl radical(s), 450–451

P

Pachygrapsus marmoratus eggs, purple carotenoproteins from, 104
  purification of, 106
Palladium-catalyzed olefination, 20–22
Palladium-coupling reaction, 22
Palm alkyl ester(s), distillation of, 165–166
Palm oil
  carotenoids in, 160–161, 350
    extraction methods for recovery of, 165–166
    high-performance liquid chromatogram of, 163
    percent composition of, 164
    regional and species variations in, 160–161, 164
  crude, carotenoid profile of, 162
  extracted from fiber, carotenoid profile of, 162–165
  extracts from, carotenoid contents of, 161
    and methods of extraction and fractionation, 161–162
  fractionation of, 162
  pressed fiber, 162–165
  second pressed fiber, 162–165

Palm oil methyl esters
  carbon adsorption of, carotenoid concentration obtained by, 165–166
  as diesel substitute, 166
Papaya(s), carotenoids in, 157, 350, 358–359
Papilioerythrinone
  absolute configuration of, determination of, 29–30
  circular dichroism spectra of, 30–31
  in *Paralithodes brevipes*, 27–28
  structure of, 28–29
Paprika, 147, 152, 274
Paralithodes brevipes, 23
  carotenoids from, 22–31
    isolated using column chromatography, 23–25
  structures of, 28, 30
  using column chromatography, 23–25
Paraquat, lipid peroxidation induced by, 406
cis-Parinaric acid, 469
  commercial source of, 461
  structure of, 470
cis-Parinaric acid fluorescence-based assay, of antioxidant radical-scavengers
  in DOPC liposomes, 463, 468
  in hexane, 462–463, 465–467
Parrot(s), 320
Parsley, carotenoids in, 249–250
Passerine(s), yellow pigments in, 320
PE. *See* Petroleum ether
Peach(es), 157–158, 350, 358
Peanut oil, 159
Peas, 148
Pectenolone
  isolated from *Paralithodes brevipes*, 27–28
  structure of, 28
Pennywort, 148
Peridinin, 14, 21, 231
  in algae, 241–243
  allenic bond in, chirality and isomerization of, 232
  all-*trans*-
    circular dichroic spectra of, 239
    in dinoflagellates, 245
    Fourier transform infrared absorption spectra of, 240
    identification criteria for, 234, 244
    visible light absorption spectra of, 243

de novo production of, 232
isolation of, 231-245
  general precautions with, 236
  procedures for, 241-244
  reagents and instrumentation for, 235-236
  strategy for, 232-234
  partition data for, in hexane-aqueous methanol, 234
  qualitative detection of, HPLC procedure for, 234
  separation of, from chlorophylls, 234
  structure of, 14, 231
  total synthesis of, 13-22
Perkin-Elmer 599 spectrophotometer, 79
Peroxyl radical(s), 454
  carotenoid quenching of, 406-407
  mechanism of, 406-407
  carotenoid scavenging of, 450-451, 454
Peroxy radical(s), deactivation of, by carotenoids, 403
Petroleum ether. *See also* Acetone-petroleum ether extraction
  as solvent in carotenoid separation, 247-249
Petroleum ether-acetonitrile-methanol extraction, 248
Petroselinum crispum. See Parsley
Phaeophyceae, 231
Phaeophyta, 171
Pheasant's eye narcissus, 147
1H-Phenalen-1-one, 421, 425
Phenol, 254
Phoenicoxanthin
  in birds, 320
  in *Paralithodes brevipes*, 26-28
  stereoisomers of, HPLC separation of, 25-26
  structure of, 28
Phosphatidylcholine
  purification of, 455-456
  soybean, 460
Phosphatidylcholine liposomes, carotenoid reaction with, 455
Phospholipid, 346
Phospholipid hydroperoxide method for, for lipid hydroperoxide assay, 457-459
C_7-Phosphonium chloride, preparation of, 18-19

Phosphonium salt(s)
  ^{14}C-labeled, 45-46, 48
    storage of, 46
  labeled, 44-46
  in synthesis of carotenoids, 44-49
  transformation of C_{20} alcohol into, 85-86
  tritium-labeled, 45-46, 48-49
    storage of, 46
Photodiode array detection. *See also* Germanium photodetector
  with HPLC, 251-265, 318, 337-338, 354, 475
  applications of, 252
  biological samples for, 252
  conditions for, 253-254
  definitive peak identification with, 256-257
  determination of peak purity with, 259-263
  instrumentation for, 253
  internal standards used in, 264
  liquid instrumentation for, 253
  preparation of biological extracts for, 254-256
  qualitative aspects of, 256-263
  quantitative aspects of, 263-265
  standard compounds for, 252-253
  tentative peak identification with, 258-259
Photodiode array detector, 74, 113, 115, 158, 210, 219, 324
  characteristics of, 251-252
Photodynamic effect, 420
Photoinduced charge separation, 452-453
Photoinhibition, protection against, 442-443
Photoisomerization
  of carotenoids in solution, 299
  iodine-catalyzed, 292, 294
Photonic flux, 422, 424
Photooxidation, type II, 420
Photoprotection, and carotenoids, 266, 305, 312, 404-405, 479-484
Photosensitization
  carotenoid protection against. *See* Photoprotection, and carotenoids
  singlet oxygen production by, 420-422
Photosynthesis
  intramolecular charge separation and

transfer in, electronic device to mimic, 452
role of carotenoids in, 420–421
Photosynthetic antennae, and carotenoporphyrins, 87
Photosynthetic energy, study of, 87
Photosystem II preparations, 36, 39–40
β-Phycoerythrin, 462
  oxidation by peroxyl radicals, 462
Phycoerythrin fluorescence-based assay, in aqueous system, of antioxidant radical-scavengers, 462–467, 469
Phycomycetes, 171
Phytoene, 11–12
  in apricots, 357
  in blueberries, 326–329
  in carrots, 326, 328
  as colorless carotenoid precursor, 480
  desaturation of, 62
  formation of, from isopentenyl diphosphate, 62
  in fruits, 158
  in human plasma, 213–214
    chromatographic profile of, 215
  intermediate accumulation of, 441
  LC-MS spectra of, 334
  in palm oil, 162, 164–165
  in papaya, 358–359
  preloading of algae with, for β-carotene biosynthesis, 441
  in pumpkin, 354, 356
  purification of, 71–72
  stereoisomers of, 439
  storage of, 71
  structure of, 217
  in tomatoes, 155, 326, 328
15-*cis*-Phytoene, 63
  radiolabeled
    biosynthesis of, 63–64
    as substrate, 64
  source of, 66
Phytofluene, 11–12
  antioxidant effects of, 413
  in apricots, 357
  in bacteria, 380
  in blueberries, 326–330
  in carrots, 326, 328, 330
  in fruits, 158
  in human plasma, 213–214
    chromatographic profile of, 215

intermediate accumulation of, 441
LC-MS spectra of, 334
in palm oil, 162, 164
in papaya, 358–359
in pumpkin, 354, 356
purification of, 71–73
stereoisomers of, 439
structure of, 217
in tomatoes, 155, 326, 328, 330
Pigeon, 320
Plant carotenoid(s), 142–167. *See also* Fruit(s), carotenoids; Vegetable carotenoid(s)
  analysis of, 144–146
  in chloroplasts, 168–169
  composition and content of, 145
  distribution of, 147
  extraction of, 274, 350–351
  in higher species, 168–169
  high-performance liquid chromatography of, 145–148, 151
  identification of, 326
  methods for separation of, 146–167
  in reproductive tissues, 169
  separation of
    historical development of, 145–146
    by mass spectrometry, 151
Plant pigments (Goodwin), 3
Plasma carotenoid(s)
  chromatographic profile of, 213–215, 278–279, 281
  chromatographic quantitation of, internal standards for, 208–210, 280
  expression of concentration of, 212
  gradient separation of, 204, 273–274
  high-performance liquid chromatography of
    for detailed separation, 212–219
    for extraction and analysis, 273–281
    internal standards for, 274
    results of, 278–280
    selection of method for, 273–274
    working standards for, 274–277
  high-resolution, 336–346
    ammonium acetate in mobile phase, 344–346
    column temperature during, 343–344
    equipment and procedures for, 337–338

equipment and solvent program for, 340
hydrolysis with microbial enzymes, 342
injection of sample in methanol, 341–342
internal standards for, 338–341
  selection of, 341
  synthesis of, 339–340
sample preparation and analysis for, 339–340
technical discussion of, 340–345
use of two columns for, 340–341
for purification, 275–276
sample extraction for, 277–278, 339
identification of, 278
major types of, 273
quantification of, 211–212
reference samples of, 208
sources of, 209
separation and quantification of, 205–219
  chromatographic conditions for, 207–208
  experimental procedures for, 207–212
  historical development of, 206
  instrumentation for, 207
spectral characteristics of, 276
types of, 213–220, 336
Plum(s), 157
Polar lipids, separation of carotenes from, 66–67
Pollen. See *Dorotheanthus bellidiformis* pollen
Polyene-chain formation, 20–21
Polyene components, coupled with chromophores, 453
Polyethylene, 233
Polymeric synthesis, 192–196, 204
Polymethacrylate, 186
Polysaccharides, depolymerization of, 405
Polysciences, Inc., 461
Polystyrene-divinyl benzene, 186
Polyunsaturated lipids, 454
Pomacea eggs, 172
Porphyrin(s), 142, 480
  in photoinduced charge separation, 452–453
  reactions of carotenoid free radical with, 449
Porphyrin derivatives

structure of, 88–89
synthesis of, 96–97
Potassium hydroxide, 254, 337
Potato(es), 159
Primate Supply Information Clearing House, 224
Procambarus clarkii, 101
  carotenoproteins from, 104–110. See also Carotenoprotein(s)
Prolycopene
  biosynthesis of, 64–65
  cyclization of, 73
  identification of, 73
  purification of, 73
  storage of, 73
  as substrate in cyclization reactions, 64–65
Proneurosporene
  identification of, 73
  purification of, 73
Protein(s), inactivation of, 404
Prune(s), 358
Prymnesiophyceae, 231
Pseudo-ionone, 58, 75
  cleavage of, 60
  kinetic comportment of, 59–60
Pseudomonas radiora, 339, 375, 385
Pulsed laser(s), in surface-enhanced Raman scattering spectroscopy, 33
Pulse radiolysis, 306, 444, 450
Pumpkin(s), 151, 350, 354
  carotenoids in, 155–156, 354, 356
  vitamin A activity of, 156
Pyramimonas parkeae, 379
Pyridine, 31, 316, 318
  preparation of, 316
Pyrrhula pyrrhula, 320
Pyrrophyta, 171

Q

Quantum Chemical, 392
Quinone(s), in photoinduced charge separation, 452–453

R

Radical scavenger(s). See Antioxidant(s)
Radioisotopic labeling, of carotenoids, 42–49

Radioprotection, 13
Raman spectroscopy, 33. *See also* Surface-enhanced Raman scattering spectroscopy
  for determination of carotenoid configurations, 298–305
    avoidance of isomerization during, 303–304
    measurements, 304–305
    procedures for, 301–305
    scope of, 305
    spectral comparison of isomers, 305
  equipment for
    conventional scanning-type, 304
    equipped with multichannel detector-305, 304
  illumination geometry schematic for, 33–34
  instrumentation for, 35
  resonance, 33
    versus surface-enhanced resonance Raman scattering spectroscopy, 39–40
  resonance Raman process, 301
Rapeseed oil, 159
Raphidophyta. *See* Chloromonadophyta
Rat(s), 267, 418, 483–484
Rayleigh scattering, 32–33
Reaction center complex(es), 36, 39–41
  carotenoid triplet states in, 308–310
  and carotenoporphyrins, 87
  of photosynthetic bacteria, 308, 311
Red bell pepper. *See* Paprika
Red blood cells, hemolysis of, carotenoid inhibition of, 416
Red chilli, 151
Redox potential
  of carotenoids, 449
  and number of conjugated double bonds, correlation between, 449
Reduction, generation of carotenoid free radical by, 445
Reference carotenes, 65–66. *See also* Carotene substrates
  incubation assays, 67–71
  purification of, 67–71
Relative retention indices, 280
Retention indices. *See* Kovats retention indices

Retention time, on high-performance liquid chromatography, 381–382
Retina(s). *See also* Macula lutea
  peripheral, carotenoids in, 227–229
    HPLC analysis of, 228–229
  primate
    carotenoids in, 220–230, 336
      chromatographic profiles of, 364–366
      extraction of, 224–225, 363
      HPLC analysis of, 226–227, 360–366
      separation and quantitation of, 363
    dissection of, 221–224, 361–362
      equipment for, 361
    distribution of carotenoids in, 360–366
    internal standards for HPLC analysis of, 360–366
    from perfusion-fixed animals, dissection of, 224
    tissue handling of, 361–362
Retinal
  commercial sources of, 274
  total ion current chromatogram of, 284, 286–287
Retinaldehyde, gas chromatography-mass spectrometry of, 282, 286–287
Retinal methoxime, syn and anti conformers of, separation of, 282
Retinal oxime, geometric isomers of, 283
Retinoic acid, 282
  commercial sources of, 462
  singlet oxygen quenching constant of, 437
Retinoic acid analog, pentafluorobenzyl ester of, 282
Retinoid(s)
  analysis of, 145
  antioxidant radical-scavenging activity of, 460–472. *See also* Antioxidant radical scavenger(s)
  chromatographic characteristics of, 281
  commercial sources of, 462
  deuterium-labeled, isotope dilution studies using, 284
  gas chromatography-mass spectrometry of, 281–290. *See also* Gas chromatography-mass spectrometry
  HPLC separation of, 274
  in human plasma, spectral characteristics of, 276

Kovats retention indices, 284, 288–289
labeled with radioisotopes, 45
phycoerythrin fluorescence assay of, 466–467
reaction stoichiometry of, with peroxyl radicals in hexane, 470–471
Retinoid radical(s), 449
Retinol. *See also* Vitamin A
commercial sources of, 462
deuterated, 282
gas chromatography-mass spectrometry of, 281–287
limitations on, 290
and glutathione radicals, reaction between, 450
in human milk. *See* Human milk
in human plasma, 211–213, 218, 273–274
chromatographic characteristics of, 281
identification and quantification of, 211–212, 278, 280
spectral characteristics of, 276
hydrogenation of, before GC analysis, 282
preparation of, by reduction of retinal, 274
standards for, preparation of, 394
total ion current chromatogram of, 284, 286–287
trimethylsilyl derivative of, 282
Retinyl acetate
gas chromatography-mass spectrometry of, 281–282, 286–287
hydrogenation of, before GC analysis, 282
total ion current chromatogram of, 284, 286–287
Retinyl esters
in human plasma, 274, 278, 280
retention indices of, 288
Retinyl hexanoate, 274, 280
Retinyl palmitate
cis-parinaric acid fluorescence assay of, 466–467
commercial sources of, 274, 462
in human plasma
chromatographic characteristics of, 281
spectral characteristics of, 276
packed-column gas chromatography of, 281–282
radical scavenging in liposomal system, 468
Retinyl stearate, antioxidant effects of, 413

Retinyl tetradecanoate, 290
Rhesus monkey(s), 267
Rhodobacter sphaeroides, 38–40
reaction centers isolated from, 308–311
Rhodococcus equi, 373
Rhodococcus erythropolis, 379
Rhodococcus fascians, 373
Rhodococcus rhodochrous
absorption parameters of, 378–380
carotenoid glycoside ester from, 366–374
position of double bonds, 373
structure of, 367, 372–374
carotenoids from
acetylation of, 369
acidic methanolysis of, 371–372
chemical modification of, 369–370
chromatographic profile of, 369
fatty acid ester composition of, analysis of, 371–372
identification of, 370–371
isolation and purification of, 368
proton nuclear magnetic resonance spectra of, 372
reduction of, 369
saponification of, 371
structures of, 370–371
sugar moiety of, 372
trimethylsilylation of, 369–370
types of, 374
cellular lipids of, structure of, 372–374
Rhodocyclus gelatinosus, 378
Rhodomicrobium vannielii, 252, 254
chromatogram of, 258, 260
major carotenoids in, 258, 260, 264–265
major xanthophylls in, 265
preparation of, 255
Rhodophyta, 171
Rhodopin, 260, 311
Rhodopseudomonas acidophila. See Rhodobacter sphaeroides
Rhodopseudomonas sphaeroides. See Rhodobacter sphaeroides
Rhodospirillaceae, 375–376
Rhodospirillum rubrum, 38
Rhodovibrin, 260
Rhodoxanthin, 168, 364
in birds, 320
Rodent(s). *See* Rat(s)
ROM 250, 140
ROM 500, 140

SUBJECT INDEX

Roots and tubers. *See also individual species*
  carotenoids in, 158–159
Rose bengal, 421, 425
Rosemary oleoresin, 139, 142
Roseobacter denitrificans, 375
Rotating cells, for avoidance of isomerization during Raman spectroscopy, 304
Rotavapor, 241–242, 245
RPLC. *See* High-performance liquid chromatography, reversed phase
RRIs. *See* Relative retention indices
Rubixanthin, 376
  in bacteria, 380
Rubrene, photooxidation of, 431

S

Saponification, 22, 317, 337, 351–352, 359, 371. *See also* Alkaline hydrolysis
  for analysis of carotenoids in breast milk, 392, 394–398
Sarcina lutea, 75
Sarcinaxanthin, 75
  (±)-all-E,*cis*-, 86
  structure of, 76
  synthesis of, 75–86
    multistep approach to, 76
    procedures for, 79–86
Semi-β-carotenone, 169
SERRS. *See* Surface-enhanced Raman scattering spectroscopy, resonance
SERS. *See* Surface-enhanced Raman scattering spectroscopy
SFV. *See* Solvent feed valve
Shimadzu CR-4A integrator, 278
Shimadzu CR 1B integrator, 58
Shimadzu IR-27G spectrophotometer, 27
Shimadzu LC-6A instrument, 27
UV-240 Shimadzu recording spectrophotometer, 324
Shimadzu SPD-6VA UV-VIS spectrophotometric detector, 27
Shimadzu UV-240 spectrophotometer, 27
Sigma Chemical Company, 176, 252, 274, 323, 362, 392, 462
Signal-to-noise ratio, 304, 325, 333
Silica columns, for chromatographic procedures, 186–187, 192, 212, 233–237, 247–248, 253, 317–318, 372
Silinol, beneficial effects of, on carotenoid separation, 192

Sinapis alba. See Mustard
Singlet energy transfer, 90
Singlet oxygen, 461
  deactivation of, in absence of physical quencher, 422
  generation and possible pathological consequences of, in biological systems, 404
  photosensitization production of, 420–422
Singlet oxygen quenching
  by carotenoids, 87, 266, 323, 386, 403–404, 421–429, 461, 472
    assay of, 429–438
      comments on, 435–438
      germanium diode setup for, 433–435
      method for, 432–435
      NDPO$_2$ preparation for, 432–433
      NDP preparation for, 432
      principles of, 431
      procedures for, 435
      reagents for, 432
    measurement of, 421–430. *See also* Near-infrared steady-state luminescence
    two reactions used for, 431
    mechanism of, 405, 421, 429–430
    rate constants for, 436–438
    solvent dependence on rate constants of, 426–429, 431
  by retinoids, 461
Singlet oxygen scavenger(s), 175, 406
Siphonaxanthin, 170
  in bacteria, 380
SL-1/SL-2. *See* Solvent loop valves
S/N ratio. *See* Signal-to-noise ratio
Sodium borohydride, 274, 318
Sodium deuteroxide, 51
Sodium hydride, 51
Sodium phosphate buffer, 462
Sodium sulfate, 299
  anhydrous, 55–56
Solvent feed valve, 247–248
Solvent loop valves, 247–248
Solvent operation valve, 247–248
Solvent radical(s), 445
Solvent system(s), for high-performance liquid chromatography, 247–248, 274, 340, 346, 394
  nonaqueous versus aqueous, 274
Sorangium compositum, 366

SOV. *See* Solvent operation valve
Soybean oil, 159
  photooxidation of, 431
Soybean phosphatidylcholine, 460
SpectrEL mass spectrometer, 284
Spherisorb AY5 alumina column, 253
Spheroidene, 39
  in bacteria, 39
  triplet state EPR signals from, 308–310
Spheroidenone, 39
Spheroplasts, 36, 38
Spinach, 147–148, 151
Spirilloxanthin(s), 4, 6, 38–39, 147, 170, 260, 376
  in benzene solution, isomerization shifts of, 301–302
  HPLC of
    eluent for, 300
    with lime column, 296
Sprague-Dawley rat(s), 267, 418
Squash(es), 152, 218, 350, 358
Squirrel monkeys, 227
SRMs. *See* Standard reference materials
Standard reference materials, 190–191
Staphylococci, 170
Staphylococcus aureus, 367
Starfruit, 157
Stereoisomerization, 291–292
Stern-Volmer plots, 424, 435
  for rate constants of carotenoids, 435–436
  of singlet oxygen quenching, by carotenoids, 426–429
Stigmatella aurantiaca, 366
Stromal proteins, 62
Sulfite, conversion to sulfate, 175
Sulfone
  in carotenoid synthesis, 15–16
  obtention of, 83–84
Sunflower seed oil, 159
Superoxide radical(s), 405, 452, 461
Surface-enhanced Raman scattering spectroscopy, 31–42
  applications of
    to membrane systems, 32
    in photosynthesis, 41
  experimental procedures with, 32–38
  illumination geometry, schematic for, 33–34
  instrumentation for, 32–35
  lasers in, 33
  limitations of, 32
  major advantages of, 32
  resonance, 32–42
    with bacterial photosynthetic systems, 38–39
    carotenoid results with, examples of, 38–41
    of carotenoids, 38–41
    critical parameters for, 37–38
    effects on protein complexes, 41
    electrochemical cell, 35–36
    with green plant photosynthetic systems, 39–41
    with photosynthetic samples, 36–37
    methods for, 37–38
    potential applications of, 41–42
    problems with, 38
    versus resonance Raman spectroscopy, 39–40
  spectrometer and detector in, 33, 35
  substrate preparation for, 33–36
    silver colloids, 36
    silver electrode, 33–36
    vacuum-deposited island films, 35–36
Sweet potato(es), 147, 158–159, 350, 354

T

Tagetes erecta. *See* French marigold
Taraxanthin, in pumpkin, 156
Taro, 159
Taurodeoxycholic acid (TDCA), 176, 462
Tenax CG, 55
Terpenoid(s), 11–12
Tert-butanol, 247
Tetraarylporphyrin(s), 90
Tetrachloroethylene, 300
Tetradehydroastaxanthin
  isolated from *Paralithodes brevipes*, 27–28
  structure of, 28
Tetrahydrofuran, 45–46, 51, 176–177, 183, 199, 248, 350
  commercial sources of, 392
Tetrapyrrole(s), cyclic, coupling with carotenoids, 91
Tetra-Z-lycopene, total synthesis of, 20–22
TFA. *See* Trifluoroacetic acid
Thermal isomerization, 291, 294–295, 300
  prevention of, 301
Thermospray, for HPLC-MS analysis, 284

THF. *See* Tetrahydrofuran
Thin-layer chromatography, 23, 114, 248, 251, 275
  for carotenoid identification, 238, 244, 313, 317–318
  for carotenoid purification, 378
  for carotenoid separation, 145–147, 151, 213, 233, 236
  equipment for, 79, 236
  reversed-phase, 318
  for separation of carotenoid stereoisomers, 293
  solvent systems for, 248
  in synthesis of deuterated β-carotene, 53
  system I, 67, 69
  system II, 67–68
Thiol(s), 461
Thylakoid membranes, carotenoids in, 39–40
TIC. *See* Total ion chromatogram
Titanium:sapphire laser(s), 42
TLC. *See* Thin-layer chromatography
TOCO 500, 140
Tocopherol(s)
  chromatographic characteristics of, 281
  commercial sources of, 274
  in human plasma, 213–215, 273–274
    chromatographic profile of, 213, 218, 281
    identification of, 278, 280
    quantification of, 211–212, 280
    spectral characteristics of, 276
α-Tocopherol, 139, 142, 176, 461, 463, 469–470, 472
  antioxidant effects of, 410–413, 451
  antioxidant radical-scavenging activity of, 460–472. *See also* Antioxidant radical scavenger(s)
  gas chromatography-mass spectrometry of, 284–285
  inhibition of MDA formation by, 415
  inhibition of red blood cell hemolysis by, 415
  interactions with carotenoids, 412
  protective effects of, on β-carotene, 176, 178–182, 185
  recycling of, 472
γ-Tocopherol, antioxidant effects of, 413
Toluene, 235
Tomato(es), 151, 265, 274, 350, 354
  carotenoids in
    analysis of, 325
    concentrations of, 154–155
    extraction of, 324
    identification of, 330–331
    UV/VIS light absorption spectra of, 326, 328, 330
  Tangerine variety of, 66, 70
    carotene fractions obtained from, 73
Torr Seal, 33–35
Torulene, 171
Total ion chromatogram, 333–335
Trephine, for retinal dissection, 223, 230
Trichloroacetic acid, 254
Triethylamine, 274
Triethyl phosphonoacetate, deuterated, 50–51
Trifluoroacetic acid, 325
Triglyceride hydrolase, 339
Triglycerides, elimination from plasma sample, 342
3,4,4'-Trihydroxy-β-carotene, structure of, 109
2,6,6-Trimethylcyclohexanone, 57
2,6,6-Trimethyl-2-cyclo-hexen-1-one, 57
Trimethylsilylation, 25
1-Trimethylsilyl-3-methyl-4-bromo-2-butene, 84
Triphenylphosphine hydrobromide, 47–48, 51
Triplet energy transfer, 90
Tris, 38
Tritium, carotenoids labeled with, 42–49
Tritium gas reduction, 44–45
Trolox, 462–463, 469
Tubers. *See* Roots and tubers
Tunaxanthin, antioxidant activity of, 411–412

U

Ubiquinol(s), 461
2070 Ultrarac II, 247
Ultrasonic agitation, 187, 190
Ultrasphere 5 CN, 236
Ultrasphere ODS column, 176
Ultraviolet light absorption spectra
  of *cis*-β-carotene, 296–297
  of carotenoids, 23, 32, 91, 134, 146, 152, 219, 291, 326, 328, 352, 475

equipment for, 79
prior to mass spectrometric analysis, 113
Upogebia pusilla, 101
Uric acid, 461
Ustilago, 482

V

Vacuum filtration, 187
Varian 3700 gas chromatograph, 56
Varian Superscan 3 spectrophotometer, 79
Varian XL-300 instrument, 27
Vegetable(s)
  green leafy
    carotenoids, 148–150, 156
      chromatographic profiles of, 353–354
      versus green nonleafy, carotenoids, 150–151
  green nonleafy, carotenoids, 149
    chromatographic profiles of, 353–354
  tuberous, carotenoids, 148–149, 151
  yellow-orange, carotenoids, 357–359
  yellow-red, carotenoids, 354–357
Vegetable carotenoid(s), 147–156. *See also* Palm oil; Vegetable oils
  chromatographic profiles of, classification of, 353
  extraction of, 350–351
  HPLC of, for separation, 353–359
  identification of, by HPLC separation in order of elution, 355
  separation of, methods for, 147–156
  standards for, source of, 350
Vegetable oils, carotenoids in, 159–167. *See also individual species*; Palm oil
Vinyl triflate, preparation of, 21
Violaxanthin, 6–7, 9, 126, 146, 358
  in algae, 242, 257, 259
  in bacteria, 380
  in chloroplasts of higher plants, 168
  commercial sources of, 114, 253
  HPLC of, composition of solvent mixtures, flow rate, and retention time for, 275
  molecular weight of, 118
  in papaya, 358–359
  standards for, plant sources of, 350
Visible light absorption spectra, 23, 91, 134, 146, 152, 219, 326, 328, 352, 475

of *cis-β*-carotene, 296–297
  for carotenoid characterization, 25–27
  for carotenoid identification, 239, 244
  equipment for, 27, 236
  prior to mass spectrometric analysis, 113
Vitamin A. *See also* Retinol
  and carotenoids, 144
  carotenoids as precursors of, 143, 175, 184
  cis-trans isomerism of, 7–8
  deficiency of, 13
  intake of, recommended safe level of, 151
  main dietary sources of, 143
  in marine animals, 172
  sources of, 143
  supply of, 13
Vitamin E. *See* α-Tocopherol; γ-Tocopherol
Volatile compounds
  formation of, by thermal degradation of carotenoids, 54–62
    kinetics of, 58–60
  identification of, 57–58
  isolation of, 55–56
  quantitation of, by gas chromatography, 58
  separation of, 56–57
Volvocales, 439
Vydac HPLC column, 326, 357

W

Water
  added to mobile phase of LC-MS, 335
  as solvent modifier, in carotenoid separation, 197
Water free radical(s), 446
Watermelon(s), 157
Waters Associates pump(s), 278, 324, 394
Waters C_{18} Resolve column, 278
Whatman filter paper, 324
Whatman Partisil ODS column, 474
WHO. *See* World Health Organization
Wittig condensation, 13, 15, 17, 90
Woodpeckers, 320
World Health Organization, 13, 151

X

Xanthine oxidase system, 405
Xanthophyll(s), 3–4, 146, 186
  in blueberries, 326–327, 329

chromatographic separation of, 192, 197, 207–208, 353
in fruits, 158
in palm oil, 162
separation of carotenes from, 66–67
singlet oxygen quenching constant of, 437
in vegetables, 149, 151
Xanthophyll regions, separation of, from chromatogram, 337, 341–342
Xanthophyta, 171
Xenon/mercury lamp, 424
X-ray crystallography, 298

Y

Yanagimoto micromelting point apparatus, 27
Yeast carotenoid(s), photoprotective activity of, 482

Z

Z-carotene, 11–12
α-Zeacarotene, in palm oil, 164
β-Zeacarotene, 153
intermediate accumulation of, 441
in palm oil, 164
stereoisomers of, 439
Zeaxanthin, 4, 6, 358, 380
in algae, 242
antioxidant activity of, 411–412
in bacteria, 376
in birds, 318
in chloroplasts of higher plants, 168

chromatographic separation of, 192, 196–197, 208, 226
commercial sources of, 462
in fruit, 158
in *Homarus americanus*, 319
HPLC of
ammonium acetate buffer in mobile phase, 344–346
composition of solvent mixtures, flow rate, and retention time for, 275
effects of column temperature on, 343
in human milk, 399
in human plasma, 206, 220, 273, 336
chromatographic profile of, 214–215, 218, 278–279, 281, 338
quantification of, 276
spectral characteristics of, 276
in human retina, 360–366, 429
chromatographic profiles of, 364–366
separation and quantitation of, 363
stereochemistry of, 363–364
isolated from *Paralithodes brevipes*, 27–28
mammalian metabolism of, 271, 337
in mustard green leaves, 274
natural occurrence of, 147
in palm oil, 162
in primate retina, 220, 225, 227–230
in pumpkin, 156
singlet oxygen quenching constant of, 437
standards for, plant sources of, 350
structure of, 28, 216
Zelnoxanthin, structure of, 216
Zorbax ODS column, 192, 206, 253, 257, 319, 321, 324–325

ISBN 0-12-182114-5